PRECALCULUS MATHEMATICS
A Functional Approach

Karl J. Smith
Santa Rosa Junior College

Brooks/Cole Publishing Company
Monterey, California
A Division of Wadsworth Publishing Company, Inc.

CONTEMPORARY UNDERGRADUATE MATHEMATICS SERIES

Consulting Editor: Robert J. Wisner

Printed in the United States of America

10 9 8 7 6 5 4 3 2

Library of Congress Cataloging in Publication Data

Smith, Karl J
    Precalculus mathematics: a functional approach.

    Includes index.
    1. Functions.   I. Title.
QA331.S617   512'.1   78–6279
ISBN 0–8185–0269–X

Acquisition Editor: *Craig F. Barth*
Manuscript Editor: *Phyllis Niklas*
Production Editor: *Micky Lawler*
Interior Design: *Jamie S. Brooks*
Cover Design: *Katherine Minerva*
Chapter-Opening Photos: *Jim Pinckney*
Illustrations: *Knapman Bartlett Ltd., Bristol, England*
Typesetting: *J. W. Arrowsmith Ltd., Bristol, England*

FUNCTIONS (continued)

Exponential function: $f(x) = b^x$, b a positive constant

EXPONENTS

Definition: if b is any nonzero real number and n is any natural number, then

$$b^n = b \cdot b \cdot b \cdots b$$

$\qquad\quad$ *n* factors

$$b^0 = 1$$

$$b^{-n} = \frac{1}{b^n}$$

$b^{1/n} = \sqrt[n]{b}$ and $b^{m/n} = \sqrt[n]{b^m} = (\sqrt[n]{b})^m$ ($b > 0$, m and n positive integers)

Laws of exponents: Let a and b be nonzero real numbers and let m and n be any real numbers, except the form $0/0$ and division by zero are excluded.

First law: $b^m \cdot b^n = b^{m+n}$ $\qquad$ Second law: $\dfrac{b^m}{b^n} = b^{m-n}$

Third law: $(b^n)^m = b^{mn}$ $\qquad$ Fourth law: $(ab)^m = a^m b^m$

Fifth law: $\left(\dfrac{a}{b}\right)^m = \dfrac{a^m}{b^m}$

Logarithmic function: $f(x) = \log_b x$ ($b > 0$, $b \neq 1$)

LAWS OF LOGARITHMS

First law: $\log_b AB = \log_b A + \log_b B$

Second law: $\log_b \dfrac{A}{B} = \log_b A - \log_b B$

Third law: $\log_b A^p = p \log_b A$

LOG OF BOTH SIDES THEOREM: $\log_b A = \log_b B \Leftrightarrow A = B$

CHANGE OF BASE THEOREM: $\log_b x = \dfrac{\log_a x}{\log_a b}$

FORMULAS

Distance between (x_1, y_1) and (x_2, y_2): $d = \sqrt{(x_2 - x_1)^2 + (y_2 - y_1)^2}$

Circle with center (h, k) and radius r: $(x - h)^2 + (y - k)^2 = r^2$

Quadratic formula: if $ax^2 + bx + c = 0$, $a \neq 0$, then $x = \dfrac{-b \pm \sqrt{b^2 - 4ac}}{2a}$

(continued inside back cover)

PRECALCULUS MATHEMATICS
A Functional Approach

Other titles by the same authors

The Nature of Modern Mathematics, Second Edition. Karl J. Smith
Introduction to Symbolic Logic. Karl J. Smith
Beginning Algebra for College Students. Karl J. Smith and Patrick J. Boyle
Study Guide for Beginning Algebra for College Students. Karl J. Smith and
Patrick J. Boyle
Intermediate Algebra for College Students. Karl J. Smith and Patrick J.
Boyle
Study Guide for Intermediate Algebra for College Students. Karl J. Smith
and Patrick J. Boyle
Algebra and Trigonometry. Karl J. Smith and Patrick J. Boyle
College Algebra. Karl J. Smith and Patrick J. Boyle
Trigonometry for College Students. Karl J. Smith
Analytic Geometry: A Refresher. Karl J. Smith
Precalculus Mathematics: A Functional Approach. Karl J. Smith

This book is dedicated with love to my daughter,
Melissa Ann

Preface

This book is designed to prepare students for the study of calculus. My main goal in writing this book is to make the material clear and interesting for the reader, and I have made every effort to point out important ideas and common pitfalls. Important ideas are enclosed in a box, examples are set off for easy reference, and I have used the margins to amplify, explain, give hints, provide historical notes, and generally enrich the exposition. A review for each chapter lists new terms and important ideas, and a sample test ends each chapter.

There seems to be a division of philosophy regarding method of presentation in recent books. Some texts take a very matter-of-fact approach and do little to motivate the reader. Others water down the material and try to be attractive by adding irrelevant illustrations and cartoons. This book takes the best features of both approaches. I have found that, if I can maintain interest by emphasizing the practicality of mathematics, my students are motivated to study the work seriously, and I've given special attention to the appearance and presentation of the material.

To make the mathematics in this book relevant to students' interests, I have given special attention to the problem sets. The exercises, which are presented in pairs of similar problems, are divided into A and B problems by order of difficulty. The answers to the odd-numbered problems are in the back of the book, and special care has been taken to make the answer section a meaningful learning aid. There are more than 260 graphs presented in the answer section alone in order to provide additional examples when students find a problem too difficult. If students are working an even-numbered problem in the book and are unsure of themselves, they might try working the preceding or following odd-numbered problem in order to check their procedure.

A wide variety of applied problems is provided, and they are indicated by the following symbols:

astronomy/space science	☾☼	physics/chemistry	⚛
aviation/navigation	(compass)	proof/theory	♋
business/economics	$	psychology	ψ

DISTINCTIVE FEATURES

1. The book is written to make the material clear and interesting for the student.

2. The text offers a good, solid treatment of mathematics.

3. An extensive answer section is provided to give students hints and additional examples.

4. A wide variety of problems is presented for each section.

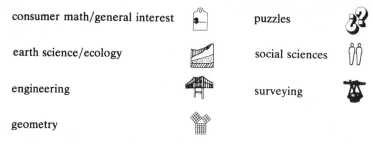

consumer math/general interest		puzzles	
earth science/ecology		social sciences	
engineering		surveying	
geometry			

The applied problems are self-contained and do not require any previous knowledge. They are integrated into nearly every problem set so that students can gain daily practice with them.

Theoretical and proof problems are specially designated. Each section also contains some problems called "Mind Bogglers." The solutions of these problems require a certain amount of ingenuity. They are not, however, just difficult problems; they are designed to provoke some additional avenues for classroom discussion. If students make a habit of at least reading them in each section, they may be interested and pleasantly surprised.

5. The main thread in the book is the idea of a function.

The topics covered in this book are those most needed by beginning calculus students. The main thread tying the material together is the notion of a function and the graphs of various functions. Although it is assumed that students have had some high school algebra, the important ideas from algebra are reviewed in Chapters 2–5. It is not necessary for students to have had trigonometry, since an entire trigonometry course is contained in Chapters 6 and 7. On the other hand, if your students are well prepared in trigonometry, you may choose to omit or pass quickly over these chapters.

6. The book has been written to allow a great deal of flexibility to suit the needs of the class.

A great deal of flexibility is possible in the selection of topics chosen for this course. For this reason, more material is provided in this book than can be used in a single semester. A map of the relationship between chapters and sections is given below.

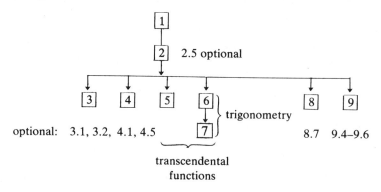

CHAPTER	NUMBER OF CLASS DAYS IF CHAPTER IS COVERED Minimum–Maximum	COMMENTS
1. *Functions and Graphs*	5–7	This chapter is required for the rest of the material in the course. However, Section 1.4 is required only for Sections 5.2 and 6.6.
2. *Linear and Quadratic Functions*	5–8	Section 2.5 is an optional review section. The remainder of this chapter is required for the rest of the material in the course.
3. *Polynomial Functions*	4–8	Sections 3.1 and 3.2 are optional review sections.
4. *Rational Functions*	4–7	Section 4.1 is optional review; Section 4.5 is optional enrichment.
5. *Exponential and Logarithmic Functions*	4–6	Calculator and logarithmic methods are compared and used in Section 5.4.
6. *Circular Functions*	6–12	Even if trigonometry has been assumed, it is suggested that Sections 6.2 and 6.3 be briefly reviewed and graphing of the circular functions in Sections 6.4 and 6.5 be covered. Sections 6.7–6.9 are optional; they require an introduction to complex numbers (Appendix A).
7. *Trigonometric identities and Applications*	6–9	Section 7.2 needs two or three class meetings if this chapter is not treated as a review.
8. *Sequences, Series, and the Binomial Theorem*	4–10	Sections 8.1–8.4 are independent of Sections 8.5–8.7. Section 8.7 is optional.
9. *Analytic Geometry*	3–10	Sections 9.4 and 9.6 require trigonometry and are optional. Section 9.5 is also optional. Most calculus books contain material in analytic geometry, but a rudimentary knowledge of the conic sections is often assumed.

7. Calculators are used wherever appropriate.

Because of the great availability of inexpensive electronic calculators, the computational aspects of the logarithmic function in Chapter 5 have been minimized. A discussion of logarithmic tables versus calculators is given in Section 5.4. This book advocates the use of calculators but gives alternative solution techniques for those who do not have a calculator. Several problems throughout the text are designated specifically as calculator problems, but many others (particularly in Chapters 5 and 6) could also be worked with a calculator, if one is available. (See also "A Word about Calculators," following this Preface.)

The following people offered helpful reviews of the manuscript: Marjorie S. Freeman of the University of Houston, Downtown College; Thomas Green of Contra Costa College; James Householder of California State University, Humboldt; Raymond McGivney of the University of Hartford; Bill Orr of San Bernardino Valley College; David Tabor of the University of Texas; and Robert Webber of Longwood College. I would also like to thank Marc Imback for checking all the problems in this book and my consulting editor, Robert J. Wisner of New Mexico State University, for all his helpful suggestions. I am grateful to the people at Brooks/Cole who worked on this book, including Phyllis Niklas, Micky Lawler, Craig Barth, and Jamie Brooks, and especially to Jack N. Thornton for assembling such a sterling staff.

Special thanks go to my wife, Linda, and our children, Melissa and Shannon, for their love and support while I was working on this book.

Karl. J. Smith

A Word about Calculators

Although a calculator is not required for this book, it would certainly be helpful if you have one. Four-function calculators are available for under $10, and calculators with the circular, logarithmic, and exponential functions can be purchased for under $20.

Three main types of logic are used on calculators: arithmetic logic, algebraic logic (recognizes the order-of-operations convention from algebra), and RPN logic. It is suggested that you use a calculator with algebraic or RPN logic. RPN calculators are characterized by an $\boxed{\text{ENTER}}$ or $\boxed{\text{SAVE}}$ key. If you're using a calculator with arithmetic logic, you must use parentheses or "pick out" the operations to be performed first. For example, consider the problem $2 + 3 \cdot 4$, which illustrates the differences among the three types of logic.

ARITHMETIC LOGIC	ALGEBRAIC LOGIC	RPN LOGIC
$\boxed{3}$	$\boxed{2}$	$\boxed{2}$
$\boxed{\times}$	$\boxed{+}$	$\boxed{\text{ENTER}}$
$\boxed{4}$	$\boxed{3}$	$\boxed{3}$
$\boxed{=}$	$\boxed{\times}$	$\boxed{\text{ENTER}}$
$\boxed{+}$	$\boxed{4}$	$\boxed{4}$
$\boxed{2}$	$\boxed{=}$	$\boxed{\times}$
$\boxed{=}$		$\boxed{+}$

If you input

$\boxed{2}\ \boxed{+}\ \boxed{3}\ \boxed{\times}\ \boxed{4}\ \boxed{=}$

on a calculator with arithmetic logic, you will obtain the incorrect answer 20. For this reason, a calculator with algebraic or RPN logic is recommended.

The correct output is 14.

In this book, we'll be using a Texas Instruments SR-50 to illustrate algebraic logic and a Hewlett-Packard HP-35 to illustrate RPN logic. However, you should always consult the owner's manual for the calculator you own.

CREDITS

This page constitutes an extension of the copyright page.

Contents

1

2

3

4

Functions and Graphs

1.1 RELATIONS AND FUNCTIONS

This book is designed to prepare you for the study of calculus. As the title suggests, the main thread that will lead you through this book is the concept of a *function*. You have, no doubt, been introduced to this idea before, probably in algebra. But functional ideas are around us in many forms.

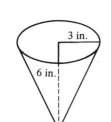

Figure 1.1   Ice cream cone

Consider an ice cream container shaped like a right circular cone, as shown in Figure 1.1. Suppose you experiment by filling the cone with different amounts of sand. You will find that the radius of the circle of sand in the cone is always $\frac{1}{2}$ the height (h) of the sand in the cone and that the volume (V) is found by

$$V = \frac{\pi h^3}{12}.$$

A table of values showing the volumes for different heights is given here.

Height	1	2	3	4	5	6
Volume	$\dfrac{\pi}{12}$	$\dfrac{2\pi}{3}$	$\dfrac{9\pi}{4}$	$\dfrac{16\pi}{3}$	$\dfrac{125\pi}{12}$	18π

Do you see how these values were obtained?
If $h = 1$, then

$$V = \frac{\pi(1)^3}{12} = \frac{\pi}{12}.$$

If $h = 2$, then

$$V = \frac{\pi(2)^3}{12} = \frac{8\pi}{12} = \frac{2\pi}{3}.$$

And so on.

INDEPENDENT VARIABLE and DEPENDENT VARIABLE of a RELATION

This table can be written as a set of *ordered pairs*: $(1, \pi/12)$, $(2, 2\pi/3)$, $(3, 9\pi/4)$, ...

A set of ordered pairs is called a *relation*. In the example, the ordered pairs are (h, V) so that $V = \pi h^3/12$. The variable associated with the first component is called the *independent variable*, and the variable associated with the second component is called the *dependent variable*. In the example, if we are free to select a replacement value for h, then the value of V *depends* on this choice.

The set of possible replacements for the independent variable is sometimes limited to a certain set of numbers. For the cone we're considering, h (the height of the sand in the cone) cannot be negative (why?), and should not be larger than 6 in., since the conical container has this height. We say that the *domain* of the function is the set of possible replacements for the independent variable. Thus, the domain, D, is

DOMAIN

$$D = \{h \mid 0 < h \le 6\}.$$

This is the so-called *set-builder notation*. It is read "the set of all h such that 0 is less than h and h is less than or equal to 6."

The set of corresponding replacements for the dependent variable is the *range* of a relation.

It is often useful to draw a graph of a relation. To do so, we set up a *rectangular coordinate system*, as shown in Figure 1.2. This is also sometimes called a *Cartesian coordinate system*.

The horizontal number line is called the *x-axis* (sometimes called the *axis of abscissas*), and *x* represents the first component of the ordered pair. The vertical number line is called the *y-axis* (sometimes called the *axis of ordinates*), and *y* represents the second component of the ordered pair. The point of intersection of the axes is the *origin*. These number lines divide the plane into four regions called *quadrants*, as shown in Figure 1.2.

By plotting the points whose coordinates are shown in the table and then connecting them with a smooth curve, we arrive at the graph of this relation, as shown in Figure 1.3.

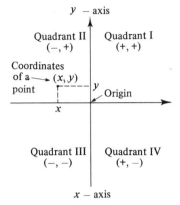

Figure 1.2 Cartesian coordinate system

Height	1	2	3	4	5	6
Volume	$\dfrac{\pi}{12}$	$\dfrac{2\pi}{3}$	$\dfrac{9\pi}{4}$	$\dfrac{16\pi}{3}$	$\dfrac{125\pi}{12}$	18π

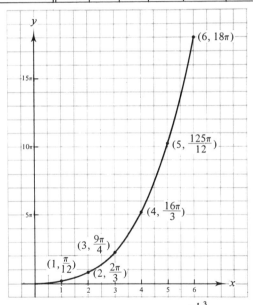

Figure 1.3 Graph of $V = \dfrac{\pi h^3}{12}$

In general, the equation of any graph sets up a *one-to-one correspondence* between points on the graph and ordered pairs satisfying the equation. A *graph of an equation* and an *equation of a graph* are closely connected: there is a one-to-one correspondence between the set of all ordered pairs (x, y) that satisfy the equation and the set of all points (x, y) that lie on the curve. The method of drawing the graph of an equation as shown here—that of plotting points—is very primitive, and one of the primary purposes of this book is to develop easier methods of graphing curves.

A *one-to-one correspondence* between the two sets is a correspondence in which each member of either set is paired with exactly one member of the other set.

Remember, the *domain* is the set of all permissible first components.

Look at the graph of Figure 1.3. Notice that for each value of x in the domain, there exists one corresponding value of y, and consequently one point on the curve, as shown in Figure 1.4.

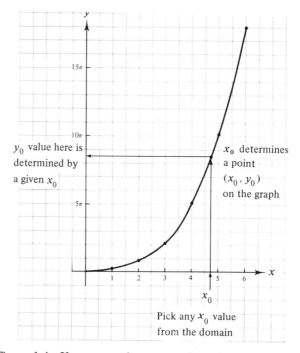

y_0 value here is determined by a given x_0

x_0 determines a point (x_0, y_0) on the graph

x_0

Pick any x_0 value from the domain

Figure 1.4   If you are given any value of x, call it x_0, a corresponding value of y, call it y_0, can be found.

Remember, the *range* is the set of all second components.

From Figure 1.4, it can be seen that the range, R, is given by

$$R = \{V \mid 0 < V \leq 18\pi\}.$$

This example illustrates a special type of relation called a *function*, which we shall describe in several ways.

The symbol x stands for a first component. The symbol $f(x)$ is read "f of x" and does not mean multiplication, but is a single symbol representing the second component of an ordered pair.

A FUNCTION IS A SET OF ORDERED PAIRS

DESCRIPTION 1: A function f is a set of ordered pairs for which each member of the domain, x, is associated with exactly one member of the range, $f(x)$.

EXAMPLES

1. $f = \{(0, 2), (1, 5), (2, 8), (3, 11)\}$
    $D = \{0, 1, 2, 3\}$ $R = \{2, 5, 8, 11\}$

D	R	
$0 \rightarrow 2$	$f(0) = 2$
$1 \rightarrow 5$	$f(1) = 5$
$2 \rightarrow 8$	$f(2) = 8$
$3 \rightarrow 11$	$f(3) = 11$

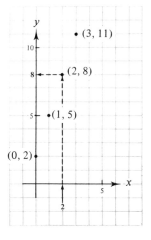

Notice that each value in the domain is associated with a single value in the range.

2. $f = \{(0, 0), (1, 1), (-1, 1), (2, 4), (-2, 4)\}$
    $D = \{0, 1, -1, 2, -2\}$ $R = \{0, 1, 4\}$

D	R	
$0 \rightarrow 0$	$f(0) = 0$
$1 \searrow$	$f(1) = 1$
$-1 \nearrow 1$	$f(-1) = 1$
$2 \searrow$	$f(2) = 4$
$-2 \nearrow 4$	$f(-2) = 4$

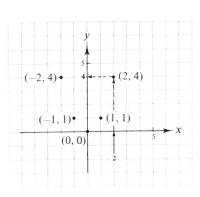

3. Not all sets of ordered pairs are functions. For example,

    $\{(0, 0), (1, 1), (1, -1), (4, 2), (4, -2)\};$
    $D = \{0, 1, 4\};$ $R = \{0, 1, -1, 2, -2\}.$

We do not use $f(x)$ notation unless the set of ordered pairs is a function.

Notice that each member of the domain is not associated with exactly one member of the range. To show that a relation is not a function, you need find only one member of the domain that is associated with more than one member of the range.

$$\begin{array}{cc} D & R \\ 0 \rightarrow & 0 \\ 1 & \begin{array}{c} 1 \\ -1 \end{array} \\ 4 & \begin{array}{c} 2 \\ -2 \end{array} \end{array}$$

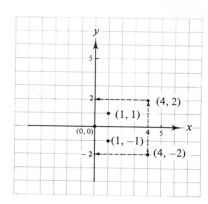

A FUNCTION IS A RULE

> DESCRIPTION 2: A function f is a rule that assigns a single real number $f(x)$ to each x in the domain of the function.

EXAMPLES

Functions do not have to be denoted by f; any letter can be used, such as f, g, h, T, f_1, f_2, ϕ, ...

4. The rule can be given verbally: T is the temperature in degrees Celsius at a certain location at some time x. Suppose the readings shown in the table are obtained (for $0 \le x < 24$; that is, for time measured on a 24-hour clock).

Time, x	Temperature, degrees Celsius	T(x)
6	8	$T(6) = 8$
10	14	$T(10) = 14$
12	17	$T(12) = 17$
14	18	$T(14) = 18$
18	14	$T(18) = 14$

The examples chosen have integral values from the domain. Can you find $V(3/2)$?

5. The rule can be given as a formula: $V = \pi h^3/12$. This formula can be restated using functional notation: $V(h) = \pi h^3/12$. From Figure 1.3, you find $V(1) = \pi/12$, $V(2) = 2\pi/3$, $V(3) = 9\pi/4$, $V(4) = 16\pi/3$, $V(5) = 125\pi/12$, and $V(6) = 18\pi$.

6. The rule can be given as an equation: $y = x^2$. This rule says "given a

Contents

1

2

3

4

5 Exponential and Logarithmic Functions   203

6 Circular Functions   245

7 Trigonometric Identities and Applications   317

8

9

Functions and Graphs

1.1 RELATIONS AND FUNCTIONS

This book is designed to prepare you for the study of calculus. As the title suggests, the main thread that will lead you through this book is the concept of a *function*. You have, no doubt, been introduced to this idea before, probably in algebra. But functional ideas are around us in many forms.

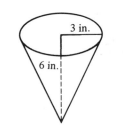

Consider an ice cream container shaped like a right circular cone, as shown in Figure 1.1. Suppose you experiment by filling the cone with different amounts of sand. You will find that the radius of the circle of sand in the cone is always $\frac{1}{2}$ the height (h) of the sand in the cone and that the volume (V) is found by

$$V = \frac{\pi h^3}{12}.$$

Figure 1.1   Ice cream cone

A table of values showing the volumes for different heights is given here.

Height	1	2	3	4	5	6
Volume	$\dfrac{\pi}{12}$	$\dfrac{2\pi}{3}$	$\dfrac{9\pi}{4}$	$\dfrac{16\pi}{3}$	$\dfrac{125\pi}{12}$	18π

Do you see how these values were obtained?
If $h = 1$, then

$$V = \frac{\pi(1)^3}{12} = \frac{\pi}{12}.$$

If $h = 2$, then

$$V = \frac{\pi(2)^3}{12} = \frac{8\pi}{12} = \frac{2\pi}{3}.$$

And so on.

INDEPENDENT
VARIABLE and
DEPENDENT VARIABLE
of a RELATION

This table can be written as a set of *ordered pairs*: $(1, \pi/12)$, $(2, 2\pi/3)$, $(3, 9\pi/4), \ldots$.

A set of ordered pairs is called a *relation*. In the example, the ordered pairs are (h, V) so that $V = \pi h^3/12$. The variable associated with the first component is called the *independent variable*, and the variable associated with the second component is called the *dependent variable*. In the example, if we are free to select a replacement value for h, then the value of V *depends* on this choice.

The set of possible replacements for the independent variable is sometimes limited to a certain set of numbers. For the cone we're considering, h (the height of the sand in the cone) cannot be negative (why?), and should not be larger than 6 in., since the conical container has this height. We say that the *domain* of the function is the set of possible replacements for the independent variable. Thus, the domain, D, is

DOMAIN

$$D = \{h \mid 0 < h \le 6\}.$$

This is the so-called *set-builder notation*. It is read "the set of all h such that 0 is less than h and h is less than or equal to 6."

The set of corresponding replacements for the dependent variable is the *range* of a relation.

It is often useful to draw a graph of a relation. To do so, we set up a *rectangular coordinate system*, as shown in Figure 1.2. This is also sometimes called a *Cartesian coordinate system*.

The horizontal number line is called the *x-axis* (sometimes called the *axis of abscissas*), and *x* represents the first component of the ordered pair. The vertical number line is called the *y-axis* (sometimes called the *axis of ordinates*), and *y* represents the second component of the ordered pair. The point of intersection of the axes is the *origin*. These number lines divide the plane into four regions called *quadrants*, as shown in Figure 1.2.

By plotting the points whose coordinates are shown in the table and then connecting them with a smooth curve, we arrive at the graph of this relation, as shown in Figure 1.3.

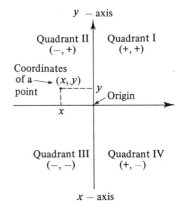

Figure 1.2   Cartesian coordinate system

Height	1	2	3	4	5	6
Volume	$\dfrac{\pi}{12}$	$\dfrac{2\pi}{3}$	$\dfrac{9\pi}{4}$	$\dfrac{16\pi}{3}$	$\dfrac{125\pi}{12}$	18π

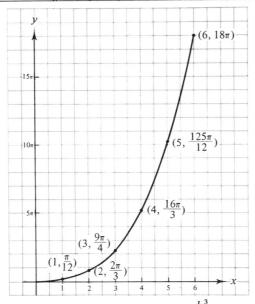

Figure 1.3   Graph of $V = \dfrac{\pi h^3}{12}$

In general, the equation of any graph sets up a *one-to-one correspondence* between points on the graph and ordered pairs satisfying the equation. A *graph of an equation* and an *equation of a graph* are closely connected: there is a one-to-one correspondence between the set of all ordered pairs (x, y) that satisfy the equation and the set of all points (x, y) that lie on the curve. The method of drawing the graph of an equation as shown here—that of plotting points—is very primitive, and one of the primary purposes of this book is to develop easier methods of graphing curves.

A *one-to-one correspondence* between the two sets is a correspondence in which each member of either set is paired with exactly one member of the other set.

Remember, the *domain* is the set of all permissible first components.

Look at the graph of Figure 1.3. Notice that for each value of x in the domain, there exists one corresponding value of y, and consequently one point on the curve, as shown in Figure 1.4.

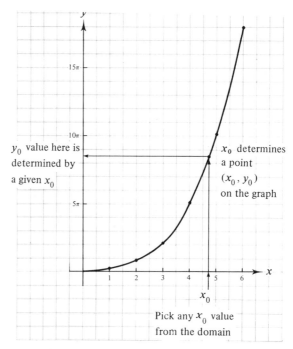

Figure 1.4 If you are given any value of x, call it x_0, a corresponding value of y, call it y_0, can be found.

Remember, the *range* is the set of all second components.

From Figure 1.4, it can be seen that the range, R, is given by

$$R = \{V \mid 0 < V \leq 18\pi\}.$$

This example illustrates a special type of relation called a *function*, which we shall describe in several ways.

The symbol x stands for a first component. The symbol $f(x)$ is read "f of x" and does not mean multiplication, but is a single symbol representing the second component of an ordered pair.

A FUNCTION IS A SET OF ORDERED PAIRS

DESCRIPTION 1: A function f is a set of ordered pairs for which each member of the domain, x, is associated with exactly one member of the range, $f(x)$.

EXAMPLES

1. $f = \{(0, 2), (1, 5), (2, 8), (3, 11)\}$
       $D = \{0, 1, 2, 3\}$ $R = \{2, 5, 8, 11\}$

    D   *R*
    $0 \to 2$ $f(0) = 2$
    $1 \to 5$ $f(1) = 5$
    $2 \to 8$ $f(2) = 8$
    $3 \to 11$ $f(3) = 11$

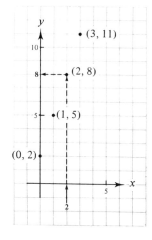

Notice that each value in the domain is associated with a single value in the range.

2. $f = \{(0, 0), (1, 1), (-1, 1), (2, 4), (-2, 4)\}$
       $D = \{0, 1, -1, 2, -2\}$ $R = \{0, 1, 4\}$

    D   *R*
    $0 \to 0$ $f(0) = 0$
    $1 \searrow$
       1 $f(1) = 1$
   $-1 \nearrow$ $f(-1) = 1$
    $2 \searrow$
       4 $f(2) = 4$
   $-2 \nearrow$ $f(-2) = 4$

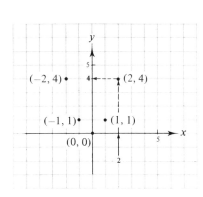

3.   Not all sets of ordered pairs are functions. For example,

    $\{(0, 0), (1, 1), (1, -1), (4, 2), (4, -2)\}$;
    $D = \{0, 1, 4\}$; $R = \{0, 1, -1, 2, -2\}$.

We do not use $f(x)$ notation unless the set of ordered pairs is a function.

Notice that each member of the domain is not associated with exactly one member of the range. To show that a relation is not a function, you need find only one member of the domain that is associated with more than one member of the range.

$$
\begin{array}{cc}
D & R \\
0 \rightarrow 0 \\
1 \begin{array}{c}\nearrow 1\\ \searrow 1\end{array} \\
4 \begin{array}{c}\nearrow 2\\ \searrow 2\end{array}
\end{array}
$$

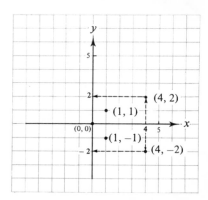

A FUNCTION IS A RULE

> DESCRIPTION 2: A function f is a rule that assigns a single real number $f(x)$ to each x in the domain of the function.

EXAMPLES

Functions do not have to be denoted by f; any letter can be used, such as f, g, h, T, f_1, f_2, ϕ, ...

4. The rule can be given verbally: T is the temperature in degrees Celsius at a certain location at some time x. Suppose the readings shown in the table are obtained (for $0 \le x < 24$; that is, for time measured on a 24-hour clock).

Time, x	Temperature, degrees Celsius	$T(x)$
6	8	$T(6) = 8$
10	14	$T(10) = 14$
12	17	$T(12) = 17$
14	18	$T(14) = 18$
18	14	$T(18) = 14$

The examples chosen have integral values from the domain. Can you find $V(3/2)$?

5. The rule can be given as a formula: $V = \pi h^3/12$. This formula can be restated using functional notation: $V(h) = \pi h^3/12$. From Figure 1.3, you find $V(1) = \pi/12$, $V(2) = 2\pi/3$, $V(3) = 9\pi/4$, $V(4) = 16\pi/3$, $V(5) = 125\pi/12$, and $V(6) = 18\pi$.

6. The rule can be given as an equation: $y = x^2$. This rule says "given a

member of the domain, x, square it to find the second component associated with x." Let $y = f(x)$. Then $f(x) = x^2$.

Replace x by 1

$f(1) = 1^2$

      $= 1$

$f(3) = 3^2$

      $= 9$

Replace x by 2

$f(2) = 2^2$

      $= 4$

$f\left(\dfrac{1}{2}\right) = \left(\dfrac{1}{2}\right)^2$

      $= \dfrac{1}{4}$

7. $g(x) = 3x^2 + 2x + 1$

Replace x by 2

$g(2) = 3 \cdot 2^2 + 2 \cdot 2 + 1$
    $= 17$

Replace x by -4

$g(-4) = 3(-4)^2 + 2(-4) + 1$
    $= 41$

8.   Not all equations represent functions. For example, compare

$$y = x^2 \qquad \text{and} \qquad x = y^2$$

For each x, there is exactly one y value as shown in Example 6. In this example, $y = f(x)$ so $f(x) = x^2$.

For some values of x, there are two values of y. For example, if $x = 4$, then $y = 2$ and $y = -2$ since $4 = 2^2$ and $4 = (-2)^2$. Thus, this is not a function so we do not use $f(x)$ notation.

Notice that each value in the domain is associated with a single value in the range.

A FUNCTION IS A MAPPING

> **DESCRIPTION 3:** A function f is a mapping that associates with each number x in the domain a unique number $f(x)$ in a set containing the range. We say that f maps a set X into a set Y.

$f(x)$ is sometimes called the *image* of x under the function f.

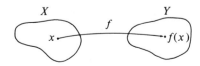

Notice that the range R of f is a subset of Y, and does not have to include all of Y. However, if the range and Y are equal, then we say X maps *onto* Y.

EXAMPLES

9.

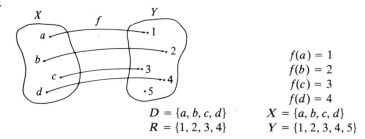

$$f(a) = 1$$
$$f(b) = 2$$
$$f(c) = 3$$
$$f(d) = 4$$

$D = \{a, b, c, d\}$ $X = \{a, b, c, d\}$
$R = \{1, 2, 3, 4\}$ $Y = \{1, 2, 3, 4, 5\}$

10.

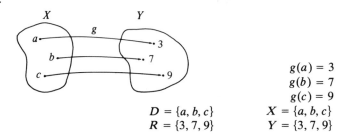

$$g(a) = 3$$
$$g(b) = 7$$
$$g(c) = 9$$

$D = \{a, b, c\}$ $X = \{a, b, c\}$
$R = \{3, 7, 9\}$ $Y = \{3, 7, 9\}$

If f maps X onto Y when x_1 and x_2 are distinct elements of X and $f(x_1) \neq f(x_2)$, then f is a *one-to-one mapping* of X onto Y. Example 11a illustrates a one-to-one mapping while 11b does not.

11. Reconsider Examples 1–3 as mappings.

a.

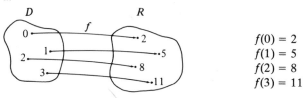

$$f(0) = 2$$
$$f(1) = 5$$
$$f(2) = 8$$
$$f(3) = 11$$

b.

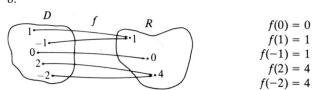

$$f(0) = 0$$
$$f(1) = 1$$
$$f(-1) = 1$$
$$f(2) = 4$$
$$f(-2) = 4$$

c.   Example 3 illustrates a mapping that is not a function.

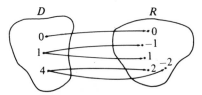

PROBLEM SET 1.1

A Problems

Classify the sets of Problems 1–14 as relations, functions, both, or neither.

1. $\{(4, 7), (3, 4), (5, 4), (6, 9)\}$
2. $\{(5, 2), (7, 3), (1, 6), (7, 4)\}$
3. $\{(8, 2), (7, 1), (6, 3), (5, 1)\}$
4. $\{6, 9, 12, 15\}$
5. $\{1, 2, 3, 4\}$
6. $\{(x, y) \mid y = 4x + 3\}$
7. $\{(x, y) \mid y \leq 4x + 3\}$
8. $\{(x, y) \mid y = -1$ if x is a rational number and $y = 1$ if x is an irrational number$\}$
9. $\{(x, y) \mid y = 1$ if x is positive and $y = -1$ if x is negative$\}$

$\$ \begin{cases} 10. & \{(x, y) \mid y = \text{closing price of IBM stock on January 2 of year } x\} \\ 11. & \{(x, y) \mid x = \text{closing price of Xerox stock on July 1 of year } y\} \end{cases}$

$\begin{cases} 12. & \{(x, y) \mid (x, y) \text{ is a point on a circle of radius 4 centered at } (2, 3)\} \\ 13. & \{(x, y) \mid (x, y) \text{ is a point on a line passing through } (2, 3) \text{ and } (4, 5)\} \\ 14. & \{(x, y) \mid (x, y) \text{ is a point on a line passing through } (4, 5) \text{ and } (-3, 5)\} \end{cases}$

For Problems 15–18, use the table below, which reflects the purchasing power of the dollar from October 1944 to October 1974 (Source: U.S. Bureau of Labor Statistics, Consumer Division). Let x represent the year; let the domain be the set {1944, 1954, 1964, 1974}; and let

$r(x) = $ *price of 1 lb of round steak,*
$s(x) = $ *price of a 5-lb bag of sugar,*
$b(x) = $ *price of a loaf of bread,*
$c(x) = $ *price of 1 lb of coffee,*
$e(x) = $ *price of a dozen eggs,*
$m(x) = $ *price of $\frac{1}{2}$ gal of milk,*
$g(x) = $ *price of 1 gal of gasoline.*

Year	Round steak (1 lb)	Sugar (5-lb bag)	Bread (loaf)	Coffee (1 lb)	Eggs (1 dozen)	Milk ($\frac{1}{2}$ gal)	Gasoline (1 gal)
1944	$0.45	$0.34	$0.09	$0.30	$0.64	$0.29	$0.21
1954	0.92	0.52	0.17	1.10	0.60	0.45	0.29
1964	1.07	0.59	0.21	0.82	0.57	0.48	0.30
1974	1.78	2.08	0.36	1.31	0.84	0.78	0.53

$\$$ 15. Find:
   a. $r(1954)$ b. $m(1954)$ c. $g(1944)$ d. $c(1974)$

$\$$ 16. Find:
   a. $s(1974) - s(1944)$ b. $b(1974) - b(1944)$

$\$$ 17. a. Find the change in the price of eggs from 1944 to 1974.
   b. Write the change in the price of eggs using functional notation.

Historical Note

The word *function* was used as early as 1694 by the great universal genius of the 17th century, Gottfried von Leibniz (1646–1716), to denote any quality connected with a curve. The notion was generalized and modified by Johann Bernoulli (1667–1748) and by Leonhard Euler (1707–1783), the most prolific mathematical writer in history. Euler was the first to use the notation $f(x)$ in 1734. The terminology we're using in this book is primarily due to the work of P. G. Lejeune-Dirichlet (1805–1859).

$ 18. a. Find $\dfrac{g(1944 + 30) - g(1944)}{30}$.

b. In words, can you attach any meaning to the number found in part a?

In Problems 19–24, let $f(x) = 3x^2$, $g(x) = 4x - 1$, $h(x) = 5 - 2x$, $k(x) = 500x$, and $V(r) = 2\pi r^3$. Find and plot the ordered pairs specified for each problem.

19. a. $f(-1)$ b. $f(0)$ c. $f(1)$ d. $f(2)$
20. a. $g(-2)$ b. $g(0)$ c. $g(2)$ d. $g(4)$
21. a. $h(5)$ b. $h(6)$ c. $h(8)$ d. $h(10)$
22. a. $k(5)$ b. $k(6)$ c. $k(7)$ d. $k(8)$
23. a. $k(0.01)$ b. $k(0.05)$ c. $k(0.08)$ d. $k(0.1)$
24. a. $V(1)$ b. $V(2)$ c. $V(3)$ d. $V(4)$

B Problems

$ 25. Use the table given for Problems 15–18.
   a. What is the average increase in the price of sugar per year from 1944 to 1954? Write this using functional notation.
   b. What is the average increase in the price of sugar per year from 1944 to 1964? Write this using functional notation.
   c. What is the average increase in the price of sugar per year from 1944 to 1974? Write this using functional notation.
   d. What is the average increase in the price of sugar per year from 1944 to 1944 + h, where h is an unspecified number of years?

$ 26. Use the table given for Problems 15–18 and answer the questions posed in Problem 25, but this time for the price of coffee.

THINK METRIC 27. *Calculator Problem.* If an object is projected vertically from the ground at 112.43 meters per second (mps), then (neglecting air resistance) the distance of the object above the ground at time *t* (in seconds) is given by the equation

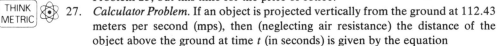

$$d = 112.43t - 4.87t^2.$$

What is the height at $t = 1.4$ sec and at $t = 2.6$ sec? If the relation is specified by (t, d), what would impose a limitation on the domain values? Specify a domain to generate nonnegative values of *d*.

28. *Calculator Problem.* The projectile of Problem 27 has velocity *v* at time *t* given by

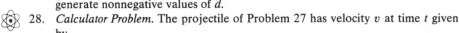

$$v = 112.43 - 9.75t.$$

What is the velocity for $t = 1.4$ sec and $t = 2.6$ sec? If the relation is specified by (t, v), what would impose a limitation on the domain values? Specify a domain to generate nonnegative values of *v*.

29. In Figure 1.5, if point *A* has coordinates $(2, f(2))$, what are the coordinates of *P* and *Q*?

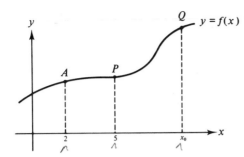

Figure 1.5 Graph of f

30. In Figure 1.6, if point A has coordinates $(3, g(3))$, what are the coordinates of P and Q?

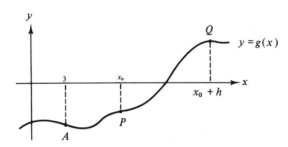

Figure 1.6 Graph of g

31. If F is a one-to-one function mapping X onto Y, and the domain of F contains exactly 5 elements, what can you conclude about the set Y?

32. If G maps X into Y, and the range of F is a set S, where S is a subset of Y, explain why f maps X onto S.

Mind Boggler

33. Functions are sometimes described as machines: a function is a machine that assigns to each "input" number a unique "output" number. A table of values from the squaring function machine is given below.

Input values	Output values
1	1
2	4
3	9
−5	25

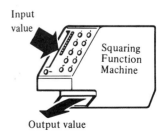

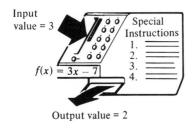

Input value = 3

$f(x) = 3x - 7$

Special Instructions
1.
2.
3.
4.

Output value = 2

a. Suppose you are given a machine that multiplies the input value by 3 and then subtracts 7. Complete the table below.

Input values	Output values
3	2
5	
0	
−3	

b. A secret machine is introduced, and the following table is found:

Smith's Secret Function Machine

?

Input values	Output values
0	3
1	5
2	7
3	9
4	11

Write a verbal description of how this secret function machine might work.

c. A super secret machine is introduced, and the following table is found:

Input values	Output values
0	5
1	6
2	9
3	14
4	21

Write a verbal description of how this machine works.

1.2 FUNCTIONAL NOTATION

The notation for functions introduced in the last section is fundamental for this course and for calculus. In this section, we'll provide additional examples and practice in using it. Since you are used to writing ordered pairs as (x, y) and plotting points (x, y), we can let $y = f(x)$ to emphasize the fact that the second component is *determined by* the first component, x.

Remember:

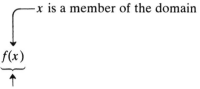

x is a member of the domain

$f(x)$

$f(x)$ is a member of the range

The *function* is denoted by f; $f(x)$ is the *number* associated with x. Sometimes functions are defined by expressions such as

$$f(x) = 3x + 2 \quad \text{or} \quad g(x) = x^2 + 4x + 3,$$

but remember that the functions are f and g (recall the three descriptions of a function from the last section).

To emphasize this difference between f and $f(x)$, some books use

$$f: x \rightarrow 3x + 2$$

to define functions, but we'll use the notation $f(x) = 3x + 2$ since it is a little more compact.

EXAMPLES: Given f and g defined by

$$f(x) = 3x + 2 \quad g(x) = x^2 + 4x + 3,$$

find the indicated values.

1. Replace x by 1 in the expression

$$f(x) = 3x + 2$$
$$\updownarrow \quad \updownarrow$$
$$f(1) = 3(1) + 2$$
$$= 5$$

2. $g(2)$: $g(x) = x^2 + 4x + 3$
$$\updownarrow \quad \updownarrow \quad \updownarrow$$
$$g(2) = (2)^2 + 4(2) + 3$$
$$= 4 \ + 8 \ + 3$$
$$= 15$$

The symbol $f(1)$ represents the second component of the ordered pair of the function f with first component 1.

3. $g(-3) = (-3)^2 + 4(-3) + 3$
$$= 9 \ - 12 \ + 3$$
$$= 0$$

4. $f(-3) = 3(-3) + 2$
$$= -9 \ + 2$$
$$= -7$$

The members of the domain of a function may also be variables, as shown in the Examples 5–12.

EXAMPLES: Let F and G be defined by

$$F(x) = x^2 + 1 \quad \text{and} \quad G(x) = (x + 1)^2.$$

5. $F(w) = w^2 + 1$

6. $G(t) = (t + 1)^2$
    $= t^2 + 2t + 1$

7. $F(w + 3) = (w + 3)^2 + 1$
    $= w^2 + 6w + 9 + 1$
    $= w^2 + 6w + 10$

8. $G(w + 3) = [(w + 3) + 1]^2$
    $= (w + 4)^2$
    $= w^2 + 8w + 16$

9. $F(x - 2) = (x - 2)^2 + 1$
    $= x^2 - 4x + 4 + 1$
    $= x^2 - 4x + 5$

10. $G(x - 2) = [(x - 2) + 1]^2$
     $= (x - 1)^2$
     $= x^2 - 2x + 1$

11. $F(w + h) = (w + h)^2 + 1$
     $= w^2 + 2wh + h^2 + 1$

12. $G(w + h) = [(w + h) + 1]^2$
     $= w^2 + wh + w + wh + h^2 + h + w + h + 1$
     $= w^2 + 2wh + h^2 + 2w + 2h + 1$

In Problem 18 of the previous section, you were asked to find the average yearly price change of gasoline from 1944 to 1974. To work this problem using functional notation, write

From the table on page 9, $g(x)$ = price of gasoline in year x:

$g(1944) = \$0.21$
$g(1954) = \$0.29$
$g(1964) = \$0.30$
$g(1974) = \$0.53$

$$\frac{g(1974) - g(1944)}{30} = \frac{\$0.53 - \$0.21}{30}$$

$$= \frac{\$0.32}{30}$$

$$\approx \$0.01.$$

Suppose that we assume that $g(1984) = \$1.53$. What is the average price change per year from 1974 to 1984?

$$\underbrace{\frac{g(1984) - g(1974)}{10}}_{\text{length of time in years}} = \frac{\$1.53 - \$0.53}{10}$$

$$= \frac{\$1.00}{10}$$

$$= \$0.10 \quad \text{or about 10¢ per year increase}$$

What would the average price change per year be in h years? Since we don't know what $g(1974 + h)$ will be, we can answer this question using functional notation:

$$\frac{g(1974 + h) - g(1974)}{h}.$$

EXAMPLES: Find $\dfrac{f(x + h) - f(x)}{h}$ for each function.

13. $f(x) = x^2$, where $x = 5$

$$\frac{f(5 + h) - f(5)}{h} = \frac{(5 + h)^2 - 5^2}{h}$$

$$= \frac{25 + 10h + h^2 - 25}{h}$$

$$= \frac{(10 + h)h}{h}$$

$$= 10 + h$$

14. $f(x) = 2x^2 + 1$, where $x = 1$

$$\frac{f(1 + h) - f(1)}{h} = \frac{[2(1 + h)^2 + 1] - [2(1)^2 + 1]}{h}$$

$$= \frac{[2(1 + 2h + h^2) + 1] - [2 + 1]}{h}$$

$$= \frac{2h^2 + 4h + 3 - 3}{h}$$

$$= 2h + 4$$

15. $f(x) = 1/x$, where $x = t$

$$\frac{f(t + h) - f(t)}{h} = \frac{\dfrac{1}{t + h} - \dfrac{1}{t}}{h}$$

$$= \frac{t - (t + h)}{t(t + h)} \cdot \frac{1}{h}$$

$$= \frac{-h}{th(t + h)}$$

$$= \frac{-1}{t(t + h)}$$

16. $f(x) = x^2 + 3x - 2$

$$\frac{f(x + h) - f(x)}{h} = \frac{[(x + h)^2 + 3(x + h) - 2] - [x^2 + 3x - 2]}{h}$$

$$= \frac{x^2 + 2xh + h^2 + 3x + 3h - 2 - x^2 - 3x + 2}{h}$$

$$= \frac{2xh + h^2 + 3h}{h}$$

$$= 2x + 3 + h$$

Functional notation can be used to work a wide variety of applied problems, as shown by Examples 17–19 and again in the problem set.

EXAMPLES

17. If an object is dropped from a certain height, it is known that it will fall a distance of s ft in t sec according to the formula

$$s(t) = 16t^2.$$

a. How far will the object fall in the first second?

$$s(1) = 16 \cdot 1^2$$
$$= 16 \quad \text{or 16 ft}$$

b. How far will it fall in the *next* 2 sec?

$$s(1 + 2) = 16 \cdot 3^2$$
$$= 144 \quad \text{or 144 ft in 3 sec}$$

So the answer to the question is

$$s(3) - s(1) = 144 - 16$$
$$= 128 \quad \text{or 128 ft}$$

c. How far will it fall in the next h sec?

$$s(1 + h) - s(1) = 16(1 + h)^2 - 16$$
$$= 16 + 32h + 16h^2 - 16$$
$$= (32h + 16h^2)\,\text{ft}$$

d. What is the average rate of change (feet per second, fps) during the time $t = 1$ sec to $t = 3$ sec?

$$\frac{s(3) - s(1)}{3 - 1} = \frac{128}{2} = 64 \text{ fps}$$

e. What is the average rate of change of distance during the time
$t = 1$ sec to $t = h$ sec?

$$\frac{s(1 + h) - s(1)}{h} = \frac{32h + 16h^2}{h}$$

$$= (32 + 16h)\,\text{fps}$$

f. What is the average rate of change of distance during the time
$t = x$ sec to $t = x + h$ sec?

$$\frac{s(x + h) - s(x)}{(x + h) - x} = \frac{s(x + h) - s(x)}{h}$$

Does this look familiar?

$$= \frac{16(x + h)^2 - 16x^2}{h}$$

$$= \frac{16x^2 + 32xh + 16h^2 - 16x^2}{h}$$

$$= (32x + 16h)\,\text{fps}$$

18. Using Example 17f, find the average velocity when $x = 2$ and h is given
below.

a. $h = 3$; $32(2) + 16(3)$ $= 64 + 48$ $= 112$ fps
b. $h = 2$; $32(2) + 16(2)$ $= 64 + 32$ $= 96$ fps
c. $h = 1$; $32(2) + 16(1)$ $= 64 + 16$ $= 80$ fps
d. $h = 0.5$; $32(2) + 16(0.5)$ $= 64 + 8$ $= 72$ fps
e. $h = 0.1$; $32(2) + 16(0.1)$ $= 64 + 1.6$ $= 65.6$ fps
f. $h = 0.01$; $32(2) + 16(0.01) = 64 + 0.16$ $= 64.16$ fps

19. A firm determines that the total cost C (in dollars) of producing x units of
a certain product is given by

$$C(x) = -0.02x^2 + 4x + 500 \quad (0 \le x \le 150).$$

a. Find $C(50)$ and $C(100)$.

$$C(50) = -0.02(50)^2 + 4(50) + 500$$
$$= 650 \quad \text{or } \$650$$
$$C(100) = -0.02(100)^2 + 4(100) + 500$$
$$= 700 \quad \text{or } \$700$$

b. What is the average cost per unit if 50 and 100 units are produced?

$$\frac{C(50)}{50} = \frac{650}{50} = \$13 \qquad \frac{C(100)}{100} = \frac{700}{100} = \$7$$

c. What is the per unit increase in cost for the 50-unit increase?

$$\frac{C(50 + 50) - C(50)}{50} = \frac{700 - 650}{50} = \$1$$

d. What is the per unit increase in cost for a 1-unit increase?

$C(51) = -0.02(51)^2 + 4(51)$
$+ 500$
$= \$651.98$

$$\frac{C(50 + 1) - C(50)}{1} = \frac{651.98 - 650}{1} = \$1.98$$

e. What is the per unit increase in cost for an h-unit increase in cost above a production level of x units?

Ah, ah—it happened again!

$$\frac{C(x + h) - C(x)}{(x + h) - x} = \frac{C(x + h) - C(x)}{h}$$

$$= \frac{[-0.02(x + h)^2 + 4(x + h) + 500] - [-0.02x^2 + 4x + 500]}{h}$$

$$= \frac{-0.02x^2 - 0.04xh - 0.02h^2 + 4x + 4h + 500 + 0.02x^2 - 4x - 500}{h}$$

$$= -0.04x - 0.02h + 4$$

We can verify this result for part d above where $x = 50$ and $h = 1$:

$$-0.04(50) - 0.02(1) + 4 = \$1.98.$$

PROBLEM SET 1.2

A Problems

In Problems 1–10, let $f(x) = 2x + 1$ and $g(x) = 2x^2 - 1$. Find the requested values.

1. a. $f(0)$ b. $f(2)$ c. $f(-3)$ d. $f(\sqrt{5})$
   e. $f(\pi)$

2. a. $f(1)$ b. $g(1)$ c. $f(\sqrt{3})$ d. $g(\sqrt{3})$
   e. $g(1 + \sqrt{3})$

3. a. $f(w)$ b. $g(w)$ c. $g(t)$ d. $g(v)$
   e. $f(m)$

4. a. $f(t)$ b. $f(p)$ c. $f(t + 1)$ d. $g(t + 1)$
   e. $f(t^2)$

5. a. $f(1 + \sqrt{2})$ b. $g(1 + \sqrt{2})$ c. $g(t + 3)$ d. $f(t^2 + 2t + 1)$
   e. $g(m - 1)$

6. a. $f(x + 2)$ b. $g(x + 2)$ c. $f(t + h)$ d. $g(t + h)$
   e. $f(x + h)$

7. a. $3f(x + 2) - 4g(x + 2)$ b. $f(t^2 + 2t + 1) - g(t + 3)$

8. a. $\dfrac{f(t + 3) - f(t)}{3}$ b. $\dfrac{g(t + 2) - g(t)}{2}$

9. a. $\dfrac{f(t + h) - f(t)}{h}$ b. $\dfrac{g(t + h) - g(t)}{h}$

10. a. $\dfrac{f(x + h) - f(x)}{h}$ b. $\dfrac{g(x + h) - g(x)}{h}$

In Problems 11–16, compute the given value where $f(x) = x^2 - 1$ *and* $g(x) = 2x + 5$.

11. a. $f(w)$ b. $f(h)$ c. $f(w + h)$ d. $f(w) + f(h)$
12. a. $g(s)$ b. $g(t)$ c. $g(s + t)$ d. $g(s) + g(t)$
13. a. $f(x^2)$ b. $f(\sqrt{x})$ c. $f(x + h)$ d. $f(-x)$
14. a. $g(x^2)$ b. $g(\pi)$ c. $g(x + \pi)$ d. $g(-x)$

15. a. $f(x + h) - f(x)$ b. $\dfrac{f(x + h) - f(x)}{h}$

16. a. $g(x + h) - g(x)$ b. $\dfrac{g(x + h) - g(x)}{h}$

B Problems

In Problems 17–22, find $\dfrac{f(x + h) - f(x)}{h}$ *for the given function f.*

17. $f(x) = 5x^2$ 18. $f(x) = 3x^2 + 2x$

19. $f(x) = \dfrac{1}{x}$ 20. $f(x) = \dfrac{x + 1}{x - 1}$

21. $f(x) = \dfrac{1}{x^2 + 1}$ 22. $f(x) = 2x^2 + 3x - 4$

\$ 23. An advertising agency conducted a survey and found that the number of units sold, N, is related to the amount, a, spent on advertising (in thousands of dollars) by the formula

$$N(a) = -15a^2 + 180a + 100 \quad (0 \le a \le 10).$$

Find $N(0)$, $N(1)$, $N(2), \ldots, N(10)$ and plot $(0, N(0))$, $(1, N(1))$, $(2, N(2)), \ldots, (10, N(10))$. Connect these points with a smooth curve.

\$ 24. For Problem 23, find the average rate of change of N as a changes from
a. 0 to 2   b. 2 to 4   c. 4 to 6   d. 6 to 8
e. Give an interpretation to the rates you found in parts a–d.

\$ 25. In Problem 23, find the average rate of change of N as a changes from 3 to
a. 6   b. 5   c. 4   d. 3.5   e. $3 + h$ Hint: $N(3.5) = 546.25$
f. Using part e, find it for 3.1.

\$ 26. In Problem 23, find the average rate of change of N as a changes from x to $x + h$.

ψ 27. In a psychology experiment, the subject was asked to memorize a list of words. The total number of words memorized, $M(t)$, is a function of the time, t (in minutes), spent in memorizing the list. The following data were obtained:

Time, t	10	20	25	30	35	40
Number of words, $M(t)$	10	19	23	25	25	25

$10 \le t \le 60$

Find the average rate of change of M per minute as t changes from
a.   10 to 20   b.   20 to 25   c.   25 to 30   d.   30 to 35
e.   Give an interpretation to the rates you found in parts a–d.

 28.   In Problem 27, find the average rate of change from
a.   10 to 30   b.   10 to 25   c.   10 to 20   d.   10 to 10 + h
e. x to $x + h$

 29.   Let d be a function that represents the distance an object falls (neglecting air resistance) in t sec. Find the distance the object falls for the intervals of time given.

Hint: $d(t) = 16t^2$

a.   From $t = 2$ to $t = 6$ b.   From $t = 2$ to $t = 4$
c.   From $t = 2$ to $t = 3$ d.   From $t = 2$ to $t = 2 + h$
e.   From $t = x$ to $t = x + h$

30.   In Problem 29, give a physical interpretation for

$$\frac{d(x + h) - d(x)}{h}.$$

31.   *Calculator Problem.* According to the U.S. Public Health Service, the number of marriages in the United States in 1970 was about 2,158,000 and in 1974 it was about 2,223,000. If $M(x)$ represents the number of marriages in year x, answer the following questions:

a.   Find $\dfrac{M(1974) - M(1970)}{4}$.

b.   Give a verbal description for the following functional expression:

$$\frac{M(1970 + h) - M(1970)}{h}.$$

32.   *Calculator Problem.* According to the U.S. Public Health Service, the number of divorces in the United States in 1970 was about 708,000 and in 1974 it was about 970,000. If $D(x)$ represents the number of divorces in year x, answer the following questions:

a.   Find $\dfrac{D(1974) - D(1970)}{4}$.

b.   Give a verbal description for the following functional expression:

$$\frac{D(1970 + h) - D(1970)}{h}.$$

33.   *Calculator Problem.* It took 4 million years to reach a worldwide human population of 1 billion by 1830. Subsequent population figures are given in the table below:

If we project the current rate of increase in population (2%) into the future, we'll see the following population changes:
    1987 (12 years); 5 billion
    1996 (9 years); 6 billion
    2004 (7½ years); 7 billion
    2010 (6 years); 8 billion

Year	Time span	World population (in billions), $P(t)$
1830	4 million years	1
1930	100 years	2
1960	30 years	3
1975	15 years	4

a. Find the growth rate (the average rate of change of the population with respect to time) for 1830 to 1930; 1930 to 1960; and 1960 to 1975.
b. Calculate the percentage rate of increases for the years 1930, 1960, and 1975 based on your calculations for part a.

Mind Boggler

34. Four snails start at the vertices of a unit square and move directly toward one another in cyclic order, at unit rate. How far will they travel before they meet?

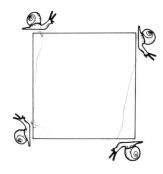

1.3   ALGEBRA OF FUNCTIONS

In algebra, you spent a great deal of time learning the algebra of real numbers. We can also consider an algebra of functions.

FUNCTION ADDITION AND SUBTRACTION

Let $f(x) = \frac{1}{2}x^2$ and $g(x) = x + 2$. These functions can be graphed by plotting points, as shown in Figure 1.7.

x	$f(x)$	$g(x)$	$(f + g)(x)$
-4	8	-2	6
-3	$\frac{9}{2}$	-1	$\frac{7}{2}$
-2	2	0	2
-1	$\frac{1}{2}$	1	$\frac{3}{2}$
0	0	2	2
1	$\frac{1}{2}$	3	$\frac{7}{2}$
2	2	4	6
3	$\frac{9}{2}$	5	$\frac{19}{2}$
4	8	6	14

Next, consider the function defined by $A(x) = f(x) + g(x)$. The function A is denoted by $f + g$.

NOTATION FOR THE
SUM OF FUNCTIONS

$$(f + g)(x) = f(x) + g(x)$$

$$= \frac{1}{2}x^2 + (x + 2)$$

$$= \frac{1}{2}x^2 + x + 2$$

This curve can be graphed by plotting points as shown in Figure 1.7 (dotted curve). Notice from the table that the values for $(f + g)(x)$ can be obtained by adding $f(x) + g(x)$ rather than by calculating $(f + g)(x) = \frac{1}{2}x^2 + x + 2$. This gives an idea about how $(f + g)(x)$ can be graphed.

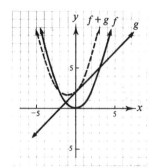

Figure 1.7   Graphs of $f(x)$ and $g(x)$

Consider a closeup portion of Figure 1.7. It is not necessary that you actually calculate the functional values. That is, choose any x_0, as shown in Figure 1.8. Then the functional values $f(x_0)$ can be represented as distances. By choosing different x values, the entire graph of $(f + g)(x)$ can be generated. This method of graphing is called *graphing by adding ordinates*.

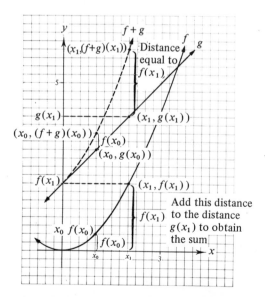

Figure 1.8   Detail of Figure 1.7 showing the procedure for graphing by adding ordinates

EXAMPLES

1. Let f and g be the functions shown in Figure 1.9. Graph $f + g$ by adding ordinates. The result is shown in Figure 1.9.

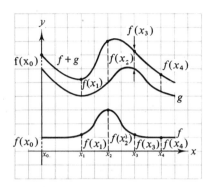

Figure 1.9   Graphs of f, g, and $f + g$

2. Let $f(x) = (1/5)x^2$ and $g(x) = (3/2)x + 2$. The graphs are shown in Figure 1.10. Graph

$$(f + g)(x) = \frac{1}{5}x^2 + \frac{3}{2}x + 2$$

by adding ordinates. The result is shown by a dotted curve in Figure 1.10.

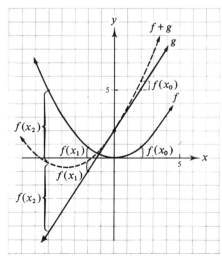

Figure 1.10   Graphs of f, g, and $f + g$

Subtraction of functions is handled similarly:

$$(f - g)(x) = f(x) - g(x).$$

If $f(x) = \frac{1}{2}x^2$ and $g(x) = x + 2$, then

$$(f - g)(x) = f(x) - g(x)$$

$$= \frac{1}{2}x^2 - (x + 2)$$

$$= \frac{1}{2}x^2 - x - 2.$$

For f and g, the domain of both f and g is the set of real numbers, so the domain of $f + g$ and $f - g$ is also the set of real numbers. In general, the domains of $f + g$ and $f - g$ are the intersection of the domains of f and g.

EXAMPLES: Find $f + g$, $f - g$, and evaluate these functions for $x = -1$ and $x = 5$.

The domain of both f and g is the set of real numbers, so the domains of $f + g$ and $f - g$ are also the set of real numbers.

3. $f(x) = x^2, \quad g(x) = x + 4$

$$(f + g)(x) = f(x) + g(x) = x^2 + x + 4$$
$$(f + g)(-1) = (-1)^2 + (-1) + 4$$
$$= 4$$
$$(f + g)(5) = 5^2 + 5 + 4$$
$$= 34$$
$$(f - g)(x) = f(x) - g(x) = x^2 - (x + 4)$$
$$= x^2 - x - 4$$
$$(f - g)(-1) = (-1)^2 - (-1) - 4$$
$$= -2$$
$$(f - g)(5) = 5^2 - 5 - 4$$
$$= 16$$

4. $f(x) = x^2 \quad$ for $-2 \le x \le 8, \quad g(x) = x + 3 \quad$ for $-3 \le x \le 4$

The intersection of $-2 \le x \le 8$ and $-3 \le x \le 4$ is $-2 \le x \le 4$, as shown below:

$$\underline{\qquad -2 \le x \le 8 \qquad}$$

$$\underline{\quad -3 \le x \le 4 \quad}$$

Intersection

$$(f + g)(x) = x^2 + x + 3 \qquad \text{for } -2 \le x \le 4$$
$$(f + g)(-1) = (-1)^2 + (-1) + 3$$
$$= 3$$

$(f + g)(5)$ is not defined since 5 is not in the domain.

$$(f - g)(x) = x^2 - x - 3 \quad \text{for } -2 \le x \le 4$$
$$(f - g)(-1) = (-1)^2 - (-1) - 3$$
$$= -1$$

$(f - g)(5)$ is not defined since 5 is not in the domain.

5. $f = \{(0, 0), (-1, 1), (2, 4), (3, 9), (5, 25)\}$
    $g = \{(0, -5), (-1, 0), (2, -3), (4, -2), (5, -1)\}$

The domain of f is $\{0, -1, 2, 3, 5\}$
The domain of g is $\{0, -1, 2, 4, 5\}$
The domain of $f + g$ and $f - g$ is the intersection: $\{0, -1, 2, 5\}$.

$$(f + g)(0) \ = f(0) + g(0) \ = 0 + (-5) \ = -5$$
$$(f + g)(-1) = f(-1) + g(-1) = 1 + 0 \ \ \ = 1$$
$$(f + g)(2) \ = f(2) + g(2) \ \ = 4 + (-3) \ = 1$$
$$(f + g)(5) \ = f(5) + g(5) \ \ = 25 + (-1) = 24$$

Thus,

$$f + g = \{(0, -5), (-1, 1), (2, 1), (5, 24)\},$$
$$f - g = \{(0, 5), (-1, 1), (2, 7), (5, 26)\}.$$

Now, by inspection,

$$(f + g)(-1) = 1 \qquad (f - g)(-1) = 1$$
$$(f + g)(5) = 24 \qquad (f - g)(5) = 26.$$

The work for $f - g$ is not shown. Can you fill in the details?

FUNCTION MULTIPLICATION AND DIVISION

Let $f(x) = 2x + 1$ and $g(x) = x^2$. The functions defined by $P(x) = f(x) \cdot g(x)$ and $Q(x) = f(x)/g(x)$ are denoted by $(fg)(x)$ and $(f/g)(x)$, respectively.

NOTATION FOR THE PRODUCT AND QUOTIENT OF FUNCTIONS

$$(fg)(x) = f(x) \cdot g(x) \qquad\qquad (f/g)(x) = \frac{f(x)}{g(x)}$$

$$= (2x + 1)(x^2) \qquad\qquad = \frac{2x + 1}{x^2} \quad (x \neq 0)$$

$$= 2x^3 + x^2$$

The domain of fg is the intersection of the domains of f and g, and the domain of f/g is the intersection of the domains of f and g for which values causing $g(x) = 0$ are excluded.

EXAMPLES: Find fg and f/g and evaluate these for $x = -1$ and $x = 5$.

6. $f(x) = x^2, g(x) = x + 4$

$$(fg)(x) = f(x) \cdot g(x)$$
$$= x^2(x + 4)$$
$$= x^3 + 4x^2$$
$$(fg)(-1) = (-1)^3 + 4(-1)^2$$
$$= 3$$
$$(fg)(5) = 5^3 + 4(5)^2$$
$$= 225$$

$$(f/g)(x) = \frac{f(x)}{g(x)}$$

$$= \frac{x^2}{x + 4} \quad (x \neq -4)$$

$$(f/g)(-1) = \frac{(-1)^2}{-1 + 4}$$

$$= \frac{1}{3}$$

$$(f/g)(5) = \frac{5^2}{5 + 4}$$

$$= \frac{25}{9}$$

7. $f = \{(0, 0), (-1, 1), (2, 4), (3, 9), (5, 25)\}$
    $g = \{(0, -5), (-1, 0), (2, -3), (4, -2), (5, -1)\}$

As in Example 5, the intersection of the domains is $\{0, -1, 2, 5\}$.

$(fg)(0)$	$= f(0) \cdot g(0)$	$= 0(-5)$	$= 0$
$(fg)(-1)$	$= f(-1) \cdot g(-1)$	$= 1(0)$	$= 0$
$(fg)(2)$	$= f(2) \cdot g(2)$	$= 4(-3)$	$= -12$
$(fg)(5)$	$= f(5) \cdot g(5)$	$= 25(-1)$	$= -25$

For f/g, we must also exclude from the domain any values of x for which g is 0; thus, -1 is excluded, and the domain of f/g is $\{0, 2, 5\}$.

In answer to part of the question, notice that

$(fg)(-1) = 0;$

$(fg)(5) = -25.$

$$(f/g)(0) = \frac{f(0)}{g(0)} = \frac{0}{-5} = 0$$

$$(f/g)(2) = \frac{f(2)}{g(2)} = \frac{4}{-3} = -\frac{4}{3}$$

$(f/g)(5) = -25$

$(f/g)(-1)$ is not defined.

$$(f/g)(5) = \frac{f(5)}{g(5)} = \frac{25}{-1} = -25$$

Thus, $f/g = \{(0, 0), (2, -4/3), (5, -25)\}$, and $(f/g)(-1)$ is not defined.

COMPOSITION OF FUNCTIONS

In addition to the obvious ways of combining functions by adding, subtracting, multiplying, and dividing, there is another way functions can be combined called *composition of functions*. As an example of composition, suppose the cost of manufacturing x calculators is $50x + 200$ (dollars). This means that the cost of each calculator is

$$c(x) = \frac{50x + 200}{x} \quad (x > 0),$$

so that the cost depends on the number of calculators produced. Suppose also that a store sells these calculators by marking up the price 20%. That is, for an item costing k dollars the price is

$$p(k) = k + 0.20k$$

$$= 1.20k.$$

Ultimately, the price depends on the number produced by finding

$$p[c(x)] = p\left[\frac{50x + 200}{x}\right]$$

$$= 1.20\left(\frac{50x + 200}{x}\right)$$

$$= 60 + \frac{240}{x}.$$

This function is called the composition of c by p and is denoted by $p \circ c$.

$p \circ c$ is read "p circle c."

DEFINITION: The composite function of f by g, denoted by $g \circ f$, is the function defined by

$$(g \circ f)(x) = g[f(x)].$$

The domain of $g \circ f$ is the subset of the domain of f containing those values for which $g \circ f$ is defined (see Figure 1.11).

DEFINITION OF
COMPOSITION

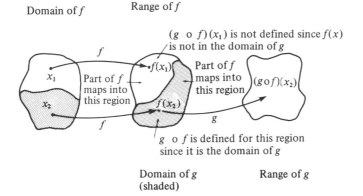

Domain of f Range of f

$(g \ o \ f)(x_1)$ is not defined since $f(x)$ is not in the domain of g

f

x_1

Part of f maps into this region

$\cdot f(x_1)$

Part of f maps into this region

$(g \circ f)(x_2)$

x_2

$f(x_2)$

f

g

$g \ o \ f$ is defined for this region since it is the domain of g

Domain of g Range of g
(shaded)

Figure 1.11   Composition, $g \circ f$, is defined for the subset of the domain of f for which $g \circ f$ is defined (shaded portion of the domain of f).

EXAMPLES

8. If $f(x) = x^2$ and $g(x) = x + 4$, find:

   a. $(g \circ f)(x)$ b. $(f \circ g)(x)$ c. $(g \circ f)(-1)$ d. $(f \circ g)(5)$

   Solution:

   a. $(g \circ f)(x) = g[f(x)]$ b. $(f \circ g)(x) = f[g(x)]$
   $= g[x^2]$ $= f[x + 4]$
   $= x^2 + 4$ $= (x + 4)^2$
   $= x^2 + 8x + 16$

   c. $(g \circ f)(-1) = (-1)^2 + 4$ d. $(f \circ g)(5) = 5^2 + 8(5) + 16$
   $= 5$ $= 81$

9. $f = \{(0, 0), (-1, 1), (-2, 4), (-3, 9), (5, 25)\}$
   $g = \{(0, -5), (-1, 0), (2, -3), (4, -2), (5, -1)\}$

f	g		g	f
$0 \rightarrow$	$0 \cdots \rightarrow 0 \rightarrow -5$		$0 \rightarrow -5 \cdots \rightarrow$ Not defined	
$-1 \rightarrow$	$1 \cdots \rightarrow$ Not defined		$-1 \rightarrow 0 \cdots \rightarrow 0 \rightarrow 0$	
$-2 \rightarrow$	$4 \cdots \rightarrow 4 \rightarrow -2$		$2 \rightarrow -3 \cdots \rightarrow -3 \rightarrow 9$	
$-3 \rightarrow$	$9 \cdots \rightarrow$ Not defined		$4 \rightarrow -2 \cdots \rightarrow -2 \rightarrow 4$	
$5 \rightarrow 25$	$\cdots \rightarrow$ Not defined		$5 \rightarrow -1 \cdots \rightarrow -1 \rightarrow 1$	

   $g \circ f = \{(0, -5), (-2, -2)\}$ $f \circ g = \{(-1, 0), (2, 9), (4, 4), (5, 1)\}$

10. For f, g, and h defined by

   $f(x) = 3x + 6, \qquad g(x) = 4x - 1, \qquad$ and $\qquad h(x) = 2x^2 + 3x + 1,$

   find:

   a. $f \circ f$ b. $g \circ g$ c. $f \circ (g \circ h)$ d. $(f \circ g) \circ h$

   Solution:

   a. $(f \circ f)(x) = f(3x + 6)$ b. $(g \circ g)(x) = g(4x - 1)$
   $= 3(3x + 6) + 6$ $= 4(4x - 1) - 1$
   $= 9x + 18 + 6$ $= 16x - 4 - 1$
   $= 9x + 24$ $= 16x - 5$

   c. Let   d. Let
   $m(x) = (g \circ h)(x)$ $t(x) = (f \circ g)(x)$
   $= g(2x^2 + 3x + 1)$ $= f(4x - 1)$
   $= 4(2x^2 + 3x + 1) - 1$ $= 3(4x - 1) + 6$
   $= 8x^2 + 12x + 3$ $= 12x + 3$

   Then   Then

   $(f \circ m)(x) = f(8x^2 + 12x + 3)$ $(t \circ h)(x) = t(2x^2 + 3x + 1)$
   $= 3(8x^2 + 12x + 3) + 6$ $= 12(2x^2 + 3x + 1) + 3$
   $= 24x^2 + 36x + 15$ $= 24x^2 + 36x + 15$

PROBLEM SET 1.3

A Problems

1. Let $f(x) = 2x - 3$ and $g(x) = x^2 + 1$. Find:
   a. $(f + g)(5)$ b. $(f - g)(3)$ c. $(fg)(2)$
   d. $(f/g)(4)$ e. $(f \circ g)(2)$

2. Let $f(x) = \dfrac{x - 2}{x + 1}$ and $g(x) = x^2 - x - 2$. Find:
   a. $(f + g)(2)$ b. $(f - g)(5)$ c. $(fg)(102)$
   d. $(f/g)(99)$ e. $(f \circ g)(1)$

3. Let $f(x) = \dfrac{2x^2 - x - 3}{x - 2}$ and $g(x) = x^2 - x - 2$. Find:
   a. $(f + g)(-1)$ b. $(f - g)(2)$ c. $(fg)(9)$
   d. $(f/g)(102)$ e. $(f \circ g)(0)$

4. Let $f = \{(0, 1), (1, 4), (2, 7), (3, 10)\}$ and $g = \{(0, -3), (1, -1), (2, 1), (3, 3)\}$.
   Find:
   a. $(f + g)(1)$ b. $(f - g)(3)$ c. $(fg)(2)$
   d. $(f/g)(0)$ e. $(f \circ g)(2)$

5. Let $f = \{(5, 3), (6, 2), (7, 9), (8, 12)\}$ and $g = \{(5, 8), (6, 5), (7, 4), (8, 3)\}$. Find:
   a. $(f + g)(6)$ b. $(f - g)(7)$ c. $(fg)(5)$
   d. $(f/g)(8)$ e. $(f \circ g)(6)$

6. Let $f = \{(5, 9), (10, 29), (15, 39), (20, 49)\}$ and $g = \{(5, 4), (10, 5), (15, 6), (20, 9)\}$. Find:
   a. $(f + g)(10)$ b. $(f - g)(15)$ c. $(fg)(15)$
   d. $(f/g)(20)$ e. $(f \circ g)(10)$

In Problems 7–10, find:
a. $(f + g)(x)$ b. $(f - g)(x)$ c. $(fg)(x)$ d. $(f/g)(x)$
e. *the domain for the functions in parts a–d*

7. $f(x) = 2x - 3$ and $g(x) = x^2 + 1$

8. $f(x) = \dfrac{x - 2}{x + 1}$ and $g(x) = x^2 - x - 2$

9. $f(x) = \dfrac{2x^2 - x - 3}{x - 2}$ and $g(x) = x^2 - x - 2$

10. $f(x) = 4x + 1$ and $g(x) = x^3 + 3$

In Problems 11–14, find:
a. $f \circ g$ b. $g \circ f$

11. $f(x) = 2x - 3$ and $g(x) = x^2 + 1$

12. $f(x) = \dfrac{x - 2}{x + 1}$ and $g(x) = x^2 - x - 2$

13. $f(x) = x^2$ and $g(x) = x^2 - x - 2$

14. $f(x) = 4x + 1$ and $g(x) = x^3 + 3$

B Problems

In Problems 15–20, let f and g be functions as shown in the given figures and graph f + g.

15.

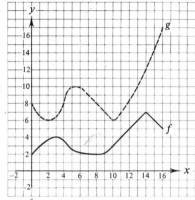

16.

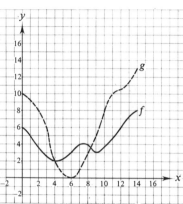

17.

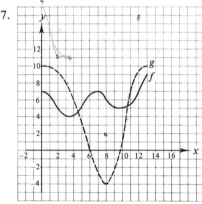

18.

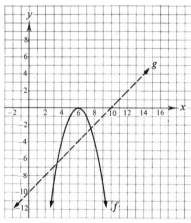

19.

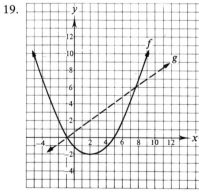

20.

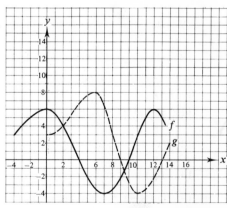

In Problems 21–27, find:

a. $f \circ g$ b. $g \circ h$ c. $(f \circ g) \circ h$ d. $f \circ (g \circ h)$

for the given functions.

21. $f(x) = x^2$, $g(x) = 2x - 1$, $h(x) = 3x + 2$
22. $f(x) = x^2$, $g(x) = 3x - 2$, $h(x) = x^2 + 1$
23. $f(x) = 2x + 4$, $g(x) = \frac{1}{2}x - 2$, $h(x) = x^2 + 1$
24. $f(x) = \sqrt{x}$, $g(x) = x^2$, $h(x) = x + 2$, $x > 0$
25. $f(x) = 3x + 2$, $g(x) = 2x - 5$, $h(x) = x + 1$
26. $f(x) = g(x) = h(x) = x$
27. $f(x) = x$, $g(x) = x^2$, $h(x) = x^3$

28. In Section 1.1, we considered the volume of a certain cone as a function of its height, namely

$$V(h) = \frac{\pi h^3}{12}.$$

Suppose the height is expressed as a function of time by $h(t) = 2t$.
a.   Find the volume for $t = 2$.
b.   Express the volume as a function of time by finding $V \circ h$.
c.   If the domain of V is $\{h | 0 < h \le 6\}$, find the domain of h; that is, what are the permissible values for t?

29. The surface area of a spherical balloon is given by

$$S(r) = 4\pi r^2.$$

Suppose the radius is expressed as a function of time by $r(t) = 3t$.
a.   Find the surface area for $t = 2$.
b.   Express the surface area as a function of time by finding $S \circ r$.
c.   If the domain of S is $\{r | 0 < r < 8\}$, find the domain of r; that is, what are the permissible values for t?

30. If $f(x) = x^2$, then

$$f\left(\frac{1}{x}\right) = \left(\frac{1}{x}\right)^2 = \frac{1}{x^2} = \frac{1}{f(x)}.$$

Give an example of a function for which

$$f\left(\frac{1}{x}\right) \ne \frac{1}{f(x)}.$$

31. If $f(x) = x^2$, then $f(-x) = (-x)^2 = x^2 = f(x)$. A function with this property, namely $f(-x) = f(x)$, is called an *even function*. Give an example of a function for which $f(-x) \ne f(x)$.   EVEN FUNCTION

32. Which of the following define an even function (see Problem 31)?

a. $f_1(x) = x^2 + 1$ b. $f_2(x) = \sqrt{x^2}$ c. $f_3(x) = \dfrac{1}{3x^3 - 4}$

d. $f_4(x) = x^3 + x$ e. $f_5(x) = \dfrac{1}{(x^3 + 3)^2}$ f. $f_6(x) = \dfrac{1}{(x^3 + x)^2}$

ODD FUNCTION

Do not assume that every function is either even or odd; for example, $f(x) = x + 3$ is neither even nor odd.

33. If $f(x) = x^3$, then $f(-x) = (-x)^3 = -x^3 = -f(x)$. A function with this property, namely $f(-x) = -f(x)$, is called an *odd function*. Give an example of a function for which $f(-x) \neq -f(x)$.

34. Which of the functions of Problem 32 define an odd function? (See also Problem 33.)

35. If $f(x) = x$, then $f(x^2) = [f(x)]^2$. Give an example of a function for which $f(x^2) \neq [f(x)]^2$.

36. If $f(x) = x^2$, then $(f \circ f)(x) = x^4 = (f \cdot f)(x)$. Give an example of a function for which $(f \circ f)(x) \neq (f \cdot f)(x)$.

Mind Bogglers

37. If $f(x) = 1 + (1/x)$, find:
    a. $(f \circ f)(x)$ b. $(f \circ f \circ f)(x)$ c. $(f \circ f \circ f \circ f)(x)$

38. *Calculator Problem*
    a. Let $f(x) = \sqrt{x}$. Choose *any* positive x. Find a numerical value for $(f \circ f)(x)$, $(f \circ f \circ f)(x)$, and $(f \circ f \circ f \circ f)(x)$. If this procedure is repeated a large number of times,

    $$(f \circ f \circ f \circ \cdots \circ f)(x),$$

    can you predict the outcome for *any* x?
    b. Let $f(x) = 2\sqrt{x}$. Answer the questions of part a.
    c. Let $f(x) = 3\sqrt{x}$. Answer the questions of part a.
    d. Let $f(x) = k\sqrt{x}$, where k is a positive integer. Answer the questions of part a.

1.4   INVERSE OF A FUNCTION

In mathematics we frequently find it necessary to "undo" a certain process or operation. For example, solving equations involves the idea of opposite operations to reverse the indicated operations of a given equation. Consider the solution of $3x = 12$. We multiply both sides by $\frac{1}{3}$.

$$\left(\frac{1}{3}\right)(3x) = \left(\frac{1}{3}\right)(12)$$

$$1(x) = 4$$

$$x = 4$$

Why do we multiply by $\frac{1}{3}$? Because $\frac{1}{3}$ is the *inverse* of 3 for multiplication. Also, recall that $\frac{1}{3}$ can be written as 3^{-1}. Thus, if we start with a number x multiplied by 3, and then we want to undo the operation of multiplying by 3, we simply multiply by the inverse of 3, namely $\frac{1}{3}$.

This simple example is introduced here because we want to generalize the idea to functions. That is, given a number x, we can evaluate some function f at x. Next, we undo the procedure to obtain x again (that is, to get back to where we started). Consider an example. Let $f(x) = 2x + 3$, where $D = \{0, 1, 3, 5\}$. Then $f = \{(0, 3), (1, 5), (3, 9), (5, 13)\}$. We can also state the relationships as a mapping:

Domain of f	Range of f
	f
0 ⟶	3
1 ⟶	5
3 ⟶	9
5 ⟶	13

To undo the results of f, we define a new relation, called the *inverse of f*, so that each element in the range maps back into the original element of the domain:

Domain of f	Range of f Domain of the inverse of f	Range of the inverse of f
0 ⟶	3 ⟶	0
1 ⟶	5 ⟶	1
3 ⟶	9 ⟶	3
5 ⟶	13 ⟶	5

The inverse can also be written $\{(3, 0), (5, 1), (9, 3), (13, 5)\}$.

Notice that the inverse of f, which is a set of ordered pairs (x, y), is the set of ordered pairs with the components interchanged—namely (y, x). In terms of the original equation, if

INVERSE OF f

$$f = \{(x, y) | y = 2x + 3\},$$

then the inverse of f is

$$\{(y, x) | y = 2x + 3\}.$$

However, because we are unaccustomed to ordered pairs of the form (y, x), we interchange the x and y values to write the inverse of f as

$$\{(x, y) | x = 2y + 3\}.$$

Solving this latter equation for y, we get

$$x = 2y + 3$$

$$x - 3 = 2y$$

$$y = \frac{x - 3}{2}.$$

If $g(x) = (x - 3)/2$, then g is called the *inverse function* for f. For example, suppose we start with the number 3 and evaluate f:

$$f(3) = 2(3) + 3$$
$$= 9.$$

Next, we apply g to this result:

$$g(9) = \frac{9 - 3}{2}$$

$$= 3.$$

This process can also be written $(g \circ f)(3)$ or $g[f(3)]$. More generally,

$$(g \circ f)(x) = g[2x + 3]$$

$$= \frac{(2x + 3) - 3}{2}$$

$$= \frac{2x}{2}$$

$$= x.$$

EXAMPLES: Find the inverse of each function.

1. $f_1 = \{(4, 7), (9, 2), (7, 3), (1, 6)\}$
   Then the inverse of f_1 is $\{(7, 4), (2, 9), (3, 7), (6, 1)\}$.

2. $f_2 = \{(6, 1), \ (7, 2), (8, 1), (9, 3)\}$
   Then the inverse of f_2 is $\{(1, 6), (2, 7), (1, 8), (3, 9)\}$.

3. $f_3 = \{(x, y)|y = 5x + 4\}$
   Then the inverse of f_3 is $\{(y, x)|y = 5x + 4\}$. But you would interchange the x and y values to write $\{(x, y)|x = 5y + 4\}$.

4. $f_4 = \{(x, y)|y = x^2\}$
   Then the inverse of f_4 is $\{(x, y)|x = y^2\}$.

The inverse of a function may or may not be a function.

As you can see from the preceding examples, the inverse of a function is not necessarily a function. The function f_2 has an inverse that is not a function, since the first component, 1, is associated with both 6 and 8. For Example 3, $f_3(x) = 5x + 4$ means that

$$y = 5x + 4.$$

The inverse is

$$x = 5y + 4.$$

To see if this is a function, solve for y:

$$y = \frac{x - 4}{5}.$$

We now see that each x value yields a single y value, so the inverse of f_3 is a function. When the inverse is a function, we write $f_3^{-1}(x) = (x - 4)/5$.

If f and g are two given functions, we can check to see whether they are inverses by checking $(f \circ g)(x)$ or $(g \circ f)(x)$. If the result of either of these operations gives x, then f and g are inverses. For Example 3:

$$f_3^{-1}[f_3(x)] = f_3^{-1}[5x + 4] \qquad \text{or} \qquad f_3[f_3^{-1}(x)] = f_3\left[\frac{x - 4}{5}\right]$$

$$= \frac{(5x + 4) - 4}{5} \qquad\qquad\qquad = 5\left(\frac{x - 4}{5}\right) + 4$$

$$= \frac{5x}{5} \qquad\qquad\qquad\qquad\qquad = (x - 4) + 4$$

$$= x \qquad\qquad\qquad\qquad\qquad\quad = x$$

The graphs of f_3 and f_3^{-1} are shown in Figure 1.12.

Be careful about this notation. The symbol f^{-1} means the inverse of the function f and does not mean 1 divided by f. Also, it is appropriate to use this notation only if the inverse is a function.

$$f_3(x) = 5x + 4$$

$$f_3^{-1}(x) = \frac{x - 4}{5}$$

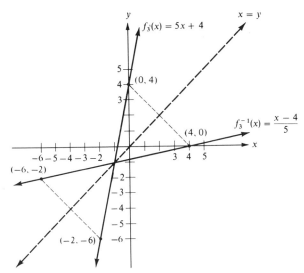

Figure 1.12 Graphs of $f_3(x) = 5x + 4$ and $f_3^{-1}(x) = (x - 4)/5$

Look at the graph of $y = x^2$ in Figure 1.13. Notice that it is not a one-to-one function. It can be shown that the inverse of a function f is a function if and only if f is a one-to-one function.

For Example 4, $f_4(x) = x^2$ means that

$$y = x^2.$$

The inverse is given by

$$x = y^2.$$

Solving for y:

$$y = \pm\sqrt{x},$$

which is not a function, because each positive value of x yields two values of y. Thus, since the inverse of f_4 is not a function, we do not write f_4^{-1}. The graphs of f_4 and its inverse are shown in Figure 1.13.

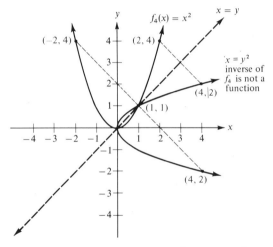

Figure 1.13   Graphs of $y = x^2$ and $x = y^2$

When we work with functions and their inverses, it is often desirable that the inverse also be a function. If we restrict the domain of a function, we frequently can force the inverse to be a function too. For example, if

$$F_4(x) = x^2 \quad (x \geq 0),$$

then

$$F_4^{-1}(x) = \sqrt{x}$$

is the inverse function. The graphs of F_4 and F_4^{-1} are shown in Figure 1.14. Notice that $f_4 \neq F_4$, since their domains are different. We will agree to use the following convention about notation: if f denotes a function with an inverse that is not a function, then F denotes the function f with restrictions on its

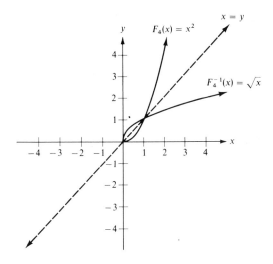

Figure 1.14 Graphs of $F_4(x) = x^2$ and $F_4^{-1}(x) = \sqrt{x}$

domain so that its inverse is a function. Consider the function from Example 2:

FUNCTION	DOMAIN	INVERSE
$f_2 = \{(6, 1), (7, 2), (8, 1), (9, 3)\}$	$\{6, 7, 8, 9\}$	$\{(1, 6), (2, 7), (1, 8), (3, 9)\}$ This is not a function.
$F_2 = \{(7, 2), (8, 1), (9, 3)\}$	$\{7, 8, 9\}$	$\{(2, 7), (1, 8), (3, 9)\}$ This is a function.

The way the domain of f_2 is limited is not unique. For this example, we could have let

$$F_2 = \{(6, 1), (7, 2), (9, 3)\},$$

with domain $\{6, 7, 9\}$. The inverse of this F_2 would also have been a function.

EXAMPLES: Given a function f, define a function F that meets the following conditions:

    a. If the inverse of f is a function, let $F = f$.
    b. If the inverse of f is not a function, restrict the domain of f in order to define a new function F so that the inverse of F is a function.

5. $f(x) = 2x + 1$, where $D = \{-2, -1, 0, 1, 2\}$
      $f = \{(-2, -3), (-1, -1), (0, 1), (1, 3), (2, 5)\}$
    inverse $= \{(-3, -2), (-1, -1), (1, 0), (3, 1), (5, 2)\}$
    The inverse is a function. Define $F = f$.

6. $f(x) = 2x^2 + 1$, where $D = \{-2, -1, 0, 1, 2\}$
      $f = \{(-2, 9), (-1, 3), (0, 1), (1, 3), (2, 9)\}$
    inverse $= \{(9, -2), (3, -1), (1, 0), (3, 1), (9, 2)\}$
    This is not a function. Thus, we restrict the domain of f (which can usually

be done in more than one way). Let $F = f$, where the domain of F is $\{0, 1, 2\}$.

$$F = \{(0, 1), (1, 3), (2, 9)\}$$
$$F^{-1} = \{(1, 0), (3, 1), (9, 2)\} \quad \text{This is a function.}$$

7. $f(x) = 2x + 1$

    The set of real numbers is understood to be the domain. Write f as a set of ordered pairs:

    $$f = \{(x, y) \mid y = 2x + 1\},$$
    $$\text{inverse} = \{(y, x) \mid y = 2x + 1\} \quad \text{or} \quad \{(x, y) \mid x = 2y + 1\}.$$

    Solving for y:

    $$x = 2y + 1$$

    $$x - 1 = 2y$$

    $$y = \frac{x - 1}{2}.$$

    We see that this is a function, so we define $F = f$.

8. $f(x) = 2x^2 + 1$

    The set of real numbers is understood to be the domain. Write f as a set of ordered pairs:

    $$f = \{(x, y) \mid y = 2x^2 + 1\},$$
    $$\text{inverse} = \{(x, y) \mid x = 2y^2 + 1\}.$$

    Solving for y:

    $$x = 2y^2 + 1$$

    $$x - 1 = 2y^2$$

    $$\frac{x - 1}{2} = y^2$$

    $$y = \pm\sqrt{\frac{x - 1}{2}}$$

    This is not a function, so we restrict the domain of f. Let $F = f$, where the domain of F is the set of nonnegative real numbers. That is,

    $$F = \{(x, y) \mid y = 2x^2 + 1, x \geq 0\},$$
    $$F^{-1} = \{(x, y) \mid x = 2y^2 + 1, y \geq 0\}.$$

    Solving for y:

    $$x = 2y^2 + 1$$

    $$y = \pm\sqrt{\frac{x - 1}{2}}$$

    └─Use the *positive value only* since $y \geq 0$.

    $$y = \sqrt{\frac{x - 1}{2}} \quad \text{This is a function.}$$

One of the main topics of this book is the graphing of functions. Sometimes when we graph the inverse of a function, our work can be simplified by using a geometric property of inverses. If (x, y) is a point on the graph of the original function, then (y, x) is a corresponding point on the graph of the inverse. Thus, the inverse is *symmetric* with respect to the line $x = y$. That is, if we were to graph the function and then place a mirror along the line $x = y$, the mirror image would represent the graph of the inverse. It was for this reason that the line $x = y$ was drawn in Figures 1.12–1.14, and, as you can see, in each figure the inverse is symmetric with respect to this line.

SYMMETRY WITH RESPECT TO A LINE

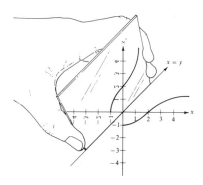

PROBLEM SET 1.4

A Problems

Suppose a mirror is held as illustrated in Problems 1–8. Draw the reflection of the curve in the mirror.

1.

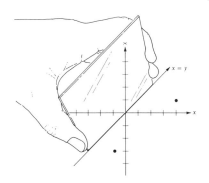

2.

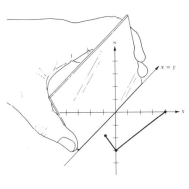

3.
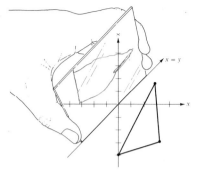

4.
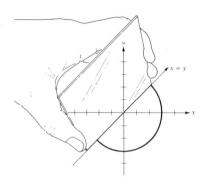

5.
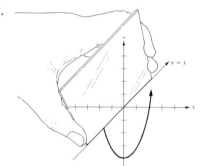

6.
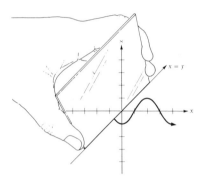

7.
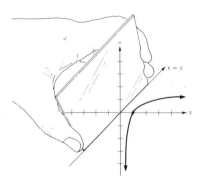

8.
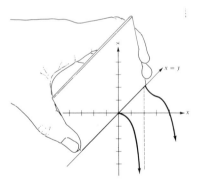

Find the inverse of each function given in Problems 9–18.

9. $f(x) = x + 3$

10. $f(x) = 2x + 3$

11. $g(x) = 5x$

12. $g(x) = \frac{1}{5}x$

13. $h(x) = x^2 - 5$

14. $h(x) = \sqrt{x} + 5$

15. $f(x) = x$

16. $f(x) = \dfrac{1}{x}$

17. $f(x) = \dfrac{1}{x - 2}$

18. $f(x) = \dfrac{2x + 1}{x}$

Determine which pairs of functions in Problems 19–24 are inverses.

EXAMPLE 9

$$f(x) = \frac{2}{5}x + 3 \qquad g(x) = \frac{5}{2}x + 5$$

$$(f \circ g)(x) = f[g(x)] = f\left(\frac{5}{2}x + 5\right)$$

$$= \frac{2}{5}\left(\frac{5}{2}x + 5\right) + 3$$

$$= x + 2 + 3$$

$$= x + 5$$

Recall that $(f \circ g)(x) = x$ if and only if f and g are inverses.

Since $(f \circ g)(x) \neq x$, then f and g are not inverses.

---

19. $f(x) = 5x + 3; g(x) = \dfrac{x - 3}{5}$

20. $f(x) = \dfrac{2}{3}x + 2; g(x) = \dfrac{3}{2}x + 3$

21. $f(x) = \dfrac{4}{5}x + 4; g(x) = \dfrac{5}{4}x + 3$

22. $f(x) = \dfrac{1}{x}, x \neq 0; g(x) = \dfrac{1}{x}, x \neq 0$

23. $f(x) = x^2, x < 0; g(x) = \sqrt{x}, x > 0$

24. $f(x) = x^2, x \geq 0; g(x) = \sqrt{x}, x \geq 0$

B Problems

Determine which pairs of functions in Problems 25–28 are inverses.

25. $f(x) = 2x^2 + 1, x \geq 0; g(x) = \dfrac{1}{2}\sqrt{2x - 2}, x \geq 1$

26. $f(x) = 2x^2 + 1, x \leq 0; g(x) = -\dfrac{1}{2}\sqrt{2x - 2}, x \geq 1$

27. $f(x) = (x + 1)^2, x \geq 1; g(x) = -1 - \sqrt{x}, x \geq 0$

28. $f(x) = (x + 1)^2, x \geq -1; g(x) = -1 + \sqrt{x}, x \geq 0$

Given the functions in Problems 29–38:
a. *Find the inverse of f.*
b. *Graph f and its inverse.*
c. *Determine whether the inverse is a function.*
d. *If the inverse is not a function, define a function F by limiting the domain of f so that F^{-1} is a function.*
e. *Graph F and F^{-1} if $F \neq f$.*

29. $f = \{(4, 5), (6, 3), (7, 1), (2, 4)\}$

30. $f = \{(1, 4), (6, 1), (4, 5), (3, 4)\}$

31. $f = \{(x, y) | y = 2x + 3\}$

32. $f = \{(x, y) | y = 3x + 1\}$

33. $f(x) = x + 5$

34. $f(x) = \frac{1}{2}x + \frac{3}{2}$

35. $f(x) = -\frac{1}{2}x^2$

36. $f(x) = 2x^2$

37. $f(x) = \sqrt{x}$

38. $f(x) = \sqrt{x + 1}$

If $y = f(x)$ *is given by the graph in Figure 1.15, define a function* $y = F(x)$ *with domain* $\{x \mid 0 \le x \le 7\}$ *by limiting the domain of* f *so that* F^{-1} *exists. Then find the values requested in Problems 39–41.*

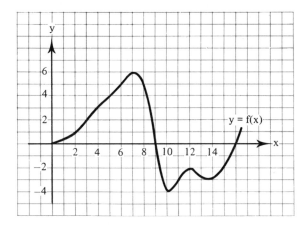

Figure 1.15   Graph of $y = f(x)$

39. a. $f(2)$ b. $f(4)$ c. $f(7)$ d. $f(10)$ e. $f(14)$
40. a. $f(0)$ b. $F(4)$ c. $F(7)$ d. $f(16)$ e. $f(0)$
41. a. $F^{-1}(0)$ b. $F^{-1}(1)$ c. $F^{-1}(5)$ d. $F^{-1}(6)$ e. $F^{-1}(3)$

If $y = f(x)$ *is given by the graph in Figure 1.16, define a function* $y = F(x)$ *with domain* $\{x \mid 0 \le x \le 6\}$ *by limiting the domain of* f *so that* F^{-1} *exists. Then find the values requested in Problems 42–45.*

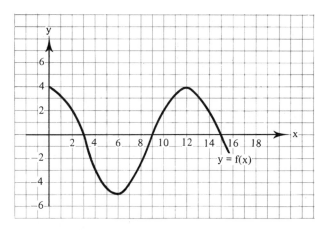

Figure 1.16   Graph of $y = f(x)$

42. a. $f(2)$ b. $F(3)$ c. $F^{-1}(2)$ d. $F^{-1}(0)$
43. a. $f(3)$ b. $f(0)$ c. $F^{-1}(4)$ d. $F^{-1}(-3)$
44. a. $f(6)$ b. $f(12)$ c. $F^{-1}(-4)$ d. $F(5)$
45. a. $f(9)$ b. $f(15)$ c. $F(7)$ d. $F^{-1}(-2)$

Mind Bogglers

46. a. Graph the inverse of f for the function shown in Figure 1.15.
     b. Graph F^{-1} for the function shown in Figure 1.15.
47. a. Graph the inverse of f for the function shown in Figure 1.16.
     b. Graph F^{-1} for the function shown in Figure 1.16.
48. A function is said to be *increasing* on an interval I if $f(x_1) < f(x_2)$ whenever $x_1 < x_2$ and x_1 and x_2 are in I. Show that if f is an increasing function, then the inverse of f is also a function.

1.5   TRANSLATION OF FUNCTIONS

Once you are given a particular function, it is possible to shift the graph of that function to other locations on the plane. For example, let $y = f(x)$ be

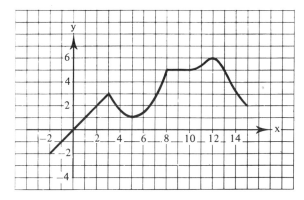

Figure 1.17   Graph of $y = f(x)$

the function shown in Figure 1.17. It is possible to shift the entire curve up, down, right, or left, as shown in Figure 1.18. In fact, this shifting can be done easily if you use tracing paper. However, since this is usually not convenient, consider the effect that shifting the coordinate axes has on the equation of a given curve. If the coordinate axes are shifted up k units, the origin of this new coordinate system would correspond to the point $(0, k)$ in the old coordinate system. If the axes are shifted to the right h units, the origin would

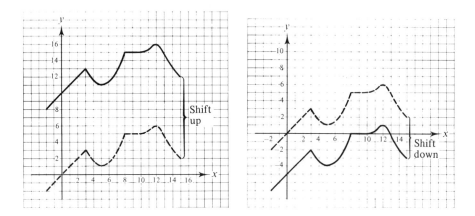

Figure 1.18 Shifting the graph of $y = f(x)$ (Part 1)

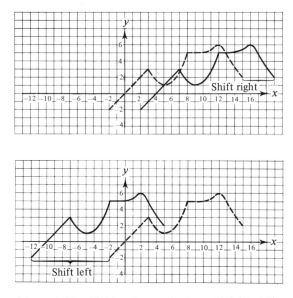

Figure 1.18 *Shifting the graph of $y = f(x)$ (Part 2)*

correspond to the point $(h, 0)$ in the old system. A horizontal shift of h units followed by a vertical shift of k units would shift the new coordinate axes so that its origin corresponds to a point (h, k) on the old axes. Suppose a *new* coordinate system with origin at (h, k) is drawn and the new axes are labeled x' and y', as shown in Figure 1.19. Every point on a given curve can now be

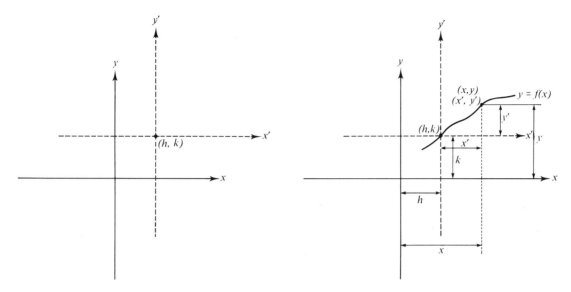

Figure 1.19   Shifting the axes to (h, k)

Figure 1.20   Comparison of coordinate axes

denoted in two ways (see Figure 1.20).

1.   As (x, y) measuring from the old origin
2.   As (x', y') measuring from the new origin

To find the relationship between (x, y) and (x', y'), consider the graph shown in Figure 1.20.

$$x = x' + h \qquad x' = x - h$$
$$\text{or}$$
$$y = y' + k \qquad y' = y - k$$

This says that if we are given any function

$$y - k = f(x - h),$$

the graph of this function is the same as

$$y' = f(x'),$$

where (x', y') are measured from the new origin located at (h, k). This can greatly simplify our work since $y' = f(x')$ is usually easier to graph than $y - k = f(x - h)$.

EXAMPLE 1: Graph $y - 3 = (x - 2)^2$:

a.   By plotting points

x	y
0	7
1	4
2	3
3	4
4	7
5	12
6	19

You will not appreciate the other method unless *you* actually *do* the arithmetic of this problem by this method.

b.   By shifting axes

$(h, k) = (2, 3)$

Plot this point. Now, plot points $y' = x'^2$.

x	y
1	1
−1	1
2	4
−2	4
3	9
−3	9

This table of values can be done mentally. Use $(2, 3)$ as the origin when counting out these values. It is not necessary to draw the $x' - y'$ axes.

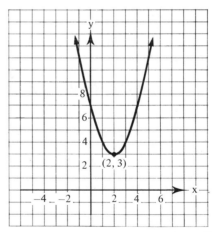

Figure 1.21   Graph of $y - 3 = (x - 2)^2$

The advantages of the shifting method become more apparent as the problems become more complicated.

EXAMPLE 2: Graph $y - \dfrac{1}{2} = \left(x - \dfrac{3}{2}\right)^2$. First, plot (3/2, 1/2). Next, *count* out the same table of values generated for part b of the previous example. Remember, you should be able to generate these mentally since you are using the equation

$$y' = x'^2.$$

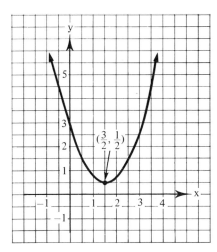

Figure 1.22   Graph of $y - \dfrac{1}{2} = \left(x - \dfrac{3}{2}\right)^2$

As you can see, the graph of $y - k = f(x - h)$ is done in two steps:

1. Plot (h, k).
2. Graph the simpler curve $y' = f(x')$ by using (h, k) as the new origin.

EXAMPLES: Find (h, k) for each equation.

3. $y - 5 = f(x - 7)$; $(h, k) = (7, 5)$

4. $y + 6 = f(x - 1)$; $(h, k) = (1, -6)$
    Notice that $y + 6$ can be written as $y - (-6)$.

5. $y + 1 = f(x + 3)$; $(h, k) = (-3, -1)$

6. $y = f(x)$; $(h, k) = (0, 0)$
    This indicates no shift.

7. $y - 15 = f\left(x + \dfrac{2}{3}\right)$; $(h, k) = \left(-\dfrac{2}{3}, 15\right)$

8. $y - 6 = f(x) + 15$
    Write the equation as $y - 21 = f(x)$; thus, $(h, k) = (0, 21)$.

SUMMARY: Given $y = f(x)$.

Right shift h units (*h* positive): $y = f(x - h)$
Left shift h units (*h* negative): $y = f(x - h)$
Up shift k units (*k* positive): $y - k = f(x)$
Down shift k units (*k* negative): $y - k = f(x)$

In general, $y - k = f(x - h)$ passes through (h, k) and is the graph of $y = f(x)$, which has been shifted *h* units horizontally and *k* units vertically.

PROBLEM SET 1.5

A Problems

For each function given in Problems 1–10:
a. Find (h, k).   b. Graph the function.

1. $y + 3 = x^2$

2. $y = (x - 6)^2$

3. $y = (x - 4)^2$

4. $y = (x + 4)^2$

5. $y - 1 = (x - 5)^2$

6. $y - 1 = (x + 3)^2$

7. $y - \dfrac{2}{3} = \left(x + \dfrac{5}{3}\right)^2$

8. $y + \dfrac{1}{2} = \left(x - \dfrac{3}{2}\right)^2$

9. $y - \dfrac{4}{7} = \left(x + \dfrac{9}{7}\right)^2$

10. $y + \dfrac{3}{8} = \left(x - \dfrac{1}{2}\right)^2$

For Problems 11–16, let f be the function shown in Figure 1.23. Graph the indicated function.

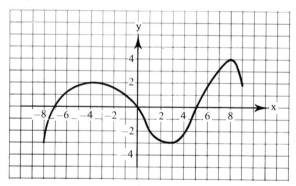

Figure 1.23   Graph of a function $y = f(x)$

11. $y + 3 = f(x)$

12. $y - 5 = f(x)$

13. $y = f(x - 4)$

14. $y = f(x + 2)$

15. $y - 2 = f(x - 1)$

16. $y + 4 = f(x - 3)$

B Problems

Graph the functions given in Problems 17–28.

17. $y - \sqrt{2} = (x + \sqrt{5})^2$

18. $y - \sqrt{2} = (x - \sqrt{3})^2$

19. $y - \pi = x^2$

20. $y = (x - \pi)^2$

21. $y - \pi = \left(x + \dfrac{\pi}{2}\right)^2$

22. $y - \pi = \left(x + \dfrac{\pi}{6}\right)^2$

23. $y + 3 = (x + 8)^2$, such that $-14 \le x \le -8$

24. $y - 2 = (x + 3)^2$, such that $-7 \le x \le -1$

25. $y - 12 = -\left(x + \dfrac{25}{2}\right)^2$, such that $y > 9$

26. $y + 3 = (x + 3)^2$, such that $y \le 9$

27. $y - 12 = -(x - 8)^2$, such that $y > 9$

28. $y + \dfrac{13}{5} = (x - 8)^2$, such that $y \le 4x - 35$

Mind Bogglers

29. Certain letters of the alphabet can be written as functions. For example,

$$m(x) = \begin{cases} 5x & \text{if} \quad 0 \le x < 1 \\ -5x + 10 & \text{if} \quad 1 \le x < 2 \\ 5x - 10 & \text{if} \quad 2 \le x < 3 \\ -5x + 20 & \text{if} \quad 3 \le x \le 4 \end{cases}$$

is the graph of the letter M, as shown in Figure 1.24. Which of the other letters of the alphabet can be written as functions?

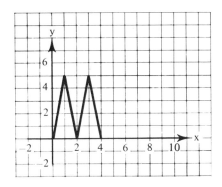

Figure 1.24   Graph of the function $m(x)$

30. The equations of some other letters (see Problem 29) are shown below.

Letter I:

$$i(x) = 5x \quad \text{if} \quad 0 \le x \le 1$$

Letter N:

$$n(x) = \begin{cases} 5x & \text{if} \quad 0 \le x < 1 \\ -5x + 10 & \text{if} \quad 1 \le x < 2 \\ 5x - 10 & \text{if} \quad 2 \le x \le 3 \end{cases}$$

Letter U:

$$u(x) = 5(x - 1)^2 \quad \text{if} \quad 0 \le x \le 2$$

Graph each of these letters.

31. a. Graph $n(x)$, $i(x - 4)$, and $m(x - 6)$ from Problems 29 and 30 on the same coordinate axes.

   b. As you can see from part a, if more than one letter is graphed, it is necessary to translate some of the letter functions. Write the equations to graph the word MINIMUM, as shown in Figure 1.25.

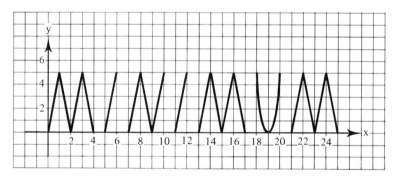

Figure 1.25   Graph of the word MINIMUM

1.6 SUMMARY AND REVIEW

TERMS

Cartesian coordinate system (1.1)
Composition of functions (1.3)
Dependent variable (1.1)
Domain (1.1)
Function (1.1)
Functional notation (1.2)
Graph (1.1)
Independent variable (1.1)
Inverse (1.4)

Inverse function (1.4)
Ordered pair (1.1)
Origin (1.1)
Quadrant (1.1)
Range (1.1)
Rectangular coordinate system (1.1)
Relation (1.1)
x-axis (1.1)
y-axis (1.1)

IMPORTANT IDEAS

(1.1) A *relation* is a set of ordered pairs. The set of first components is called the *domain* and the set of second components is called the *range*.

(1.1) A *function* is a relation for which each member of the domain is associated with exactly one member of the range. It can also be described as a rule or a mapping.

(1.1, 1.2) Functional notation.

(1.3) Adding, subtracting, multiplying, dividing, and composing functions.

(1.3) Graphing functions by adding ordinates.

(1.4) The inverse of a function is found by interchanging the x and y components. Two functions f and g are inverses if and only if $(f \circ g)(x) = x$ or $(g \circ f)(x) = x$. The graphs of a function and its inverse are symmetric with respect to the line $x = y$.

(1.5) Shifting a function $y = f(x)$:

$$y = f(x - h) \quad \text{Right shift if } h \text{ is positive}$$
$$\text{Left shift if } h \text{ is negative}$$
$$y - k = f(x) \quad \text{Up shift if } k \text{ is positive}$$
$$\text{Down shift if } k \text{ is negative}$$

In general, $y - k = f(x - h)$ is the same as the curve $y' = f(x')$, where the origin has been translated to the point (h, k).

REVIEW PROBLEMS

1. (1.1) What is a function?
2. (1.1) Classify the given set as a relation, a function, both, or neither.
   a. $\{5, 6, 7, 8\}$
   b. $\{(x, y) | (x, y) \text{ is a point on the line passing through } (7, 8) \text{ and } (7, 2)\}$
   c. $\{(x, y) | x = \text{ the temperature at a particular location at time } y\}$
   d. $\{(x, y) | y = \text{ the temperature at a particular location at time } x\}$
   e. $\{(x, y) | y = x \text{ if } x \text{ is nonnegative and } y = -x \text{ if } x \text{ is negative}\}$
3. (1.2) If $f(x) = 3x + 2$, find:
   a. $f(-3)$ b. $f(w)$ c. $f(m^2 + n^2)$ d. $f(x^2 + 2x - 1)$
   e. $\dfrac{f(x + h) - f(x)}{h}$
4. (1.2, 1.3) If $g(x) = 5x^2 + x + 1$, find:
   a. $\dfrac{g(x + h) - g(x)}{h}$ b. $(g \circ g)(x)$
5. (1.3) For $f(x) = x^2$ and $g(x) = 5x - 2$, find:
   a. $(f + g)(x)$ b. $(f - g)(x)$ c. $(fg)(x)$ d. $(f/g)(x)$
   e. $(f \circ g)(x)$
6. (1.3) For f and g of Problem 5 and $h(x) = 2x + 1$, find:
   a. $g \circ f$ b. $f \circ f$ c. $(g \circ f) \circ h$ d. $g \circ (f \circ h)$
   e. $h \circ h \circ h$
7. (1.3) Let f and g be functions as shown in Figure 1.26, and graph $f + g$.

The answer to Problem 1 should be given from memory.

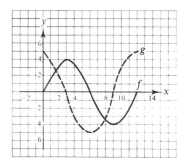

Figure 1.26   Graphs of f and g

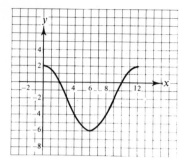

Figure 1.27   Graph of f

8. (1.4) Find the inverse of each function. If the inverse is not a function, limit the domain of f to define a function F so that F^{-1} is also a function.

a. $f = \{(0, 1), (1, 4), (2, 9), (-2, 9)\}$ b. $f(x) = -\dfrac{3}{2}x$

c. $f(x) = -\dfrac{3}{2}x + 3$ d. $f(x) = -\dfrac{3}{2}x^2$

e. $f(x) = -\dfrac{3}{2}x^2 + 3$

9. (1.4) Let f be the function shown in Figure 1.27. Define a function $y = F(x)$ with domain $\{x \mid 0 \le x \le 6\}$ by limiting the domain of f so that the inverse of F exists.

a. Graph F.   b. Graph the inverse of f.
c. Graph the inverse of F.   d. Find $F^{-1}(2)$.
e. Find $F^{-1}(-2)$.

10. (1.5) Let f be the function shown in Figure 1.27. Graph

a. $y + 2 = f(x)$ b. $y = f(x - 3)$ c. $y - 4 = f(x + 2)$

Linear and Quadratic Functions

2

2.1 LINEAR FUNCTIONS

This chapter is concerned with two special types of functions from algebra: linear and quadratic functions.

DEFINITION OF LINEAR AND QUADRATIC FUNCTIONS

> DEFINITION OF LINEAR AND QUADRATIC FUNCTIONS: A function f is called
>
> 1. *linear* if $f(x) = mx + b$;
> 2. *quadratic* if $f(x) = ax^2 + bx + c, \quad a \neq 0$.

In Section 1.1, we spoke of the graph of a relation. This is an essential notion in mathematics—one that relates the set of ordered pairs of a relation to the set of ordered pairs of a graph in a one-to-one fashion. Thus, graphs that are simple curves can usually be determined by plotting points. But plotting points at random is usually not efficient.

In algebra, you studied several forms of the equation of a line. The derivation of some of these is reviewed in the problems, and the forms are stated below for review.

> FORMS OF A LINEAR EQUATION
>
> STANDARD FORM OF THE EQUATION OF A LINE
>
> STANDARD FORM: $Ax + By + C = 0$, where (x, y) is any point on the line and A, B, and C are constants, A and B not both 0.
>
> SLOPE-INTERCEPT FORM
>
> SLOPE-INTERCEPT FORM: $y = mx + b$, where m is the slope and b is the y-intercept.
>
> POINT-SLOPE FORM
>
> POINT-SLOPE FORM: $y - k = m(x - h)$, where m is the slope and (h, k) is a point on the line.
>
> TWO-POINT FORM
>
> TWO-POINT FORM: $y - y_1 = \left(\dfrac{y_2 - y_1}{x_2 - x_1}\right)(x - x_1)$, where (x_1, y_1) and (x_2, y_2) are points on the line.
>
> INTERCEPT FORM
>
> INTERCEPT FORM: $\dfrac{x}{a} + \dfrac{y}{b} = 1$, where $(a, 0)$ and $(0, b)$ are the x- and y-intercepts, respectively.

SLOPE

The slope of a line represents its "steepness." It is defined by selecting two points on the line, say (x_1, y_1) and (x_2, y_2), where $x_1 \neq x_2$, as shown in

Figure 2.1. Then

Using functional notation for the points $(x_1, f(x_1))$ and $(x_2, f(x_2))$, the slope is

$$\text{slope} = \frac{\text{vertical change}}{\text{horizontal change}} = \frac{y_2 - y_1}{x_2 - x_1} = \frac{\text{rise}}{\text{run}}.$$

$$\frac{f(x_2) - f(x_1)}{x_2 - x_1}.$$

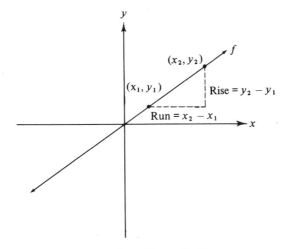

Figure 2.1   Slope of a line

Consider the function

$$f(x) = mx.$$

Since $f(0) = 0$, $f(1) = m$, $f(2) = 2m, \ldots,$ the points $(0, 0)$, $(1, m)$, $(2, 2m), \ldots,$ lie on this line. Let $(x_1, f(x_1))$ and $(x_2, f(x_2))$ be two distinct points on this line. Then (x_1, mx_1) and (x_2, mx_2) lie on the line. The slope is

$$\frac{\text{vertical change}}{\text{horizontal change}} = \frac{mx_2 - mx_1}{x_2 - x_1}$$

$$= \frac{m(x_2 - x_1)}{x_2 - x_1}.$$

$$= m.$$

The slope of a line is denoted by m.

Thus, the slope of the function $f(x) = mx$ is m.

EXAMPLES

1. Find the slope of the line passing through $(2, -3)$ and $(-1, 2)$.

$$m = \frac{2 - (-3)}{-1 - 2} = \frac{5}{-3} = -\frac{5}{3}$$

Recall from algebra that

$$-\frac{p}{q} = \frac{-p}{q} = \frac{p}{-q}.$$

If p and q are positive, then the first two are the preferred forms.

2. Find the slope of the line passing through $(x_0, f(x_0))$ and $(x_0 + h, f(x_0 + h))$.

$$m = \frac{f(x_0 + h) - f(x_0)}{x_0 + h - x_0} = \frac{f(x_0 + h) - f(x_0)}{h}$$

Remember the applied problems of Section 1.2? Does this look familiar?

3. Graph $y = (-5/3)x$. This line passes through the origin and has slope $m = -5/3$. To graph, we start at the known point (the origin) and count out the slope using rise $= -5$ and run $= 3$, as shown in Figure 2.2.

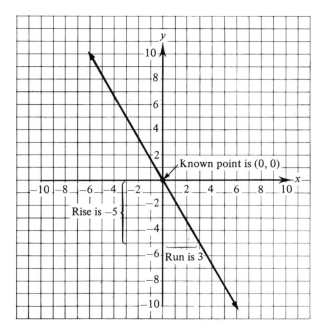

Figure 2.2 Negative slope: $y = -\frac{5}{3}x$

Notice that the same line is obtained if you use rise $= 5$ and run $= -3$ since

$$-\frac{5}{3} = \frac{-5}{3} = \frac{5}{-3}.$$

4. Graph $6x - 2y = 0$. Solve for y:

$$y = 3x.$$

This line passes through the origin and has slope $m = 3$. Since $3 = 3/1$, we use rise $= 3$ and run $= 1$, as shown in Figure 2.3.

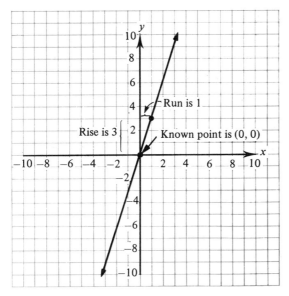

Figure 2.3   Positive slope: $y = 3x$

5. Graph $y = \frac{2}{3}x$ on coordinate axes with the following scales:

a. 1 square = 1 unit   b. 1 square = $\frac{1}{5}$ unit   c. 1 square = 500 units

The solutions are shown in Figure 2.4.

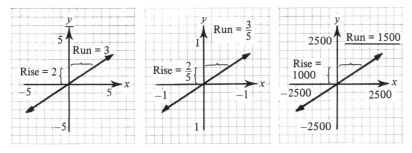

Figure 2.4   Graph of $y = \frac{2}{3}x$ with different scales

For Example 5, notice that $m = 2/3$. For part a, we used rise = 2, run = 3; for part b, rise = 2/5, run = 3/5; and for part c, rise = 1000, run = 1500. For each part, the *ratio*

$$\frac{\text{rise}}{\text{run}} = \frac{2}{3} = \frac{2/5}{3/5} = \frac{1000}{1500}$$

is the same. Thus, if you are given a ratio

$$\frac{y}{x} = \frac{2}{3},$$

you *cannot* say $y = 2$ and $x = 3$, since there are *many* choices of x and y that satisfy this ratio.

y-INTERCEPT

The general form

$$y = mx + b$$

can be compared with $f(x) = mx$, since

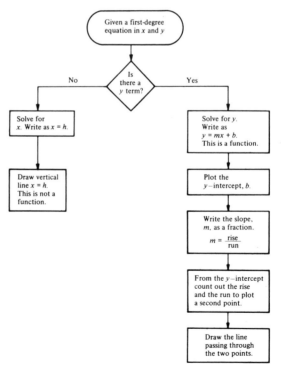

Figure. 2.5   Procedure for graphing a line by using the slope-intercept form

$$y = mx + b$$
$$y = f(x) + b$$
$$y - b = f(x),$$

which means the function has been translated to the point $(0, b)$. This point $(0, b)$ is called the *y-intercept*. The slope of a line is not affected by a translation. This means that to graph a line, we can follow the flowchart shown in Figure 2.5.

y-INTERCEPT

EXAMPLES

6. $y = \frac{1}{2}x + 3$

   The *y*-intercept is 3 and the slope is $\frac{1}{2}$; the line is graphed as shown in Figure 2.6.

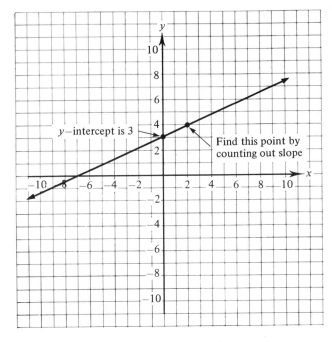

Figure 2.6 Graph of $y = \frac{1}{2}x + 3$

7. $2x + 3y - 6 = 0$

   Solve for y:

$$3y = -2x + 6$$

$$y = -\frac{2}{3}x + 2$$

The y-intercept is 2 and the slope is $-2/3$; the line is graphed in Figure 2.7.

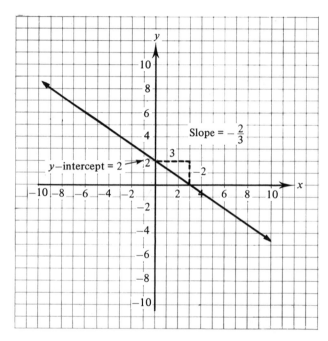

Figure 2.7 Graph of $2x + 3y - 6 = 0$

8. $4x + 2y - 5 = 0$ for $-1 \le x \le 3$
   Solve for y:

$$2y = -4x + 5$$

$$y = -2x + \frac{5}{2}$$

The y-intercept is $5/2$ and the slope is -2; this line is shown as a dotted line in Figure 2.8. However, we have a restriction on the domain, so we draw the part of the line with x values between -1 and 3 (inclusive) as shown by a solid line segment in Figure 2.8.

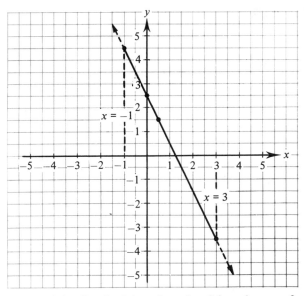

Figure 2.8 Graph of $4x + 2y - 5 = 0$ for $-1 \le x \le 3$

9. When given the standard form of the equation of a line, the best method for graphing the line is usually the slope-intercept method described by Figure 2.5. However, if given the point-slope form, it is easier to translate the function. For example, compare

$$y - 3 = \frac{2}{5}(x + 2) \qquad \text{with} \qquad y - k = \frac{2}{5}(x - h);$$

$(h, k) = (-2, 3)$ and $m = 2/5$, so this line is graphed by translation, as shown in Figure 2.9.

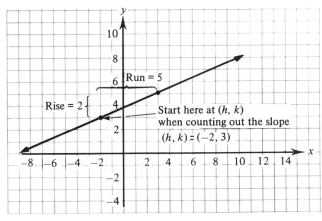

Figure 2.9 Graph of $y - 3 = \frac{2}{5}(x + 2)$

PROBLEM SET 2.1

A Problems

Graph the line segments given in Problems 1–8 on the same coordinate axes by plotting the given point and sketching the line through that point with the given slope. Apply the stated restrictions.

When Problems 1–8 are finished, you will have drawn a picture.

1. $(0, 0)$, $m = 0$; $-8 \leq x \leq 10$
2. $(0, 7)$, $m = 1$; $1 \leq y \leq 11$
3. $(4, 11)$, $m = -\frac{5}{2}$; $4 \leq x \leq 8$
4. $(7, -3)$, $m = 1$; $-3 \leq y \leq 0$
5. $(-9, 3)$, $m = -3$; $-3 \leq y \leq 0$
6. $(0, -3)$, $m = 0$; $-7 \leq x \leq 7$
7. $(8, 1)$, $m = 0$; $-6 \leq x \leq 8$
8. $(4, 5)$, no slope; $0 \leq y \leq 11$

Graph the lines given in Problems 9–22 by finding the slope and y-intercept.

9. $y = 3x + 3$
10. $y = -4x - 1$
11. $y = \frac{2}{3}x + \frac{4}{3}$
12. $y = \frac{1}{5}x - \frac{6}{5}$
13. $y = 40x$
14. $y = 300x$
15. $-y + 2x = 1$
16. $3x + y = 10$
17. $x - 4 = 0$
18. $y + 2 = 0$
19. $5x - 4y - 8 = 0$
20. $x - 3y + 2 = 0$
21. $100x - 275y + 500 = 0$
22. $2x - 5y - 1200 = 0$

Graph the line segments in Problems 23–30 on the same coordinate axes by sketching the lines and noting restrictions on either x or y. Set up your axes so that $-15 \leq x \leq 15$ and $-15 \leq y \leq 15$.

Be careful! Some restrictions are on x and others are on y.

23. $y - 7 = \frac{5}{2}(x + 7)$; $-11 \leq x \leq -7$
24. $y + 7 = -\frac{2}{5}(x + 1)$; $-11 \leq y \leq -3$
25. $y + 2 = \frac{1}{12}(x - 1)$; $1 \leq x \leq 13$
26. $y - 1 = \frac{4}{9}(x + 2)$; $-3 \leq y \leq 1$
27. $y - 1 = -\frac{2}{5}(x - 8)$; $-7 \leq x \leq 13$
28. $y + 2 = -\frac{9}{8}(x - 1)$; $-7 \leq x \leq 1$
29. $y + 6 = \frac{5}{2}(x - 11)$; $-11 \leq y \leq -1$
30. $y + 11 = \frac{9}{4}(x - 9)$; $5 \leq x \leq 9$

B Problems

Find the equation of the line satisfying the given conditions in Problems 31–43. Leave your answer in standard form.

EXAMPLE 10: y-intercept, 4; slope, 6

Solution: Use $y = mx + b$; $b = 4$, $m = 6$.
$$y = 6x + 4$$
In standard form, $6x - y + 4 = 0$.

---

31. y-intercept, 6; slope, 5
32. y-intercept, -3; slope, -2
33. y-intercept, 0; slope, 0
34. y-intercept, 5; slope, 0

EXAMPLE 11: Slope, 4; passing through $(-3, 2)$

Solution: Use $y = mx + b$; $m = 4$, b is not given. However, since $(-3, 2)$ is on the curve, it satisfies the equation

$$2 = 4(-3) + b$$
$$2 = -12 + b$$
$$14 = b.$$

See Problem 44 for another solution to this problem.

Thus, $y = 4x + 14$. In standard form,

$$4x - y + 14 = 0.$$

35. Slope, 3; passing through $(2, 3)$
36. slope, -1; passing through $(-4, 5)$
37. Slope, $\frac{1}{2}$; passing through $(3, 3)$
38. Slope, $\frac{2}{5}$; passing through $(5, -2)$

EXAMPLE 12: Passing through $(2, 3)$ and $(5, 7)$

Solution: Use $y = mx + b$; m and b are both unknown. However, since $(2, 3)$ is on the curve:

$$3 = 2m + b.$$

See Problem 45 for another solution to this problem.

And, since $(5, 7)$ is on the curve:

$$7 = 5m + b.$$

We can solve these equations simultaneously.

$$\begin{cases} 2m + b = 3 \\ 5m + b = 7 \end{cases} \qquad b = 3 - 2m$$
$$\overline{-3m = -4} \qquad = 3 - 2\left(\frac{4}{3}\right)$$
$$m = \frac{4}{3} \qquad = \frac{9}{3} - \frac{8}{3}$$
$$= \frac{1}{3}$$

Thus, $y = \frac{4}{3}x + \frac{1}{3}$. In standard form,

$$3y = 4x + 1$$
$$4x - 3y + 1 = 0.$$

39. Passing through $(-4, -1)$ and $(4, 3)$
40. Passing through $(6, -3)$ and $(-2, -4)$
41. Passing through $(4, -2)$ and $(4, 5)$
42. Passing through $(5, 6)$ and $(1, -2)$

43. Passing through $(5, 6)$ and $(7, 6)$

44. Example 11 used the slope-intercept form. However, we could have used the point-slope form,

$$y - k = m(x - h).$$

For example, if $m = 4$ and $(h, k) = (-3, 2)$, then

$$y - 2 = 4(x + 3)$$
$$y - 2 = 4x + 12$$
$$4x - y + 14 = 0.$$

Derive the point-slope form by using the method for Problem 35 with slope m and point (h, k).

45. Example 12 used the slope-intercept form. However, we could have used the two-point form,

$$y - y_1 = \left(\frac{y_2 - y_1}{x_2 - x_1}\right)(x - x_1).$$

For example, given $(2, 3)$ and $(5, 7)$:

$$y - 3 = \left(\frac{7 - 3}{5 - 2}\right)(x - 2)$$

$$y - 3 = \frac{4}{3}(x - 2)$$

$$3y - 9 = 4x - 8$$

$$4x - 3y + 1 = 0.$$

Derive the two-point form by using the method for Problem 39 with the points (x_1, y_1) and (x_2, y_2).

46. Consider Figure 2.10.

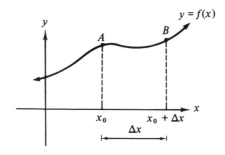

Figure 2.10

a. What are the coordinates of A and B?
b. What is the slope of the line passing through A and B?

47. Consider Figure 2.11.
    a. What are the coordinates of A and B?
    b. What is the slope of the line passing through A and B?
48. Answer Problem 47b where $x_0 = 1$ and Δx is given below.
    a. $\Delta x = 2$ b. $\Delta x = 1$ c. $\Delta x = 0.5$ d. $\Delta x = 0.25$
    e. $\Delta x = 0.1$ f. $\Delta x = 0.01$ g. $\Delta x = 0.001$
49. Consider Figure 2.12.
    a. What are the coordinates of A and B?
    b. What is the slope of the line passing through A and B?
50. The population of New York State in 1960 was roughly 16.8 million and in 1970 was about 18.2 million. Assuming that the increase will continue linearly, write an equation relating the population to the date. Make the base year 1950 ($x = 0$ represents 1950).

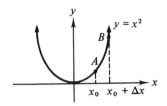

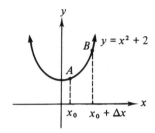

Figure 2.11

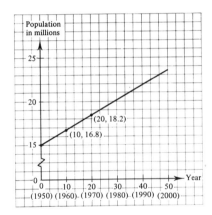

Figure 2.12

Note: even though the number of people is a positive integer, the graph is drawn as a line and then the answers are interpreted appropriately. The graph showing the population of New York State from 1950 to 2000 is shown at left:

51. Use the results of Problem 50 to estimate the population of New York State in 1980 and in 2000.
52. The demand for a certain product is related to its price. Suppose we are manufacturing a new line of stationery and decide to test the market at two stores. We find that we can sell 170 boxes in a month if they are priced at $1.00 and 110 boxes if we price them at $3.00. Assuming that the demand is linear, write an equation expressing this relationship for p (the price), where $0 \le p \le 3$. Let n, representing the demand, be the dependent variable. Also, draw the graph.
53. The demand for a product is related to its supply. The supply of stationery (see Problem 52) is also related to the price. At $1.00 per box we could supply 50 boxes, and at $2.00 a box we could supply 110 boxes. Assuming that the supply is linear, write an equation expressing this relationship for p (the price), where $0 \le p \le 3$. Let n, representing the supply, be the dependent variable. Also, draw the graph.

$ 54. Suppose you intend to manufacture and sell rings. You find that you can sell 200 rings if the price is $10.00, but you can sell 2200 rings if they are priced at $1.00. Assuming that the demand is linear, write an equation expressing this relationship for p (the price), where $1 \le p \le 10$. Let n, representing the demand, be the dependent variable. Also, draw the graph.

$ 55. Suppose you produce rings to sell (see Problem 54). You find that you can supply 700 rings if they sell for a price of $1.00 and 3700 if they sell at $10.00. Assuming that the supply is linear, write an equation expressing this relationship for p (the price), where $1 \le p \le 10$. Let n, representing the supply, be the dependent variable. Also, draw the graph.

Mind Boggler

$ 56. *Calculator Problem.* Sometimes a set of data is not linear, but can be approximated by a straight line. Suppose we have some rather incomplete sales figures, as shown by the following chart:

Year	Sales (in thousands of dollars)
1955	160
1956	310
1965	220
1968	360
1970	450

Making 1940 the base year (that is, 1940 = 0, 1955 = 15, 1956 = 16, and so on), plot these points on a set of coordinate axes. The line that best approximates the data is called the *best-fitting line.* It is found that the best-fitting line passes through the points (12, 173) and (28, 361). Write the equation of this line. Now complete the following table:

Year	Actual	Sales predicted (from equation)	Sales difference (actual − predicted)	Squares of sales difference
1955				
1956				
1965				
1968				
1970				

Add the squares of the differences. This should be the smallest possible sum if you found the best-fitting line. Use the equation you found to predict the sales in 1975.

2.2 ABSOLUTE VALUE

In the last section, the distance between two points was found while calculating the slope. Although the procedure for finding the distance between two points on a number line is relatively easy, certain observations should be made, as illustrated by Examples 1–3.

EXAMPLES: Find the distance between the given points.

1. (46) and (12)
    This distance can be found on a number line, as shown in Figure 2.13.

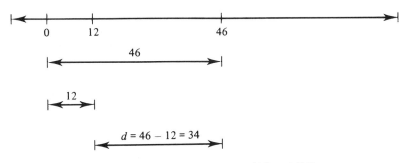

Figure 2.13   Distance between (46) and (12)

Thus, the distance, *d*, is $46 - 12 = 34$.

2. (146) and (−38)
    The distance is shown in Figure 2.14.

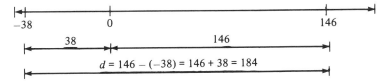

Figure 2.14   Distance between (146) and (−38)

With most numbers it is neither convenient nor necessary to draw the graph. It is obvious that the distance is $146 - (-38) = 184$.

3. (x_1) and (x_2)
    The distance is $x_1 - x_2$ or $x_2 - x_1$, depending on whether x_1 or x_2 is further from the origin, as shown in Figure 2.15.

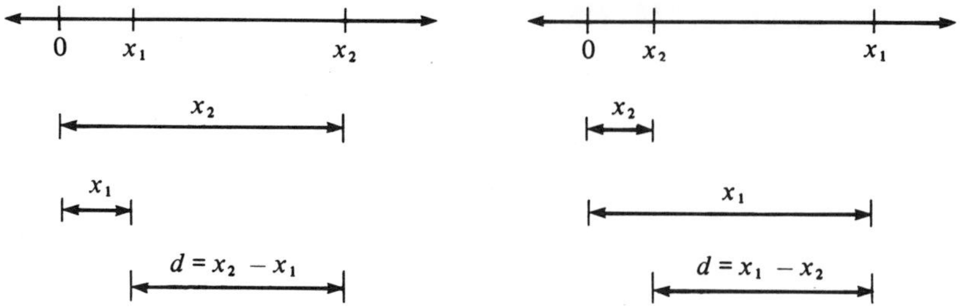

Figure 2.15 Distance between (x_1) and (x_2)

The situation shown in Example 3 is not very satisfactory, so we introduce a notion called *absolute value* to mean the distance of a given point from the origin.

DEFINITION OF
ABSOLUTE VALUE

DEFINITION: The *absolute value* of a, written $|a|$, is defined by

$|a| = a$ if $a \geq 0$
$|a| = -a$ if $a < 0$.

EXAMPLES

4. $|5| = 5$, since $5 \geq 0$
   Notice that the point (5) is 5 units from the origin.

5. $|-5| = -(-5)$
       $= 5$, since $-5 < 0$
   Notice that the point (-5) is 5 units from the origin.

6. $|5^2 - 3| = 5^2 - 3$, since $5^2 - 3 \geq 0$

7. $|5^2 - 125| = -(5^2 - 125)$
             $= 125 - 5^2$, since $5^2 - 125 < 0$

8. $|\pi - 3| = \pi - 3$, since $\pi - 3 \geq 0$

9. $|\pi - 4| = -(\pi - 4)$, since $\pi - 4 < 0$
         $= 4 - \pi$

10. $|w^2 + 1| = w^2 + 1$, since $w^2 + 1 \geq 0$

11. $|-4 - t^2| = -(-4 - t^2)$
             $= t^2 + 4$, since $-4 - t^2 < 0$

We can now use absolute value to obtain a more satisfactory answer for Example 3.

ONE-DIMENSIONAL
DISTANCE FORMULA

The distance between (x_1) and (x_2) is

$$d = |x_2 - x_1|.$$

Notice that this is the same as $d = |x_1 - x_2|$.

Some properties of absolute value with which you will have to be familiar are now stated.

Properties of absolute value:

1. $|a| \geq 0$

   1. Absolute value is nonnegative

2. $|-a| = |a|$

   2. Absolute values of opposites are equal

3. $|a|^2 = a^2$

   3. Squares of absolute values

4. $|a||b| = |ab|$

   4. Multiplying absolute values

5. $\dfrac{|a|}{|b|} = \left|\dfrac{a}{b}\right|$

   5. Dividing absolute values

6. $|a + b| \leq |a| + |b|$

   6. Adding absolute values; this is called the *triangle inequality*

7. $|a| = |b|$ if and only if $a = \pm b$

   7. Absolute value equations

8. For positive numbers b, $|a| < b$ if and only if $-b < a < b$ (also, $|a| \leq b$ if and only if $-b \leq a \leq b$).

   8. Absolute value inequalities

The proofs of the first three properties follow directly from the definition of absolute value. The next three can be proved by carefully considering cases. Property 7 can be used when solving absolute value equations. We will prove this property.

i. *Prove:* If $|a| = |b|$, then $a = \pm b$.
   If $|a| = |b|$, then $|a|^2 = |b|^2$ and
   $$a^2 = b^2 \quad \text{by Property 3}$$
   $$a = \pm b.$$

ii. *Prove:* If $a = \pm b$, then $|a| = |b|$.
   If $a = \pm b$, then $|a| = |\pm b|$ and by Property 2, $|b| = |-b|$.
   Thus, $\qquad |a| = |b|$.

$a = \pm b$ is the same as $\pm a = \pm b$, since $-a = \pm b$ is equivalent to $a = \pm b$ if you multiply both sides by (-1).

EXAMPLES: Solve the given equations.

12. $|x + 5| = 2$
    By Property 7: $x + 5 = 2 \quad$ or $\quad x + 5 = -2$
    $\qquad\qquad\qquad\quad x = -3 \quad$ or $\qquad x = -7$
    $$\{-3, -7\}$$

13. $|x + 5| = -2$

    By Property 1, $|x + 5| \geq 0$, so the solution set is empty.

14. $|x + 5| = |3x - 4|$

    By Property 7:

    $x + 5 = 3x - 4 \quad$ or $\quad x + 5 = -(3x - 4)$

    $-2x = -9 \qquad\qquad x + 5 = -3x + 4$

    $x = \dfrac{9}{2} \qquad\qquad\qquad 4x = -1$

    $\qquad\qquad\qquad\qquad x = -\dfrac{1}{4}$

    $\left\{\dfrac{9}{2}, -\dfrac{1}{4}\right\}$

Property 8 can be used when solving absolute value inequalities. This property is also proved in two parts, but we will prove only the part we expect to use when solving absolute value inequalities.

Prove: If $|a| < b$, then $-b < a < b$.

By definition:

$$|a| = a \quad \text{if} \quad a \geq 0$$
$$|a| = -a \quad \text{if} \quad a < 0.$$

Case i: if $a \geq 0$

$|a| < b \quad$ given

$a < b \quad$ by substitution, $|a| = a$

Case ii: if $a < 0$

$|a| < b \qquad$ given

$-a < b \qquad$ by substitution, $|a| = -a$

$0 < a + b \quad$ add a to both sides

$-b < a \qquad$ subtract b from both sides

Therefore, $-b < a$ and $a < b$ from Cases i and ii. This is the same as saying $-b < a < b$.

EXAMPLES: Solve the given inequalities.

See Appendix B for a review of solving inequalities.

15. $|2x - 3| \leq 4$

    By Property 8:

    $-4 \leq 2x - 3 \leq 4$

    $-4 + \mathbf{3} \leq 2x - 3 + \mathbf{3} \leq 4 + \mathbf{3}$

    $-1 \leq 2x \leq 7$

    $-\dfrac{1}{2} \leq x \leq \dfrac{7}{2}$

16. $|4 - x| < -3$

By Property 1, $|4 - x| \geq 0$, so the solution set is empty.

17. $|4 - x| < 3$

By Property 8:

$$-3 < 4 - x < 3$$
$$-3 - \mathbf{4} < 4 - x - \mathbf{4} < 3 - \mathbf{4}$$
$$-7 < -x < -1$$
$$7 > x > 1$$
$$\text{or}$$
$$1 < x < 7$$

18. $|2x + 2| > 4$

To solve this, we have to be a little tricky in order to apply Property 8. Every real number is a member of one of the following sets:

$|2x + 2| > 4$	$|2x + 2| \not> 4$
↑	↑
We want the solution for this inequality.	But we solve this one instead.

$$|2x + 2| \not> 4$$
$$|2x + 2| \leq 4$$
$$-4 \leq 2x + 2 \leq 4$$
$$-6 \leq 2x \leq 2$$
$$-3 \leq x \leq 1$$

Thus, the set we are looking for is the set of x values *not* included in $-3 \leq x \leq 1$. The solution is $x < -3$ or $x > 1$, as shown in Figure 2.16.

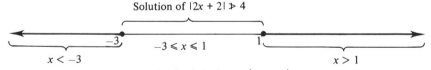

Solution of $|2x + 2| \not> 4$

$x < -3$ $-3 \leq x \leq 1$ $x > 1$

Figure 2.16   Solution of $|2x + 2| > 4$

Two approaches can be taken to graphing absolute value functions. The first uses the definition of absolute value, and the second uses the idea of translation, which was discussed in Section 1.5.

EXAMPLES: Graph the following functions and find $f(1)$, $f(-3)$, and $f(5)$.

19. $f(x) = |x|$

Solution: First, apply the definition of absolute value:

$$f(x) = \begin{cases} x & \text{if} \quad x \geq 0 \\ -x & \text{if} \quad x < 0 \end{cases}$$

Graph $f(x) = x$, $x \geq 0$; this is shown in Figure 2.17a. Next, graph $f(x) = -x$, $x < 0$; this is shown in Figure 2.17b. Finally, the union of the two parts is the graph of $f(x)$, as shown in Figure 2.17c.

In practice, you shouldn't draw three separate graphs but should instead draw all the parts on one set of coordinate axes.

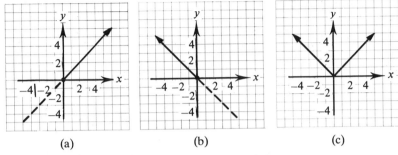

(a)   (b)   (c)

Figure 2.17   Graph of $f(x) = |x|$

$f(1) = |1| = 1; f(-3) = |-3| = 3; f(5) = |5| = 5$

20. $f(x) = |x - 3| + 2$

Solution: This can be rewritten as

$y - 2 = |x - 3|$,

which is the graph of $y = |x|$ on the coordinate axes translated to (3, 2), as shown in Figure 2.18.

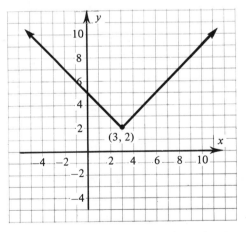

Figure 2.18   Graph of $f(x) = |x - 3| + 2$

Verify that these points are on the graph shown in Figure 2.18.

$f(1) = |1 - 3| + 2 = 4; f(-3) = |-3 - 3| + 2 = 8; f(5) = |5 - 3| + 2 = 4$

21. $f(x) = |x - 2| + |x|$

Solution: This cannot be written as a simple translation, so we'll work it by the method of adding the ordinates introduced in Section 1.3. First, graph $f_1(x) = |x|$ and $f_2(x) = |x - 2|$, as shown in Figure 2.19a. Next, add some

$$f(1) = |1 - 2| + |1| = 2$$
$$f(2) = |-3 - 2| + |-3| = 8$$
$$f(5) = |5 - 2| + |5| = 8$$

Verify that the points $(1, 2), (-3, 8), (5, 8)$ are on the graph shown in Figure 2.19c.

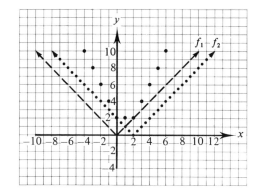

(a)

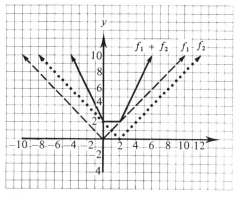

(b)

(c)

Figure 2.19   Graph of $f(x) = |x - 2| + |x|$

convenient ordinates of

$$f = f_1 + f_2,$$

as shown in Figure 2.19b. Finally, connect these points to complete the graph, as shown in Figure 2.19c.

PROBLEM SET 2.2

A Problems

Find the distance between the points given in Problems 1–4.

· 1. (45) and (88)　　　　　　　　　2. (16) and (−14)
  3. (−53) and (23)　　　　　　　　　4. (−28) and (−103)

Write the expressions in Problems 5–14 without using absolute value notation.

5. a. $|8|$　　b. $|-7|$　　c. $|15|$　　　　d. $|-25|$　　e. $|0|$
6. a. $|x^2|$　　b. $|x|^2$　　c. $|-6.28 \times 10^4|$　　d. $|10^3 - 46|$　　e. $|10^3 - 1462|$
7. $|u^2 + 7|$　　　　　　　　　　　8. $|-y^2 - 4|$
9. $|\pi^2 - 9|$　　　　　　　　　　10. $|\pi^2 - 10|$
11. $|5 - 2\pi|$　　　　　　　　　　12. $|7 - 2\pi|$
13. $|m|$　　　　　　　　　　　　　14. $|m + 1|$

Solve the equations or inequalities in Problems 15–20.

15. $|2x + 3| = 7$　　　　　　　　　16. $|5x + 1| = -3$
17. $|4 - x| = 5$　　　　　　　　　18. $|x - 2| < 0.01$
19. $|3x - 2| \le 5$　　　　　　　　20. $|x - 3| < -2$

B Problems

Solve the equations or inequalities in Problems 21–26.

21. $|x + 2| = |2x - 4|$　　　　　　22. $|x - 2| = |2 - x|$
23. $|5x - 2| > 8$　　　　　　　　　24. $|x - 3| > 0.001$
25. $|x - 1| + x = 3$　　　　　　　26. $|x + 1| + 2x = 7$

Graph the equations in Problems 27–38.

27. $y = |x| + 1$　　　　　　　　　28. $y = |x| + 3$
29. $y = |x - 1|$　　　　　　　　　30. $y + 2 = |x - 4|$
31. $f(x) = |x - 2| + 3$　　　　　　32. $f(x) = |x - 3| - 4$
33. $|x| = 3$, where $-3 \le y \le -1$　　34. $|y| = 2$, where $-1 \le x \le 4$
35. $y = |x| + 1$, where $-3 \le x \le 3$　　36. $y = |x| - 5$, where $|x| \le 3$
37. $f(x) = |x - 4| + |x|$　　　　　38. $f(x) = |x + 1| - |x|$

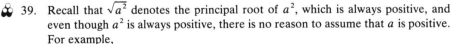 39. Recall that $\sqrt{a^2}$ denotes the principal root of a^2, which is always positive, and even though a^2 is always positive, there is no reason to assume that a is positive. For example,

if $a = 2$, $\sqrt{2^2} = \sqrt{4} = 2$;　　and　　if $a = -2$, $\sqrt{(-2)^2} = \sqrt{4} = 2$.

This shows that $\sqrt{a^2} \neq a$ for all values of a. In particular, we define

$$\sqrt{a^2} = a \quad \text{if} \quad a \geq 0$$

$$\sqrt{a^2} = -a \quad \text{if} \quad a < 0.$$

But this is the same as the definition of absolute value. Thus,

$$\sqrt{a^2} = |a|.$$

Use this fact and the properties of square root to prove

$$|ab| = |a||b|.$$

40. Prove $\left|\dfrac{a}{b}\right| = \dfrac{|a|}{|b|}$ by using the information given in Problem 39.
41. Let $f(x) = 2x + 1$. Show that if $|x - 3| < 1$, then $5 < f(x) < 9$.
42. Let $f(x) = 2x + 1$. Show that if $|x - 3| < 1/2$, then $6 < f(x) < 8$.
43. Let $f(x) = 2x + 1$. Show that if $|x - 3| < 1/10$, then $6.8 < f(x) < 7.2$.
44. If $f(x) = 2x + 1$, find d such that $|f(x) - 7| < 1$ whenever $|x - 3| < d$.
45. If $f(x) = 2x + 1$, find d such that $|f(x) - 7| < \frac{1}{100}$ whenever $|x - 3| < d$.
46. Show that the statement $|x + y| \leq |x| + |y|$ is true when
  a. $x = 2$ and $y = 1$ b. $x = -2$ and $y = 1$
  c. $x = 2$ and $y = -1$ d. $x = -2$ and $y = -1$

Mind Bogglers

47. Solve $|2x - 1| + |x| - 3 = 0$.
48. Graph $y = ||||x - 2| - 2| - 2| - 2|$.
49. Set up axes so that $-10 < x < 10$ and $-10 < y < 10$ in order to accommodate the full graph. Graph the following equations on the same coordinate axes:

  a. $y = x + 5$, where $-6 \leq x \leq -3$
  b. $y = -x + 5$, where $3 \leq x \leq 6$
  c. $y = x - 10$, where $3 \leq x \leq 6$
  d. $y = -x - 10$, where $-6 \leq x \leq -3$
  e. $|x| = 3$, where $-7 \leq y \leq -4$ or $2 \leq y \leq 5$
  f. $|x| = 6$, where $-4 \leq y \leq -1$
  g. $y + 4 = |x + 3|$, where $-6 \leq x \leq 0$
  h. $y + 4 = |x - 3|$, where $0 \leq x \leq 6$
  i. $y + 1 = |x|$, where $|x| \leq 3$
  j. $y - 2 = |x|$, where $|x| \leq 3$
  k. $|x| + y = 8$, where $|x| \leq 3$
  l. $|x| + y = -4$, where $|x| \leq 3$
  m. $x = 0$, where $|y + 1| \leq 3$

2.3 DISTANCE FORMULA AND CIRCLES

Finding the distance between points in two dimensions requires the Pythagorean Theorem. Suppose we wish to find the distance between $A(8, 2)$

Recall the Pythagorean Theorem: the square of the hypotenuse of a right triangle is equal to the sum of the squares of its legs.

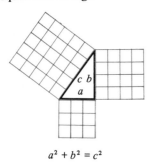

$$a^2 + b^2 = c^2$$

𝕳istorical 𝕹ote

The Greek Pythagoras (ca. 540 B.C.) founded the famous Pythagorean School, which developed into a closely knit brotherhood with secret rites and observances. Although the Egyptians and the Chinese knew of the Pythagorean Theorem before the Pythagoreans, the brotherhood did make significant contributions to mathematics. A story about Pythagoras is told by Howard Eves in his book *In Mathematical Circles* (Boston: Prindle, Weber, & Schmidt, 1969). Pythagoras was trying to find some students so he offered to pay a fellow one coin for each geometric theorem he could teach him. Since this amounted to more money than the fellow could earn working, he agreed, and soon he had earned quite a pile of coins. However, in spite of

and $B(5, 6)$, as shown in Figure 2.20. Draw a line through A parallel to the x-axis and a line through B parallel to the y-axis, as shown in Figure 2.20.

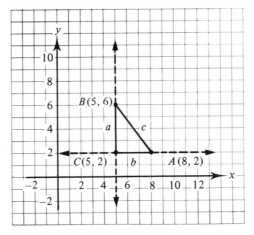

Figure 2.20   Distance between $A(8, 2)$ and $B(5, 6)$

These lines intersect at point C with coordinates $(5, 2)$. We are looking for the length of side c. The Pythagorean Theorem gives us the following equation:

$$c = \sqrt{a^2 + b^2}.$$

The lengths of sides a and b can be measured as distances on the y-axis and x-axis, respectively.

$$a = |6 - 2| \qquad b = |8 - 5|$$
$$= 4 \qquad\qquad = 3$$

Therefore,

$$c = \sqrt{4^2 + 3^2}$$
$$= \sqrt{16 + 9}$$
$$= \sqrt{25}$$
$$= 5.$$

EXAMPLES: Find the distances between the given points.

1. $A(4, 1)$ and $B(7, 6)$: the point C has coordinates $(7, 1)$

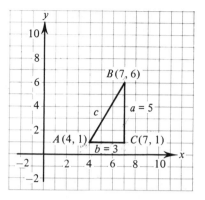

$$c = \sqrt{a^2 + b^2}$$
$$= \sqrt{|6-1|^2 + |7-4|^2}$$
$$= \sqrt{5^2 + 3^2}$$
$$= \sqrt{34}$$

himself, the student found that he had become interested in his studies and asked Pythagoras to go faster. Pythagoras said he could go faster provided the student now pay him a coin for each new theorem mastered. By the time the student finished his study of geometry, Pythagoras had gained back all his coins.

2. $A(5, 8)$ and $B(3, 2)$: the point C has coordinates $(3, 8)$

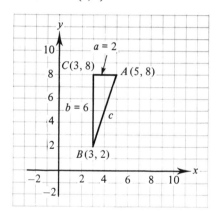

$$c = \sqrt{a^2 + b^2}$$
$$= \sqrt{|3-5|^2 + |2-8|^2}$$
$$= \sqrt{|-2|^2 + |-6|^2}$$
$$= \sqrt{4 + 36}$$
$$= 2\sqrt{10}$$

3. $A(x_1, y_1)$ and $B(x_2, y_2)$: the point C has coordinates (x_2, y_1)

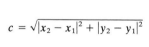

$$c = \sqrt{|x_2 - x_1|^2 + |y_2 - y_1|^2}$$

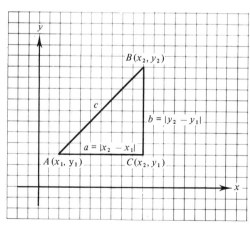

Notice that if we take the absolute value of a number and square it, the result is the same as simply squaring the number (Property 3). This and the general form illustrated by Example 3 lead us to a general formula called the *distance formula*.

DISTANCE FORMULA

This formula gives the distance between any two points in a plane.

If $P(x_1, y_1)$ and $Q(x_2, y_2)$ are any two points, then the distance, d, from P to Q is
$$d = \sqrt{(x_2 - x_1)^2 + (y_2 - y_1)^2}.$$

Notice that when we use the distance formula, we don't have to go through the steps of finding point C and applying the Pythagorean Theorem, as we did in the previous examples.

EXAMPLES: Find the distances between the given points.

4.   (6, 3) and (4, 7)
$$d = \sqrt{(4 - 6)^2 + (7 - 3)^2}$$
$$= \sqrt{4 + 16}$$
$$= 2\sqrt{5}$$

5.   (−3, 2) and (−1, −6)
$$d = \sqrt{[(-1) - (-3)]^2 + (-6 - 2)^2}$$
$$= \sqrt{4 + 64}$$
$$= 2\sqrt{17}$$

A *circle* is defined as the set of all points in a plane at a given distance from a given point. The given distance is called the *radius* and the given point is called the *center*.

The distance formula leads to the equation of a *circle*. Consider a circle with center (h, k) and radius r. Then from the definition of a circle, the distance from (h, k) to *any point* (x, y) on the circle is equal to r. Thus,

$$r = \sqrt{(x - h)^2 + (y - k)^2}$$

or
$$r^2 = (x - h)^2 + (y - k)^2.$$

EQUATION OF A CIRCLE

The graph of the equation
$$(x - h)^2 + (y - k)^2 = r^2$$
is a circle with center (h, k) and radius r.

EXAMPLES: Graph each equation.

6. $(x - 2)^2 + (y - 3)^2 = 5$

Solution: By inspection, the center is $(2, 3)$ and the radius is $\sqrt{5}$. The graph is shown below.

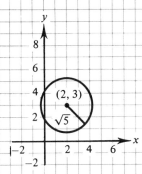

7. $(x - 2)^2 + y^2 = 9$

Solution: By inspection, the center is $(-2, 0)$ and the radius is 3. The graph is shown below.

Notice that a circle is not a function. Can you explain why?

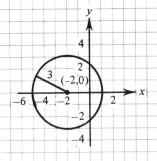

8. $(x - 4)^2 + (y - 1)^2 = 0$

Solution: By inspection, the center is $(4, 1)$ and the radius is 0. The graph is the single point $(4, 1)$.

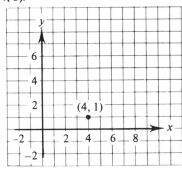

PROBLEM SET 2.3

A Problems

Graph the equations in Problems 1–10.

1. $x^2 + y^2 = 4$
2. $x^2 + y^2 = 9$
3. $x^2 + y^2 = 25$
4. $x^2 + y^2 = 100$
5. $x^2 = 16 - y^2$
6. $y^2 = 36 - x^2$
7. $x^2 + y^2 + 3 = 0$
8. $x^2 + y^2 - 3 = 0$
9. $x^2 - 4 = -y^2$
10. $x^2 + 4 = -y^2$

Find the distance between the points given in Problems 11–16.

11. $(5, 1)$ and $(8, 5)$
12. $(1, 4)$ and $(13, 9)$
13. $(-2, 4)$ and $(0, 0)$
14. $(0, 0)$ and $(5, -2)$
15. $(4, 5)$ and $(3, -1)$
16. $(-2, 1)$ and $(-1, -5)$

B Problems

Graph the equations in Problems 17–28.

17. $(x - 1)^2 + (y - 4)^2 = 49$
18. $(x - 5)^2 + (y - 3)^2 = 64$
19. $(x - 1)^2 + (y - 2)^2 = \frac{1}{4}$
20. $(x - 2)^2 + (y - 1)^2 = \frac{1}{9}$
21. $(y - 2)^2 + (x + 4)^2 = 10$
22. $(y + 4)^2 + (x - 7)^2 = 8$
23. $(x - 4)^2 + (y - 2)^2 = 7$
24. $(x + 1)^2 + (y + 2)^2 = 0$
25. $(x - 1)^2 + (y + 5)^2 + 10 = 0$
26. $(x - 1)^2 + (y + 5)^2 - 10 = 0$
27. $(x + 4) + (y - 3) = 7$
28. $(x + 2) + (y + 7) = 4$

Find the distance between the points given in Problems 29–36.

29. (a, b) and (c, d)
30. $(4x, 5x)$ and $(-3x, 2x)$, $x > 0$
31. $(x, f(x))$ and $(2x, 0)$
32. $(5, 3)$ and $(x, f(x))$
33. $(\sqrt{12}, \sqrt{50})$ and $(1/\sqrt{3}, 10/\sqrt{18})$

In Problem 34, we mean to find the distance between $(4, 5)$ and a point P on the x-axis so that the segment from the given point to P is perpendicular to the x-axis.

34. $(4, 5)$ and the x-axis
35. $(-3, -4)$ and the x-axis
36. $(-2, -5)$ and the y-axis
37. Find the set of points (x, y) for which the distance from (x, y) to $(2, 3)$ is 7.
38. Find the set of points (x, y) for which the distance from (x, y) to $(-3, -5)$ is 5.
39. Find the set of points (x, y) for which the distance from (x, y) to $(3, 0)$ plus the distance from (x, y) to $(-3, 0)$ equals 10.
40. Find the set of points (x, y) for which the distance from (x, y) to $(4, 0)$ plus the distance from (x, y) to $(-4, 0)$ equals 10.
41. Find the set of points (x, y) for which the distance from (x, y) to $(3, 0)$ minus the distance from (x, y) to $(-3, 0)$, in absolute value, is 1.
42. Find the set of points (x, y) for which the distance from (x, y) to $(4, 0)$ minus the distance from (x, y) to $(-4, 0)$, in absolute value, is 1.

Mind Bogglers

$\begin{smallmatrix}\infty\\\infty\end{smallmatrix}$ 43. Find the set of points (x, y) for which the distance from (x, y) to $(c, 0)$ plus the distance from (x, y) to $(-c, 0)$ equals $2a$. Let $b^2 = a^2 - c^2$ when simplifying the equation.

$\begin{smallmatrix}\infty\\\infty\end{smallmatrix}$ 44. Find the set of points (x, y) for which the distance from (x, y) to $(c, 0)$ minus the distance from (x, y) to $(-c, 0)$, in absolute value, equals $2a$. Let $b^2 = c^2 - a^2$ when simplifying the equation.

> The set of points satisfying the conditions of Problem 43 is called an *ellipse*.

> The set of points satisfying the conditions of Problem 44 is called a *hyperbola*.

2.4 QUADRATIC FUNCTIONS

The second function to be considered in this chapter is the quadratic function $f(x) = ax^2 + bx + c, a \neq 0$. Consider a special case, namely

$$y = x^2.$$

By plotting points, you can find the graph of this curve, as shown in Figure 2.21.

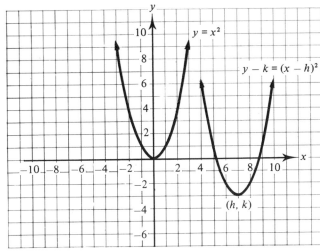

Figure 2.21   Graphs of $y = x^2$ and $y - k = (x - h)^2$

A curve of this shape is called a *parabola*. Recall that in Section 1.5, we graphed $y = x^2$ as well as $(y - k) = (x - h)^2$ by shifting the axes as shown in Figure 2.21.

Next, compare $y = x^2$ and $y = ax^2$. Suppose $a > 0$; then the graph of $y = ax^2$ may be obtained from $y = x^2$ by stretching or shrinking the curve by the factor of a in the y direction only. For example, consider the following table:

PARABOLA

	x	$y = x^2$	$y = 2x^2$	$y = \frac{1}{4}x^2$
Positive x-values	1	1	2	$\frac{1}{4}$
	2	4	8	1
	3	9	18	$\frac{9}{4}$
	4	16	32	4
Negative x-values	−1	1	2	$\frac{1}{4}$
	−2	4	8	1
	−3	9	18	$\frac{9}{4}$
	−4	16	32	4
			↑	↑
			$a = 2$	$a = \frac{1}{4}$

Each entry in these columns is found by multiplying the corresponding entry in the $y = x^2$ column by a.

These parabolas are shown in Figure 2.22.

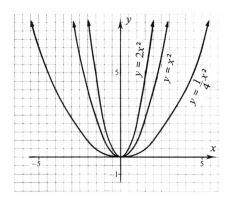

Figure 2.22   Graphs of parabolas for which $a > 0$

If a is <0, the parabolas are the same except that they open down, as shown in Figure 2.23.

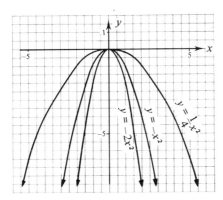

Figure 2.23   Graphs of parabolas for which $a < 0$

We can make four observations based on the examples.

1. If $a > 0$, the parabola opens up; if $a < 0$, the parabola opens down.
2. The point $(0, 0)$ is the lowest point if the parabola opens up; $(0, 0)$ is the highest point if the parabola opens down.
3. The vertical line passing through the vertex is the y-axis. The points to the left of this line form a mirror image of the points to the right of it, so we say that the curve is *symmetric* with respect to this line.
4. Relative to a fixed scale, the magnitude of a determines the "fatness" of the parabola; small values of $|a|$ yield "fat" parabolas, and large values of $|a|$ yield "skinny" parabolas.

Properties of parabolas of the form $y = ax^2$.

For graphs of parabolas of the form

$$y - k = a(x - h)^2,$$

we simply translate the axes to the point (h, k) and then graph the parabola $y' = ax'^2$, as shown in Examples 1 and 2.

General form of the equation for a parabola; the *vertex* of the parabola is the point (h, k). Using functional notation, this can be written

$$f(x) = a(x - h)^2 + k.$$

EXAMPLES: Graph each of the given curves.

1. $y = 3x^2$

    This parabola opens up since $a = 3 > 0$. The vertex is $(0, 0)$, and if

    $x = 1,$ then $y = 3,$
    $x = 2,$ then $y = 12,$
    $\vdots$

    Using symmetry, this parabola is plotted as shown in Figure 2.24.

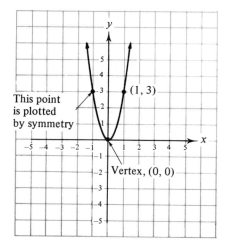

Figure 2.24   Graph of $y = 3x^2$

2. $y + 5 = 3(x + 2)^2$
    By inspection, $(h, k) = (-2, -5)$ and the standard-position parabola is $y = 3x^2$. The graph is shown in Figure 2.25.

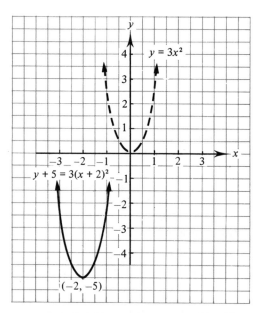

Figure 2.25   Graph of $y + 5 = 3(x + 2)^2$

COMPLETING THE SQUARE

Now we are ready to consider the general quadratic function

$$y = ax^2 + bx + c \quad (a \neq 0).$$

Consider Example 2 above: $y + 5 = 3(x + 2)^2$. This was graphed by translating the axes to $(-2, -5)$ and then considering $y' = 3x'^2$. Suppose we rewrite the given equation as

$$y + 5 = 3(x + 2)^2$$
$$y + 5 = 3(x^2 + 4x + 4)$$
$$y + 5 = 3x^2 + 12x + 12$$
$$y = 3x^2 + 12x + 7.$$

The last form is the general quadratic form, where $a = 3$, $b = 12$, and $c = 7$. Suppose you are given this form and are asked to graph the curve. The process whereby the steps shown above are reversed is called *completing the square*. In general,

$$y = ax^2 + bx + c$$

can be put into the form

$$y - \text{some number} = a(x - ?)^2.$$

To complete the square, follow these steps:

Step 1. Subtract c (the constant term) from both sides.

$$y = ax^2 + bx + c \qquad y = 3x^2 + 12x + 7$$
$$y - c = ax^2 + bx \qquad y - 7 = 3x^2 + 12x$$

Step 2. Factor the a term from the expression on the right (remember $a \neq 0$).

$$y - c = a\left(x^2 + \frac{b}{a}x\right) \qquad y - 7 = 3(x^2 + 4x)$$

Step 3. To complete the square

$$x^2 + \frac{b}{a}x + ? = (x + ?)^2,$$

square $\frac{1}{2}$ the coefficient of the x term.

$$\left(\frac{b}{2a}\right)^2 = \frac{b^2}{4a^2}$$

Then add a times this number to both sides.

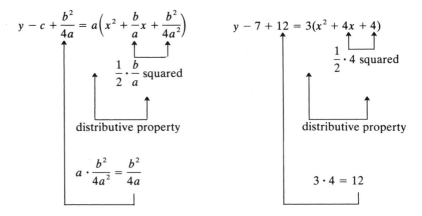

$$y - c + \frac{b^2}{4a} = a\left(x^2 + \frac{b}{a}x + \frac{b^2}{4a^2}\right) \qquad y - 7 + 12 = 3(x^2 + 4x + 4)$$

$$\frac{1}{2} \cdot \frac{b}{a} \text{ squared} \qquad\qquad \frac{1}{2} \cdot 4 \text{ squared}$$

distributive property   distributive property

$$a \cdot \frac{b^2}{4a^2} = \frac{b^2}{4a}$$

$$3 \cdot 4 = 12$$

$$y + \frac{-4ac}{4a} + \frac{b^2}{4a} = a\left(x + \frac{b}{2a}\right)^2 \qquad y - 7 + 12 = 3(x + 2)^2$$

common denominator if
fractions are involved

$$\left(y + \frac{b^2 - 4ac}{4a}\right) = a\left(x + \frac{b}{2a}\right)^2 \qquad y + 5 = 3(x + 2)^2$$

The general form may look complicated, but remember that actual calculations will look like the numerical example.

EXAMPLES: Graph the given curves.

3. $y = x^2 + 6x + 10$

Solution: Complete the square.

$$y - 10 = x^2 + 6x$$

Since $\frac{1}{2} \cdot 6 = 3$ and $3^2 = 9$, we add 9 to both sides.

$$y - 10 + \mathbf{9} = x^2 + 6x + \mathbf{9}$$

$$\frac{1}{2} \cdot 6 \text{ squared}$$

$$y - 1 = (x + 3)^2$$

The vertex is at $(-3, 1)$, the parabola opens up, and the graph is shown in Figure 2.26.

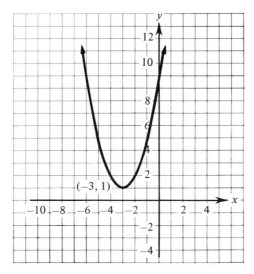

Figure 2.26   Graph of $y = x^2 + 6x + 10$

4. $2y = x^2 - 10x + 29$

    Solution:

    $$2y - 29 = x^2 - 10x$$

    $$2y - 29 + \mathbf{25} = x^2 - 10x + \mathbf{25} \qquad \tfrac{1}{2} \cdot 10 = 5 \text{ and } 5^2 = 25.$$

    $$2y - 4 = (x - 5)^2$$

    $$y - 2 = \frac{1}{2}(x - 5)^2$$

    The vertex is (5, 2), the parabola opens up, and the graph is shown in Figure 2.27.

5. $y = 1 - 5x - 2x^2$

    Solution:
    $$y - 1 = -2x^2 - 5x$$

    $$y - 1 = -2\left(x^2 + \frac{5}{2}x\right)$$

    $$y - 1 - \frac{\mathbf{25}}{\mathbf{8}} = -2\left(x^2 + \frac{5}{2}x + \frac{\mathbf{25}}{\mathbf{16}}\right)$$

    $$y - \frac{33}{8} = -2\left(x + \frac{5}{4}\right)^2$$

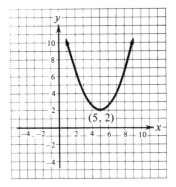

Figure 2.27   Graph of $2y = x^2 - 10x + 29$

$$\frac{1}{2} \cdot \frac{5}{2} = \frac{5}{4} \text{ and } \left(\frac{5}{4}\right)^2 = \frac{25}{16}$$

We add $-2(25/16) = -25/8$ to both sides.

The vertex is $(-5/4, 33/8)$, the parabola opens down, and the graph is shown in Figure 2.28. For fractions, you can sometimes choose a scale that is more convenient than one square per unit. Notice in Figure 2.28 that there are four squares per unit. Also, remember that once you have found the vertex, you simply have to graph $y = -2x^2$ translated to the point $(-5/4, 33/8)$.

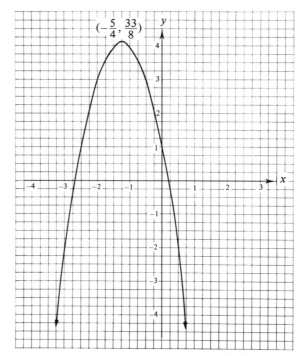

Figure 2.28   Graph of $y = 1 - 5x - 2x^2$

6. $y = 3x^2 + 5x - 4$

Solution:

$$y + 4 = 3x^2 + 5x$$

$$y + 4 = 3\left(x^2 + \frac{5}{3}x\right)$$

$$\frac{1}{2} \cdot \frac{5}{3} = \frac{5}{6} \text{ and } \left(\frac{5}{6}\right)^2 = \frac{25}{36}$$

$$y + 4 + \frac{25}{12} = 3\left(x^2 + \frac{5}{3}x + \frac{25}{36}\right)$$

We added $3(25/36) = 25/12$ to both sides.

$$y + \frac{73}{12} = 3\left(x + \frac{5}{6}\right)^2$$

The vertex is $(-5/6, -73/12)$, the parabola opens up, and the graph is shown in Figure 2.29. Don't let the fractions bother you. Once you have plotted the vertex, you can compare the given curve with $y = 3x^2$, which has been translated to the point $(-5/6, -73/12)$. Also notice that it was not particularly advantageous to change the scale for this problem.

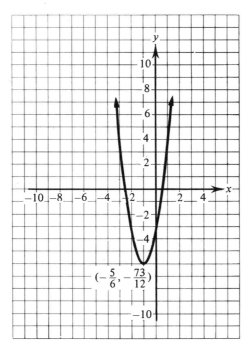

Figure 2.29   Graph of $y = 3x^2 + 5x - 4$

PROBLEM SET 2.4

A Problems

Graph the curves given in Problems 1–18.

1. $y = (x - 1)^2$
2. $y = -2(x - 1)^2$
3. $y = \frac{1}{4}(x - 1)^2$
4. $y - 2 = (x - 1)^2$
5. $y - 2 = -\frac{3}{5}(x - 1)^2$
6. $y - 2 = 3(x + 2)^2$
7. $y = x^2 + 4x + 4$
8. $y = x^2 + 6x + 9$
9. $y = x^2 + 2x - 3$
10. $y = x^2 - 5x + 2$
11. $x^2 + x - y + 2 = 0$
12. $2x^2 - x - y + 3 = 0$
13. $y = 2x^2 - 4x + 5$
14. $3y = x^2 + 6x + 15$
15. $2x^2 - 8x + 3y + 20 = 0$
16. $4x^2 - 20x - 16y + 33 = 0$
17. $4x^2 + 24x - 27y - 17 = 0$
18. $25x^2 - 30x - 5y + 2 = 0$

B Problems

Graph the curves with the restrictions given in Problems 19–26.

19. $y = 2x^2 - 4x + 4; \; y \le 5$
20. $y = 2x^2 + 8x + 5; \; -5 \le x \le 0$
21. $y = 2x^2 + 8x + 5; \; |x - 1| \le 1$
22. $x^2 + 2x + 2y - 3 = 0; \; y \ge -1$
23. $y = x^2; \; |x| \le \frac{1}{10}$
24. $y = -2x^2; \; |x| \le \frac{1}{100}$
25. $y - 3 = (x - 4)^2; \; y \le 9 - x$
26. $y + 4 = -\frac{1}{2}(x + 2)^2; \; 5y \ge 6x - 30$

Find the maximum value for the functions given in Problems 27–32.

EXAMPLE 7: $y = -5x^2 + 10x - 2$

Solution: We recognize this as a parabola that opens down. The maximum value will occur at its vertex. Complete the square to find the vertex.

$$y + 2 = -5x^2 + 10x$$
$$y + 2 = -5(x^2 - 2x)$$
$$y + 2 - \mathbf{5} = -5(x^2 - 2x + \mathbf{1})$$
$$y - 3 = -5(x - 1)^2$$

The vertex is $(1, 3)$ so the maximum value $y = 3$ occurs when $x = 1$.

---

27. $y = -4x^2 - 8x - 1$ 28. $y = -3x^2 + 6x - 5$
29. $10x^2 - 160x + y + 655 = 0$ 30. $6x^2 + 84x + y + 302 = 0$
31. $9x^2 + 6x + 81y - 53 = 0$ 32. $100x^2 - 120x + 25y + 41 = 0$

33. The highest bridge in the world is the bridge over the Royal Gorge of the Arkansas River in Colorado. It is 1053 ft above the water. If a rock is projected vertically upward from this bridge with an initial velocity of 64 fps, the height, h, of the object above the river at any time t is described by the function

$$h(t) = -16t^2 + 64t + 1053.$$

What is the maximum height possible for a rock projected vertically upward from the bridge with an initial velocity of 64 fps? After how many seconds does it reach that height?

34. *Calculator Problem.* In 1974, Evel Knievel attempted a skycycle ride across the Snake River. If the path of the skycycle is given by the equation

$$d(x) = -0.0005x^2 + 2.39x + 600,$$

where $d(x)$ is the height above the canyon floor for a horizontal distance of x units from the launching ramp, what was Knievel's maximum height?

35. A small manufacturer of quality citizen-band radios determines that the price of each item is related to the number of items produced. If x items are produced per

day, and the maximum number that can be produced is 10 items, then the price should be

$400 - 25x$ dollars.

It is also determined that the overhead (the cost of producing x items) is

$5x^2 + 40x + 600$ dollars.

The daily profit is then found by subtracting the overhead from the revenue.

$$
\begin{aligned}
\text{profit} &= \text{revenue} - \text{cost} \\
&= (\text{number of items})(\text{price per item}) - \text{cost} \\
&= x(400 - 25x) - (5x^2 + 40x + 600) \\
&= 400x - 25x^2 - 5x^2 - 40x - 600 \\
&= -30x^2 + 360x - 600
\end{aligned}
$$

How many items should the manufacturer produce per day, and what should be the expected daily profit in order to maximize the profit?

$ 36. Graph $p = -30x^2 + 360x - 600$ from Problem 35. *From your graph* estimate the answers to the following questions:

a. What is the domain for x?

b. For what values of the domain does the manufacturer show a profit?

c. Does the manufacturer ever show a loss? For what values?

d. If there were a strike, what would be the value of x? What would be the profit or loss for this situation?

37. *Calculator Problem*

a. In most states, drivers are required to know approximately how long it takes to stop their cars at various speeds. Suppose you estimate three car lengths for 30 mph and six car lengths for 60 mph. One car length per 10 mph assumes a linear relationship between speed and distance covered. Write a linear equation where x is the speed of the car and y is the distance traveled by the car in ft. (Assume that one car length = 20 ft.)

b. The scheme for the stopping distance of a car given in part a is convenient but not accurate. The stopping distance is more accurately approximated by the quadratic equation

$y = 0.071x^2$.

This says that your car requires four times as many feet to stop at 60 mph as at 30 mph. Doubling your speed quadruples the braking distance. Graph this quadratic equation and the linear equation you found in part a on the same coordinate axes.

c. Comment on the results from part b. At about what speed are the two measures the same?

Mind Bogglers

38. Find the equation of the parabola passing through $(-3, -1)$, $(-6, 2)$, and $(-1, 7)$.

39. Suppose the area under the arc of the parabola

$$y = Ax^2 + Bx + C$$

between $x = -h$ and $x = +h$ $(h > 0)$ is denoted by A_p. If

$$A_p = \frac{2Ah^3}{3} + 2Ch,$$

show that

$$A_p = \frac{h}{3}(y_0 + 4y_1 + y_2)$$

by noting that the curve passes through $(-h, y_0)$, $(0, y_1)$, and (h, y_2), as shown in Figure 2.30.

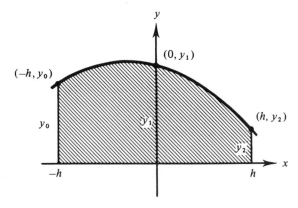

Figure 2.30 Area under the arc of the parabola $y = Ax^2 + Bx + C$ between $x = -h$ and $x = +h$

2.5 ROOTS AND THE DISCRIMINANT

In Problem 33 of the previous section, it was stated that if a rock is projected vertically upward with an initial velocity of 64 fps from the world's highest bridge over the Royal Gorge of the Arkansas River, then its height y at time x is described by the function

$$y = -16x^2 + 64x + 1053.$$

The graph was found by completing the square:

$$y - 1053 - \mathbf{64} = -16(x^2 - 4x + \mathbf{4})$$
$$y - 1117 = -16(x - 2)^2.$$

It is shown in Figure 2.31. We now wish to answer two questions:

1. What is the height of the bridge?
2. How long does it take the rock to hit the river below?

This is not the path of the rock but the height of the rock with respect to a given time x.

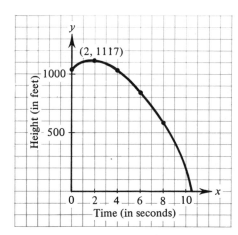

Figure 2.31 Graph of $y = -16x^2 + 64x + 1053$, $x \geq 0$, $y \geq 0$

The height of the bridge is the value of y when $x = 0$. This is the *y-intercept*:

$$y = -16(0)^2 + 64(0) + 1053$$
$$= 1053 \text{ ft.}$$

The second question asks for the length of time it takes the rock to hit the river below; this is found by finding the value(s) of x for which $y = 0$. These values are called the *roots*, or *zeros* of the function:

ROOTS, ZEROS

$$0 = -16x^2 + 64x + 1053.$$

To solve this, divide by -16 and complete the square.

$$x^2 - 4x - \frac{1053}{16} = 0$$

$$x^2 - 4x + 4 = \frac{1053}{16} + \frac{64}{16}$$

$$(x - 2)^2 = \frac{1117}{16}$$

$$\sqrt{(x - 2)^2} = \sqrt{\frac{1117}{16}}$$

$$|x - 2| = \sqrt{\frac{1117}{16}}$$

$$x - 2 = \pm \sqrt{\frac{1117}{16}}$$

$$x = 2 \pm \sqrt{\frac{1117}{16}}$$

$$x = \frac{8 \pm \sqrt{1117}}{4}$$

Since x must be positive (why?),

$$x = \frac{8 + \sqrt{1117}}{4} \approx 10.36 \text{ sec.}$$

We frequently find it necessary to find the zeros of a function, so we will examine the question more closely. Consider the equation

$$f(x) = ax^2 + bx + c \quad (a \neq 0).$$

We wish to find x so that $f(x) = 0$; this is accomplished by completing the square.

$$ax^2 + bx + c = 0$$

$$x^2 + \frac{b}{a}x = -\frac{c}{a} \quad \text{since } a \neq 0$$

$$x^2 + \frac{b}{a}x + \left(\frac{b}{2a}\right)^2 = -\frac{c}{a} + \frac{b^2}{4a^2}$$

$$\left(x + \frac{b}{2a}\right)^2 = \frac{-4ac + b^2}{4a^2}.$$

DISCRIMINANT

Let $d = b^2 - 4ac$. Then d is called the *discriminant* because it discriminates the different types of solutions obtained when solving a quadratic equation. If $d > 0$, then

$$\left(x + \frac{b}{2a}\right)^2 = \frac{d}{4a^2}$$

Since $d > 0$ and $4a^2 > 0$, we see that the radicand $d/4a^2 > 0$.

$$\sqrt{\left(x + \frac{b}{2a}\right)^2} = \sqrt{\frac{d}{4a^2}}$$

$$\left|x + \frac{b}{2a}\right| = \sqrt{\frac{d}{4a^2}}$$

$$x + \frac{b}{2a} = \pm \sqrt{\frac{d}{4a^2}}$$

$$x = -\frac{b}{2a} \pm \frac{\sqrt{d}}{2a}$$

$$x = \frac{-b \pm \sqrt{d}}{2a}.$$

This represents two real numbers,

$$\frac{-b + \sqrt{d}}{2a} \quad \text{and} \quad \frac{-b - \sqrt{d}}{2a}.$$

On the other hand, if $d < 0$, then $-d > 0$ and $d = (-d)(-1)$. So

$$x = \frac{-b \pm \sqrt{d}}{2a}$$

Since $d < 0$, $\sqrt{d}$ is not a real number; however, $\sqrt{-d}$ is a real number.

$$= \frac{-b \pm \sqrt{(-d)(-1)}}{2a}$$

$$= \frac{-b \pm \sqrt{-d} \cdot \sqrt{-1}}{2a}$$

$$= \frac{-b \pm \sqrt{-d}\, i}{2a} \quad \text{where } i = \sqrt{-1}.$$

The number i, along with a review of algebra involving i, is given in Appendix A.

If $d = 0$, then

$$\left(x + \frac{b}{2a}\right)^2 = 0$$

$$x + \frac{b}{2a} = 0$$

$$x = -\frac{b}{2a}.$$

Given $ax^2 + bx + c = 0$, $a \neq 0$, and $d = b^2 - 4ac$,

if $d > 0$,   then there are two real roots:

$$x = \frac{-b \pm \sqrt{d}}{2a};$$

if $d = 0$,   then there is one real root:

$$x = \frac{-b}{2a};$$

if $d < 0$,   then there are no real roots:

$$x = \frac{-b \pm \sqrt{-d}\, i}{2a}.$$

If we state this as

$$x = \frac{-b \pm \sqrt{b^2 - 4ac}}{2a},$$

then the result is called the *Quadratic Formula*.

The nature of the roots of a quadratic equation is determined by the discriminant.

QUADRATIC FORMULA

This situation can be illustrated graphically. If the zeros of a function are the real numbers, then they are the x-intercepts, as shown in Figure 2.32.

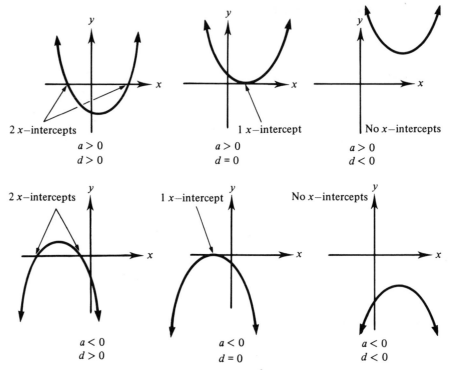

2 x-intercepts
$$a > 0$$
$$d > 0$$

1 x-intercept
$$a > 0$$
$$d = 0$$

No x-intercepts
$$a > 0$$
$$d < 0$$

2 x-intercepts
$$a < 0$$
$$d > 0$$

1 x-intercept
$$a < 0$$
$$d = 0$$

No x-intercepts
$$a < 0$$
$$d < 0$$

Figure 2.32 Graphs of $y = ax^2 + bx + c, a \neq 0$

EXAMPLES: Solve for x.

1. $5x^2 + 2x - 3 = 0$

$$x = \frac{-2 \pm \sqrt{4 - 4(5)(-3)}}{2(5)}$$

$$= \frac{-2 \pm 2\sqrt{1 + 15}}{2(5)}$$

$$= \frac{-2 \pm 8}{10}$$

$$= -1 \quad \text{or} \quad \frac{3}{5}$$

2. $5x^2 + 2x - 2 = 0$

$$x = \frac{-2 \pm \sqrt{4 - 4(5)(-2)}}{2(5)}$$

$$= \frac{-2 \pm 2\sqrt{1 + 10}}{2(5)}$$

$$= \frac{-1 \pm \sqrt{11}}{5}$$

3. $5x^2 + 2x + 2 = 0$

$$x = \frac{-2 \pm \sqrt{4 - 4(5)(2)}}{2(5)}$$

$$= \frac{-2 \pm 2\sqrt{-9}}{10}$$

$$= \frac{-1 \pm 3i}{5}$$

A review of complex numbers is given in Appendix A.

4. $5x^2 + 2x - (w + 4) = 0$

$$x = \frac{-2 \pm \sqrt{4 - 4(5)(-w - 4)}}{2(5)}$$

$$= \frac{-2 \pm 2\sqrt{5w + 21}}{2(5)}$$

$$= \frac{-1 \pm \sqrt{5w + 21}}{5}$$

$4 - 4(5)(-w - 4)$
$\quad = 4(1 + 5w + 20)$
$\quad = 4(5w + 21)$

5. $5x^2 + (1 - w)x - (w + 4) = 0$

$$x = \frac{-(1 - w) \pm \sqrt{(1 - w)^2 - 4(5)(-w - 4)}}{2(5)}$$

$$= \frac{-1 + w \pm \sqrt{1 - 2w + w^2 + 20w + 80}}{10}$$

$$= \frac{(w - 1) \pm \sqrt{w^2 + 18w + 81}}{10}$$

$$= \frac{(w - 1) \pm (w + 9)}{10}$$

$$= \frac{2w + 8}{10}, \ \frac{-10}{10}$$

$$= \frac{w + 4}{5}, \ -1$$

Notice that
$$\sqrt{w^2 + 18w + 81}$$
$$= \sqrt{(w + 9)^2}$$
$$= |w + 9|$$
$$= \pm(w + 9).$$

Suppose you wish to check your answers for the above examples, say Example 2. One method is to substitute the roots into the original equation. This may involve considerable effort:

$$x = \frac{-1 \pm \sqrt{11}}{5}.$$

Substituting, we get

$$5\left(\frac{-1 + \sqrt{11}}{5}\right)^2 + 2\left(\frac{-1 + \sqrt{11}}{5}\right) - 2 \overset{?}{=} 0$$

and

$$5\left(\frac{-1 - \sqrt{11}}{5}\right)^2 + 2\left(\frac{-1 - \sqrt{11}}{5}\right) - 2 \overset{?}{=} 0.$$

Instead, suppose we represent the roots by r_1 and r_2. Then:

$$x = r_1 \quad \text{or} \quad x = r_2$$
$$x - r_1 = 0 \quad \text{or} \quad x - r_2 = 0$$
$$(x - r_1)(x - r_2) = 0$$
$$x^2 - r_1 x - r_2 x + r_1 r_2 = 0$$
$$x^2 - (r_1 + r_2)x + r_1 r_2 = 0$$

$$x^2 - \left(\begin{matrix}\text{sum of}\\\text{the roots}\end{matrix}\right)x + \left(\begin{matrix}\text{product of}\\\text{the roots}\end{matrix}\right) = 0$$

This result provides an easy method for checking your answers. That is,

$$r_1 + r_2 = -\frac{b}{a}$$

$$r_1 r_2 = \frac{c}{a}.$$

It also gives a method for finding an equation if you are given the roots.

Comparing this result to $ax^2 + bx + c = 0$ or

$$x^2 + \frac{b}{a}x + \frac{c}{a} = 0,$$

we can see that

$$\textbf{sum of roots} = -\frac{b}{a} \quad \text{and} \quad \textbf{product of roots} = \frac{c}{a}.$$

For Example 2:

$$5x^2 + 2x - 2 = 0$$

$$x^2 + \frac{2}{5}x - \frac{2}{5} = 0$$

sum of roots:

$$\frac{-1 + \sqrt{11}}{5} + \frac{-1 - \sqrt{11}}{5} = -\frac{2}{5}$$

product of roots:

$$\left(\frac{-1 + \sqrt{11}}{5}\right)\left(\frac{-1 - \sqrt{11}}{5}\right) = \frac{1 - 11}{25} = \frac{-10}{25} = -\frac{2}{5}$$

The answer checks since

$$x^2 + \frac{2}{5}x - \frac{2}{5}.$$

$\text{opposite of the}\!\!\!\uparrow$ $\;\;\uparrow\!\!\!\text{same as the product}$
$\;\;\;\;\;$ sum of roots $\quad\quad$ of roots

EXAMPLES

6. Check the answer for Example 5 using the sum and product method.

$$5x^2 + (1 - w)x - (w + 4) = 0$$

$$x^2 + \frac{1 - w}{5}x + \frac{-w - 4}{5} = 0$$

opposite of sum of roots   product of roots

$$r_1 = \frac{w + 4}{5} \quad\quad r_2 = -1 \quad r_1 + r_2 = \frac{w + 4}{5} - 1 \quad\quad r_1 r_2 = \left(\frac{w + 4}{5}\right)(-1)$$

$$= \frac{w + 4 - 5}{5} \quad\quad = \frac{-w - 4}{5}$$

$$= \frac{w - 1}{5}$$

The answer checks.

7. Find a quadratic equation with roots

$$\frac{2 + \sqrt{5}}{3} \quad \text{and} \quad \frac{2 - \sqrt{5}}{3}.$$

sum of roots: $\dfrac{2 + \sqrt{5}}{3} + \dfrac{2 - \sqrt{5}}{3} = \dfrac{4}{3}$

product of roots: $\left(\dfrac{2 + \sqrt{5}}{3}\right)\left(\dfrac{2 - \sqrt{5}}{3}\right) = \dfrac{4 - 5}{9} = -\dfrac{1}{9}$

equation: $x^2 - \dfrac{4}{3}x - \dfrac{1}{9} = 0$

$$9x^2 - 12x - 1 = 0$$

8. Find a quadratic equation with roots $3 + 2i$ and $3 - 2i$.

Remember: $i^2 = -1$.

sum of roots: $(3 + 2i) + (3 - 2i) = 6$
product of roots: $(3 + 2i)(3 - 2i) = 9 - 4i^2$
$$= 9 + 4$$
$$= 13$$

equation: $x^2 - 6x + 13 = 0$

PROBLEM SET 2.5

A Problems

In Problems 1–10, find the discriminant and determine the number and type of roots for the given equation.

1. $x^2 + 5x - 6 = 0$ 2. $x^2 + 5x + 6 = 0$
3. $x^2 - 10x + 25 = 0$ 4. $x^2 + 6x + 9 = 0$
5. $3x^2 - 2 = 5x$ 6. $5x^2 = 3x - 4$
7. $3x^2 = 7x$ 8. $7x^2 = 3$
9. $x = \sqrt{2} - 2x^2$ 10. $\sqrt{5} - 4x^2 = 3x$

Solve each equation in Problems 11–20 over the set of complex numbers.

11. $2x^2 + x - 15 = 0$ 12. $12x^2 + 5x - 2 = 0$
13. $4x^2 - 5 = 0$ 14. $3x^2 - 1 = 0$
15. $2x^2 - 6x + 5 = 0$ 16. $5x^2 - 4x + 1 = 0$
17. $4x^2 = 12x - 9$ 18. $3x^2 = 5x + 2$
19. $3x = 1 - 2x^2$ 20. $5x = 3 - 4x^2$

B Problems

Solve each equation in Problems 21–30 for x in terms of the other variable.

21. $2x^2 + x - w = 0$ 22. $2x^2 + wx + 5 = 0$
23. $3x^2 + 2x + (y + 2) = 0$ 24. $3x^2 + 5x + (4 - y) = 0$
25. $4x^2 - 4x + (1 + t^2) = 0$
26. $4x^2 - (3t + 10)x + (6t + 4) = 0, \ t > 2$
27. $2x^2 + 3x + 4 - y = 0$ 28. $y = 2x^2 + 3x + 4$
29. $(x - 3)^2 + (y - 2)^2 = 4$ 30. $x^2 - 6x + y^2 - 4y + 9 = 0$

31. If $f(x) = ax^2 + bx + c$, $a \neq 0$, with roots r_1 and r_2, then show that $f(x) = a(x - r_1)(x - r_2)$.

32. If $f(x) = ax^2 + bx + c$, show that the following are roots:

$$r_1 = \frac{2c}{-b + \sqrt{b^2 - 4ac}} \quad \text{and} \quad r_2 = \frac{2c}{-b - \sqrt{b^2 - 4ac}}.$$

$ 33. In Problem 35 of Problem Set 2.4, it was determined that a manufacturer of citizen-band radios has a profit (in dollars) given by

$$p = -30x^2 + 360x - 600,$$

where x is the number of radios produced per day. What is the number of radios produced if there is no profit?

$ 34. An investor owns 100 shares of stock valued at $30 per share. She also plans to purchase 10 shares per week until she decides to sell all the stock. If the value of the stock is expected to decrease by $1.50 per share per week, the equation representing the value, y (in dollars), after x weeks is

$$y = \begin{pmatrix} \text{value of} \\ \text{stock} \end{pmatrix} \begin{pmatrix} \text{number of} \\ \text{shares} \end{pmatrix}.$$

a. In how many weeks should the investor sell the stock to realize maximum income from sale?
b. What is the maximum income from sale?
c. In how many weeks will the stock be worthless?

35. *Calculator Problem.* If you were to throw a rock straight up at 80 fps from the world's tallest dam in Rojunsky, U.S.S.R., the height y in feet from the ground after x seconds is given by

$$y = -16x^2 + 80x + 1066.$$

a. What is the height of the dam?
b. How long will the rock take to hit the bottom?
c. Graph the height of the rock as a function of time.

36. *Calculator Problem.* If you throw a rock straight up at 48 fps from the top of the Sears Tower in Chicago, the height in feet, y, from the ground after x seconds is given by

$$y = -16x^2 + 48x + 784.$$

a. What is the height of the Sears Tower?
b. How long will it take the rock to hit the ground?
c. Graph the height of the rock as a function of time.

37. *Calculator Problem.* Many materials, such as brick, steel, aluminum, and concrete, will expand due to increases in temperature. This is why fillers are placed between the cement slabs in sidewalks. Suppose you have a 100-ft beam securely fastened at both ends, and let the height of the buckle be x feet. If the percentage of swelling is y, then

$$\text{new length} = \text{old length} + \text{change in length}$$

$$= 50 + (\text{percentage})(\text{length})$$

$$= 50 + \left(\frac{y}{100}\right)50$$

$$= 50 + \frac{y}{2}.$$

These relationships are shown in Figure 2.33.

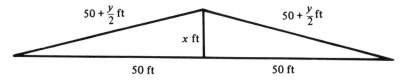

Figure 2.33   Expansion of a 100-ft beam

Then by the Pythagorean Theorem,

$$x^2 + 50^2 = \left(50 + \frac{y}{2}\right)^2$$

$$x^2 + 50^2 = \frac{(100 + y)^2}{4}$$

$$4x^2 + 4 \cdot 50^2 = 100^2 + 200y + y^2$$

$$4x^2 - y^2 - 200y = 0.$$

Solve this equation for x and then calculate the amount of buckling (in inches) for the following materials.

Material	y
Brick	0.03
Steel	0.06
Aluminum	0.12
Concrete	0.05

38. *Calculator Problem.* Suppose a model rocket weighs $\frac{1}{4}$ lb. Its engine propels it vertically to a height of 52 ft and a speed of 120 fps at burnout. If its parachute fails to open, what will be the approximate time to fall to Earth, according to the following equation for free-fall in vacuum:

$$h(t) = h_0 + v_0 t - \frac{1}{2}gt^2,$$

For this problem, $h_0 = 52$ ft and $v_0 = 120$ fps.

where $h(t)$ is the height (in feet) at time t, h_0 and v_0 are the height (in feet) and velocity (in feet per second) at the time selected at $t = 0$, and g is approximately 32 ft/sec/sec?

Mind Bogglers

39. A farmer wants to use a 120-ft natural barrier, as shown in Figure 2.34, and 300 ft of fence to construct a simple rectangular enclosure of maximum area. What are the dimensions of the enclosure?

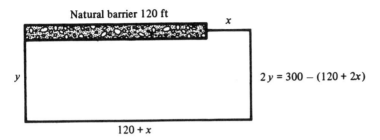

Natural barrier 120 ft

x

y

$2y = 300 - (120 + 2x)$

$120 + x$

Figure 2.34 Enclosure with 300 ft of fence

40. A space capsule is in the shape of the frustrum of a cone, as shown in Figure 2.35. Given that its volume is 19,792 cu ft and its height is 12 ft, what is the diameter of the base if it is twice the diameter of the top?

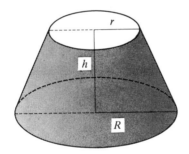

Figure 2.35 Shape of a space capsule

2.6 SUMMARY AND REVIEW

TERMS

Absolute value (2.2)
Circle (2.3)
Completing the square (2.5)
Discriminant (2.5)
Distance formula (2.3)
Linear function (2.1)
Parabola (2.4)
Quadratic equation (2.5)

Quadratic function (2.1, 2.4)
Root (2.5)
Slope of a line (2.1)
Triangle inequality (2.2)
x-intercepts of a parabola (2.5)
y-intercept of a line (2.1)
y-intercept of a parabola (2.5)
Zeros (2.5)

IMPORTANT IDEAS

(2.1)   A function is linear if $f(x) = mx + b$ and is quadratic if $f(x) = ax^2 + bx + c$, $a \neq 0$.

(2.1)   Forms of a linear equation:

standard form: $Ax + By + C = 0$

slope-intercept form: $y = mx + b$

point-slope form: $y - k = m(x - h)$

two-point form: $y - y_1 = \dfrac{(y_2 - y_1)}{x_2 - x_1}(x - x_1)$

Be able to find the equation for a line.

(2.1)   Graphing a line: solve for y and use slope and intercept (see Figure 2.5, on page 58).

(2.2)   The absolute value of a, written $|a|$, is

$$|a| = a \quad \text{if} \quad a \geq 0$$
$$|a| = -a \quad \text{if} \quad a < 0.$$

(2.2, 2.3)   The distance between points in one dimension and in two dimensions is

$$d = |x_2 - x_1| \quad \text{for points } (x_1) \text{ and } (x_2);$$
$$d = \sqrt{(x_2 - x_1)^2 + (y_2 - y_1)^2} \quad \text{for points } (x_1, y_1) \text{ and } (x_2, y_2).$$

(2.2)   Solving absolute value equations and inequalities.

(2.3)   The equation of a circle with center (h, k) and radius r is

$$(x - h)^2 + (y - k)^2 = r^2.$$

(2.4)   Graph the parabola $y = ax^2 + bx + c$ by completing the square to obtain $y - k = a(x - h)^2$. The parabola opens up if $a > 0$ and down if $a < 0$. It is the same as the parabola $y = ax^2$, translated to the point (h, k).

(2.4)   To complete the square for $y = ax^2 + bx + c$:
1.   Subtract the constant term from both sides.
2.   Factor a.
3.   Add the square of $\frac{1}{2}$ the coefficient of the x term to both sides.

(2.5)   Quadratic Formula: if $ax^2 + bx + c = 0$, $a \neq 0$, then

$$x = \frac{-b \pm \sqrt{b^2 - 4ac}}{2a}.$$

The sum of the roots is $-b/a$ and the product of the roots is c/a. The discriminant, $d = b^2 - 4ac$, determines the nature of the roots:

if $d > 0$,   then there are two real roots;
if $d = 0$,   then there is one real root;
if $d < 0$,   then there are no real roots.

REVIEW PROBLEMS

1. Graph:
    a. (2.1) $y = -3x + 2$ for $-1 \le x \le 2$
    b. (2.3) $(x - 3)^2 + (y + 2)^2 = 16$
2. Graph:
    a. (2.2) $y = |2x - 1|$ b. (2.2) $y = |x + 3| + 1$
3. (2.5) Graph:

    a. $y = \dfrac{-1}{3}x^2$ for $x \le 3$ b. $y = \dfrac{-1}{3}(x + 2)^2$

    c. $y = \dfrac{-1}{3}(x + 2)^2 + 3$

4. (2.1) Find the standard form equation of the line satisfying the given conditions.
    a. y-intercept, $1/2$; slope, $-5/2$
    b. Slope, 2; passing through $(-6, -5)$
    c. Passing through $(-1, 2)$ and $(6, -4)$
    d. No slope; passing through $(4, 6)$
5. (2.2) Solve the given equations and inequalities.
    a. $|x - 3| = |2x - 5|$
    b. $|3x + 2| \le 5$
    c. $|2 - x| < 4$
    d. $|4x + 1| > 8$
6. (2.2, 2.3) Find the distance between the given points.
    a. $(\sqrt{8})$ and $(1/\sqrt{2})$
    b. $(-1, 2)$ and $(6, -4)$
    c. (α, β) and (γ, δ)
    d. $(a, f(a))$ and $(b, f(b))$
$\$$ 7. (2.4, 2.5) A person selling hand-made shirts produces x shirts per day and prices them at $15 - x$ dollars each. The overhead is found to be $9x^2 - 85x + 205$. Then

$$\begin{aligned} \text{profit} &= \text{revenue} - \text{cost} \\ &= x(15 - x) - (9x^2 - 85x + 205) \\ &= 15x - x^2 - 9x^2 + 85x - 205 \\ &= -10x^2 + 100x - 205. \end{aligned}$$

    a. Graph $p = -10x^2 + 100x - 205$.
    b. How many shirts should be made to maximize the profit?
    c. What is the maximum profit?
    d. For how many shirts is there a loss?
$\$$ 8. (2.5) The number of telephones, x, handled by a switchboard is related to the number of connections, y, by the equation

$$2y = x^2 - x.$$

If a switchboard has a capacity of 780 connections, how many telephones can be handled?

9. (2.5) Solve:
     a. $2x^2 + 7x - 15 = 0$ b. $3x^2 - 10x - 6 = 0$
10. (2.5) Solve for x:
     a. $4x^2 - 5x + 6 = 0$ b. $3x^2 + (1 + 6t)x + (-2 - 4t) = 0, t \geq 0$

Polynomial Functions

3

3.1 EXPONENTS AND SIMPLIFYING EXPRESSIONS

ALGEBRAIC EXPRESSION

POLYNOMIAL

EXAMPLES:

Polynomials	Non-polynomials
5	$\dfrac{5}{0}$
$x, \dfrac{1}{3}x$	$\dfrac{1}{x}$
$2x + 3$	$2\sqrt{x} + 3$
$5x^2 + x + \sqrt{3}$	$\dfrac{x^2 + x + 2}{x - 2}$

An *algebraic expression* is any expression containing only algebraic symbols and operations such as addition, subtraction, multiplication, nonzero division, extraction of roots, and raising to integral or fractional powers. The simplest type of algebraic expression is a *polynomial*, which is a finite sum of terms with whole number exponents on the variables. A polynomial is written in functional form as

$$P(x) = a_n x^n + a_{n-1} x^{n-1} + a_{n-2} x^{n-2} + \cdots + a_1 x + a_0,$$

where n is an integer greater than or equal to 0, and the coefficients $a_0, a_1, a_2, \ldots, a_{n-1}, a_n$ are real numbers. For example, if

$$P(x) = 4x - 5, \quad Q(x) = 5x^2 + 2x + 1, \qquad R(x) = 3x^3 - 4x^2 + 3x - 2,$$

then:

$$
\begin{aligned}
P(2) &= 4(2) - 5 & Q(2) &= 5(2)^2 + 2(2) + 1 & R(2) &= 3(2)^3 - 4(2)^2 + 3(2) - 2 \\
&= 8 - 5 & &= 20 + 4 + 1 & &= 24 - 16 + 6 - 2 \\
&= 3 & &= 25 & &= 12
\end{aligned}
$$

Some additional terminology concerning polynomials is reviewed in Problems 1–20 of Problem Set 3.1.

The key to working with polynomials is the proper handling of exponents.

DEFINITION OF EXPONENTS
b is called the *base*
n is called the *exponent*
b^n is called a *power*

If b is any real number and n is any natural number, then

$$b^n = \underbrace{b \cdot b \cdot b \cdots b}_{n \text{ factors}}.$$

And further, if $b \neq 0$, then

$$b^0 = 1, \qquad b^{-n} = \frac{1}{b^n}.$$

ADDITION AND SUBTRACTION OF POLYNOMIALS

Similar terms are sometimes called *like terms*.

The distributive law shows you how to add polynomials instead of combining *similar terms*. For example

$$
\begin{aligned}
Q(x) + R(x) &= (5x^2 + 2x + 1) + (3x^3 - 4x^2 + 3x - 2) \\
&= 3x^3 + \underbrace{5x^2 + (-4)x^2}_{\substack{\text{similar} \\ \text{terms}}} + \underbrace{2x + 3x}_{\substack{\text{similar} \\ \text{terms}}} + \underbrace{1 + (-2)}_{\substack{\text{similar} \\ \text{terms}}} \\
&= 3x^3 + x^2 + 5x - 1
\end{aligned}
$$

Remember that terms containing the same power (or powers) of the variables are similar terms. This means that only the numerical coefficients may be different. We can apply the distributive property and combine the numerical coefficients:

$$5x^2 + (-4)x^2 = [5 + (-4)]x^2 = (5 - 4)x^2 = 1 \cdot x^2 = x^2.$$

EXAMPLES

1. Find $P(x) + Q(x)$.
   Solution: $(4x - 5) + (5x^2 + 2x + 1) = 5x^2 + (4x + 2x) + (-5 + 1)$
   $$= 5x^2 + 6x - 4$$

2. Find $P(x) + R(x)$.
   Solution: $(4x - 5) + (3x^3 - 4x^2 + 3x - 2) = 3x^3 - 4x^2 + 7x - 7$

3. Find $Q(x) - R(x)$.
   Solution: $(5x^2 + 2x + 1) - (3x^3 - 4x^2 + 3x - 2)$
   $$= 5x^2 + 2x + 1 + (-1)(3x^3 - 4x^2 + 3x - 2)$$
   $$= 5x^2 + 2x + 1 + (-3x^3) + 4x^2 + (-3x) + 2$$
   $$= -3x^3 + 9x^2 - x + 3$$

4. Find $4P(x) - 2Q(x)$.
   Solution: $4(4x - 5) - 2(5x^2 + 2x + 1) = 16x - 20 - 10x^2 - 4x - 2$
   $$= -10x^2 + 12x - 22$$

Remember, on p. 108 we let
$$P(x) = 4x - 5,$$
$$Q(x) = 5x^2 + 2x + 1,$$
$$R(x) = 3x^3 - 4x^2 + 3x - 2.$$
It is customary to rearrange the terms in the order of highest to lowest power.

Notice that many of the steps can be done mentally.

← This step would probably be done mentally, but we wanted to emphasize the step where we handled the subtraction. Remember,

$$Q(x) - R(x)$$
$$= Q(x) + (-1)R(x)$$

and then distribute the -1, being careful to use the properties of signs correctly.

MULTIPLICATION OF POLYNOMIALS

To understand the multiplication of polynomials, you must understand the multiplication of monomials, as illustrated by the following examples.

EXAMPLES: Multiply and simplify the given monomials.

5. $x^2(x^3) = xx(xxx)$
   $$= x^5$$

6. $(x^2)^3 = (x^2)(x^2)(x^2)$
   $$= (xx)(xx)(xx)$$
   $$= x^6$$

7. $(x^2y^3)^4 = (x^2y^3)(x^2y^3)(x^2y^3)(x^2y^3)$
   $$= (x^2x^2x^2x^2)(y^3y^3y^3y^3)$$
   $$= x^8y^{12}$$

Example 5 suggests that
$$b^m \cdot b^n = b^{m+n}.$$

Example 6 suggests that
$$(b^n)^m = b^{mn}.$$

Example 7 suggests that
$$(ab)^m = a^m b^m,$$
or, as shown,
$$(x^2y^3)^4 = (x^2)^4(y^3)^4$$
$$= x^8y^{12}.$$

Example 8 suggests that

$$\left(\frac{a}{b}\right)^m = \frac{a^m}{b^m}.$$

8. $\left(\dfrac{x}{y}\right)^3 = \left(\dfrac{x}{y}\right)\left(\dfrac{x}{y}\right)\left(\dfrac{x}{y}\right)$

$$= \frac{xxx}{yyy}$$

$$= \frac{x^3}{y^3}$$

Notice in Example 9 that

$5 - 3 = 2.$

9. $\dfrac{x^5}{x^3} = \dfrac{xxxxx}{xxx}$

$$= xx$$

$$= x^2$$

Notice in Example 10 that

$3 - 5 = -2.$

The last two examples suggest that

$$\frac{b^m}{b^n} = b^{m-n}.$$

10. $\dfrac{x^3}{x^5} = \dfrac{xxx}{xxxxx}$

$$= \frac{1}{x^2}$$

$$= x^{-2}$$

These examples illustrate five laws of exponents that are used to simplify expressions.

LAWS OF EXPONENTS

> Let a and b be real numbers and let m and n be any integers. Then the five rules listed below govern the use of exponents except that the form $0/0$ and division by zero are excluded.
>
> First Law: $b^m \cdot b^n = b^{m+n}$
>
> Second Law: $\dfrac{b^m}{b^n} = b^{m-n}$
>
> Third Law: $(b^n)^m = b^{mn}$
>
> Fourth Law: $(ab)^m = a^m b^m$
>
> Fifth Law: $\left(\dfrac{a}{b}\right)^m = \dfrac{a^m}{b^m}$

These laws of exponents are used together with the distributive property when we multiply polynomials. Here's an example:

$$\begin{aligned}
(4x - 5)(5x^2 + 2x + 1) &= (4x - 5)5x^2 + (4x - 5)2x + (4x - 5) \cdot 1 \\
&= 20x^3 - 25x^2 + 8x^2 - 10x + 4x - 5 \\
&= 20x^3 - 17x^2 - 6x - 5
\end{aligned}$$

EXAMPLES

11.　$(2x - 3)(x + 4) = (2x - 3)x + (2x - 3)4$
$$= 2x^2 - 3x + 8x - 12$$
$$= 2x^2 + 5x - 12$$

This step could be done mentally.

12.　$(2x - 3)(x^2 - x - 2) = (2x - 3)x^2 + (2x - 3)(-x) + (2x - 3)(-2)$
$$= 2x^3 - 3x^2 - 2x^2 + 3x - 4x + 6$$
$$= 2x^3 - 5x^2 - x + 6$$

13.　$(2x - 3)(x + 3)(3x - 4) = [(2x - 3)x + (2x - 3)3](3x - 4)$
$$= [2x^2 - 3x + 6x - 9](3x - 4)$$
$$= (2x^2 + 3x - 9)(3x - 4)$$
$$= (2x^2 + 3x - 9)3x + (2x^2 + 3x - 9)(-4)$$
$$= 6x^3 + 9x^2 - 27x - 8x^2 - 12x + 36$$
$$= 6x^3 + x^2 - 39x + 36$$

We *associated* the first two polynomials in this example.

Products of binomials are common, and for this reason a special pattern called *FOIL* is observed. In Example 11, there are four pairs of terms to be multiplied.

FOIL stands for the product of:

First terms (see ①)
+
Outer terms (see ②)
+
Inner terms (see ③)
+
Last terms (see ④).

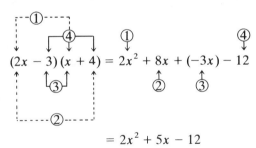

$$= 2x^2 + 5x - 12$$

Steps 2 and 3 are usually added mentally.

EXAMPLES: Multiply mentally.

14.　$(2x - 3)(x + 3) = 2x^2 + 3x - 9$

Mentally

$6x + (-3x)$
② ③

$$\overset{\textcircled{1}}{\downarrow} \qquad \overset{\textcircled{4}}{\downarrow}$$

15. $(x + 3)(3x - 4) = 3x^2 + 5x - 12$

$$\underset{\textcircled{2} + \textcircled{3}}{\underbrace{\qquad\qquad}}$$

16. $(5x - 2)(3x + 4) = 15x^2 + 14x - 8$

Division of polynomials will be considered in Section 3.3.

PROBLEM SET 3.1

A Problems

In your previous algebra courses, you were introduced to the basic terminology involving polynomials. Problems 1–20 were designed to help you review this terminology. In Problems 1–10, match the term at the left with its definition at the right. In Problems 11–20, match the description at the left with an example from the right.

1. Term
2. Degree of a term
3. Polynomial
4. Degree of a polynomial
5. Coefficient
6. Numerical coefficient
7. Monomial
8. Binomial
9. Trinomial
10. Real polynomial
11. A quadratic binomial in x and y
12. A quadratic trinomial over the rationals
13. A monomial that is second degree in x
14. A linear monomial in x
15. A zero-degree polynomial
16. A linear polynomial in x over the integers
17. A second-degree monomial in x and y
18. A third-degree polynomial in x and y over the integers
19. A fifth-degree polynomial that contains a third-degree term

A. A finite sum of terms with whole number exponents on the variables
B. An indicated product of numbers and variables
C. A polynomial with one term
D. A polynomial with coefficients and variables that are restricted to real numbers
E. A polynomial with three terms
F. The exponent of the variable or the sum of the exponents of the variables if more than one variable
G. A polynomial with two terms
H. The product of all the factors of a term except a certain one or ones
I. The numerical factor of a term
J. An expression with more than three terms
K. A polynomial with fewer than three terms
L. The degree of its highest-degree term
M. $\frac{2}{3}$
N. $\frac{5}{2}x$
P. $3x + 2$
Q. $4x^3y^3$
R. $\sqrt{3}\,xy$
S. $2xy^2$
T. x^2y^2
U. $x^2 + y^2$

20. A fifth-degree polynomial over the integers

V. $\frac{2}{3}x^2 + 5x - \frac{1}{2}$
W. $x^2y^3 + 2xy^3 + y^4$
X. $\sqrt{5}\,x^2 - x + 1$
Y. $x^2y^3 + \frac{1}{2}x^2y^2 + xy^2$
Z. $\frac{1}{3}x^2 + y^3$

21. Use Problems 1–20 to complete the quotation. Substitute the letter for the corresponding problem number in parentheses to find the quotation. Notice that the letters O and K have already been filled in for you. For example, the fourth word is (O)(2), which translates to "OF" since the answer to Problem 2 is F.

(15)(3)(13)(5)(9)(15)(3)(13)(6)(7)(18)-(13)(5)(9)
(11)(14)(18)(5)(3)(K)(9)(14)
(2)(O)(11)(14)(10)(3)(13)(6)(O)(14)　　　(O)(2)
(18)(7)(6)(9)(14)(7)(9)(18),　　　(3)(14)(10)　　　(13)(5)(9)
(16)(4)(9)(14)(13)(6)(2)(11)(4)　　　(2)(O)(11)(14)(13)(3)(6)(14)
(O)(2)　　　(3)(10)(12)(3)(14)(13)(3)(8)(9)　　　(13)(O)
(5)(11)(15)(3)(14)　　　(3)(2)(2)(3)(6)(17)(18).

Historical Quote

This is a quotation of Isaac Barrow (1630–1677). Calculus was invented by Newton and Leibniz, but Barrow came very close to a breakthrough in the study of calculus on his own. He preceded Newton in the mathematics chair at Cambridge University.

Multiply the expressions in Problems 22–34 mentally.

22. a. $(x + 2)(x + 1)$　　b. $(y - 2)(y + 3)$
23. a. $(x + 1)(x - 2)$　　b. $(y - 3)(y + 2)$
24. a. $(a - 5)(a - 3)$　　b. $(b + 3)(b - 4)$
25. a. $(c + 1)(c - 7)$　　b. $(z - 3)(z + 5)$
26. a. $(2x + 1)(x - 1)$　　b. $(2x - 3)(x - 1)$
27. a. $(x + 1)(3x + 1)$　　b. $(x + 1)(3x + 2)$
28. a. $(2a + 3)(3a - 2)$　　b. $(2a + 3)(3a + 2)$
29. a. $(x + y)(x + y)$　　b. $(x - y)(x - y)$
30. a. $(x + y)(x - y)$　　b. $(a + b)(a - b)$
31. a. $(5x - 4)(5x + 4)$　　b. $(3y - 2)(3y + 2)$
32. a. $(a + 2)^2$　　b. $(b - 2)^2$
33. a. $(x + 4)^2$　　b. $(y - 3)^2$
34. a. $(s + t)^2$　　b. $(u - v)^2$

B Problems

Let $P(x) = 5x + 1$, $Q(x) = 3x^2 - 5x + 2$, and $R(x) = x^3 - 4x^2 + x - 4$ in Problems 35–42. Find the required result.

35. a. $P(2)$　　b. $Q(2)$　　c. $R(2)$
36. a. $P(3)$　　b. $Q(0)$　　c. $R(1)$
37. a. $P(-1)$　　b. $Q(-1)$　　c. $R(-1)$
38. a. $P(5)$　　b. $Q(1)$　　c. $R(0)$
39. a. $P(x) + Q(x)$　　b. $Q(x) + R(x)$
40. a. $P(x) + R(x)$　　b. $Q(x) - P(x)$
41. a. $R(x) - Q(x)$　　b. $P(x)Q(x)$
42. a. $P(x)Q(x) - R(x)$　　b. $3Q(x) - 4R(x)$

Simplify the expressions in Problems 43–52.

43. $(3x - 1)(x^2 + 3x - 2)$

44. $(5x + 1)(x^3 - 2x^2 + 3x - 5)$

45. $(x + 1)(x - 3)(2x + 1)$

46. $(2x - 1)(x + 3)(3x + 1)$

47. $(x - 2)^2(x + 1)$

48. $(x - 2)(x + 1)^2$

49. $(x - 3)^3$

50. $(2x - 1)^3$

51. $(2x - 1)^2(3x^4 - 2x^3 + 3x^2 - 5x + 12)$

52. $(x - 3)^3(2x^3 - 5x^2 + 4x - 7)$

Mind Boggler

 53. Fill in the blanks with whole numbers less than 10 so that each horizontal, vertical, and diagonal equation is true. The answer is not unique.

The * means multiplication.

3	+		=	
+	*	+	⟍	+
	+		=	
‖		‖	⟍	‖
	+		=	

3.2 FACTORING

FACTORING

The process of reversing multiplication to find divisors of a given number is called *factoring*. You probably spent a great deal of time learning

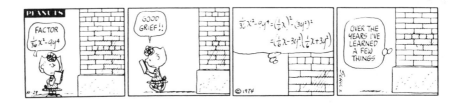

the process of factoring, and hopefully your sentiments are closer to those of the school building in the cartoon than those expressed by Sally. The point to remember when factoring an expression is to classify it into one of several types. In the cartoon, the expression

$$\frac{1}{36}x^2 - 9y^4$$

is of the type called *the difference of squares* since it can be written as shown in the third frame of the cartoon:

$$\left(\frac{x}{6}\right)^2 - (3y^2)^2 = \left(\frac{x}{6} - 3y^2\right)\left(\frac{x}{6} + 3y^2\right).$$

The types of factoring problems usually encountered are summarized in Table 3.1, and they are listed in order of difficulty. For example, the expression

$$6x^2 + 3x - 18$$

is properly factored by *first* finding the common factor (type 1) and then using FOIL (type 7):

$$6x^2 + 3x - 18 = 3(2x^2 + x - 6)$$
$$= 3(2x - 3)(x + 2).$$

We usually factor *over the set of integers*, which means that all the numerical coefficients are integers. If the original problem involves fractions (as in the example from the cartoon), we will factor out the fractional part first:

$$\frac{1}{36}x^2 - 9y^4 = \frac{1}{36}(x^2 - 6^2 \cdot 3^2 y^4)$$

$$= \frac{1}{36}[x^2 - (18y^2)^2]$$

$$= \frac{1}{36}(x - 18y^2)(x + 18y^2).$$

Factoring over the set of integers also rules out factoring

$$x^2 - 3 = (x - \sqrt{3})(x + \sqrt{3}),$$

since the factors do not have integer coefficients. We call an expression *completely factored* or *irreducible* if all fractions are eliminated by common factoring and if no further factoring is possible over the set of integers.

TABLE 3.1 Factoring Types

Type	Form	Comments
1. Common factors	$ax + ay + az$ $= a(x + y + z)$	This is simply using the distributive property. It can be applied with any number of terms.
2. Difference of squares	$x^2 - y^2 = (x - y)(x + y)$	This is the type shown in the cartoon. The *sum* of two squares cannot be factored in the set of real numbers.
3. Difference of cubes	$x^3 - y^3$ $= (x - y)(x^2 + xy + y^2)$	This is similar to the difference of squares and can be proved by multiplying the factors.
4. Sum of cubes	$x^3 + y^3$ $= (x + y)(x^2 - xy + y^2)$	Unlike the sum of squares, the sum of cubes can be factored.
5. Perfect square	$x^2 + 2xy + y^2 = (x + y)^2$ $x^2 - 2xy + y^2 = (x - y)^2$	The middle term is twice the product of xy.
6. Perfect cube	$x^3 + 3x^2y + 3xy^2 + y^3$ $= (x + y)^3$ $x^3 - 3x^2y + 3xy^2 - y^3$ $= (x - y)^3$	The numerical coefficients of the terms are: 1 3 3 1 or 1 − 3 3 − 1
7. FOIL	See examples in the text.	This trial-and-error procedure is used with a trinomial. It should be used after types 1–6 have been checked.
8. Grouping	See examples in the text.	After types 1–7 have been checked, you can factor some expressions by proper grouping.
9. Irreducible	Examples arise in every factoring situation. Expressions such as $x + 4$ and $x^2 + y^2$ cannot be factored further.	When factoring, you are not through until all the factors are irreducible.

EXAMPLES: Completely factor.

1. $2x + 3xy = x(2 + 3y)$ Common factor x

2. $4 - 20x = 4(1 - 5x)$ Common factor 4

3. $2xz + yz = (2x + y)z$ Common factor z

4. $2x(2a - b) + y(2a - b) = (2x + y)(2a - b)$ Common factor $(2a - b)$

5. $x^2 - 9 = (x - 3)(x + 3)$ Difference of squares

6. $3x^2 - 75 = 3(x^2 - 25)$ Common factor first
$\qquad\qquad = 3(x - 5)(x + 5)$ Difference of squares

7. $16x^4 - 1 = (4x^2 - 1)(4x^2 + 1)$ Difference of squares
$\qquad\qquad = (2x - 1)(2x + 1)(4x^2 + 1)$ Difference of squares

8. $(x + 3y)^2 - 1 = [(x + 3y) - 1][(x + 3y) + 1]$ Difference of squares
$\qquad\qquad = (x + 3y - 1)(x + 3y + 1)$ Simplify

9. $\dfrac{9a^2}{b^2} - (a + 3b)^2 = \dfrac{1}{b^2}[9a^2 - b^2(a + 3b)^2]$ Common factor to eliminate fractions

$\qquad\qquad = \dfrac{1}{b^2}\{(3a)^2 - [b(a + 3b)]^2\}$ This step can be done mentally.

$\qquad\qquad = \dfrac{1}{b^2}[3a - b(a + 3b)][3a + b(a + 3b)]$ Difference of squares

$\qquad\qquad = \dfrac{1}{b^2}(3a - ab - 3b^2)(3a + ab + 3b^2)$ Simplify

10. $x^2 - 8x + 15$
This is a trinomial, so we use FOIL (a trial-and-error procedure) to try to factor it into two binomials:
$x^2 - 8x + 15 = (x - 5)(x - 3)$

11. $6x^2 + x - 12 = (2x + 3)(3x - 4)$ FOIL—you may have to try several possibilities before you find the correct one.

12. $6x^2 - 9x - 15 = 3(2x^2 - 3x - 5)$
$\qquad\qquad = 3(2x - 5)(x + 1)$

13. $4x^4 - 13x^2y^2 + 9y^4$
This is a trinomial, so we use FOIL (a trial-and-error procedure) to try to factor it into two binomials.
$4x^4 - 13x^2y^2 + 9y^4 = (x^2 - y^2)(4x^2 - 9y^2)$
$\qquad\qquad = (x - y)(x + y)(2x - 3y)(2x + 3y)$ You might try several possibilities before finding this one. Remember to factor completely; several types may be combined in one problem.

14. $9x^3 + 18x^2 - x - 2$
None of the types 1–7 from Table 3.1 seem to apply. Thus, we try grouping the terms. Some groupings may lead to a factorable form; others may not.

Grouping

Common factor

Common factor

Difference of squares

$$9x^3 + 18x^2 - x - 2 = (9x^3 + 18x^2) - (x + 2)$$
$$= 9x^2(x + 2) - (x + 2)$$
$$= (9x^2 - 1)(x + 2)$$
$$= (3x - 1)(3x + 1)(x + 2)$$

GRAPHS OF EQUATIONS EXPRESSIBLE IN FACTORED FORM

Suppose we wish to graph the curve

$$x^2 - y^2 = 0.$$

Since this equation can be factored as

$$(x - y)(x + y) = 0,$$

we can use the result from algebra, which says that the solution is found by setting each factor equal to zero. That is,

Recall, if $A \cdot B = 0$, then $A = 0$ or $B = 0$ (perhaps both). The important thing to remember here is that this procedure works only if the product is equal to zero.

$$x - y = 0 \qquad \text{or} \qquad x + y = 0$$

so the graph of $x^2 - y^2 = 0$ is shown in Figure 3.1.

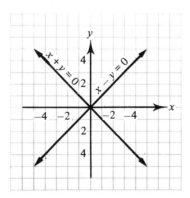

Figure 3.1   Graph of $x^2 - y^2 = 0$

EXAMPLE 15: Graph $3x^3 - 3xy - 2x^2y + 2y^2 = 0$.

Solution: Factor by grouping:

$$3x(x^2 - y) - 2y(x^2 - y) = 0$$
$$(x^2 - y)(3x - 2y) = 0.$$

Thus,

$$x^2 - y = 0 \quad \text{or} \quad 3x - 2y = 0.$$

The graph is shown in Figure 3.2.

Notice the use of the word *or*. If we used *and*, then we would mean the simultaneous solution or intersection of graphs. Instead, we want *all* points on both graphs.

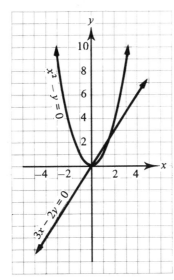

Figure 3.2 Graph of $3x^3 - 3xy - 2x^2y + 2y^2 = 0$

PROBLEM SET 3.2

A Problems

Factor completely, if possible, the expressions in Problems 1–24.

1. $me + mi + my$
2. $a^2 - b^2$
3. $a^2 + b^2$
4. $a^3 - b^3$
5. $a^3 + b^3$
6. $s^2 + 2st + t^2$
7. $m^2 - 2mn + n^2$
8. $u^2 + 2uv + v^2$
9. $a^3 + 3a^2b + 3ab^2 + b^3$
10. $p^3 - 3p^2q + 3pq^2 - q^3$
11. $-c^3 + 3c^2d - 3cd^2 + d^3$
12. $x^2y + xy^2$
13. $(a + b)x + (a + b)y$
14. $(4x - 1)x + (4x - 1)3$
15. $x^2 - 2x - 35$
16. $2x^2 + 7x - 15$
17. $3x^2 - 5x - 2$
18. $6y^2 - 7y + 2$
19. $8a^2b + 10ab - 3b$
20. $2s^2 - 10s - 48$
21. $4y^3 + y^2 - 21y$
22. $12m^2 - 7m - 12$
23. $12p^4 + 11p^3 - 15p^2$
24. $9x^2y + 15xy - 14y$

B Problems

Factor completely, if possible, the expressions in Problems 25–58.

25. $(x - y)^2 - 1$ 26. $(2x + 3)^2 - 1$
27. $(5a - 2)^2 - 9$ 28. $(3p - 2)^2 - 16$

29. $\dfrac{4x^2}{9} - (x + y)^2$ 30. $\dfrac{4}{25}x^2 - (x + 2)^2$

31. $\dfrac{x^6}{y^8} - 169$ 32. $(a + b)^2 - (x + y)^2$

33. $(m - 2)^2 - (m + 1)^2$ 34. $x^{2n} - y^{2n}$
35. $x^{3n} - y^{3n}$ 36. $x^{3n} + y^{3n}$
37. $x^{2n} - 2x^n y^n + y^{2n}$ 38. $8(x - 2)^3 - 27(x + 2)^3$
39. $(x - 2)^2 - (x - 2) - 6$ 40. $(x + 3)^2 - (x + 3) - 6$
41. $4x^4 - 17x^2 + 4$ 42. $4x^4 - 45x^2 + 81$
43. $z^6 - 64$ 44. $x^8 - y^8$
45. $z^5 - 8z^2 - 4z^3 + 32$ 46. $x^5 + 8x^2 - x^3 - 8$
47. $(x^3 + \frac{1}{27})(x^3 - \frac{1}{8})$ 48. $(x^3 + \frac{1}{8})(x^2 - \frac{1}{4})$
49. $x^6 - 6x^3 - 16$ 50. $x^6 + 9x^3 + 8$
51. $x^2 - 2xy + y^2 - a^2 - 2ab - b^2$ 52. $x^2 + 2xy + y^2 - a^2 - 2ab - b^2$
53. $x^2 + y^2 - a^2 - b^2 - 2xy + 2ab$
54. $x^3 + 3x^2 y + 3xy^2 + y^3 + a^3 + 3a^2 b + 3ab^2 + b^3$
55. $(x + y + 2z)^2 - (x - y + 2z)^2$
56. $(x^2 - 3x - 6)^2 - 4$
57. $2(x + y)^2 - 5(x + y)(a + b) - 3(a + b)^2$
58. $2(s + t)^2 + 3(s + t)(s + 2t) - 2(s + 2t)^2$

Graph the curves in Problems 59–68 by factoring.

59. $\dfrac{x^2}{9} - \dfrac{y^2}{4} = 0$ 60. $\dfrac{x^2}{16} - \dfrac{y^2}{25} = 0$

61. $\dfrac{y^2}{49} = \dfrac{x^2}{36}$ 62. $\dfrac{y^2}{25} - \dfrac{x^2}{36} = 0$

63. $2x^3 - 2xy + x^2 y - y^2 = 0$ 64. $4x^3 - 8xy + 3x^2 y - 6y^2 = 0$
65. $(x - y - 1)(3x + 2y - 4) = 0$ 66. $(5x - 2y - 6)(x + y + 2) = 0$
67. $(x - y)(x + 2y)(x^2 - y) = 0$ 68. $(x^2 - y^2)(2x + y - 3) = 0$

Mind Bogglers

69. In Problem 29 of Problem Set 1.5 (page 49), we wrote equations for the letters M, I, N, and U. Let
    A: $(y - 2)(2x - y)(2x + y - 8) = 0$,
    E: $xy(y - 4)(y - 2) = 0$,
    J: $[(x - 2)^2 - 4y](x - 4) = 0$,
    K: $x(x + y - 2)(2x - 3y + 6) = 0$,

where the domain and range for each equation is $0 \le x \le 4$, $0 \le y \le 4$. Use these letters as examples to help you write the equations for the letters F, I, and T.

70. We can handle messages using the letters from Problem 69 by translating the functions as necessary or by enclosing each letter in brackets and assuming that the axes are shifted five units to the right each time a bracket is encountered. With this second method, graph the following message using the restrictions for the domain and range given in Problem 69.

$[y(x - 2)(y - 4)][x - 5][(y - 2)(2x - y)(2x + y - 8)]$
$[(x - 2)(y - 4)][xy(y - 4)(y - 2)][y + 2]$
$[(x - 4)((x - 2)^2 - 4y)][(y - 2)(2x - y)(2x + y - 8)]$
$[(4x - y)(4x - y - 8)(4x + y - 8)(4x + y - 16)]$
$[x^2 + (y - 1)^2] = 0.$

71. Using the ideas of Problems 69 and 70, write equations for the following message: A NEAT FEAT.

72. Using the ideas of Problems 69 and 70, write equations for the following message: A FINE VINE.

73. *Calculator Problem.* The number 126 can be factored into two factors, 6 and 21, that have the same digits $(1, 2, 6)$ appearing in the original number. Many numbers have this property. For example, $(5829)(63,741) = 371,546,289$. The smallest nine-digit number containing each nonzero digit once that can be factored as above (in more than one way) is 129,378,546. What are the factors?

3.3 SYNTHETIC DIVISION

In Section 3.1, we considered sums, differences, and products of polynomials. In this section, we'll consider quotients of polynomials and will see how this division can be quickly and easily accomplished. This will, in turn, lead us to some methods for finding the roots of certain polynomials in the next sections.

We'll begin by considering the positive integers P and D. If the result of P divided by D is an integer Q, then D is a factor of P. For example, if $P = 15$ and $D = 3$, then $P/D = 5$, and 5 is called the *quotient*. If D is not a factor of P, then we will obtain a quotient Q plus a remainder R which is less than D so that

$$\frac{P}{D} = Q + \frac{R}{D}.$$

For example, if $P = 17$ and $D = 3$, then $P/D = 5 + \frac{2}{3}$.

Division may be checked by multiplying:

if $\dfrac{P}{D} = Q$, then $P = QD$;

if $\dfrac{P}{D} = Q + \dfrac{R}{D}$, then $P = QD + R$ $(R < D)$.

For example,

if $\dfrac{15}{3} = 5$, then

$15 = 5 \cdot 3$;

if $\dfrac{17}{3} = 5 + \dfrac{2}{3}$, then

$17 = 5 \cdot 3 + 2.$

Division of polynomials is similar and leads to a result called the *Division Algorithm.*

DIVISION ALGORITHM
This result can be checked by multiplication:

$$P(x) = Q(x)D(x) + R(x).$$

If $P(x)$ and $D(x)$ are polynomials $[D(x) \neq 0]$, then

$$\frac{P(x)}{D(x)} = Q(x) + \frac{R(x)}{D(x)},$$

where $Q(x)$ is a unique polynomial and $R(x)$ is a polynomial such that the degree of $R(x)$ is less than the degree of $D(x)$.

1. If $R(x) = 0$, then $D(x)$ is a factor of $P(x)$.
2. If the degree of $D(x)$ is greater than the degree of $P(x)$, then $Q(x) = 0$.

The question to be considered is how to *find* $Q(x)$ and $R(x)$. The first method is by long division. Let $P(x) = 3x^4 + 7x^3 + 2x^2 + 3x + 5$ and $D(x) = x + 1$.

Multiply $3x^3(x + 1)$
$= 3x^4 + 3x^3$ and write the answer so that similar terms are aligned.

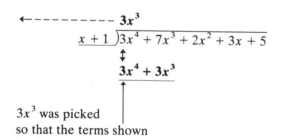

$3x^3$ was picked
so that the terms shown
by the arrow are *identical.*

Next, subtract (or add the opposite).

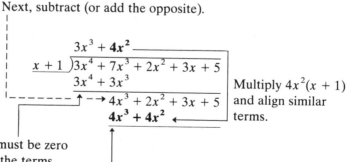

Multiply $4x^2(x + 1)$
and align similar
terms.

This must be zero
since the terms
were identical.

$4x^2$ was picked so that these terms
are identical.

Now, subtract and repeat the procedure.

$$
\begin{array}{r}
3x^3 + 4x^2 - 2x\ + 5 \\
x + 1\ \overline{)\,3x^4 + 7x^3 + 2x^2 + 3x + 5} \\
3x^4 + 3x^3 \\
\hline
4x^3 + 2x^2 + 3x + 5 \\
4x^3 + 4x^2 \\
\hline
-2x^2 + 3x + 5 \\
-2x^2 - 2x \\
\hline
5x + 5 \\
5x + 5 \\
\hline
0
\end{array}
$$

Don't forget to subtract:
$$2x^2 - 4x^2 = -2x^2.$$

subtract $3x - (-2x) = 5x$

$0 \leftarrow 5 - 5 = 0$

The remainder is 0, so $x + 1$ is a factor of $3x^4 + 7x^3 + 2x^2 + 3x + 5$.

We can check this by verifying $P(x) = Q(x)D(x)$:

$Q(x)D(x)$

$= (3x^3 + 4x^2 - 2x + 5)(x + 1)$

$= 3x^4 + 4x^3 - 2x^2 + 5x + 3x^3 + 4x^2 - 2x + 5$

$= 3x^4 + 7x^3 + 2x^2 + 3x + 5.$

Since this is $P(x)$, the result checks.

EXAMPLE 1: Let $P(x) = 4x^4 - 6x^2 - 10x + 3$ and $D(x) = x - 2$; find $P(x)/D(x)$.

Solution:

$$
\begin{array}{r}
4x^3 + 8x^2 + 10x\ + 10 \\
x - 2\ \overline{)\,4x^4 \qquad - 6x^2 - 10x + 3} \\
4x^4 - 8x^3 \\
\hline
8x^3 - 6x^2 - 10x + 3 \\
8x^3 - 16x^2 \\
\hline
10x^2 - 10x + 3 \\
10x^2 - 20x \\
\hline
10x + 3 \\
10x - 20 \\
\hline
23
\end{array}
$$

Notice that there is no x^3 term, but we left a space for this "missing" term.

The remainder is 23. Thus,

$$\frac{P(x)}{D(x)} = 4x^3 + 8x^2 + 10x + 10 + \frac{23}{x - 2}.$$

Check this by verifying

$$P(x) = Q(x)D(x) + R(x).$$

The process of long division is indeed *long* because of the duplication of symbols when carrying out this process. Consider the following example:

Notice that each number in boldface type is a repetition of the number directly above so we could eliminate writing it down.

$$\begin{array}{r}
2x^3 - 4x^2 - x + 3 \\
x - 1 \overline{)2x^4 - 6x^3 + 3x^2 + 4x - 3} \\
\mathbf{2x^4 - 2x^3} \\
\hline
-4x^3 + 3x^2 + 4x - 3 \\
\mathbf{-4x^3 + 4x^2} \\
\hline
-x^2 + 4x - 3 \\
\mathbf{-x^2 + x} \\
\hline
3x - 3 \\
3x - 3 \\
\hline
0
\end{array}$$

This is of the first degree.

The degree of the first term is one less than the degree of the given polynomial.

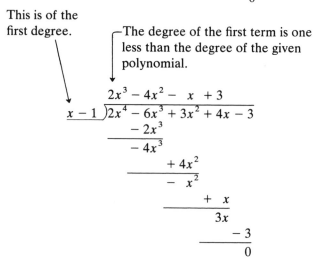

Next, the position of the term indicates the degree, so it is not necessary to write down the variable.

Interpret this answer by recognizing that it is of degree one less than the degree of the dividend. This means the answer is $2x^3 - 4x^2 - x + 3$.

$$\begin{array}{r}
2 - 4 - 1 \quad 3 \\
-1 \overline{)2 - 6 + 3 + 4 - 3} \\
-2 \\
\hline
-4 \\
+4 \\
\hline
-1 \\
+1 \\
\hline
3 \\
-3 \\
\hline
0
\end{array}$$

There is no reason to spread out the array; it can be compressed into a more efficient form.

$$
\begin{array}{r}
\textcircled{2}\quad -4 \quad -1 \quad 3 \qquad \leftarrow\text{Delete these terms.}\\
\hspace{3em}\leftarrow\text{This line is no longer needed.}\\
-1\,\overline{)\,2\quad -6\quad +3\quad +4\quad -3}\\
-2\quad +4\quad +1\quad -3\\
\hline
-4\quad -1\quad 3\qquad 0 \leftarrow\text{This is the remainder.}
\end{array}
$$

That's better, but the top line is the same as the bottom line if we bring down the leading coefficient.

$$
\begin{array}{r}
-1\,\bigg|\ 2\quad -6\quad +3\quad +4\quad -3\\
-2\quad +4\quad +1\quad -3\\
\hline
2\quad -4\quad -1\quad 3\qquad 0
\end{array}
$$

These are the coefficients of the quotient (which begins with a degree one less than that of the given polynomial).

This is the remainder.

The process is now fairly compact:

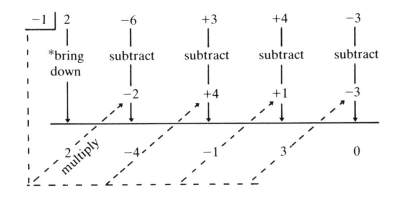

*Start here and carry out the zig-zag process:

bring down the 2;
multiply by (-1);
subtract: $-6 - (-2) = -4$;
multiply by (-1);
subtract: $3 - 4 = -1$;
and so on.

However, it is usually easier to add the opposite than to subtract, so change the sign of the divisor (-1 to $+1$) and add:

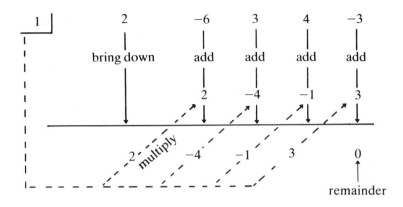

This condensed form of the division of a polynomial by $x - r$ is called *synthetic division*.

SYNTHETIC DIVISION
Notice that synthetic division is used only when the divisor is of the form $x - r$ (r positive or negative). If the divisor is not linear, then long division must be used.

EXAMPLES

2. Divide $x^4 + 3x^3 - 12x^2 + 5x - 2$ by $x - 2$.

opposite of -2; $r = 2$

$$
\begin{array}{r|rrrrr}
2 & 1 & 3 & -12 & 5 & -2 \quad \leftarrow\text{coefficients of given polynomial} \\
  &   & 2 & 10 & -4 & 2 \\
\hline
  & 1 & 5 & -2 & 1 & 0 \\
\end{array}
$$

quotient   remainder

degree one less than the given polynomial

$$\frac{x^4 + 3x^3 - 12x^2 + 5x - 2}{x - 2} = x^3 + 5x^2 - 2x + 1$$

3. Divide $x^4 - 2x^3 - 10x^2 + 26x + 3$ by $x + 3$.

opposite of $+3$

$$
\begin{array}{r|rrrrr}
-3 & 1 & -2 & -10 & 26 & 3 \quad \leftarrow\text{coefficients} \\
   &   & -3 & 15 & -15 & -33 \\
\hline
   & 1 & -5 & 5 & 11 & -30 \\
\end{array}
$$

remainder

The quotient is $x^3 - 5x^2 + 5x + 11 + \dfrac{-30}{x + 3}$.

4. Divide $x^5 - 3x^2 + 1$ by $x - 2$.

$$\underline{2}\,\underline{|\ 1 \quad 0 \quad 0 \quad -3 \quad 0 \quad 1}$$

$$\quad\quad\quad 2 \quad 4 \quad 8 \quad 10 \quad 20$$

$$\overline{\quad 1 \quad 2 \quad 4 \quad\ 5 \quad 10 \quad 21}\ (R)$$

The quotient is $x^4 + 2x^3 + 4x^2 + 5x + 10 + \dfrac{21}{x-2}$.

Be sure to include zero coefficients for missing terms.

PROBLEM SET 3.3

A Problems

Fill in the blanks for the synthetic divisions in Problems 1–10.

1. $\dfrac{3x^3 - 9x^2 + 11x - 10}{x - 2}$ $\underline{2}\,\underline{|\quad 3 \quad \underline{\quad} \quad 11 \quad -10}$

$$\quad\quad\quad\quad\quad\quad\quad\quad 6 \quad \underline{\quad} \quad 10$$

$$\quad\quad\quad\quad\quad\overline{\underline{\quad} \quad -3 \quad\ 5 \quad\ 0}$$

The quotient is $\underline{\quad\quad} - 3x + 5$.

2. $\dfrac{4x^3 - 6x^2 - 8x - 5}{x - 3}$ $\underline{\quad}\,\underline{|\quad 4 \quad -6 \quad \underline{\quad} \quad \underline{\quad}}$

$$\quad\quad\quad\quad\quad\quad\quad\quad \underline{\quad} \quad \underline{\quad} \quad \underline{\quad}$$

$$\quad\quad\quad\quad\quad\overline{\quad 4 \quad\ 6 \quad\ 10 \quad 25}$$

The quotient is $4\underline{\quad} + 6\underline{\quad} + 10 + \dfrac{25}{x-3}$.

3. $\dfrac{2x^3 - 4x^2 - 10x + 12}{x + 2}$ $\underline{\quad}\,\underline{|\quad 2 \quad -4 \quad \underline{\quad} \quad \underline{\quad}}$

$$\quad\quad\quad\quad\quad\quad\quad\quad \underline{\quad} \quad \underline{\quad} \quad \underline{\quad}$$

$$\quad\quad\quad\quad\quad\overline{\quad 2 \quad -8 \quad\ 6 \quad\ 0}$$

The quotient is $2\underline{\quad} - 8\underline{\quad} + 6$.

4. $\dfrac{5x^3 + 12x^2 + 2x - 4}{x + 3}$ $\underline{\quad}\,\underline{|\quad 5 \quad\ 12 \quad \underline{\quad} \quad \underline{\quad}}$

$$\quad\quad\quad\quad\quad\quad\quad\quad -15 \quad 9 \quad -33$$

$$\quad\quad\quad\quad\quad\overline{\underline{\quad} \quad \underline{\quad} \quad \underline{\quad} \quad \underline{\quad}}$$

The quotient is $\underline{\quad\quad\quad}$.

5. $\dfrac{x^3 + 7x^2 + 8x - 20}{x + 4}$ $\underline{\quad}\,\underline{|\quad 1 \quad\ 7 \quad \underline{\quad} \quad \underline{\quad}}$

$$\quad\quad\quad\quad\quad\quad\quad\quad -4 \quad -12 \quad 16$$

$$\quad\quad\quad\quad\quad\overline{\underline{\quad} \quad \underline{\quad} \quad \underline{\quad} \quad \underline{\quad}}$$

The quotient is $\underline{\quad\quad\quad}$.

6. $\dfrac{x^4 - 3x^2 + 2x - 7}{x - 1}$

	1	1	-2	0
1	1	-2		

The quotient is _____.

7. $\dfrac{x^4 + 2x^3 - 5x + 2}{x - 1}$

	1	2	0	-5	2
	1	3	3	-2	
1	3	3	-6		

The quotient is _____.

8. $\dfrac{x^5 - 1}{x - 1}$

	1	1	1	1	1
1	1	1	1		

The quotient is _____.

9. $\dfrac{x^5 - 32}{x + 2}$

		-2	4	-8		
1	-2	4				

The quotient is _____.

10. $\dfrac{x^4 + 3x^3 - 2x^2}{x + 3}$

		-3			
1					

The quotient is _____.

Use synthetic division to find the quotients in Problems 11–20.

11. $\dfrac{x^3 + 5x^2 - 2x - 24}{x + 4}$

12. $\dfrac{x^3 + 3x^2 - 6x - 8}{x - 2}$

13. $\dfrac{x^3 - 4x^2 - 17x + 60}{x - 5}$

14. $\dfrac{x^5 - 3x^4 + x - 3}{x - 3}$

15. $\dfrac{4x^4 + 4x^3 - 15x^2 + 7}{x - 1}$

16. $\dfrac{5x^3 - 21x^2 - 13x - 35}{x - 5}$

17. $\dfrac{x^4 - x^3 + x^2 - x - 4}{x + 1}$

18. $\dfrac{2x^4 + 6x^3 - 4x^2 - 11x + 3}{x + 3}$

19. $\dfrac{3x^4 + 10x^3 - 8x^2 - 5x - 20}{x + 4}$

20. $\dfrac{x^4 - 9x^3 + 20x^2 - 15x + 18}{x - 6}$

B Problems

In Problems 21–25, use long division to find $Q(x)$ and $R(x)$ if $P(x)$ is divided by $D(x)$.

21. $P(x) = 4x^4 - 14x^3 + 10x^2 - 6x + 2; D(x) = 2x - 1$

22. $P(x) = x^5 - x^3 + x^2 + x - 1; D(x) = x^2 + x$
23. $P(x) = x^5 - x^3 + x^2 + 1; D(x) = x^2 + x$
24. $P(x) = x^2 + 2x + 1; D(x) = x^3$
25. $P(x) = 6x^4 - 11x^3 + 6x^2 - 2x + 5; D(x) = 2x^2 + x + 1$

In Problems 26–35, use synthetic division to find the quotient.

26. $\dfrac{4x^3 - x^2 + 2x - 1}{x + 1}$

27. $\dfrac{6x^4 - x^2 + 1}{x - 3}$

28. $\dfrac{5x^5 - 2x + 1}{x + 2}$

29. $\dfrac{5x^4 + 7x^3 - 27x^2 + 14x - 12}{x - 1}$

30. $\dfrac{x^5 + 3x^3 - 3x^4 - 16x^2 + 21x - 6}{x - 3}$

31. $\dfrac{5x^4 - x + x^5 - 1}{x + 5}$

32. $\dfrac{x^4 - 12x^2 + 4x + 15}{(x + 1)(x - 3)}$

33. $\dfrac{x^4 - x^3 - 12x^2 + 28x - 16}{(x - 2)(x + 4)}$

34. $\dfrac{x^4 + 7x^3 + 5x^2 - 23x + 10}{x^2 + 4x - 5}$

35. $\dfrac{2x^4 - 7x^3 - 4x^2 + 27x - 18}{x^2 - x - 6}$

Hint for Problems 34 and 35: see Problems 32 and 33.

Mind Bogglers

36. Find K so that $x^4 + Kx^3 + 7x^2 - 2x + 8$ has no remainder when divided by $x + 2$.
37. Find h and k so that $x^4 + hx^3 - kx + 15$ has no remainder when divided by $x - 1$ and $x + 3$.
38. Let $P(x) = 2x^3 - x^2 + 5x + 3$.
    a. Divide $P(x)$ by $2x + 1$ using long division.
    b. Divide $P(x)$ by $x + \frac{1}{2}$ using synthetic division.
    c. Compare your answers to parts a and b.
    d. Notice that the divisor of part a was divided by 2. If a divisor in any division problem is divided by 2, how will the quotient be affected?
    e. Write out a general procedure for dividing $P(x)$ by $ax + b$ synthetically.
39. Check the procedure you outlined in Problem 38e by applying it to the following problems (divide synthetically):
    a. $2x^3 + 7x^2 + x - 1$ divided by $2x + 1$
    b. $6x^3 - 13x^2 + 14x - 12$ divided by $2x - 3$
    c. $6x^3 + x^2 + 3x + 1$ divided by $2x + 1$
    d. $3x^3 - 10x^2 - 19x + 5$ divided by $3x + 5$

Check your answer by using long division.

Did the procedure you outlined in Problem 38 take the possibility of remainders into account?

3.4   GRAPHING POLYNOMIAL FUNCTIONS

In Chapter 2, we considered the graphs of linear and quadratic functions. Suppose we wish to graph the cubic function

$$f(x) = 2x^3 - 3x^2 - 12x + 17.$$

Since there is a one-to-one correspondence between points on the graph and ordered pairs satisfying this equation, we can begin by plotting points.

$$f(0) = 2 \cdot 0^3 - 3 \cdot 0^2 - 12 \cdot 0 + 17$$
$$= 17 \qquad\qquad (0, 17) \text{ is on the graph}$$
$$f(1) = 2 \cdot 1^3 - 3 \cdot 1^2 - 12 \cdot 1 + 17$$
$$= 2 - 3 - 12 + 17$$
$$= 4 \qquad\qquad (1, 4) \text{ is on the graph}$$
$$f(-1) = 2(-1)^3 - 3(-1)^2 - 12(-1) + 17$$
$$= -2 - 3 + 12 + 17$$
$$= 24 \qquad\qquad (-1, 24) \text{ is on the graph}$$
$$f(2) = 2 \cdot 2^3 - 3 \cdot 2^2 - 12 \cdot 2 + 17$$
$$= 16 - 12 - 24 + 17$$
$$= -3 \qquad\qquad (2, -3) \text{ is on the graph}$$
$$f(-2) = 2(-2)^3 - 3(-2)^2 - 12(-2) + 17$$
$$= -16 - 12 + 24 + 17$$
$$= 13 \qquad\qquad (-2, 13) \text{ is on the graph}$$

For other values, this procedure becomes quite tedious. However, let's try the problem synthetically.

Divide $f(x) = 2x^3 - 3x^2 - 12x + 17$ by $x - 1$:

$$
\begin{array}{r|rrrr}
1 & 2 & -3 & -12 & 17 \\
  &   & 2 & -1 & -13 \\
\hline
  & 2 & -1 & -13 & 4 \ (R)
\end{array}
$$

Notice that the remainder is the same as $f(1)$.

Divide $f(x)$ by $x - 2$:

$$
\begin{array}{r|rrrr}
2 & 2 & -3 & -12 & 17 \\
  &   & 4 & 2   & -20 \\
\hline
  & 2 & 1 & -10 & -3 \ (R)
\end{array}
$$

This remainder is the same as $f(2)$.

Divide $f(x)$ by $x + 2$:

$$
\begin{array}{r|rrrr}
-2 & 2 & -3 & -12 & 17 \\
   &   & -4 & 14 & -4 \\
\hline
   & 2 & -7 & 2   & 13 \ (R)
\end{array}
$$

The remainder is $f(-2)$.

It would appear that the remainder when dividing $f(x)$ by $x - r$ is the same as $f(r)$.

REMAINDER THEOREM

> **REMAINDER THEOREM:** If a polynomial $f(x)$ is divided by $x - r$ until a remainder independent of x is obtained, then the remainder is equal to $f(r)$.

To verify this theorem, recall the Division Algorithm from Section 3.3:

$$\frac{P(x)}{D(x)} = Q(x) + \frac{R(x)}{D(x)} \quad \text{or} \quad P(x) = Q(x)D(x) + R(x).$$

In this context, $P(x) = f(x)$, $D(x) = x - r$, and $R(x)$ is a constant since the degree of $R(x)$ must be less than the degree of $D(x)$, which is 1. The Division Algorithm can be restated as

$$f(x) = Q(x)(x - r) + R,$$

where R represents the remainder. Now,

$$\begin{aligned} f(r) &= Q(r)(r - r) + R \\ &= R \end{aligned}$$

since $Q(r)(r - r) = Q(r) \cdot 0 = 0$. Points on the curve can therefore be found by synthetic division. Also, since the same polynomial is to be evaluated repeatedly, it is not necessary to recopy it each time, and the work can be arranged as follows:

	2	−3	−12	17	point
1	2	−1	−13	4	(1, 4)
−1	2	−5	−7	24	(−1, 24)
2	2	1	−10	−3	(2, −3)
−2	2	−7	2	13	(−2, 13)
0				17	(0, 17)
3	2	3	−3	8	(3, 8)
−3	2	−9	15	−28	(−3, −28)
4	2	5	8	49	(4, 49)
−4	2	−11	32	−111	(−4, −111)

Try to do the work mentally:

$$1 \cdot 2 + (-3) = -1;$$
$$1 \cdot (-1) + (-12) = -13;$$

and so on.

Why is $f(0)$ always equal to the constant term?

Now, plot the points from the table, paying attention to the scales on the axes so that they accommodate most of the values obtained (see Figure 3.3).

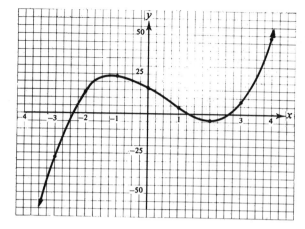

Figure 3.3   Graph of $f(x) = 2x^3 - 3x^2 - 12x + 17$

Connect the points to draw a smooth curve.

EXAMPLES: Graph the given polynomial functions by plotting points.

1. $f(x) = 3x^4 - 8x^3 - 30x^2 + 72x + 47$

	3	-8	-30	72	47	
-3	3	-17	21	9	20	(-3, 20)
-2	3	-14	-2	76	-105	(-2, -105)
-1	3	-11	-19	91	-44	(-1, -44)
0					47	(0, 47)
1	3	-5	-35	37	84	(1, 84)
2	3	-2	-34	4	55	(2, 55)
3	3	1	-27	-9	20	(3, 20)
4	3	4	-14	16	111	(4, 111)

You select the integers that are convenient.

Plot the points and draw a smooth curve, as shown in Figure 3.4.

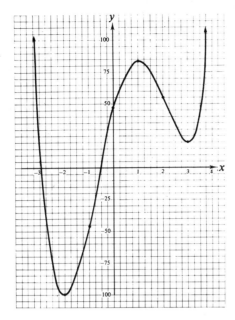

Figure 3.4   Graph of $f(x) = 3x^4 - 8x^3 - 30x^2 + 72x + 47$

2. $f(x) = 8x^3 - 12x^2 + x - 15$

	8	−12	1	−15	
−3	8	−36	109	−342	(−3, −342)
−2	8	−28	57	−129	(−2, −129)
−1	8	−20	21	−36	(−1, −36)
0				−15	(0, −15)
1	8	−4	−3	−18	(1, −18)
2	8	4	9	3	(2, 3)
3	8	12	37	96	(3, 96)

Plot the points as shown in Figure 3.5. Plotting these points does not give us enough information about the behavior of the graph, so we need more points, particularly between −1 and 2.

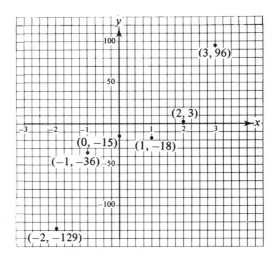

Figure 3.5 Points from synthetic division

	8	−12	1	−15	
−0.5	8	−16	9	−19.5	(−0.5, −19.5)
0.5	8	−8	−3	−16.5	(0.5, −16.5)
1.5	8	0	1	−13.5	(1.5, −13.5)

Notice that decimals were used instead of fractions. You may wish to use a calculator.

The graph is shown in Figure 3.6.

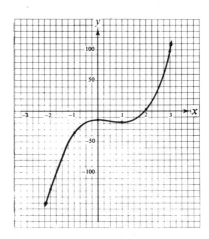

Figure 3.6 Graph of $f(x) = 8x^3 − 12x^2 + x − 15$

However, it looks as if the interesting portion is between -1 and 2, so we can focus in on this part of the graph by changing the scale, as shown in Figure 3.7.

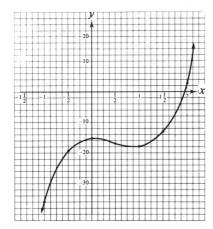

Figure 3.7   Graph of $f(x) = 8x^3 - 12x^2 + x - 15$

Two questions come to mind when we graph polynomial functions. The first is "How many points are enough?" There is no pat answer to this question. For example, the curve

$$y = 8x^3 + 12x^2 - 26x - 15$$

passes through the points $(-2, 21)$, $(-1, 15)$, $(0, -15)$, $(1, -21)$, and $(2, 45)$. Are these points sufficient to graph the curve between -2 and 2? Three possibilities are shown in Figure 3.8. To decide which curve is the correct one really requires calculus, but by plotting additional points, we can obtain a better approximation of the graph.

	8	12	−26	−15	
−1.5	8	0	−26	24	$(-1.5, 24)$
−0.5	8	8	−30	0	$(-0.5, 0)$
0.5	8	16	−18	−24	$(0.5, -24)$

curves are estimated

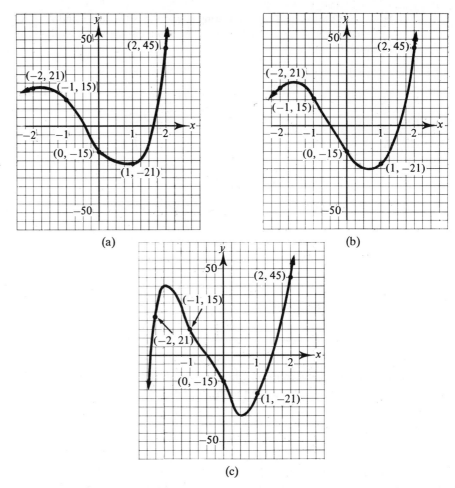

(a)   (b)

(c)

Figure 3.8   Possible graphs of $y = 8x^3 + 12x^2 - 26x - 15$

Using this additional information, we can see that Figure 3.8b is a good approximation to the curve.

The second question that comes up when we graph polynomial functions concerns the places where the graph of the polynomial crosses the *x*-axis. The terminology for polynomials is the same as was used for quadratics. Values of *x* that satisfy the equation

$$P(x) = 0$$

ROOT or SOLUTION of a polynomial equation.

are called the *roots* or *solutions* of the polynomial equation. If *x* is a real number, then it is also an *x-intercept*. If it is easy to find the roots of an equation, we can use this information to help us draw the graph. On the other hand, if it is difficult to find the roots, we can use the graph to help us approximate the roots.

It is sometimes helpful to know whether we have found all the negative or positive zeros. *Descartes' Rule of Signs* gives this result.

We'll consider the question of finding the roots of an equation in the next section.

DESCARTES' RULE OF SIGNS: Let $P(x)$ be a polynomial written in descending powers of x. Count the number of sign changes in the signs of the coefficients.

1. The number of positive real roots is equal to the number of sign changes or is equal to that number decreased by an even integer.

2. The number of negative real roots is equal to the number of sign changes in $P(-x)$ or is equal to that number decreased by an even integer.

DESCARTES' RULE OF SIGNS
A root of multiplicity m is counted as m roots.

EXAMPLES

3.
$$f(x) = 3x^4 - 8x^3 - 30x^2 + 72x + 47$$

number of sign changes: 1   2

Two sign changes occur, so there are two or no positive roots (see Figure 3.4).

$$f(-x) = 3(-x)^4 - 8(-x)^3 - 30(-x)^2 + 72(-x) + 47$$
$$= 3x^4 + 8x^3 - 30x^2 - 72x + 47$$

number of sign changes: 1   2

Two sign changes occur, so there are two or no negative roots (see Figure 3.4).

This is the same as Example 1; its graph is shown in Figure 3.4, on page 133. Notice that there were no positive roots and two negative roots.

4.
$$f(x) = 8x^3 - 12x^2 + x - 15$$

sign changes: 1   2   3

Three sign changes occur, so there are three or one positive root(s).

$$f(-x) = -8x^3 - 12x^2 - x - 15$$

sign changes: none

There are no sign changes, so there are no negative roots.

This is the same as Example 2; its graph is shown in Figure 3.7, on page 135. Notice that there was one positive root and there were no negative roots.

PROBLEM SET 3.4

A Problems

Use synthetic division to find the values specified for the functions given in Problems 1–9.

1. $f(x) = 5x^3 - 7x^2 + 3x - 4$
   a. $f(1)$ b. $f(-1)$ c. $f(0)$ d. $f(6)$ e. $f(-4)$
2. $g(x) = 4x^3 + 10x^2 - 120x - 350$
   a. $g(1)$ b. $g(-1)$ c. $g(0)$ d. $g(5)$ e. $g(-5)$
3. $P(x) = x^4 - 10x^3 + 20x^2 - 23x - 812$
   a. $P(1)$ b. $P(-1)$ c. $P(0)$ d. $P(3)$ e. $P(7)$
4. $f(t) = 3t^4 + 5t^3 - 8t^2 - 3t$
   a. $f(0)$ b. $f(1)$ c. $f(2)$ d. $f(-2)$ e. $f(-4)$
5. $g(t) = 4t^4 - 3t^3 + 5t - 10$
   a. $g(0)$ b. $g(-1)$ c. $g(-2)$ d. $g(5)$ e. $g(-3)$
6. $P(t) = t^4 - t^3 - 39t - 70$
   a. $P(0)$ b. $P(1)$ c. $P(-1)$ d. $P(-5)$ e. $P(7)$
7. $f(h) = 8h^4 - 6h^3 + 5h^2 + 4h - 3$
   a. $f(0)$ b. $f(1)$ c. $f(\frac{1}{2})$ d. $f(-\frac{1}{2})$ e. $f(-3)$
8. $g(h) = 16h^4 + 64h^3 + 19h^2 - 81h + 18$
   a. $g(0)$ b. $g(-2)$ c. $g(-3)$ d. $g(\frac{1}{4})$ e. $g(\frac{3}{4})$
9. $P(x) = 4x^4 - 8x^3 - 43x^2 + 29x + 60$
   a. $P(-1)$ b. $P(1)$ c. $P(4)$ d. $P(\frac{3}{2})$ e. $P(-\frac{5}{2})$

*Sketch the graph of each polynomial function in Problems 10–13. Also, use Descartes'
Rule of Signs to determine the number of positive and negative roots.*

10. $f(x) = x^3 - 3x^2 + 10$
11. $f(x) = x^3 + 3x^2 + 11$
12. $f(x) = 2x^3 - 3x^2 - 12x + 3$
13. $f(x) = 2x^3 + 3x^2 - 12x + 48$

B Problems

*Sketch the graph of each polynomial function in Problems 14–25. Also, use Descartes'
Rule of Signs to determine the number of positive and negative roots.*

14. $f(x) = x^3 - 6x^2 + 9x - 9$
15. $f(x) = 2x^3 - 3x^2 - 36x + 78$
16. $f(x) = 4x^4 - 8x^3 - 43x^2 + 29x + 60$
17. $f(x) = 16x^4 + 64x^3 + 19x^2 - 81x + 18$
18. $f(x) = x^4 - 7x^2 - 2x + 2$
19. $f(x) = x^4 - 14x^3 + 58x^2 - 46x - 9$
20. $f(x) = x^6 - 4x^4 - 4x^2 + 4$
21. $f(x) = x^5 + 2x^4 - 5x^3 - 10x^2 + 4x + 8$
22. *Calculator Problem.* $f(x) = 3x^4 - 7x^3 + 5x^2 + x - 10$
23. *Calculator Problem.* $f(x) = 8x^4 + 12x^3 - 3x^2 + 4x + 20$
24. *Calculator Problem.* $f(x) = x^5 - 3x^4 + 2x^3 - 7x + 15$
25. *Calculator Problem.* $f(x) = x^5 - 5x^4 + 3x^3 + x^2$

26. Graph on the same axes:
    a. $y = x^3$ b. $y = -x^3$ c. $y = (x - 1)^3$ d. $y - 2 = x^3$
    e. $y - 2 = (x - 1)^3$
27. Graph on the same axes:
    a. $y = x^3$ b. $y = \frac{1}{2}x^3$ c. $y = -\frac{1}{2}x^3$ d. $y = \frac{1}{10}x^3$
    e. $y = -\frac{1}{100}x^3$
28. Using Problems 26 and 27, discuss the graph of $y - k = a(x - h)^3$ by comparing it with the graph of $y = x^3$.
29. Graph on the same axes:
    a. $y = x^4$ b. $y = -\frac{1}{2}x^4$ c. $y = -\frac{1}{2}(x - 4)^4$ d. $y - 3 = \frac{1}{2}x^4$
    e. $y - 3 = \frac{1}{2}(x - 4)^4$
30. Using Problem 29, discuss the graph of $y - k = a(x - h)^4$ by comparing it with the graph of $y = x^4$.

Mind Bogglers

31. *Calculator Problem.* On the same axes, graph $y = x^n$ for $-1 \le x \le 1$, where n is equal to:
    a. 0   b. 2   c. 4   d. 6
32. *Calculator Problem.* Repeat Problem 31 for $-4 \le x \le 4$ and $-500 \le y \le 500$.
33. *Calculator Problem.* On the same axes, graph $y = x^n$ for $-1 \le x \le 1$, where n is equal to:
    a. 1   b. 3   c. 5   d. 7
34. *Calculator Problem.* Repeat Problem 33 for $-4 \le x \le 4$ and $-2000 \le y \le 2000$.
35. Using the results from Problems 31–34, make some conjectures concerning the graph of $y = x^n$.
36. Using the results from Problems 31–35, make a conjecture about the graph of $y - k = (x - h)^n$.

3.5 ROOTS OF POLYNOMIAL FUNCTIONS

In this chapter, we have been considering polynomial functions of the form

$$P(x) = a_n x^n + a_{n-1} x^{n-1} + \cdots + a_2 x^2 + a_1 x + a_0 \quad (a_n \ne 0),$$

and you have learned that the roots of this equation are values of x for which $P(x) = 0$. Such an equation is called a *polynomial equation*.
    Consider the polynomial equation

$$4x^4 - 8x^3 - 43x^2 + 29x + 60 = 0.$$

To solve this equation, we have to find the values of x that make it true. As a

POLYNOMIAL EQUATION

This is the function that was graphed in Problem 9 of the previous problem set.

first step, we try to write the equation in factored form, because (from Section 3.4)

$$P(x) = Q(x)(x - r) + R,$$

Remember the Division Algorithm:

$$P(x) = Q(x)D(x) + R(x).$$

And for $D(x) = x - r$, we have

$$P(x) = Q(x)(x - r) + R,$$

where R is a constant.

and if $R = 0$, then $P(x) = Q(x)(x - r)$. This says that $x - r$ is a factor of $P(x)$. But we are looking for values of x such that $P(x) = 0$, so

$$0 = Q(x)(x - r),$$

which is true for $x = r$. We summarize this with a result called the *Factor Theorem*.

FACTOR THEOREM

> If r is a root of the polynomial $P(x)$, then $x - r$ is a factor of $P(x)$. Also, if $x - r$ is a factor of $P(x)$, then r is a root of the polynomial $P(x)$.

The Factor Theorem is a generalization of the method of solving quadratics by factoring. Remember, if $P \cdot Q = 0$, then $P = 0$ or $Q = 0$ (perhaps both). Thus, if you wish to solve

$$x^2 - x - 2 = 0,$$

you can do so by factoring:

$$(x - 2)(x + 1) = 0.$$

Then each factor is set equal to zero to find

$$x = 2, -1.$$

This relationship between roots and factors can tell us how many roots to expect for a polynomial function. Suppose $P(x)$ is a third-degree polynomial. It is impossible to have more than three roots, because if it had more, say four roots (r_1, r_2, r_3, r_4), then the factor theorem provides

$$P(x) = a_4(x - r_1)(x - r_2)(x - r_3)(x - r_4),$$

which gives a *fourth*-degree polynomial. Thus, a third-degree polynomial cannot have more than three roots. But does it necessarily have three roots? Consider

$$x^3 + x = 0$$
$$x(x^2 + 1) = 0,$$

which has three complex roots: one real root, $x = 0$, and two imaginary roots, $x = i$, $x = -i$. We now answer the question about the number of roots with a theorem that is so important it is called the *Fundamental Theorem of Algebra*.

FUNDAMENTAL THEOREM OF ALGEBRA

> If $P(x)$ is a polynomial of degree $n \geq 1$ with complex coefficients, then $P(x) = 0$ has at least one complex root.

Remember that the real numbers form a subset of the complex numbers, so this theorem applies if the coefficients are real.

This theorem tells us there is at least one root, and it leads to another theorem that tells us just how many roots to expect if we first make an agreement about counting multiple roots. Consider

$$x^2 - 2x + 1 = 0$$
$$(x - 1)(x - 1) = 0.$$

The solution set, $\{1\}$, has one element, but in factored form the factor $x - 1$

appears *twice*. The root is therefore called a *root of multiplicity two*. For the fourth-degree polynomial

$$(x - 2)(x + 1)(x - 2)(x - 2) = 0,$$

the solution set is $\{-1, 2\}$ and 2 is a root of multiplicity three.

If we count roots of multiplicity, then the second-degree equation

$$x^2 - 2x + 1 = 0$$

has two roots and the fourth-degree equation

$$(x - 2)(x + 1)(x - 2)(x - 2) = 0$$

has four roots (even though the solution set is $\{-1, 2\}$).

NUMBER OF ROOTS THEOREM: If $P(x)$ is a polynomial of degree $n \geq 1$, with complex coefficients, then $P(x) = 0$ has exactly n roots (counting roots of multiplicity).

NUMBER OF ROOTS
THEOREM

You can use synthetic division to test a particular value to see if it is a root of a polynomial. If the remainder when dividing the polynomial by $x - r$ is zero, then $x - r$ is a factor and r is a root. Consider

$$4x^4 - 8x^3 - 43x^2 + 29x + 60 = 0.$$

$$
\begin{array}{r|rrrrr}
 & 4 & -8 & -43 & 29 & 60 \\
\hline
-1 & 4 & -12 & -31 & 60 & 0 \\
\end{array}
$$

↑

The remainder is 0
so $x + 1$ is a factor
and -1 is a root.

This means we can write

$$4x^4 - 8x^3 - 43x^2 + 29x + 60 = 0,$$

as

$$(x + 1)(4x^3 - 12x^2 - 31x + 60) = 0.$$

Now we can focus attention on factoring $4x^3 - 12x^2 - 31x + 60$. The equation $4x^3 - 12x^2 - 31x + 60 = 0$ is called the *depressed equation*.

$$
\begin{array}{r|rrrr}
 & 4 & -12 & -31 & 60 \\
\hline
4 & 4 & 4 & -15 & 0 \\
\end{array}
$$

↑

This shows that $x - 4$
is a factor.

Historical Note

Carl Gauss (1777–1855) proved the Fundamental Theorem of Algebra in 1797, at the age of 20. Along with Archimedes and Newton, Gauss is considered to be one of the three greatest mathematicians of all time. On March 30, 1796, an event occurred that made Gauss decide on a career in mathematics. He discovered that regular polygons having a prime number of sides p can be constructed with a straightedge and compass if and only if p is of the form $2^{2^n} + 1$. He was so proud of this discovery that he decided on mathematics as a career and requested that a regular polygon of 17 sides be inscribed on his tombstone.

Thus,

$$(x + 1)(4x^3 - 12x^2 - 31x + 60) = 0$$
$$(x + 1)(x - 4)(4x^2 + 4x - 15) = 0$$
$$(x + 1)(x - 4)(2x + 5)(2x - 3) = 0,$$

and the roots are $\{-1, 4, -5/2, 3/2\}$.

Now we focus attention on $4x^2 + 4x - 15$, but since this is quadratic, we can try to factor it directly without further synthetic division.

EXAMPLE 1: Solve $x^4 + 5x^3 + 5x^2 - 5x - 6 = 0$.

	1	5	5	−5	−6
1	1	6	11	6	0 ← $x - 1$ is a factor:
−2	1	4	3	0	$(x - 1)(x^3 + 6x^2 + 11x + 6) = 0$

This represents a second-degree equation, so we can stop synthetic division.

$x + 2$ is a factor

$x^3 + 6x^2 + 11x + 6 = 0$ is the *depressed equation.* Notice that it is not necessary to rewrite these numbers for the synthetic division

$$(x - 1)(x + 2)(x^2 + 4x + 3) = 0$$
$$(x - 1)(x + 2)(x + 1)(x + 3) = 0$$

Solution: $\{1, -2, -1, -3\}$

Of course, we solved the previous two polynomial equations, but you should view these solutions very suspiciously. How was it that we just happened to pick the values that gave us a zero remainder? Suppose you are asked to solve

$$4x^4 + 16x^3 - 21x^2 - 4x + 5 = 0$$

and are told that the roots can be found in the set

$$\left\{1, -1, 5, -5, \frac{1}{2}, -\frac{1}{2}, \frac{1}{4}, -\frac{1}{4}, \frac{5}{2}, -\frac{5}{2}, \frac{5}{4}, -\frac{5}{4}\right\}.$$

How could you proceed? By trial and error using synthetic division:

	4	16	−21	−4	5
1	4	20	−1	−5	0 ← $x - 1$ is a factor
−1	4	16	−17	−12	← $x + 1$ is not a factor

Next, try 5, but rather than recopy your work, just disregard the part of your work that is shaded below.

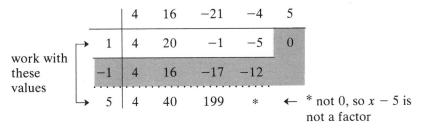

Next, try −5.

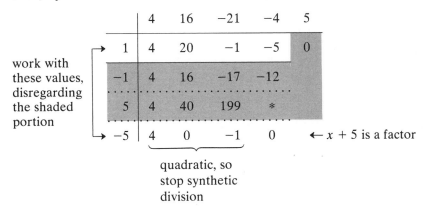

The resulting depressed equation is now quadratic and is solved by factoring or by the quadratic formula.

$$(x - 1)(x + 5)(4x^2 - 1) = 0$$
$$(x - 1)(x + 5)(2x - 1)(2x + 1) = 0$$

Solution: $\{1, -5, \frac{1}{2}, -\frac{1}{2}\}$

Even this example should be viewed with suspicion, since you were given a list of possible roots. However, mathematicians have developed a method giving you such a list for every polynomial equation that has rational roots.

If $P(x) = a_n x^n + a_{n-1} x^{n-1} + \cdots + a_1 x + a_0$ has integer coefficients and has a rational root p/q (where p/q is reduced), then p is a factor of a_0 and q is a factor of a_n.

RATIONAL ROOT THEOREM

To use this theorem, make a list of *all* factors of a_0 and divide these integers by the factors of a_n.

EXAMPLES: List all possible rational roots of the given polynomial equations.

Shorten these lists by using the ± sign:

p: ±1, ±2, ±4
q: ±1

If $a_n = \pm 1$, then the Rational Root Theorem implies that any rational root is an integer.

2. $P(x) = x^3 - x^2 - 4x + 4 = 0$
p: $a_0 = 4$, with factors 1, −1, 2, −2, 4, −4
q: $a_n = 1$, with factors 1, −1
Form all possible fractions

$$\frac{p}{q}: \ \frac{1}{1}, \frac{1}{-1}, \frac{-1}{1}, \frac{-1}{-1}, \frac{2}{1}, \frac{2}{-1}, \frac{-2}{1}, \frac{-2}{-1}, \frac{4}{1}, \frac{4}{-1}, \frac{-4}{1}, \frac{-4}{-1}.$$

Simplifying and not bothering to rewrite those that are repeated, we have:
±1, ±2, ±4.

3. $P(x) = 2x^3 - x^2 - 18x + 9$
p: $a_0 = 9$, with factors ±1, ±3, ±9
q: $a_n = 2$, with factors ±1, ±2
Possible rational roots $\dfrac{p}{q}$: ±1, ±$\dfrac{1}{2}$, ±3, ±$\dfrac{3}{2}$, ±9, ±$\dfrac{9}{2}$

4. $P(x) = x^3 - 2x^2 - 9x + 18$
p: $a_0 = 18$: ±1, ±2, ±3, ±6, ±9, ±18
q: $a_n = 1$: ±1
Possible $\dfrac{p}{q}$: ±1, ±2, ±3, ±6, ±9, ±18

5. $P(x) = 4x^4 - 8x^3 - 43x^2 + 29x + 60$
p ($a_0 = 60$): ±1, ±2, ±3, ±4, ±5, ±6, ±10, ±12, ±15, ±20, ±30, ±60
q ($a_n = 4$): ±1, ±2, ±4

$$\frac{p}{q}: \pm1, \pm\frac{1}{2}, \pm\frac{1}{4}, \pm2, \pm3, \pm\frac{3}{2}, \pm\frac{3}{4}, \pm4, \pm5, \pm\frac{5}{2}, \pm\frac{5}{4}, \pm6, \pm10, \pm12, \pm15,$$
$$\pm\frac{15}{2}, \pm\frac{15}{4}, \pm20, \pm30, \pm60$$

Note: if a factor is already listed, it is not repeated. For example, ±$\frac{2}{4}$ was not listed separately from ±$\frac{1}{2}$.

You can see from the examples that the list of possible rational roots may be quite large, but it is a *finite* list, so with enough time and effort the entire list could be checked. Usually, this is not necessary if you first pick those values that are easiest to check. (That is, don't start with the fractions or large numbers.) For example, the polynomial given in Example 5 was the first polynomial given in this section, and we solved it by dividing only two values synthetically. There is also another theorem, called the *Upper and Lower Bound Theorem*, that helps to rule out many of the values listed with the Rational Root Theorem.

> If $a > 0$ and all the numbers in the last row have the *same sign* in the synthetic division of $P(x)$ by $x - a$, then a is an *upper bound* for the roots of $P(x) = 0$.
>
> If $b < 0$ and the numbers in the last row *alternate in sign* in the synthetic division of $P(x)$ by $x - b$, then b is a *lower bound* for the roots of $P(x) = 0$.

UPPER AND LOWER BOUND THEOREM

An *upper bound* means there are no larger roots. Thus, once you've found an upper bound you need not check any larger number in your list of possible roots. A *lower bound* means there are no smaller roots.

EXAMPLES: Solve the given polynomial equation.

6. $2x^4 - 5x^3 - 8x^2 + 25x - 10 = 0$

p $(a_0 = -10)$: $\pm 1, \pm 2, \pm 5, \pm 10$

q $(a_n = 2)$: $\pm 1, \pm 2$

$\dfrac{p}{q}$: $\pm 1, \pm \dfrac{1}{2}, \pm 2, \pm 5, \pm \dfrac{5}{2}, \pm 10$

	2	−5	−8	25	−10
1	2	−3	−11	14	−4
−1	2	−7	−1	24	−34
2	2	−1	−10	5	0	← $x - 2$ is a factor
−2	2	−5	0	5	
5	2	9	35	180	
−5	2	−11	45	−220	
$\frac{1}{2}$	2	0	−10	0	

← $x - 2$ is a factor

← All sums are positive, so 5 is an upper bound. No larger values need be checked.

← Sums have alternating signs, so −5 is a lower bound. No smaller values need be checked.

← $x - \frac{1}{2}$ is a factor

$$(x - 2)\left(x - \frac{1}{2}\right)(2x^2 - 10) = 0$$

$$2(x - 2)\left(x - \frac{1}{2}\right)(x^2 - 5) = 0$$

Solution: $\{2, \frac{1}{2}, \sqrt{5}, -\sqrt{5}\}$

7. $8x^3 - 12x^2 - 26x + 15 = 0$

p $(a_0 = 15)$: $\pm 1, \pm 3, \pm 5, \pm 15$

Note: the resulting quadratic may not have rational roots. In fact, you may need the quadratic formula for its solution.

$q\ (a_n = 8)$: $\pm1, \pm2, \pm4, \pm8$

$\dfrac{p}{q}$: $\pm1, \pm\dfrac{1}{2}, \pm\dfrac{1}{4}, \pm\dfrac{1}{8}$

$\pm3, \pm\dfrac{3}{2}, \pm\dfrac{3}{4}, \pm\dfrac{3}{8}$

$\pm5, \pm\dfrac{5}{2}, \pm\dfrac{5}{4}, \pm\dfrac{5}{8}$

$\pm15, \pm\dfrac{15}{2}, \pm\dfrac{15}{4}, \pm\dfrac{15}{8}$

	8	−12	−26	15
1	8	−4	−30	−15
−1	8	−20	−6	21
5	8	28	114	585 ← an upper bound
3	8	12	10	45 ← an upper bound
−3	8	−36	82	−231 ← a lower bound; we know the roots are between −3 and 3
$\frac{1}{2}$	8	−8	−30	0 ← $x - \frac{1}{2}$ is a factor

Notice that there may be many upper bounds. This is why it is better to start with ±1 and work your way to larger and smaller numbers.

$$\left(x - \frac{1}{2}\right)(8x^2 - 8x - 30) = 0$$

$$2\left(x - \frac{1}{2}\right)(4x^2 - 4x - 15) = 0$$

$$2\left(x - \frac{1}{2}\right)(2x - 5)(2x + 3) = 0$$

Solution: $\{\frac{1}{2}, \frac{5}{2}, -\frac{3}{2}\}$

8. $8x^5 - 44x^4 + 86x^3 - 73x^2 + 28x - 4 = 0$
   $p\ (a_0 = -4)$: $\pm1, \pm2, \pm4$
   $q\ (a_n = 8)$: $\pm1, \pm2, \pm4, \pm8$

   $\dfrac{p}{q}$: $\pm1, \pm\dfrac{1}{2}, \pm\dfrac{1}{4}, \pm\dfrac{1}{8}, \pm2, \pm4$

	8	−44	86	−73	28	−4
1	8	−36	50	−23	5	1
−1	8	−52	138	−211	239	−243 ← lower bound
2	8	−28	30	−13	2	0 ← $x - 2$ is a factor
4	8	4	46	171	686	← upper bound
2	8	−12	6	−1	0	← $x - 2$ is a factor
2	8	4	14	27		← upper bound; this is a new upper bound because we are working with the depressed equation
$\frac{1}{2}$	8	−8	2	0		← $x - \frac{1}{2}$ is a factor

Note: don't forget possible multiple roots. Just because $x - 2$ was a factor once doesn't mean it can't be again.

$$(x - 2)(x - 2)\left(x - \frac{1}{2}\right)(8x^2 - 8x + 2) = 0$$

$$2(x - 2)(x - 2)\left(x - \frac{1}{2}\right)(4x^2 - 4x + 1) = 0$$

$$2(x - 2)(x - 2)\left(x - \frac{1}{2}\right)(2x - 1)(2x - 1) = 0$$

Solution: $\{2, \frac{1}{2}\}$; 2 is a root of multiplicity two and $\frac{1}{2}$ is a root of multiplicity three.

9. $x^4 - 3x^2 - 6x - 2 = 0$
$p \ (a_0 = -2)$: $\pm 1, \pm 2$
$q \ (a_n = 1)$: ± 1

$\dfrac{p}{q}$: $\pm 1, \pm 2$

	1	0	−3	−6	−2
1	1	1	−2	−8	−10
2	1	2	1	−4	−10
−1	1	−1	−2	−4	2
−2	1	−2	1	−8	14 ← lower bound

There are no rational roots. However, we can find some additional points and draw a sketch, as shown in Figure 3.9.

	1	0	−3	−6	−2	
0					−2	
−3	1	−3	6	−24	70	
3	1	3	6	12	34	← upper bound
2.5	1	2.5	3.25	2.125	3.3125	
1.5	1	1.5	−0.75	−7.125	−12.6875	

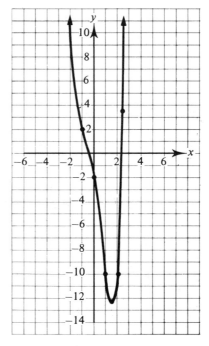

Figure 3.9 Graph of $y = x^4 - 3x^2 - 6x - 2$

Verify the type of roots using Descartes' Rule of Signs:

$$f(x) = x^4 - 3x^2 - 6x - 2$$
$$1$$

one positive root;

From the graph, we would expect roots between −1 and 0 as well as between 2 and 3. Also, because of the upper and lower bounds, we would expect these to be the only real roots. Even though we cannot find those real roots exactly, we can approximate them to any desired degree of accuracy by using synthetic division. For the root between −1 and 0:

	1	0	−3	−6	−2	
−0.5	1	−0.5	−2.75	−4.625	0.3125	← root between −0.5
−0.4	1	−0.4	−2.84	−4.864	−0.0544	← and −0.4 since one is negative and the other is positive
−0.41	1	−0.41	−2.8319	−4.8389	−0.0161	←
−0.42	1	−0.42	−2.8236	−4.8141	0.0219	←

root between −0.41 and −0.42

$$f(-x) = x^4 - 3x^2 + 6x - 2$$
$$1 \quad\; 2 \quad\; 3$$

three or one negative root(s).

A calculator would be very helpful.

Continue in this fashion to approximate the root to any degree of accuracy desired. (This would be a good problem for a computer.) Repeat the procedure for the root between 2 and 3 (it is about 2.41). Thus, the real roots (to the nearest tenth) are: {−0.4, 2.4}.

From Example 9, we can see the relationship between the roots and the graph and we have found a method for approximating the roots even when they are not rational. Notice that the crux of the argument used in Example 9 was that in each case we found two points for which the y values had opposite signs. We summarize this with a result called the *Location Theorem*.

If $P(x)$ is a polynomial and $P(a)$ and $P(b)$ are opposite in sign, then there is at least one zero on the interval between a and b.

LOCATION THEOREM

This says that if a polynomial is positive at a and negative at b, then somewhere between a and b it must cross the x-axis (it may cross more than once).

For Example 9, the Number of Roots Theorem demands four roots, but from Figure 3.9 there seem to be only two intercepts. This is because there are two real and two complex solutions. One result that is sometimes helpful when finding roots is the *Conjugate Pair Theorem*.

CONJUGATE PAIR THEOREM

1. If $P(x)$ is a polynomial with rational coefficients and $m + \sqrt{n}$ is a root, then $m - \sqrt{n}$ is also a root.
2. If $P(x)$ is a polynomial with real coefficients and $a + bi$ is a root, then $a - bi$ is also a root.

CONJUGATE PAIR THEOREM

m and n are rational and $\sqrt{n}$ is irrational

a and b are real numbers

EXAMPLES

10. For Example 9, solve $x^4 - 3x^2 - 6x - 2 = 0$ given that $-1 + i$ is a root.
Solution:

	1	0	-3	-6	-2
$-1 + i$	1	$-1 + i$	$(-1 + i)$ $\cdot (-1 + i) - 3$ $= 1 - 2i + i^2 - 3$	$(-1 + i)$ $\cdot (-3 - 2i) - 6$ $= 3 - i - 2i^2 - 6$	$(-1 + i)$ $\cdot (-1 - i) - 2$ $= 1 - i^2 - 2$
	$= 1$	$= -1 + i$	$= -3 - 2i$	$= -1 - i$	$= 0$
$-1 - i$	1	$(-1 - i)$ $+ (-1 + i)$	$-2(-1 - i)$ $+ (-3 - 2i)$ $= 2 + 2i - 3 - 2i$	$-1(-1 - i)$ $+ (-1 - i)$ $= 1 + i - 1 - i$	
	$= 1$	$= -2$	$= -1$	$= 0$	

The depressed equation is $x^2 - 2x - 1 = 0$, which is solved by the quadratic formula.

$$x = \frac{2 \pm \sqrt{4 - 4(1)(-1)}}{2}$$

$$= \frac{2 \pm 2\sqrt{2}}{2}$$

$$= 1 \pm \sqrt{2}$$

The roots are $\{-1 \pm i, 1 \pm \sqrt{2}\}$.

Notice that $1 \pm \sqrt{2}$ corresponds to the approximate values we found in Example 9, namely -0.4, 2.4.

11. Solve $x^4 - 4x - 1 = 4x^3$ given that $2 + \sqrt{5}$ is a root.
Solution:

$$x^4 - 4x^3 - 4x - 1 = 0$$

$2 + \sqrt{5}$	1	-4	0	-4	-1
		$2 + \sqrt{5}$	1	$2 + \sqrt{5}$	1
	1	$-2 + \sqrt{5}$	1	$-2 + \sqrt{5}$	0

and $2 - \sqrt{5}$ must be a root also:

$2 - \sqrt{5}$	1	$-2 + \sqrt{5}$	1	$-2 + \sqrt{5}$
		$2 - \sqrt{5}$	0	$2 - \sqrt{5}$
	1	0	1	0

The depressed equation is

$$x^2 + 1 = 0;$$
$$x = \pm i.$$

The roots are $\{2 \pm \sqrt{5}, \pm i\}$.

PROBLEM SET 3.5

A Problems

Find the possible rational roots for the polynomial equations in Problems 1–10.

1. $x^4 - 3x^3 + 7x^2 - 19x + 15 = 0$
2. $3x^3 - 7x^2 + 5x + 7 = 0$
3. $2x^5 + 6x^4 - 3x + 12 = 0$
4. $x^3 + 2x^2 - 5x - 6 = 0$
5. $x^3 + 3x^2 - 4x - 12 = 0$
6. $2x^3 + x^2 - 13x + 6 = 0$
7. $2x^3 - 3x^2 - 32x - 15 = 0$
8. $x^4 - 12x^3 + 54x^2 - 108x + 81 = 0$
9. $x^4 + 3x^3 - 20x^2 - 3x + 18 = 0$
10. $x^4 - 13x^2 + 36 = 0$

Solve the polynomial equations in Problems 11–20.

11. $x^3 - x^2 - 4x + 4 = 0$
12. $2x^3 - x^2 - 18x + 9 = 0$
13. $x^3 - 2x^2 - 9x + 18 = 0$
14. $x^3 + 2x^2 - 5x - 6 = 0$
15. $x^3 + 3x^2 - 4x - 12 = 0$
16. $2x^3 + x^2 - 13x + 6 = 0$
17. $2x^3 - 3x^2 - 32x - 15 = 0$
18. $x^4 - 12x^3 + 54x^2 - 108x + 81 = 0$
19. $x^4 + 3x^3 - 19x^2 - 3x + 18 = 0$
20. $x^4 - 13x^2 + 36 = 0$

B Problems

Solve the polynomial equations in Problems 21–30.

21. $x^3 + 15x^2 + 71x + 105 = 0$
22. $x^3 - 15x^2 + 74x - 120 = 0$
23. $8x^3 - 12x^2 - 66x + 35 = 0$
24. $12x^3 + 16x^2 - 7x - 6 = 0$
25. $x^5 + 8x^4 + 10x^3 - 60x^2 - 171x - 108 = 0$
26. $x^5 + 6x^4 + x^3 - 48x^2 - 92x - 48 = 0$
27. $x^4 - 7x^3 + 14x^2 + 2x - 20 = 0$
28. $x^4 - 2x^3 + 4x^2 + 2x - 5 = 0$
29. $4x^4 - 10x^3 + 10x^2 - 5x + 1 = 0$
30. $27x^4 - 180x^3 + 213x^2 + 62x - 10 = 0$

Solve the equations in Problems 31–39 if the given value is a root.

31. $x^3 - 2x^2 + 4x - 8 = 0; 2i$
32. $x^4 + 13x^2 + 36 = 0; -3i$

Historical Note

You may have wondered whether there is a formula similar to the Quadratic Formula for solving higher-degree polynomials. The answer is yes; there is a cubic and a quartic formula, but Niels H. Abel (1802–1829) proved at age 22 that no such formula exists for polynomials of degree higher than four. However, his proof contained a slight error and Evariste Galois (1811–1832) was the first to publish a correct proof. Galois was killed in a duel on his 21st birthday. The night before he died, he spent the entire night writing out this and other results that kept mathematicians busy for years after.

33. $x^4 - 6x^2 + 25 = 0; 2 + i$
34. $x^4 - 4x^3 + 3x^2 + 8x - 10 = 0; 2 + i$
35. $2x^4 - 5x^3 + 9x^2 - 15x + 9 = 0; i\sqrt{3}$
36. $2x^4 - x^3 - 13x^2 + 5x + 15 = 0; -\sqrt{5}$
37. $3x^5 + 10x^4 - 8x^3 + 12x^2 - 11x + 2 = 0; -2 + \sqrt{5}$
38. $2x^5 + 9x^4 - 3x^2 - 8x - 42 = 0; i\sqrt{2}$
39. $x^5 - 11x^4 + 24x^3 + 16x^2 - 17x + 3 = 0; 2 + \sqrt{3}$
40. Solve $x^6 + x^5 - 3x^4 - 4x^3 + 4x + 4 = 0$ if $-\sqrt{2}$ is a multiple root.
41. Verify that $f(x) = 4x^4 - 16x^3 + 23x^2 - 14x - 2$ has at least one root between -1 and 0.
42. Verify that $g(x) = 9x^4 - 27x^3 - 10x^2 + 37x + 19$ has at least one root between 2 and 3.
43. Does there exist a real number that exceeds its cube by 1?
44. The dimensions of a rectangular box are consecutive integers, and its volume is 720 cu cm. What are the dimensions of the box?
45. A 2-cm-thick slice is cut from a cube, leaving a volume of 384 cu cm. What is the length of a side of the original cube?
46. A rectangular sheet of tin with dimensions 3×5 m has equal squares cut from its four corners, as shown in Figure 3.10. The resulting sheet is folded up on the sides to form a topless box. Find all possible dimensions of the cutout square to the nearest 0.1 m such that the box has a volume of 1 cu m.

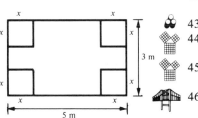

Figure 3.10 Problem 46

THINK METRIC

Mind Bogglers

In Problems 47–50, you are not looking for the roots that are not real.

Italian mathematicians discovered the algebraic solution of cubic and quartic equations in the 16th century. At that time they would challenge one another to solve certain equations. Problems 47–50 were such challenge problems. Find the real roots in each problem to the nearest tenth.

47. In 1515, Scipione del Ferro solved the cubic equation $x^3 + mx + n = 0$ and revealed his secret to his pupil Antonio Fior. At about the same time, Tartaglia solved the equation $x^3 + px^2 = n$. Fior thought Tartaglia was bluffing and challenged him to a public contest of solving cubic equations. According to the historian Howard Eves, Tartaglia triumphed completely. Solve the cubic $x^3 + px^2 = n$, where $p = 5$ and $n = 21$.

48. Girolamo Cardano stole the solution of the cubic equation from Tartaglia and published it in his *Ars magna*. Tartaglia protested, but Cardano's pupil, Ludovico Ferrari (who solved the biquadratic equation), claimed that both Cardano and Tartaglia stole it from del Ferro. According to historian Howard Eves, there was a dispute from which Tartaglia was lucky to have escaped alive. One of the problems in *Ars magna* was $x^3 - 63x = 162$. Solve this cubic.

49. *Calculator Problem.* Cardano solved the quartic $13x^2 = x^4 + 2x^3 + 2x + 1$. Find the real roots for this equation.

50. *Calculator Problem.* In 1540, Cardano was given the problem "Divide 10 into three parts such that they shall be in continued proportion and that the product

of the first two shall be 6." This information yields the equation

$$z^4 + 6z^2 + 36 = 60z.$$

Find the real roots for this equation.

3.6 SUMMARY AND REVIEW

TERMS

Algebraic expression (3.1)
Base (3.1)
Common factor (3.2)
Completely factored (3.2)
Difference of cubes (3.2)
Difference of squares (3.2)
Exponent (3.1)
FOIL (3.1)
Multiplicity of roots (3.5)
Perfect cube (3.2)

Perfect square (3.2)
Polynomial (3.1)
Power (3.1)
Roots (3.4)
Similar terms (3.1)
Solution (3.4)
Sum of cubes (3.2)
Synthetic division (3.3)
Zeros (3.4)

Derivation of Problem 50: let x, y, and z be the three parts. Then

$$x + y + z = 10.$$

Also,

$$\frac{x}{y} = \frac{y}{z}$$

and

$$xy = 6.$$

Eliminating x and y, you obtain

$$z^4 + 6z^2 + 36 = 60z.$$

IMPORTANT IDEAS

(3.1) Definition of exponents: if b is any nonzero real number and n is any natural number, then

$$b^n = \underbrace{b \cdot b \cdot b \cdots b,}$$

$$n \text{ factors}$$

$$b^0 = 1,$$

$$b^{-n} = \frac{1}{b^n}.$$

(3.1) Laws of exponents: let a and b be real numbers and let m and n be any integers except that the form $0/0$ and division by zero are excluded.

First Law: $b^m \cdot b^n = b^{m+n}$

Second Law: $\dfrac{b^m}{b^n} = b^{m-n}$

Third Law: $(b^n)^m = b^{mn}$

Fourth Law: $(ab)^m = a^m b^m$

Fifth Law: $\left(\dfrac{a}{b}\right)^m = \dfrac{a^m}{b^m}$

(3.1) Addition, subtraction, multiplication, and evaluation of polynomials
(3.2) Factoring—see Table 3.1, on page 116.
(3.2) Graphing by factoring

(3.3) Division algorithm: if $P(x)$ and $D(x)$ are polynomials $[D(x) \neq 0]$, then

$$\frac{P(x)}{D(x)} = Q(x) + \frac{R(x)}{D(x)},$$

where $Q(x)$ is a unique polynomial, and $R(x)$ is a polynomial with degree less than the degree of $D(x)$.

(3.3) Synthetic division

(3.4) Graphing polynomial functions by plotting points

(3.4) Remainder Theorem: if a polynomial $P(x)$ is divided by $x - r$ until a remainder independent of x is obtained, then the remainder is equal to $P(r)$.

(3.4) Descartes' Rule of Signs: let $P(x)$ be a polynomial written in descending powers of x. Count the number of sign changes in the signs of the coefficients.

   1. The number of positive real roots is equal to the number of sign changes or to that number decreased by an even integer.

   2. The number of negative real roots is equal to the number of sign changes in $P(-x)$ or to that number decreased by an even integer.

(3.5) Factor Theorem: if r is a root of the polynomial $P(x)$, then $x - r$ is a factor of $P(x)$. Also, if $x - r$ is a factor of $P(x)$, then r is a root of the polynomial $P(x)$.

(3.5) Fundamental Theorem of Algebra: if $P(x)$ is a polynomial of degree $n \geq 1$ with complex coefficients, then $P(x) = 0$ has at least one complex root.

(3.5) Number of Roots Theorem: if $P(x)$ is a polynomial of degree $n \geq 1$, with complex coefficients, then $P(x) = 0$ has exactly n roots (counting roots of multiplicity).

(3.5) Rational Root Theorem: if $P(x) = a_n x^n + a_{n-1} x^{n-1} + \cdots + a_1 x + a_0$ has a rational root p/q, then p is a factor of a_0 and q is a factor of a_n.

(3.5) Upper and Lower Bound Theorem: if $a > 0$ and all the sums are the *same sign* in the synthetic division of $P(x)$ by $x - a$, then a is an *upper bound* for the roots of $P(x) = 0$.

   If $b < 0$ and all the sums *alternate in sign* in the synthetic division of $P(x)$ by $x - b$, then b is a *lower bound* for the roots of $P(x) = 0$.

(3.5) Location Theorem: if $P(x)$ is a polynomial and $P(a)$ and $P(b)$ are opposite in sign, then there is at least one zero on the interval between a and b.

(3.5) Conjugate Pair Theorem:

   1. If $P(x)$ is a polynomial with rational coefficients and $m + \sqrt{n}$ is a root (m and n are rational and $\sqrt{n}$ is irrational), then $m - \sqrt{n}$ is also a root.

   2. If $P(x)$ is a polynomial with real coefficients and $a + bi$ is a root (a and b are real numbers), then $a - bi$ is also a root.

REVIEW PROBLEMS

Let $P(x) = 3x^3 + 4x^2 - 35x - 12$ and $D(x) = 3x + 1$ in Problems 1–4.

  1. (3.1) Find:

     a. $P(x)D(x)$ b. $P(t) + D(t)$ c. $\dfrac{P(x)}{D(x)}$ d. $[D(w)]^3$

e. $\dfrac{D(x + h) - D(x)}{h}$

2. (3.4) Find:
   a. $P(0)$ b. $P(1)$ c. $P(2)$ d. $P(-3)$ e. $P(-\frac{1}{3})$

3. (3.4) Graph $P(x)$.

4. (3.5) Solve $P(x) = 0$.

5. (3.2) Factor:

   a. $(3x + 2)^2 - 4$ b. $x^4 - 26x^2 + 25$ c. $\dfrac{4x^2}{y^2} - (2x + y)^2$

   d. $4x^3 + 8x^2 - x - 2$ e. $(x^3 - \frac{1}{8})(8x^3 + 8)$

6. (3.3) Use long division to find $\dfrac{12x^4 + 22x^3 + 7x + 4}{3x + 1}$.

7. (3.4) Graph $f(x) = 3x^4 - 8x^3 - 48x^2 + 492$.

8. (3.5) Solve $x^4 - 4x^3 + 8x^2 - 8x - 5$ if $1 + 2i$ is a root.

9. (3.5) Solve $x^5 - 4x^4 + 7x^3 - 7x^2 + 4x - 1 = 0$.

10. (3.5) Does there exist a real number that exceeds its fifth power by 1?

Rational Functions

4

4.1 RATIONAL EXPRESSIONS

In order to evaluate and graph polynomial functions, we used synthetic division, which was motivated by looking at the division process and the Division Algorithm. That is, we considered $P(x)/D(x)$ for polynomials $P(x)$ and $D(x)$, where $D(x) \neq 0$. Now, let's look at this problem from another viewpoint. An expression of the form $P(x)/D(x)$ is called a *rational function*.

RATIONAL FUNCTION

> DEFINITION: A *rational function* is the quotient of polynomial functions $P(x)$ and $D(x)$; that is,
>
> $$f(x) = \frac{P(x)}{D(x)}, \quad \text{where } D(x) \neq 0.$$

The domain of a rational function is the set of all real x such that $D(x) \neq 0$. When writing a rational function, we will assume this domain. That is, if

$$f(x) = \frac{x + 3}{x - 2},$$

it is not necessary to write $x \neq 2$ since this is implied in the definition. However, this can become a little tricky when simplifying rational expressions, as illustrated by the following examples.

Remember, when we are stating a rational function, the domain does not have to be stated but rather is implied. The purpose of these examples and the problems like them is to make sure you understand what the implied domain is when you are stating rational functions.

EXAMPLES: State the domain for each function.

1. $M(x) = \dfrac{2x^2 + 3x - 5}{(5x - 1)(x + 1)}$

    The domain is the set of all real numbers except $x = 1/5$, $x = -1$.

2. $N(x) = \dfrac{x^2 - 1}{x^2 + 1}$

    The domain is the set of all real numbers.

3. $f(x) = 2x - 1$

    The domain is the set of all real numbers.

4. $g(x) = \dfrac{(2x - 1)(x + 3)}{(x + 3)}$

    The domain is the set of all real numbers except $x = -3$.

Compare Examples 3 and 4 and notice that Example 4 can be simplified. Can we say $f = g$? The answer is not as obvious as you may think. A precise definition of the equality of functions is needed before we can answer this question.

DEFINITION: Two functions f and g are *equal* if and only if

1. f and g have the same domain; and
2. $f(x) = g(x)$ for all x in the domain.

EQUALITY FOR FUNCTIONS

Using this definition, we see that $f \neq g$ in Examples 3 and 4 since the domains are not the same. But, can't we write

$$g(x) = \frac{(2x - 1)(x + 3)}{x + 3} = 2x - 1?$$

Not if $g \neq f$. What can we do? We can write

$$g(x) = 2x - 1 \quad (x \neq -3).$$

Notice that the domain for g remains the same, but $f(x) = 2x - 1$ is changed when the condition $x \neq -3$ is added.

Remember from Section 1.4 that, when we deal with inverse functions, a function F is defined so that $F = f$ *except* that the domain of F may be different from the domain of f. If the domains of F and f are not the same, then $F \neq f$.

EXAMPLES: State whether the pairs of functions are equal.

5. $f(x) = \dfrac{(x - 3)(x + 5)}{x + 5}$ $g(x) = x - 3$

   $f \neq g$ since the domain of f is all reals except -5, and the domain of g is all real numbers.

6. $f(x) = \dfrac{(2x - 5)(x + 1)}{x + 1}$ $g(x) = 2x - 5 \quad (x \neq -1)$

   $f = g$ since the domain of both f and g is all reals except $x = -1$ and $f(x) = g(x)$ for all x in the domain.

7. $f(x) = \dfrac{x - 3x^2}{x}$ $g(x) = \dfrac{2x^3 - 6x^4 - 3x^2 + x}{2x^3 + x}$

   The domain of f is all reals except 0; the domain of g is all reals except 0. Also,

   $$f(x) = \frac{x - 3x^2}{x} = 1 - 3x \quad (x \neq 0);$$

$$g(x) = \frac{2x^3 - 6x^4 - 3x^2 + x}{2x^3 + x} = \frac{x(-6x^3 + 2x^2 - 3x + 1)}{x(2x^2 + 1)}$$

$$= \frac{x(-3x + 1)(2x^2 + 1)}{x(2x^2 + 1)}$$

$$= -3x + 1 \quad (x \neq 0).$$

Thus, since f and g have the same domain and $f(x) = g(x)$ for all x in the domain, $f = g$.

You probably remember how to perform the basic operations with rational expressions, but the procedures are summarized below for convenience.

PROPERTIES OF
RATIONAL EXPRESSIONS

By a *nonzero polynomial* we mean that values in the domain that cause the polynomial to be zero are excluded. In this context, as before, values that cause division by zero are excluded.

PROPERTIES OF RATIONAL EXPRESSIONS: Let P, Q, R, S, and K represent any nonzero real polynomials.

Equality: $\dfrac{P}{Q} = \dfrac{R}{S}$ if and only if $PS = QR$

Fundamental property: $\dfrac{PK}{QK} = \dfrac{P}{Q}$

Properties of signs: $\dfrac{P}{Q} = \dfrac{-P}{-Q} = -\dfrac{-P}{Q} = -\dfrac{P}{-Q}$

$$-\dfrac{P}{Q} = \dfrac{-P}{Q} = \dfrac{P}{-Q}$$

$$\dfrac{P - Q}{Q - P} = -1, \quad \text{since} \quad -(P - Q) = Q - P$$

Addition: $\dfrac{P}{Q} + \dfrac{R}{S} = \dfrac{PS + QR}{QS}$

Subtraction: $\dfrac{P}{Q} - \dfrac{R}{S} = \dfrac{PS - QR}{QS}$

Multiplication: $\dfrac{P}{Q} \cdot \dfrac{R}{S} = \dfrac{PR}{QS}$

Division: $\dfrac{P}{Q} \div \dfrac{R}{S} = \dfrac{PS}{QR}$

Note that addition and subtraction of rational expressions may be handled as a factoring problem. Recall that, when factoring

$$x^2y^3 + xy^4 + x^3y^3 = xy^3(\qquad),$$

xy^3 is the common factor since each term is factored and the representative of each common factor with the smallest exponent is selected. The factoring is then completed via multiplication and the distributive property:

$$x^2y^3 + xy^4 + x^3y^3 = xy^3(x + y + x^2).$$

Consider

$$\frac{y^2}{x^2} + xy + y^3.$$

It is understood that values that cause division by zero are excluded. This means, for this example, that $x \neq 0$.

Write this as $x^{-2}y^2 + xy + x^0y^3$ and select $x^{-2}y$ as the common factor—remember, use the smallest exponent of each different factor:

$$x^{-2}y^2 + xy + x^0y^3 = x^{-2}y(y + x^3 + x^2y^2),$$

or

$$\frac{y(y + x^3 + x^2y^2)}{x^2}.$$

EXAMPLES: Perform the indicated operations.

8. $\dfrac{a + b}{a - b} + \dfrac{a - 2b}{2a + b} = \dfrac{(a + b)(2a + b) + (a - b)(a - 2b)}{(a - b)(2a + b)}$

Definition of addition

$$= \frac{2a^2 + 3ab + b^2 + a^2 - 3ab + 2b^2}{(a - b)(2a + b)}$$

$$= \frac{3a^2 + 3b^2}{(a - b)(2a + b)}$$

$$= \frac{3(a^2 + b^2)}{(a - b)(2a + b)}$$

9. $\dfrac{x + 1}{x} - \dfrac{x + 2}{x^2} + \dfrac{x + 3}{x^3} = x^{-1}(x + 1) - x^{-2}(x + 2) + x^{-3}(x + 3)$

You can, of course, work this by obtaining a common denominator. This example illustrates how to handle this problem by factoring.

$$= x^{-3}[x^2(x + 1) - x(x + 2) + (x + 3)]$$

$$= x^{-3}[x^3 + x^2 - x^2 - 2x + x + 3]$$

$$= x^{-3}(x^3 - x + 3)$$

$$= \frac{x^3 - x + 3}{x^3}$$

10. $\left[\dfrac{x^2 + 5x + 6}{2x^2 - x - 1} \cdot \dfrac{2x^2 - 9x - 5}{x^2 + 7x + 12}\right] \div \dfrac{2x^2 - 13x + 15}{x^2 + 3x - 4}$

The key to working problems involving multiplication and division of rational expressions is in the proper factoring of each polynomial.

$= \left[\dfrac{(x + 2)(x + 3)}{(x - 1)(2x + 1)} \cdot \dfrac{(2x + 1)(x - 5)}{(x + 3)(x + 4)}\right] \div \dfrac{(x - 5)(2x - 3)}{(x - 1)(x + 4)}$

$= \dfrac{(x + 2)\cancel{(x + 3)}\cancel{(2x + 1)}\cancel{(x - 5)}\cancel{(x - 1)}\cancel{(x + 4)}}{\cancel{(x - 1)}\cancel{(2x + 1)}\cancel{(x + 3)}\cancel{(x + 4)}\cancel{(x - 5)}(2x - 3)}$

$= \dfrac{x + 2}{2x - 3}$

PROBLEM SET 4.1

A Problems

1. State the domain for the rational function

$$\frac{3x^2 - 4x - 4}{x^2 - 4}.$$

2. State the domain for the rational function

$$\frac{x}{x - 2} + \frac{x + 1}{2 - x}.$$

3. State the domain for the rational function

$$\frac{4x - 12}{x^2 - 49} \div \frac{18 - 2x^2}{x^2 - 4x - 21}.$$

In Problems 4–10, state whether the given functions are equal.

4. $f(x) = \dfrac{2x^2 + x}{x};$ $\quad g(x) = 2x + 1$

5. $f(x) = \dfrac{2x^2 + x}{x};$ $\quad g(x) = 2x + 1, \quad x \neq 0$

6. $f(x) = \dfrac{(3x + 1)(x - 2)}{x - 2}, \quad x \neq 6;$ $\quad g(x) = \dfrac{(3x + 1)(x - 6)}{x - 6}, \quad x \neq 2$

7. $f(x) = \dfrac{2x^2 + x - 6}{x - 2};$ $\quad g(x) = 2x + 3, \quad x \neq 2$

8. $f(x) = \dfrac{3x^2 - 7x - 6}{x - 3};$ $\quad g(x) = 3x + 2, \quad x \neq 3$

9. $f(x) = \dfrac{x^3 - 2x^2 - x + 2}{x - 1};$ $\quad g(x) = x^2 - x - 2, \quad x \neq -1$

10. $f(x) = \dfrac{x^3 - 3x^2 + 3x - 1}{x - 1};$ $\quad g(x) = x^2 - 2x + 1, \quad x \neq 1$

Simplify the expressions in Problems 11–23.

11. $\dfrac{1}{x} + \dfrac{1}{y}$

12. $x^{-1} + y^{-1}$

13. $2x^{-1} + 3y^{-1}$

Remember, denominators of rational expressions can be written by using negative exponents.

14. $(2x)^{-1} + (3y)^{-1}$

15. $2(x + y)^{-1} + 3$

16. $\dfrac{x^2 - y^2}{2x + 2y}$

17. $\dfrac{3x^2 - 4x - 4}{x^2 - 4}$

18. $\dfrac{3}{x + y} + \dfrac{5}{2x + y}$

19. $\dfrac{2}{x - y} + \dfrac{5}{y - x}$

20. $\dfrac{a}{a - 1} + \dfrac{a - 3}{1 - a}$

21. $\dfrac{x}{x - 2} + \dfrac{x + 1}{2 - x}$

22. $(x^2 - 36) \cdot \dfrac{3x + 1}{x + 6}$

23. $(y^2 - 9) \div \dfrac{y + 3}{y - 3}$

B Problems

Factor the expressions in Problems 24–36. Factor out all fractions.

24. $x^{-3} + 2x^{-2}y^{-1} + x^{-1}y^{-2}$

25. $xy^{-1} + 2 + x^{-1}y$

26. $x^{-3}y^{-2} + x^{-1}y + 2xy$

27. $\frac{1}{3}x^{-1}y^{-2} + xy + x^{-3}y^{-1}$

28. $x^{-2} + y^{-2} + 1$

29. $(x^2 + 1) - 4x^2(x^2 + 1)^{-1}$

30. $\frac{1}{2}y^{-1} + x^{-1} + \frac{1}{2}x^{-2}y$

31. $(x^2 + 1)^{-1} - x^2(x^2 + 1)^{-2}$

32. $(2x - 3)(3)(1 - x)^2(-1) - (1 - x)^3(-3)$

33. $4(x + 5)^3(x^2 - 2)^3 + (x + 5)^4(3)(x^2 - 2)^2(2x)$

34. $5(x - 2)^4(x^2 + 1)^3 + (x - 2)^5(3)(x^2 + 1)^2(2x)$

35. $x(\frac{1}{3})(x^2 - 3)^{-2}(2x) - (x^2 - 3)^{-1}$

36. $(x^2 - 4) - (x^2 - x - 2)(x^2 - 4)^{-1}$

Simplify the expressions in Problems 37–53.

37. $x^{-1}(x + 1) + x^{-2}(2 - x)$

38. $\dfrac{x + 3}{x} + \dfrac{3 - x}{x^2}$

39. $\dfrac{2x + 1}{x^2} + \dfrac{3 - x}{x}$

40. $x^{-2}(2x + 3) + x^{-1}(3 - x)$

41. $x^{-2}(2x^2 + x - 1) + x^{-3}(x^3 - 2x^2 + x - 1)$

42. $(a + b)^{-2}(2a + b) + (a + b)^{-3}(a^2 + 2ab + b^2)$

43. $\dfrac{2s + t}{(s + t)^2} + \dfrac{s^2 - 2t^2}{(s + t)^3}$

44. $\dfrac{4x - 12}{x^2 - 49} \div \dfrac{18 - 2x^2}{x^2 - 4x - 21}$

45. $\dfrac{36 - 9y}{3y^2 - 48} \div \dfrac{15 + 13y + 2y^2}{12 + 11y + 2y^2}$

46. $\dfrac{x^6 - y^6}{x^2 + xy + y^2} \cdot (x^2 - xy + y^2)^{-1}$

47. $\dfrac{x^6 - y^6}{x^2 - y^2} \div (x^4 + x^2y^2 + y^4)$

48. $\dfrac{4}{x^2 - 2x} + \dfrac{8}{4x - x^3} - \dfrac{-4}{x + 2}$

49. $\dfrac{6x}{2x + 1} - \dfrac{2x}{x - 3} + \dfrac{4x^2}{2x^2 - 5x - 3}$

50. $\dfrac{xy^{-1} + x^2y^{-2}}{yx^{-2} + 2x^{-1} + y^{-1}}$

51. $\dfrac{\dfrac{m}{n} + \dfrac{m^2}{n^2}}{\dfrac{m}{n^2} - m + \dfrac{1}{n}}$

52. $1 + \dfrac{x}{x + \dfrac{1}{1 + x}}$

53. $\dfrac{x}{1 + \dfrac{x}{x + \dfrac{1}{x + 1}}}$

4.2 LIMITS

One of the primary considerations in this book is graphing various types of functions. In this chapter, we'll consider several techniques for graphing rational functions. The simplest types are those that reduce to polynomial functions, or polynomial functions with deleted points. For example, if a function f is defined by

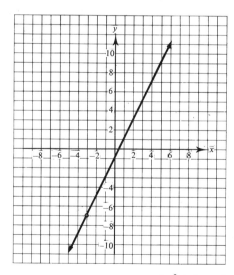

Figure 4.1   Graph of $f(x) = \dfrac{2x^2 + 5x - 3}{x + 3}$

$$f(x) = \frac{2x^2 + 5x - 3}{x + 3}$$

$$= \frac{(2x - 1)(x + 3)}{x + 3}$$

$$= 2x - 1 \quad (x \neq -3),$$

then the graph of f is the same as for the linear function $y = 2x - 1$ with the point at $x = -3$ deleted from the domain, as shown in Figure 4.1.

EXAMPLE 1: Graph $f(x) = \dfrac{x^3 + 4x^2 + 7x + 6}{x + 2}$.

Solution: Simplify, if possible:

These factors were obtained by synthetic division:

$$f(x) = \frac{(x + 2)(x^2 + 2x + 3)}{x + 2}$$

$$\begin{array}{r|rrrr} -2 & 1 & 4 & 7 & 6 \\ & & -2 & -4 & -6 \\ \hline & 1 & 2 & 3 & 0 \end{array}$$

$$= x^2 + 2x + 3 \quad (x \neq -2)$$

The graph of f is the same as for the quadratic function $y = x^2 + 2x + 3$ with the point at $x = -2$ deleted from the domain. To sketch this parabola, complete the square.

$$y - 3 = x^2 + 2x$$

$$y - 3 + 1 = x^2 + 2x + 1$$

$$y - 2 = (x + 1)^2$$

The vertex is at $(-1, 2)$ and the parabola opens up. It is drawn with the point where $x = -2$ deleted, as shown in Figure 4.2.

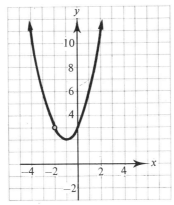

Figure 4.2   Graph of $f(x) = \dfrac{x^3 + 4x^2 + 7x + 6}{x + 2}$

For rational functions that do not reduce to polynomials so easily, we need additional information to sketch the function. In this and the next sections, two ideas that are essential in mathematics will be introduced from an intuitive standpoint and then will be used for graphing rational functions. These are the idea of *limit* and the idea of *continuity*.

Consider the function f defined by

$$f(x) = \frac{x^3 + 4x^2 + 7x + 6}{x + 2}.$$

Notice that -2 is not in the domain of f. But what about values of "x close to -2"? In mathematics, this question is phrased another way by asking what happens to $f(x)$ as x *approaches* -2. Consider the table of values given here (found by using a calculator).

We use an arrow $\rightarrow$ to mean *approaches*.

$x \rightarrow -2^+$ means x is approaching -2 from the right on the number line.

$x \rightarrow -2^-$ means x is approaching -2 from the left on the number line.

The concept of a *limit of a function* is rather sophisticated, and we are considering it only from an intuitive standpoint.

x	$f(x)$	x	$f(x)$
-1	2	-3	6
-1.5	2.25	-2.5	4.25
-1.9	2.81	-2.1	3.21
-1.99	2.9801	-2.01	3.0201
-1.999	2.9981	-2.001	3.002001
$x \rightarrow -2^+$	$f(x) \rightarrow 3$	$x \rightarrow -2^-$	$f(x) \rightarrow 3$

Notice that f is *not defined* for $x = -2$, but as the value of x gets closer to -2, the value of $f(x)$ approaches 3. We denote this relationship by

$$\lim_{x \to -2} f(x) = 3,$$

which is read "the *limit* of f of x as x approaches -2 is 3."

EXAMPLE 2: Find $\lim_{x \to 4} x^2$.

x	$f(x)$	x	$f(x)$
3	9	5	25
3.5	12.25	4.5	20.25
3.9	15.21	4.1	16.81
3.99	15.9201	4.01	16.0801
3.999	15.992001	4.001	16.008001
$x \rightarrow 4^-$	$f(x) \rightarrow 16$	$x \rightarrow 4^+$	$f(x) \rightarrow 16$

Therefore, $\lim_{x \to 4} x^2 = 16$.

Notice that the limit in Example 2 is the same as $f(4)$. But we are not concerned with the value *at $x = 4$*; we care only about the values of f *close to* $x = 4$. Compare this with Example 1, where the function

$$f(x) = \frac{x^3 + 4x^2 + 7x + 6}{x + 2}$$

is not defined at $x = -2$. The function of Example 2 is a *polynomial function*, whereas the function of Example 1 is a *rational function*. Since polynomial functions are defined for all real values, there is a useful result concerning limits of polynomial functions.

LIMIT OF A POLYNOMIAL FUNCTION: If f is any polynomial function, then

$$\lim_{x \to c} f(x) = f(c) \qquad \text{for any real number } c.$$

LIMIT OF A
POLYNOMIAL
FUNCTION

EXAMPLE 3: Find $\lim_{x \to 1} f(x)$, where

$$f(x) = 4x^3 - 2x^2 + x - 1.$$

Since f is a polynomial function,

$$\lim_{x \to 1} f(x) = f(1) = 4(1)^3 - 2(1)^2 + 1 - 1$$

$$= 2.$$

EXAMPLE 4: Let $f(x) = 5$.

a. $\lim_{x \to 2} f(x) = 5$ b. $\lim_{x \to -3} f(x) = 5$ c. $\lim_{x \to a} f(x) = 5$

The limit of a constant is that constant.

EXAMPLE 5: Find

$$\lim_{x \to 1} \frac{x^2 - 1}{x - 1}.$$

We could proceed with a table of values, but instead, consider the following argument. Notice that the function is *not defined* at $x = 1$, but since we are concerned with values *near $x = 1$*, we can say $x \neq 1$. But if $x \neq 1$, then

$$\frac{x^2 - 1}{x - 1} = \frac{(x - 1)(x + 1)}{x - 1}$$

$$= x + 1$$

Historical Note

The notion of a *limit* or a *limiting line* gave many mathematicians trouble until the 19th century. Some mathematicians worked with "infinitely short" line segments or "infinitely short" periods of time. Limits are studied in calculus, which was developed independently by two famous mathematicians, Isaac Newton (1645–1727) and Gottfried Leibniz (1646–1716).

and for values near 1 it is easy to see that $x + 1$ is near 2. Thus,

$$\lim_{x \to 1} \frac{x^2 - 1}{x - 1} = 2.$$

EXAMPLE 6: Find

$$\lim_{x \to 1} \frac{x + 1}{x^2 - 1}.$$

If $x \neq 1$, then

$$\frac{x + 1}{x^2 - 1} = \frac{x + 1}{(x - 1)(x + 1)} = \frac{1}{x - 1}.$$

Consider a table of values for $1/(x - 1)$ as x approaches 1.

x	$f(x)$	x	$f(x)$
2	1	0.5	−2
1.5	2	0.9	−10
1.1	10	0.99	−100
1.01	100	0.999	−1,000
1.001	1,000	0.9999	−10,000
1.0001	10,000	0.99999	−100,000

Example 6 shows that some limits may not exist.

It appears from the table that $f(x)$ increases without bound as x approaches 1 from the right or decreases without bound as x approaches 1 from the left. In such cases, we say that the limit doesn't exist.

EXAMPLE 7: Find the value of $1/x$ as x increases without bound.

x	$\dfrac{1}{x}$
1	1
2	0.5
10	0.1
100	0.01
1,000	0.001
10,000	0.0001

We say that $1/x \to 0$ as x increases without bound and symbolize this by $\lim_{x \to \infty} (1/x) = 0$.

EXAMPLE 8: Find the value of $1/x$ as x decreases without bound.

x	$\dfrac{1}{x}$
-1	-1
-2	-0.5
-10	-0.1
-100	-0.01
$-1,000$	-0.001
$-10,000$	-0.0001

We say that $1/x \to 0$ as x decreases without bound and symbolize this by
$$\lim_{x \to -\infty} (1/x) = 0.$$

Since $1/x \to 0$ for both $x \to \infty$ and $x \to -\infty$ in Examples 7 and 8, we say $1/x \to 0$ as x increases or decreases without bound and symbolize this by $\lim_{|x| \to \infty} (1/x) = 0$. We can also show that $\lim_{|x| \to \infty} (k/x) = 0$ for any constant k. And $\lim_{|x| \to \infty} (1/x^n) = 0$, $n > 0$, and $\lim_{|x| \to \infty} (k/x^n) = 0$.

When you see the symbol
$$\lim_{x \to \infty} \frac{1}{x} = 0,$$
it means that
$$\frac{1}{x} \to 0$$
as x increases without bound. Don't read more into the symbol ∞.

EXAMPLE 9: Find
$$\lim_{x \to \infty} \frac{x}{2x + 1}.$$

Solution: We could consider a table of values. Instead, suppose we multiply the rational expression by 1, written as $(1/x)/(1/x)$:

$$\frac{x}{2x + 1} \cdot \frac{1/x}{1/x} = \frac{1}{2 + (1/x)}.$$

The values of x that are not permitted in Example 9 are $x = -\frac{1}{2}$ and $x = 0$, but these values are excluded as x becomes large.

Now, since
$$\lim_{x \to \infty} \frac{1}{x} = 0,$$

we see that
$$\lim_{x \to \infty} \frac{1}{2 + (1/x)} = \frac{1}{2 + 0} = \frac{1}{2}.$$

EXAMPLE 10: Find
$$\lim_{x \to \infty} \frac{3x^2 - 7x + 2}{7x^2 + 2x + 5}$$

The procedure illustrated by Examples 9 and 10 should be applied only when $x \to \infty$ or $x \to -\infty$. It does not apply when $x \to a$ (a some real number).

Notice that the largest power of x in the expression is x^2, so we multiply the numerator and denominator by $1/x^2$.

$$\lim_{x \to \infty} \left[\frac{3x^2 - 7x + 2}{7x^2 + 2x + 5} \cdot \frac{1/x^2}{1/x^2} \right]$$

Now, since k/x and k/x^2 both approach 0 as x increases without bound, we have

$$= \lim_{x \to \infty} \frac{3 - \dfrac{7}{x} + \dfrac{2}{x^2}}{7 + \dfrac{2}{x} + \dfrac{5}{x^2}} = \frac{3}{7}.$$

PROBLEM SET 4.2

A Problems

Evaluate the limits in Problems 1–25.

1. $A = \lim_{x \to 0} (x^3 + 3x^2 + 12x - 5)$

2. $B = \lim_{x \to 5} (4x^5 - 25x^4 + 30x^3 - 25x^2 + 10x - 40)$

3. $C = \lim_{x \to -3} (5x^5 + 21x^4 + 15x^3 + x^2 + 25x - 6)$

4. $D = \lim_{x \to 3} \dfrac{x^2 + 3x - 10}{x - 2}$

5. $E = \lim_{x \to 2} \dfrac{x^3 - 8}{x - 2}$

6. $F = \lim_{x \to -5} \dfrac{x^2 + 3x - 10}{x + 5}$

7. $G = \lim_{x \to 3} \dfrac{x^2 - 8x + 15}{x - 3}$

8. $H = \lim_{x \to 4} \dfrac{2x^2 - 5x - 12}{x - 4}$

9. $I = \lim_{x \to 2} \dfrac{x^2 - 1}{x - 2}$

10. $J = \lim_{x \to 2} \dfrac{x^2 + 2x + 4}{x^3 - 8}$

11. $K = \lim_{x \to 2} \dfrac{x^3 - 8}{x^2 + 2x + 4}$

12. $L = \lim_{x \to 2} \dfrac{6 - x}{2x - 15}$

13. $M = \lim_{x \to 2} \dfrac{x + 2}{x^3 + 8}$

14. $N = \lim_{|x| \to \infty} \dfrac{3x - 1}{2x + 3}$

15. $P = \lim_{|x| \to \infty} \dfrac{2x^2 - 5x - 3}{x^2 - 9}$

16. $Q = \lim_{|x| \to \infty} \dfrac{x^2 + 6x + 9}{x + 3}$

17. $R = \lim_{|x| \to \infty} \dfrac{(x + 1)(x - 2)(x + 4)}{3x^3 + 2x^2 - x + 5}$

18. $S = \lim_{|x| \to \infty} \dfrac{(2x + 1)(x - 5)}{(3x - 4)(x + 6)}$

19. $T = \lim_{|x| \to \infty} \dfrac{6x^2 - 5x + 2}{2x^2 + 5x + 1}$

20. $U = \lim\limits_{x \to \infty} \dfrac{5x + 10{,}000}{x - 1}$

21. $V = \lim\limits_{x \to \infty} \left(x + 2 + \dfrac{3}{x - 1} \right)$

22. $W = \lim\limits_{x \to -\infty} \dfrac{4x + 10^5}{x + 1}$

23. $X = \lim\limits_{x \to -\infty} \left(2x - 3 + \dfrac{4}{x + 2} \right)$

24. $Y = \lim\limits_{x \to -\infty} \dfrac{4x^4 - 3x^3 + 2x + 1}{3x^4 - 9}$

25. $Z = \lim\limits_{x \to 1} \dfrac{x^2 + x + 1}{x^3 \div 1}$

26. Replace the value within each set of parentheses with the corresponding variable for the limits of Problems 1–25. The letter "O" has already been filled in.

$(\tfrac{2}{3})(-5)(\tfrac{4}{3})$ $(4)(11)(-5)(3)$ $(\tfrac{4}{3})(O)(5)$

$(0)(\tfrac{2}{3})(O)(4),$ $(8)(O)$ $(4)(11)(-5)(3)$

$(\tfrac{4}{3})(O)(5)$ $(\tfrac{1}{4})(5)(\tfrac{2}{3})(3),$ $(9)(O)(\tfrac{1}{4})(12)$

$(4)(11)(-5)(3)$ $(\tfrac{1}{4})(-5)(\tfrac{4}{3})$

B Problems

Graph the rational functions given in Problems 27–36.

27. $f(x) = \dfrac{x^2 - x - 6}{x + 2}$

28. $f(x) = \dfrac{x^2 - x - 12}{x + 3}$

29. $f(x) = \dfrac{2x^2 - 13x + 15}{x - 5}$

30. $f(x) = \dfrac{6x^2 - 5x - 4}{2x + 1}$

31. $y = \dfrac{2x^3 - 3x^2 - 32x - 15}{x^2 - 2x - 15}$

32. $y = \dfrac{3x^3 + 5x^2 - 26x + 8}{x^2 + 2x - 8}$

33. $y = \dfrac{x^3 + 6x^2 + 10x + 4}{x + 2}$

34. $y = \dfrac{x^3 + 9x^2 + 15x - 9}{x + 3}$

35. $y = \dfrac{x^3 + 12x^2 + 40x + 40}{x + 2}$

36. $y = \dfrac{x^3 + 5x^2 + 6x}{x + 3}$

Mind Boggler

37. The notion of a limit is difficult to express. If

$$\lim_{x \to 4} (x^2 - 4x + 3) = 3,$$

we say "$x^2 - 4x + 3$ is close to 3 if x is close to 4." Try to formulate this idea of closeness by drawing the graph of $f(x) = x^2 - 4x + 3$ and developing a geometric argument of what *you* would mean by close.

4.3 CONTINUITY

You may remember the following puzzle from elementary school: Can you draw the illustration in Figure 4.3 without lifting your pencil or retracing

Historical Quote

The quotation given in Problem 26 was the motto of one of the most famous women in mathematics, Sonja Kovalevsky (1850–1891). She was born in Russia, but since Russian universities were closed to women, she went to Germany to study under the famous mathematician Karl Weierstrass (1815–1897). After arriving in Germany, she found that the Berlin University also did not accept women. However, Weierstrass gave her private lessons, and she obtained a degree from Göttingen University in absentia. In 1888, she was awarded a prize for a paper submitted to the *Institut de France*, which included the quotation of Problem 26. The paper was so outstanding that the judges raised the prize money from 3000 to 5000 francs.

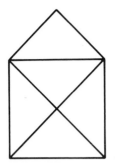

Figure 4.3

any of the lines? In calculus, we will be concerned with figures that can be drawn without lifting a pencil from the paper, but, instead of the example in Figure 4.3, we will focus our attention on functions. Which of the functions in Figure 4.4 do you think you can draw without lifting your pencil from the paper? Intuitively, we say that if you can draw the graph of the function without lifting your pencil from the paper, then it is a *continuous function*. However, since continuity is a very important idea in calculus, we will discuss the concept with some care. We begin with a discussion of the *continuity at a point*, and then we will talk about *continuity on an interval*.

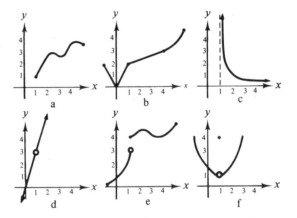

Figure 4.4   Parts c and d are not defined at the value $x = 1$.

CONTINUITY AT A POINT

There are two essential conditions for a function f to be continuous at a point c. First, $f(c)$ must be defined. For example, the functions

$$f_1(x) = \frac{3x^2 - x + 2}{x} \qquad f_2(x) = \frac{x^2 - 5x}{x} \qquad f_3(x) = \begin{cases} x + 3 & \text{if } x > 0 \\ x^2 & \text{if } x < 0 \end{cases}$$

can't be continuous at $x = 0$ since none of these functions is defined at this point.

A second condition for continuity at a point $x = c$ is that the function makes no jumps at this point. Looking at Figure 4.4, we see that the graphs in parts d, e, and f make jumps at the point $x = 1$. This means that if "x is close to c," then "$f(x)$ must be close to $f(c)$." We recognize this as the concept of limit, and now define the concept of continuity at a point.

A function f is continuous at $x = c$ if

1. $f(c)$ exists; and
2. $\lim\limits_{x \to c} f(x) = f(c)$.

𝕳istorical 𝕹ote

The idea of *continuity* evolved from the notion of a curve "without breaks or jumps" to a rigorous definition given by Karl Weierstrass. Galileo and Leibniz had thought of continuity in terms of the density of points on a curve, but they were in error since the rational numbers have this property of denseness yet do not form a continuous curve. Another mathematician, J. W. R. Dedekind (1831–1916), took an entirely different approach and concluded that continuity is due to the division of a segment into two parts by a point on the segment. Dedekind visualized the division of a curve into two parts so that there is one and only one point that makes this division. As Dedekind wrote, "By this commonplace remark, the secret of continuity is to be revealed" [from Carl Boyer, *A History of Mathematics* (New York: Wiley, 1968), p. 607]. Boyer adds "Commonplace the remark may have been, but the author [Dedekind] seems to have had some qualms about it for he hesitated for some years before committing himself in print."

EXAMPLES: Test the continuity of each function at the point $x = 1$.

1. $f(x) = \dfrac{x + 1}{x - 1}$

   Solution: Not continuous at $x = 1$ since f is not defined at this point.

2. $f(x) = \dfrac{x^2 + 2x - 3}{x - 1}$

   Solution: Not continuous at $x = 1$ since f is not defined at this point.

3. $f(x) = \begin{cases} \dfrac{x^2 + 2x - 3}{x - 1} & \text{if } x \neq 1 \\ 4 & \text{if } x = 1 \end{cases}$

   Solution:

   (1) $f(1) = 4$, so f is defined at $x = 1$.

   (2) $\lim\limits_{x \to 1} f(x) = \lim\limits_{x \to 1} \dfrac{x^2 + 2x - 3}{x - 1}$

   $\qquad\qquad = \lim\limits_{x \to 1} \dfrac{(x - 1)(x + 3)}{x - 1}$

   $\qquad\qquad = \lim\limits_{x \to 1} (x + 3)$

   $\qquad\qquad = 4$

   Since $\lim\limits_{x \to 1} f(x) = f(1)$, the function is continuous at $x = 1$.

4. $f(x) = 3x^3 + 5x^2 - 4x + 1$

   Solution:

   (1) $f(1) = 3(1)^3 + 5(1)^2 - 4(1) + 1$

   $\qquad = 5$

   So f is defined at $x = 1$.

   (2) $\lim\limits_{x \to 1} f(x) = f(1)$

   Thus, the function is continuous at $x = 1$.

Notice that Example 4 is a polynomial, and from the previous section we know that if f is any polynomial function, then

$$\lim_{x \to c} f(x) = f(c)$$

for any real number c. Thus, we immediately have the following result:

Every polynomial function is continuous at every point in its domain.

CONTINUITY OVER AN INTERVAL

If a function is continuous for every point in a given interval, then we say the function is continuous over that interval. For example, polynomials are continuous for all real numbers. On the other hand,

$$f(x) = \frac{x + 1}{x - 1}$$

is continuous for $-1 \le x \le 0$, but not for $0 \le x \le 2$ since f is undefined at $x = 1$. The function f from Example 3,

$$f(x) = \begin{cases} \dfrac{x^2 + 2x - 3}{x - 1} & \text{if } x \ne 1 \\ 4 & \text{if } x = 1, \end{cases}$$

is continuous for all real numbers. As you can see from the above examples, we are usually concerned with finding the points of discontinuity. The following result summarizes the properties of continuity; these are usually proved in a calculus course.

CONTINUITY THEOREM

The condition $g \ne 0$ means all values of x that cause $g(x) = 0$ are excluded from the domain.

CONTINUITY THEOREM: Let f and g be continuous functions at $x = c$. Then the following functions are also continuous at $x = c$:

1. $f + g$ 2. $f - g$

3. fg 4. $\dfrac{f}{g}$ $(g \ne 0)$

5. $f \circ g$ 6. $g \circ f$

SUSPICIOUS POINTS

Since most of the functions we will consider in this book are continuous over certain intervals, our task will be to look for points of discontinuity. These points may be values for which the definition of the function changes or values that cause division by zero. We will call such points *suspicious points*.

We then check the continuity at the suspicious points and use the Continuity Theorem for the rest of the points in the interval.

EXAMPLES

5. $f_1(x) = \dfrac{x^2 + 3x - 10}{x - 2}$

Solution: The suspicious point is $x = 2$, since this is a value for which the function $(x - 2)$ is equal to zero. The function f is not defined at this point, so it is *discontinuous at* $x = 2$. By the Continuity Theorem (part 4), it is continuous at all other points.

A point for which the function is not continuous is said to be *discontinuous* at that point.

6. $f_2(x) = \begin{cases} \dfrac{x^2 + 3x - 10}{x - 2} & \text{if} \quad x \neq 2 \\ 6 & \text{if} \quad x = 2 \end{cases}$

Solution: The suspicious point is $x = 2$, since it is a value for which the definition of the function changes. Checking we get:
(1) $f_2(2) = 6$, so the function is defined at $x = 2$.

$$
\begin{aligned}
(2) \quad \lim_{x \to 2} f_2(x) &= \lim_{x \to 2} \frac{x^2 + 3x - 10}{x - 2} \\
&= \lim_{x \to 2} \frac{(x + 5)(x - 2)}{x - 2} \\
&= \lim_{x \to 2} (x + 5) \\
&= 7
\end{aligned}
$$

But $f_2(2) = 6$ and thus $\lim_{x \to 2} f_2(x) \neq f_2(2)$, so the function is *discontinuous at* $x = 2$.

7. $f_3(x) = \begin{cases} \dfrac{x^2 + 3x - 10}{x - 2} & \text{if} \quad x \neq 2 \\ 7 & \text{if} \quad x = 2 \end{cases}$

Solution: Notice that we have simply redefined the functional value of $f_2(x)$ of Example 6 at a single point. Thus, $f_3(x) = 7$ and

$$\lim_{x \to 2} f_3(x) = f_3(2)$$

so the function is continuous for all real numbers.

8. $f_4(x) = \begin{cases} -x + 3 & \text{if} \quad -5 \leq x < 2 \\ x - 2 & \text{if} \quad 2 \leq x \leq 5 \end{cases}$

Solution: The suspicious point is $x = 2$, since it is a value for which the definition of the function changes.

Notice, we did not check the continuity at the other points of the domain. The other points are not suspicious points and we apply the Continuity Theorem for all of these points.

(1) f_4 is defined at $x = 2$ since $f_4(2) = 0$.
(2) $\lim_{x \to 2} f_4(x)$ is found by checking left- and right-hand limits:

$$\lim_{x \to 2^-} f_4(x) = \lim_{x \to 2^-} (-x + 3) = 1$$

$$\lim_{x \to 2^+} f_4(x) = \lim_{x \to 2^+} (x - 2) = 0$$

If you draw the graph of f_4, you can see the discontinuity at $x = 2$.

Since the left- and right-hand limits are different, $\lim_{x \to 2} f_4(x)$ doesn't exist, so the function is discontinuous at $x = 2$.

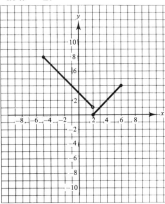

9. $f_5(x) = \begin{cases} -x + 2 & \text{if } -5 \le x < 2 \\ x - 2 & \text{if } 2 \le x < 5 \end{cases}$

Solution: The suspicious point is $x = 2$.
(1) f_5 is defined at $x = 2$ since $f_5(2) = 0$
(2) $\lim_{x \to 2^-} f_5(x) = \lim_{x \to 2^+} (-x + 2) = 0$

$$\lim_{x \to 2^+} f_5(x) = \lim_{x \to 2^+} (x - 2) = 0$$

Thus, $\lim_{x \to 2} f_5(x) = 0$, since the left- and right-hand limits are the same. Since $\lim_{x \to 2} f_5(x) = f_5(2)$, we see that it is continuous at $x = 2$. By the Continuity Theorem, it is continuous everywhere on the interval, so f is continuous over $-5 \le x \le 5$.

Notice that f_4 is continuous over $-5 \le x < 2$ and over $2 \le x \le 5$ but not over $-5 \le x \le 5$.

By looking at the graph of f_5, you can see that it is continuous over $-5 \le x \le 5$.

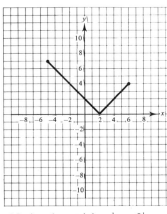

Notice that $f_5(x) = |x - 2|$.

PROBLEM SET 4.3

A Problems

In Problems 1–15, a function is defined over a certain domain. State whether the function is continuous at all points in this domain. Give the points of discontinuity.

1. $f(x) = \dfrac{1}{x^2 + 5}, \quad -5 \le x \le 5$

2. $f(x) = \dfrac{1}{x^2 - 5}, \quad -5 \le x \le 5$

3. $f(x) = \dfrac{1}{x^2 - 9}, \quad 5 \le x \le 10$

4. $f(x) = \dfrac{1}{x^2 - 9}, \quad -5 \le x \le 5$

5. $f(x) = \begin{cases} \dfrac{1}{x - 3} & -5 \le x \le 5, \ x \ne 3 \\ 4 & x = 3 \end{cases}$

6. $f(x) = \dfrac{x^2 - x - 6}{x + 2}, \quad -5 \le x \le 5$

7. $f(x) = \begin{cases} \dfrac{x^2 - x - 6}{x + 2} & -5 \le x \le 5, \ x \ne -2 \\ -4 & x = -2 \end{cases}$

8. $f(x) = \begin{cases} \dfrac{x^2 - x - 6}{x + 2} & -5 \le x \le 5, \quad x \ne -2 \\ -5 & x = -2 \end{cases}$

[handwritten: make disc = x = -2]

9. $f(x) = \dfrac{x^2 - 3x - 10}{x + 2}, \quad 0 \le x \le 5$

[handwritten: examine x = -2]

10. $f(x) = \dfrac{x^2 - 3x - 10}{x + 2}, \quad -5 \le x \le 5$

11. $f(x) = \begin{cases} \dfrac{x^2 - 3x - 10}{x + 2} & -5 \le x \le 5, \quad x \ne -2 \\ -3 & x = -2 \end{cases}$

[handwritten: examine x = -2]

12. $f(x) = \begin{cases} \dfrac{x^2 + x + 1}{x^3 - 1} & 0 < x < 5, \quad x \ne 1 \\ 1 & x = 1 \end{cases}$

[handwritten: examine]

13. $f(x) = \begin{cases} 1 & \text{if } x \text{ is rational} \\ -1 & \text{if } x \text{ is irrational} \end{cases}$

[handwritten: as x approaches a rational, f will be irration... discontinuous]

14. $f(x) = |x|, \quad -5 \le x \le 5$

15. $f(x) = |x - 2|, \quad -5 \le x \le 5$

B Problems

16. On July 10, 1913, the hourly temperatures at Death Valley, California, were recorded:

8 A.M.:	90°F
10 A.M.:	115°F
12 Noon:	123°F
1 P.M.:	134°F
3 P.M.:	130°F
5 P.M.:	128°F
7 P.M.:	105°F

This was the highest temperature ever officially recorded in North America.

If we let T represent the temperature in degrees Fahrenheit and assume that T is a function of time (measured on a 24-hour clock), we can write $T(8) = 90$; $T(10) = 115$; $T(12) = 123$; $T(13) = 134$; ...; $T(19) = 105$.
a. Is T a continuous function? What is the domain?
b. On the day in question, what is the minimum number of times during the day that the temperature was
   i. 100°F?   ii. 110°F?   iii. 80°F?

Which of the functions described in Problems 17–20 represent continuous functions? State the domain, if possible, for each example.

17. The humidity on a certain day at a given location considered as a function of time.

$ 18. The selling price of IBM stock on a certain day considered as a function of time.

19. The number of people who were unemployed in the United States during the month of December 1976 considered as a function of the date.

20. The charges for a telephone call from Los Angeles to New York considered as a function of time.

21. Give an example of a function defined on $1 \leq x \leq 3$ that is discontinuous at $x = 2$ and continuous elsewhere on the interval.

22. Give an example of a function defined on $3 \leq x \leq 7$ that is discontinuous at $x = 4$ and $x = 5$ and continuous elsewhere on the interval.

Mind Bogglers

23. On Saturday evenings, a favorite pastime of some Santa Rosa high school students is to cruise the main streets of town. The commonly followed streets are shown on the map given in Figure 4.5. Is it possible to choose a route so that all

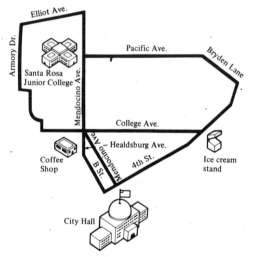

Figure 4.5   Santa Rosa street problem

the permitted streets are traveled exactly once? The intersections are identified by the following buildings: Santa Rosa Junior College (Mendocino and Pacific); coffee shop (Mendocino, College, and Healdsburg); City Hall (Mendocino and 4th); ice cream stand (College and 4th).

24. a. Give an example of a function that is discontinuous at every integral value but continuous everywhere else.

b. Give an example of a function that is discontinuous everywhere.

4.4 GRAPHS OF RATIONAL FUNCTIONS

As an aid to sketching certain rational functions, we will need to consider the notion of an *asymptote*. An asymptote is a line having the property that the distance from a point P on the curve to the line approaches zero as the distance from P to the origin increases without bound and P is *on a suitable part of the curve.* This last phrase (in italics) is best illustrated by considering Figure 4.6, where L is an asymptote for the function f. Consider P

ASYMPTOTE

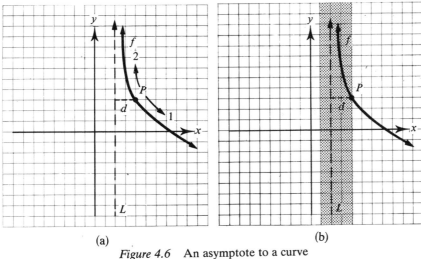

(a)   (b)

Figure 4.6   An asymptote to a curve

and d, the distance from P to the line L, as shown in Figure 4.6a. Now, the distance from P to the origin can increase two ways depending on whether P moves along the curve in direction 1 or direction 2. In direction 1, the distance d increases without bound, but in direction 2, the distance d approaches zero. Thus, if we consider the portion of the curve in the shaded region of Figure 4.6b, we see that the conditions of the definition of an asymptote apply. Even though some rational functions may not have asymptotes, there are three types of asymptotes that occur frequently enough to merit consideration. These are *vertical, horizontal,* and *slant asymptotes*. An example of each of these is shown in Figure 4.7.

In this section, we consider first the mechanics for finding these asymptotes and then the graphing of rational functions using asymptotes.

VERTICAL ASYMPTOTES

Of the rational functions that have asymptotes, the easiest to find are the vertical asymptotes. If $f(x) = P(x)/D(x)$, where $P(x)$ and $D(x)$ have no common factors, then $|f(x)|$ must get large as x gets close to any value for

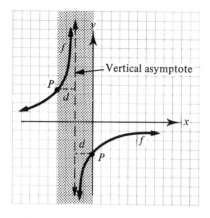 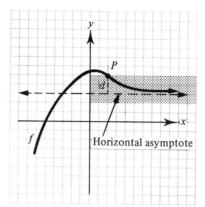 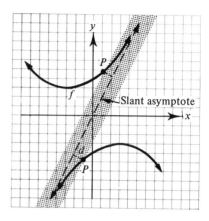

Figure 4.7 Examples of asymptotes

which $D(x) = 0$ and $P(x) \neq 0$. This means that if r is a root of $D(x) = 0$, then $x = r$ is the equation of a vertical asymptote.

EXAMPLES: Find the vertical asymptotes of the given functions.

1. $f(x) = \dfrac{1}{x - 3}$

Later in this section we'll show how a knowledge of the asymptotes can help sketch these functions.

We see that $x - 3 = 0$ if $x = 3$, which is the equation of a vertical line. This line is a vertical asymptote.

2. $f(x) = \dfrac{2x^2 - 3x + 5}{x^2 - x - 2}$

There are no common factors in the numerator and denominator of f so solve

$$x^2 - x - 2 = 0$$
$$(x - 2)(x + 1) = 0$$
$$x = 2 \quad \text{or} \quad -1$$

The vertical asymptotes are given by $x = 2$ and $x = -1$ since neither of these values also causes the numerator to be zero.

Be careful with functions like the one in Example 3. If you did not notice the common factor, you might be led to the incorrect conclusion that $x = 2$ and $x = -2$ are the equations of the vertical asymptotes.

3. $f(x) = \dfrac{x^2 + 2x - 8}{x^2 - 4}$

$$f(x) = \dfrac{x^2 + 2x - 8}{x^2 - 4}$$

$$= \dfrac{(x - 2)(x + 4)}{(x - 2)(x + 2)}$$

$$= \frac{x+4}{x+2} \quad (x \neq 2)$$

From this simplified form, you can see that $x = -2$ is the equation of the vertical asymptote.

Notice that since division by zero would be involved, the graph of a function can never cross a vertical asymptote.

HORIZONTAL ASYMPTOTES

If the equation can be solved easily for x so that

$$x = \frac{p(y)}{d(y)},$$

where $p(y)$ and $d(y)$ have no common factors, and if r is a root of $d(y) = 0$, then the line with the equation $y = r$ is a horizontal asymptote. Unfortunately, it is not always easy to solve for x, so consider $f(x) = P(x)/D(x)$, which is the usual form for a rational function, where $P(x)$ and $D(x)$ have no common factors. If $f(x)$ approaches some number k as $|x|$ becomes large without bound, then the line with the equation $y = k$ is a horizontal asymptote. Remember that the distance between a point on the graph of $y = f(x)$ and the horizontal line $y = k$ must approach zero as $|x|$ gets large. Thus,

$$\lim_{|x| \to \infty} \underbrace{[f(x) - k]}_{\substack{\text{the distance between} \\ f(x) \text{ and the line} \\ \text{given by } y = k}} = 0$$

This is the same as saying

$$\lim_{|x| \to \infty} f(x) = k.$$

EXAMPLES: Find the horizontal asymptotes of the given functions.

4. $f(x) = \dfrac{1}{x-3}$

Find

$$\lim_{|x| \to \infty} \frac{1}{x-3} = \lim_{|x| \to \infty} \frac{1}{x-3} \cdot \frac{\dfrac{1}{x}}{\dfrac{1}{x}}$$

Notice that this function is simple enough that we could have solved for x to find the horizontal asymptotes.

$$y = \frac{1}{x - 3}$$

$$y(x - 3) = 1$$

$$xy - 3y = 1$$

$$xy = 1 + 3y$$

$$x = \frac{1 + 3y}{y}$$

Thus, the line given by $y = 0$ is an asymptote.

$$= \lim_{|x| \to \infty} \frac{\dfrac{1}{x}}{1 - \dfrac{3}{x}}$$

$$= 0$$

Thus, the line given by $y = 0$ is a horizontal asymptote.

5. $f(x) = \dfrac{2x^2 - 3x + 5}{x^2 - x - 2}$

Find

$$\lim_{|x| \to \infty} f(x) = \lim_{|x| \to \infty} \frac{2x^2 - 3x + 5}{x^2 - x - 2} \cdot \frac{\dfrac{1}{x^2}}{\dfrac{1}{x^2}}$$

$$= \lim_{|x| \to \infty} \frac{2 - \dfrac{3}{x} + \dfrac{5}{x^2}}{1 - \dfrac{1}{x} - \dfrac{2}{x^2}}$$

$$= 2$$

Thus, the line given by $y = 2$ is a horizontal asymptote.

6. $f(x) = \dfrac{x^2 + 2x - 8}{x^2 - 4}$

Rewrite f:

$$\frac{(x - 2)(x + 4)}{(x - 2)(x + 2)} = \frac{x + 4}{x + 2} \cdot \frac{1/x}{1/x}$$

$$= \frac{1 + \dfrac{4}{x}}{1 + \dfrac{2}{x}}$$

Now, find

$$\lim_{|x| \to \infty} f(x) = \lim_{|x| \to \infty} \frac{1 + \dfrac{4}{x}}{1 + \dfrac{2}{x}}$$

$$= 1$$

Thus, the line given by $y = 1$ is a horizontal asymptote.

For vertical asymptotes, we noted that the graph of the function and the asymptotes could not intersect. This is not the case for horizontal asymptotes. For Example 5, the horizontal asymptote is $y = 2$ for the curve given by

$$y = \frac{2x^2 - 3x + 5}{x^2 - x - 2}.$$

If we solve these equations simultaneously, we find

$$\frac{2x^2 - 3x + 5}{x^2 - x - 2} = 2$$

$$2x^2 - 3x + 5 = 2(x^2 - x - 2)$$

$$2x^2 - 3x + 5 = 2x^2 - 2x - 4$$

$$x = 9$$

Thus, the point $(9, 2)$ is a point of intersection for the function and the asymptote.

This is graphed later in this section (Figure 4.11).

SLANT ASYMPTOTES

Linear asymptotes that are neither horizontal nor vertical are called *slant asymptotes*. Consider $f(x) = P(x)/D(x)$, where $P(x)$ and $D(x)$ have no common factors; there are three possibilities:

1. The degree of $P(x)$ is less than or equal to the degree of $D(x)$.
2. The degree of $P(x)$ is one more than the degree of $D(x)$.
3. The degree of $P(x)$ exceeds the degree of $D(x)$ by more than one.

Consider $\lim\limits_{|x| \to \infty} f(x)$ for these three possibilities.

1. If the degree of $P(x)$ is less than the degree of $D(x)$, then

$$\lim_{|x| \to \infty} \frac{P(x)}{D(x)} = 0.$$

This gives the horizontal asymptote represented by $y = 0$. If the degree of $P(x)$ is the same as the degree of $D(x)$, we have the situation discussed previously for other horizontal asymptotes.

2. If the degree of $P(x)$ is one more than the degree of $D(x)$, then

$$f(x) = \frac{P(x)}{D(x)} = mx + b + \frac{R(x)}{D(x)},$$

where the degree of $R(x)$ is less than the degree of $D(x)$. Then

$$\lim_{|x| \to \infty} \frac{R(x)}{D(x)} = 0,$$

which means that for large values of $|x|$, $f(x)$ is near the line given by $y = mx + b$. This says that the line with the equation $y = mx + b$ is a slant asymptote for the curve given by $y = f(x)$.

Note that $R(x) \ne 0$, because if $R(x) = 0$, then $P(x)$ and $D(x)$ have at least one common factor—which is contrary to the hypothesis.

3. If the degree of $P(x)$ exceeds the degree of $D(x)$ by more than one, then the quotient is no longer linear; and since asymptotes are lines, we see that there will be no slant asymptotes.

EXAMPLES: Find the slant asymptotes of the given functions.

7. The functions given in Examples 4, 5, and 6 cannot have slant asymptotes since the degrees of the numerators do not exceed the degrees of the denominators by one.

8. $f(x) = \dfrac{2x^2 - 5x + 1}{x - 3}$

    Divide synthetically to find

    $$f(x) = 2x + 1 + \frac{4}{x - 3}.$$

    Thus, the slant asymptote is given by $y = 2x + 1$.

$$
\begin{array}{r|rrr}
3 & 2 & -5 & 1 \\
  &   & 6 & 3 \\
\hline
  & 2 & 1 & 4
\end{array}
$$

9. $f(x) = \dfrac{3x^3 - 2x^2 + x - 5}{x^2 + 3}$

    Divide:

$$
\require{enclose}
\begin{array}{r}
3x - 2 \\
x^2 + 3 \enclose{longdiv}{3x^3 - 2x^2 + x - 5} \\
\underline{3x^3 + 9x } \\
-2x^2 - 8x - 5 \\
\underline{-2x^2 - 6} \\
-8x + 1
\end{array}
$$

    Thus,

$$f(x) = 3x - 2 + \frac{-8x + 1}{x^2 + 3},$$

    and the slant asymptote is given by $y = 3x - 2$.

As with horizontal asymptotes, the graph of a function can pass through a slant asymptote. For Example 9:

$$\underbrace{\frac{3x^3 - 2x^2 + x - 5}{x^2 + 3}}_{f(x)} = \overbrace{3x - 2}^{\text{slant asymptote}}$$

$$3x^3 - 2x^2 + x - 5 = (3x - 2)(x^2 + 3)$$

$$3x^3 - 2x^2 + x - 5 = 3x^3 - 2x^2 + 9x - 6$$

$$8x = 1$$

$$x = \frac{1}{8}$$

And $y = 3x - 2 = 3(1/8) - 2 = -13/8$, so the curve intersects the slant asymptote at the point $(1/8, -13/8)$.

SUMMARY

Let $f(x) = P(x)/D(x)$, where $P(x)$ and $D(x)$ have no common factors. Then

1. the *vertical line* given by $x = r$ is an asymptote if $D(r) = 0$;
2. the *horizontal line* given by $y = k$ is an asymptote if

$$\lim_{|x| \to \infty} f(x) = k;$$

3. the *slant line* given by $y = mx + b$ is an asymptote if the degree of $P(x)$ is one more than the degree of $D(x)$ and

$$\frac{P(x)}{D(x)} = mx + b + \frac{R(x)}{D(x)}.$$

ASYMPTOTES FOR A RATIONAL FUNCTION $f(x)$

Use synthetic division if $D(x)$ is of the form $x - r$; otherwise, use long division. Here $R(x) \neq 0$ and is of a degree less than the degree of $D(x)$.

GRAPHING RATIONAL FUNCTIONS

We now know a great deal about the asymptotes of certain rational functions but how can we put this information to use in order to graph the function? Generally, we follow the procedure summarized below.

1. Find and graph the asymptotes, if any.
2. Find and plot any points where the graph passes through an asymptote or the coordinate axes.
3. The plane is now divided into one or more regions with asymptotes as boundaries. Plot selected points within each region to determine the shape of the graph within that region. Remember to make use of the fact that the boundaries of these regions are asymptotes and that the distance between the curve and the asymptote must decrease as you move out from the origin.

Everyone knows what a curve is, until he has studied enough mathematics to become confused through the countless number of possible exceptions.

Felix Kline

Examples 10–13 illustrate this procedure.

EXAMPLES: Graph each rational function.

10. Let the function f be defined by the equation

$$y = \frac{1}{x - 3}.$$

Step 1. The asymptotes are given by $x = 3$ and $y = 0$ (from Examples 1, 4, and 7).

Step 2. The curve passes through $(0, -\frac{1}{3})$ and does not pass through any asymptotes since $1/(x - 3) = 0$ has an empty solution set. The asymptotes and this point are shown in Figure 4.8. Also notice

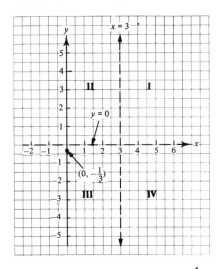

Figure 4.8 Asymptotes for $y = \dfrac{1}{x - 3}$

that the only point of discontinuity is $x = 3$. The plane is divided into Regions I, II, III, and IV.

Step 3. Region I

x	y
3.5	2
4	1
5	$\frac{1}{2}$
6	$\frac{1}{3}$

Plot the points corresponding to $x = 3.5, 4, 5,$ and 6. Draw the curve in this region, as shown in Figure 4.9. Use the facts that the lines given by $x = 3$ and $y = 0$ are asymptotes and that the graph of the function doesn't cross these lines.

Region II: Since f is a function and its graph does not cross the asymptote given by $y = 0$, and since there is a point $(0, -\frac{1}{3})$ below this line, there are no points in Region II.

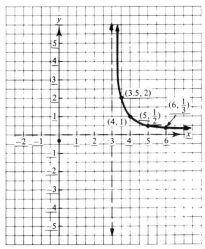

Figure 4.9 Part of the graph of $y = \dfrac{1}{x - 3}$

Notice that f is discontinuous at $x = 3$.

Region III

x	y
-1	$-\frac{1}{4}$
1	$-\frac{1}{2}$
2	-1
2.5	-2

Draw the curve in this region, as shown in Figure 4.10. Use the

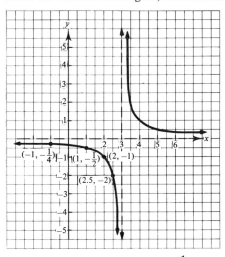

Figure 4.10 Graph of $y = \dfrac{1}{x - 3}$

facts that the lines given by $x = 3$ and $y = 0$ are asymptotes and that the curve doesn't cross these lines.

Region IV: Since the graph does not cross the asymptote given by $y = 0$, there are no points in Region IV. The graph is complete as shown in Figure 4.10.

11.   Let f be a function defined by the equation

$$y = \frac{2x^2 - 3x + 5}{x^2 - x - 2}.$$

Step 1.   The asymptotes are given by $x = 2$, $x = -1$, and $y = 2$ (from Examples 2, 5, and 7).

For the intercepts, let

$x = 0$:
$$y = \frac{2 \cdot 0^2 - 3 \cdot 0 + 5}{0^2 - 0 - 2}$$

$$= -\frac{5}{2}$$

Step 2.   The curve passes through $(0, -5/2)$ and also passes through the horizontal asymptote at $(9, 2)$. Label Regions I, II, III, IV, V, and VI, as shown in Figure 4.11. The only points of discontinuity are at the vertical asymptotes, namely $x = -1$ and $x = 2$.

Step 3.   Plot points; do each region separately.

and

$y = 0$: $0 = \dfrac{2x^2 - 3x + 5}{x^2 - x - 2}$

$2x^2 - 3x + 5 = 0$

No real roots since $b^2 - 4ac = 9 - 4 \cdot 2 \cdot 5 < 0$. For the asymptotes see pages 180 and 182.

Note: the order in which you plot points in these regions doesn't matter.

Region I

x	y
3	3.5
4	2.5
9	2

← crosses asymptote at this point and crosses into Region VI

Region VI

x	y
9	2
12	1.98

← crosses asymptote

Region II is empty since f is a function and its graph does not cross $y = 2$ between $x = -1$ and $x = 2$.

Region III

x	y
-2	4.75
-4	2.72

Region V

x	y
1	-2
1.5	-4
0.5	-1.8
0	$-\frac{5}{2}$

← y-intercept

Region IV is empty since the graph does not cross $y = 2$ for $x < -1$ and does not cross $x = -1$.

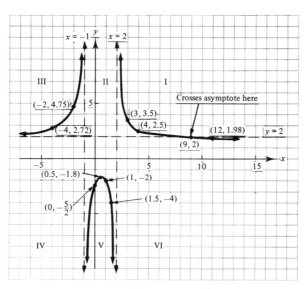

Figure 4.11   Graph of $y = \dfrac{2x^2 - 3x + 5}{x^2 - x - 2}$

12.   Let f be a function defined by

$$y = \frac{x^2 + 2x - 8}{x^2 - 4}.$$

Simplify first:

$$y = \frac{x^2 + 2x - 8}{x^2 - 4}$$

$$= \frac{(x - 2)(x + 4)}{(x - 2)(x + 2)}$$

$$= \frac{x + 4}{x + 2} \quad (x \neq 2)$$

Step 1.   The asymptotes are given by $x = -2$ and $y = 1$ (from Examples 3, 6, and 7).

Step 2.   The curve passes through $(0, 2)$ and $(-4, 0)$ and does not pass through the asymptotes. Label Regions I, II, III, and IV, as shown in Figure 4.12. The only points of discontinuity are $x = 2$ and $x = -2$.

Step 3.   Plot points.

For the intercepts, let

$$x = 0: \quad y = \frac{0 + 4}{0 + 2}$$

$$= 2$$

$$y = 0: \quad 0 = \frac{x + 4}{x + 2}$$

$$-4 = x$$

To see if the curve crosses the horizontal asymptote, solve

$$\frac{x + 4}{x + 2} = 1.$$

But this gives

$$x + 4 = x + 2$$

$$4 = 2$$

and this is impossible; therefore, the curve does not intersect the asymptote.

Region I

x	y
-1	3
1	$\frac{5}{3}$
2	$\frac{3}{2}$ $\;\leftarrow$ deleted point since $x \neq 2$
4	$\frac{4}{3}$

Region II is empty since f is a function and its graph does not cross the line $y = 1$.

Region III

x	y
-6	$\frac{1}{2}$
-5	$\frac{1}{3}$
-3	-1

Region IV is empty since f is a function and its graph does not cross the line $y = 1$.

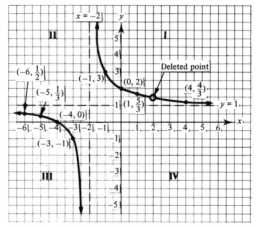

Figure 4.12 Graph of $y = \dfrac{x^2 + 2x - 8}{x^2 - 4}$

13. Let a function f be defined by

$$y = \frac{2x^2 - 5x + 1}{x - 3}.$$

Step 1. The vertical asymptote is given by $x = 3$ (by inspection). The slant asymptote is $y = 2x + 1$ (from Example 8). By considering the degree of the numerator and denominator, we find that there is no horizontal asymptote.

Step 2. The curve passes through $(0, -\frac{1}{3})$, $(2.28, 0)$, and $(0.2, 0)$. The only point of discontinuity is $x = 3$. Label Regions I, II, III, and IV, as shown in Figure 4.13.

The y-intercept is found by inspection. For the x-intercepts, solve

$$\frac{2x^2 - 5x + 1}{x - 3} = 0$$

$$2x^2 - 5x + 1 = 0$$

$$x = \frac{5 \pm \sqrt{17}}{4}$$

$$\approx 2.28, 0.2$$

Region II

x	y
3.5	16
4	13
4.5	12.67
5	13
6	14.33

Region IV

x	y	
0	−0.33	← y-intercept
0.2	0	
2.28	0	x-intercepts
1	1	
2	1	
1.5	1.33	

Since the graph is a function and does not cross the asymptotes, Regions I and III are empty.

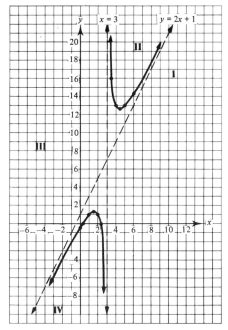

Figure 4.13 Graph of $f(x) = \dfrac{2x^2 - 5x + 1}{x - 3}$

PROBLEM SET 4.4

A Problems

Find horizontal, vertical, and slant asymptotes for the functions given in Problems 1–16.

1. $y = \dfrac{1}{x}$

2. $y = \dfrac{1}{x} + 2$

3. $y = -\dfrac{1}{x} + 1$

4. $y = \dfrac{4}{x^2}$ $y=0$

5. $y = \dfrac{2x^2 + 2}{x^2}$ $y=2$ hor. asy

6. $y = \dfrac{1}{x - 4}$ $y=0$

7. $y = \dfrac{-1}{x + 3}$

8. $y = \dfrac{4x}{x^2 - 2}$ $y=0$

9. $y = \dfrac{x^2}{x - 4}$

10. $y = \dfrac{x^3}{(x - 1)^2}$

11. $y = \dfrac{-x^2}{x - 1}$

12. $y = \dfrac{x^2 + x - 2}{x - 1}$

13. $y = \dfrac{x^2 + x - 6}{x + 3}$

14. $y = \dfrac{x^3 - 2x^2 + x - 2}{(x - 2)(x^2 + 1)}$

15. $y = \dfrac{(x - 3)(x^2 + 1)}{x^3 + 3x^2 + x + 3}$

16. $y = \dfrac{x^2 + 3x - 2}{x^2 + 2x - 8}$

B Problems

Graph the functions given in Problems 17–32. (These functions are the same as those in Problems 1–16 for which you've found the asymptotes.)

17. $y = \dfrac{1}{x}$

18. $y = \dfrac{1}{x} + 2$

19. $y = -\dfrac{1}{x} + 1$

20. $y = \dfrac{4}{x^2}$

21. $y = \dfrac{2x^2 + 2}{x^2}$

22. $y = \dfrac{1}{x - 4}$

23. $y = \dfrac{-1}{x + 3}$

24. $y = \dfrac{4x}{x^2 - 2}$

25. $y = \dfrac{x^2}{x - 4}$

26. $y = \dfrac{x^3}{(x - 1)^2}$

27. $y = \dfrac{-x^2}{x - 1}$

28. $y = \dfrac{x^2 + x - 2}{x - 1}$

29. $y = \dfrac{x^2 + x - 6}{x + 3}$

30. $y = \dfrac{x^3 - 2x^2 + x - 2}{(x - 2)(x^2 + 1)}$

31. $y = \dfrac{(x - 3)(x^2 + 1)}{x^3 + 3x^2 + x + 3}$

32. $y = \dfrac{x^2 + 3x - 2}{x^2 + 2x - 8}$

Mind Bogglers

33. Graph:

a. $y = \dfrac{x}{x^2 - 1}$ b. $y = \dfrac{x^2}{x^2 - 1}$

c. $y = \dfrac{x^3}{x^2 - 1}$ d. $y = \dfrac{x^4}{x^2 - 1}$

e. Can you see a pattern in parts a–d and describe it?

34. Problem 33d does not have a linear asymptote. Define what you might mean by a quadratic asymptote. What is the quadratic asymptote for Problem 33d? Illustrate with another example of a function with a quadratic asymptote.

4.5 PARTIAL FRACTIONS

In Section 4.1, rational expressions were added by finding common denominators. For example,

$$\frac{5}{x - 2} + \frac{3}{x + 1} = \frac{5(x + 1) + 3(x - 2)}{(x - 2)(x + 1)}$$

$$= \frac{8x - 1}{(x - 2)(x + 1)}.$$

However, in calculus it is sometimes necessary to break apart the expression

$$\frac{8x - 1}{(x - 2)(x + 1)}$$

into two fractions with denominators that are each linear. The technique for doing this is called the *method of partial fractions*. Thus, in general, if

$$\frac{ax + b}{(x - r)(x - s)}$$

is given, we wish to find A and B so that

$$\frac{ax + b}{(x - r)(x - s)} = \frac{A}{x - r} + \frac{B}{x - s}.$$

For the example above,

$$\frac{8x - 1}{(x - 2)(x + 1)} = \frac{A}{x - 2} + \frac{B}{x + 1}.$$

Pretend you don't know $A = 5$ and $B = 3$. That is, given *this* problem, find A and B.

To find A and B, add the fractions on the right:

$$\frac{A}{x - 2} + \frac{B}{x + 1} = \frac{Ax + A + Bx - 2B}{(x - 2)(x + 1)}$$

$$= \frac{(A + B)x + (A - 2B)}{(x - 2)(x + 1)}.$$

But this must equal

$$\frac{8x - 1}{(x - 2)(x + 1)}$$

Since

$$(A + B)x + (A - 2B)$$
$$= \quad 8 \quad x + \quad (-1)$$
$$\uparrow \qquad\qquad \uparrow$$

coefficients   constant
of x must   terms must
be the same   be the same

which means

$$\begin{cases} A + B = 8 & \leftarrow \text{coefficients of } x \\ A - 2B = -1 & \leftarrow \text{constant terms} \end{cases}$$

Solving simultaneously, we get

$$3B = 9$$
$$B = 3,$$

and

$$A = -1 + 2B$$
$$= -1 + 2(3)$$
$$= 5.$$

Thus,

$$\frac{8x - 1}{(x - 2)(x + 1)} = \frac{5}{x - 2} + \frac{3}{x + 1}.$$

Notice that the degree of the numerator is less than the degree of the denominator. If it is not, then use the Division Algorithm and focus only on the remainder.

The Division Algorithm calls for a remainder term where the degree of the numerator is less than the degree of the denominator.

EXAMPLES: Decompose the given rational functions into partial fractions.

1. $\dfrac{2x + 10}{(x + 3)(x + 4)} = \dfrac{A}{x + 3} + \dfrac{B}{x + 4}$

$$= \frac{Ax + 4A + Bx + 3B}{(x + 3)(x + 4)}$$

$$= \frac{(A + B)x + (4A + 3B)}{(x + 3)(x + 4)}$$

Thus,

$$\begin{cases} A + B = 2 \\ 4A + 3B = 10. \end{cases}$$

Solving simultaneously, we find $A = 4$, $B = -2$, and

$$\frac{2x + 10}{(x + 3)(x + 4)} = \frac{4}{x + 3} + \frac{-2}{x + 4}.$$

2. $\dfrac{1}{x^2 - x - 2} = \dfrac{1}{(x - 2)(x + 1)}$

$$= \frac{A}{x - 2} + \frac{B}{x + 1}$$

$$= \frac{Ax + A + Bx - 2B}{(x - 2)(x + 1)}$$

$$= \frac{(A + B)x + (A - 2B)}{(x - 2)(x + 1)}$$

Thus,

$$\begin{cases} A + B = 0 \\ A - 2B = 1, \end{cases}$$

and $A = \frac{1}{3}$, $B = -\frac{1}{3}$, so

$$\frac{1}{x^2 - x - 2} = \frac{1/3}{x - 2} + \frac{-1/3}{x + 1}.$$

$$\begin{cases} A + B = 2 \\ 4A + 3B = 10 \end{cases}$$

$$+ \begin{cases} -3A - 3B = -6 \\ 4A + 3B = 10 \end{cases}$$

$$\overline{\qquad\qquad A = 4}$$

If $A = 4$, then $B = -2$.

$$\begin{cases} A + B = 0 \\ A - 2B = 1 \end{cases}$$

$$3B = -1$$

$$B = -\frac{1}{3}, \quad A = \frac{1}{3}$$

If the degree of the denominator is larger than two, but can still be factored into linear factors, the situation is similar, except we have more terms.

Let $f(x) = P(x)/D(x)$, where $P(x)$ and $D(x)$ have no common factors and the degree of $P(x)$ is less than the degree of $D(x)$. Also, suppose $D(x) = (x - r)^n$. Then $f(x)$ can be decomposed into partial fractions,

$$\frac{A_1}{x - r} + \frac{A_2}{(x - r)^2} + \cdots + \frac{A_n}{(x - r)^n}.$$

EXAMPLES: Decompose the given rational functions into partial fractions.

3. $f(x) = \dfrac{x^2 - 6x + 3}{(x - 2)^3}$

$$\frac{x^2 - 6x + 3}{(x - 2)^3} = \frac{A}{x - 2} + \frac{B}{(x - 2)^2} + \frac{C}{(x - 2)^3}$$

$$= \frac{A(x - 2)^2 + B(x - 2) + C}{(x - 2)^3}$$

$$= \frac{Ax^2 - 4Ax + 4A + Bx - 2B + C}{(x - 2)^3}$$

$$= \frac{Ax^2 + (-4A + B)x + (4A - 2B + C)}{(x - 2)^3}$$

Thus,

$$\begin{cases} A = 1 \\ -4A + B = -6 \\ 4A - 2B + C = 3 \end{cases}$$

$$-4(1) + B = -6$$
$$B = -2$$
$$4(1) - 2(-2) + C = 3$$
$$8 + C = 3$$
$$C = -5$$

$$\begin{cases} A = 1 \\ -4A + B = -6 \\ 4A - 2B + C = 3. \end{cases}$$

Solve simultaneously to find $A = 1$, $B = -2$, and $C = -5$. So

$$\frac{x^2 - 6x + 3}{(x - 2)^3} = \frac{1}{x - 2} + \frac{-2}{(x - 2)^2} + \frac{-5}{(x - 2)^3}.$$

4. $f(x) = \dfrac{4x^2}{(x - 1)(x + 1)^2}$

$$\frac{4x^2}{(x - 1)(x + 1)^2} = \frac{A}{x - 1} + \frac{B}{x + 1} + \frac{C}{(x + 1)^2}$$

$$= \frac{A(x + 1)^2 + B(x + 1)(x - 1) + C(x - 1)}{(x - 1)(x + 1)^2}$$

$$= \frac{Ax^2 + 2Ax + A + Bx^2 - B + Cx - C}{(x - 1)(x + 1)^2}$$

$$= \frac{(A + B)x^2 + (2A + C)x + (A - B - C)}{(x - 1)(x + 1)^2}$$

Thus,

$$\begin{cases} A + B = 4 \\ 2A + C = 0 \\ A - B - C = 0. \end{cases}$$

Solving simultaneously, we find $A = 1$, $B = 3$, and $C = -2$. So

$$\frac{4x^2}{(x - 1)(x + 1)^2} = \frac{1}{x - 1} + \frac{3}{x + 1} + \frac{-2}{(x + 1)^2}.$$

Theoretically, any polynomial with real coefficients can be expressed as a product of real linear and quadratic factors. Therefore, the next step is to consider quadratic factors.

It is often difficult to find these factors. In this book, however, we will consider only those problems for which the denominators can be easily factored.

Let $f(x) = P(x)/D(x)$, where $P(x)$ and $D(x)$ have no common factors and where the degree of $P(x)$ is less than the degree of $D(x)$. Also, if $D(x) = (x^2 + sx + t)^m$, then $f(x)$ can be decomposed into partial fractions,

$$\frac{A_1x + B_1}{x^2 + sx + t} + \frac{A_2x + B_2}{(x^2 + sx + t)^2} + \cdots + \frac{A_mx + B_m}{(x^2 + sx + t)^m}.$$

EXAMPLES: Decompose the given rational functions into the sum of partial fractions.

5. $f(x) = \dfrac{2x^3 + 3x^2 + 3x + 2}{(x^2 + 1)^2}$

$$\frac{2x^3 + 3x^2 + 3x + 2}{(x^2 + 1)^2} = \frac{Ax + B}{x^2 + 1} + \frac{Cx + D}{(x^2 + 1)^2}$$

$$= \frac{(Ax + B)(x^2 + 1) + Cx + D}{(x^2 + 1)^2}$$

$$= \frac{Ax^3 + Bx^2 + Ax + B + Cx + D}{(x^2 + 1)^2}$$

$$= \frac{Ax^3 + Bx^2 + (A + C)x + (B + D)}{(x^2 + 1)^2}$$

Thus,

$$\begin{cases} A = 2 \\ B = 3 \\ A + C = 3 \\ B + D = 2. \end{cases}$$

Solve simultaneously to find $A = 2$, $B = 3$, $C = 1$, and $D = -1$. So

$$\frac{2x^3 + 3x^2 + 3x + 2}{(x^2 + 1)^2} = \frac{2x + 3}{x^2 + 1} + \frac{x - 1}{(x^2 + 1)^2}.$$

6. $f(x) = \dfrac{x^5 - x^4 + 8x^3 - 17x^2 + 14x - 11}{x^4 - 2x^3 + 3x^2 - 4x + 2}$

First notice that the degree of the numerator is larger than the degree of the denominator. Therefore, we divide:

$$x^4 - 2x^3 + 3x^2 - 4x + 2 \overline{\smash{)}x^5 - x^4 + 8x^3 - 17x^2 + 14x - 11}$$

$$\begin{array}{r} x + 1 \\ \hline x^5 - 2x^4 + 3x^3 - 4x^2 + 2x \\ \hline x^4 + 5x^3 - 13x^2 + 12x - 11 \\ x^4 - 2x^3 + 3x^2 - 4x + 2 \\ \hline 7x^3 - 16x^2 + 16x - 13 \end{array}$$

The result is

$$f(x) = x + 1 + \frac{7x^3 - 16x^2 + 16x - 13}{x^4 - 2x^3 + 3x^2 - 4x + 2}.$$

The next step is to factor the denominator using the Factor Theorem:

$$p \ (a_0 = 2): \pm 1, \pm 2$$
$$q \ (a_n = 1): \pm 1$$
$$\frac{p}{q}: \pm 1, \pm 2$$

$$\begin{array}{r|rrrrr} & 1 & -2 & 3 & -4 & 2 \\ \hline 1 & 1 & -1 & 2 & -2 & 0 \\ \hline 1 & 1 & 0 & 2 & 0 \end{array}$$

Thus, $x^4 - 2x^3 + 3x^2 - 4x + 2 = (x - 1)^2(x^2 + 2)$. Consider

$$\frac{7x^3 - 16x^2 + 16x - 13}{(x - 1)^2(x^2 + 2)}$$

$$= \frac{Ax + B}{x^2 + 2} + \frac{C}{x - 1} + \frac{D}{(x - 1)^2}$$

$$= \frac{(Ax + B)(x - 1)^2 + C(x - 1)(x^2 + 2) + D(x^2 + 2)}{(x - 1)^2(x^2 + 2)}$$

$$= \frac{(Ax + B)(x^2 - 2x + 1) + C(x^3 - x^2 + 2x - 2) + D(x^2 + 2)}{(x - 1)^2(x^2 + 2)}$$

$$= \frac{Ax^3 - 2Ax^2 + Ax + Bx^2 - 2Bx + B + Cx^3 - Cx^2 + 2Cx - 2C + Dx^2 + 2D}{(x - 1)^2(x^2 + 2)}$$

$$= \frac{(A + C)x^3 + (-2A + B - C + D)x^2 + (A - 2B + 2C)x + (B - 2C + 2D)}{(x - 1)^2(x^2 + 2)}$$

Thus,

$$\begin{cases} A + C = 7 \\ -2A + B - C + D = -16 \\ A - 2B + 2C = 16 \\ B - 2C + 2D = -13. \end{cases}$$

Solving simultaneously, we get $A = 4$, $B = -3$, $C = 3$, and $D = -2$. Therefore,

$$f(x) = x + 1 + \frac{4x - 3}{x^2 + 2} + \frac{3}{x - 1} + \frac{-2}{(x - 1)^2}.$$

PROBLEM SET 4.5

A Problems

Decompose the rational functions given in Problems 1–14 into the sum of partial fractions.

1. $\dfrac{x^2 + 2x + 5}{x^3}$

2. $\dfrac{3x^2 - 2x + 1}{x^3}$

3. $\dfrac{2x - 1}{(x - 2)^2}$

4. $\dfrac{4x - 22}{(x - 5)^2}$

5. $\dfrac{2x^2 + 7x + 2}{(x + 1)^3}$

6. $\dfrac{7x - 3x^2}{(x - 2)^3}$

7. $\dfrac{7x - 10}{(x - 2)(x - 1)}$

8. $\dfrac{11x - 1}{(x - 1)(x + 1)}$

9. $\dfrac{5x^2 - 2x + 2}{x(x - 1)^2}$

10. $\dfrac{x^2 + 5x + 1}{x(x + 1)^2}$

11. $\dfrac{x^3}{(x - 1)^2}$

12. $\dfrac{x^3}{(x + 1)^2}$

13. $\dfrac{1}{(x + 2)(x + 3)}$

14. $\dfrac{1}{(x - 4)(x + 5)}$

B Problems

Decompose the rational functions given in Problems 15–30 into the sum of partial fractions.

15. $\dfrac{x}{x^2 + 4x - 5}$

16. $\dfrac{x}{x^2 - 2x - 3}$

17. $\dfrac{4(x - 1)}{x^2 - 4}$

18. $\dfrac{10x^2 - 11x - 6}{x^3 - x^2 - 2x}$

19. $\dfrac{5x^2 - 5x - 4}{x^3 - x}$

20. $\dfrac{x^2}{(x + 1)(x^2 + 1)}$

21. $\dfrac{5x^2 - 6x + 7}{(x - 1)(x^2 + 1)}$

22. $\dfrac{34 - 5x}{48 - 14x + x^2}$

23. $\dfrac{x - 7}{20 - 9x + x^2}$

24. $\dfrac{2x^3 - 3x^2 + 6x - 1}{1 - x^4}$

25. $\dfrac{3x^3 + 5x^2 - 8x - 4}{(x + 1)^2(x^2 + 5x + 1)}$

26. $\dfrac{10x^3 - 44x^2 + 84x - 77}{(x - 2)^2(x^2 - 2x + 5)}$

Hint for Problems 27–32: use the results of Chapter 3 to factor the denominators.

27. $\dfrac{8x^3 - 12x^2 + 8x - 6}{x^4 - 2x^3 + 2x^2 - 2x + 1}$

28. $\dfrac{3x^3 - 8x^2 - 5x - 3}{x^4 - 3x^3 + x^2 + 4}$

29. $\dfrac{2x^5 + 5x^4 + 13x^3 + 16x^2 + 13x + 5}{x^4 + 2x^3 + 2x^2 + 2x + 1}$

30. $\dfrac{2x^5 + 3x^4 - 8x^3 + x^2 + 23x - 12}{x^4 - 3x^3 - x^2 + 8x - 4}$

Mind Bogglers

Decompose the rational functions given in Problems 31 and 32 into the sum of partial fractions.

31. $\dfrac{6x^6 + 21x^5 + 29x^4 + 39x^3 + 27x^2 + 12x - 1}{x^7 + 3x^6 + 5x^5 + 7x^3 + 5x^2 + 3x + 1}$

Hint for Problem 32: $x^2 + 2x + 5$ is a factor of the denominator.

32. $\dfrac{2x^6 - 11x^5 + 37x^4 - 94x^3 + 212x^2 - 471x + 661}{x^7 - 5x^6 + 5x^5 - 25x^4 + 115x^3 - 63x^2 + 135x - 675}$

4.6 SUMMARY AND REVIEW

TERMS

Asymptote (4.2)
Continuity (4.3)
Equal functions (4.1)
Horizontal asymptote (4.4)
Limit (4.2)

Partial fractions (4.5)
Rational function (4.1)
Slant asymptote (4.4)
Vertical asymptote (4.4)

IMPORTANT IDEAS

(4.1)   Properties of rational expressions: fundamental property, equality, addition, subtraction, multiplication, and division.

(4.2) $\lim\limits_{|x|\to\infty} (k/x) = 0$ and $\lim\limits_{|x|\to\infty} (k/x^2) = 0$ for any constant k

(4.2)   Limit of a polynomial function: if f is any polynomial function, then

$$\lim_{x\to c} f(x) = f(c) \qquad \text{for any real number } c.$$

(4.3)   Continuity: a function f is continuous at $x = c$ if
1. $f(c)$ exists; and
2. $\lim\limits_{x\to c} f(x) = f(c)$.

(4.3)   Continuity Theorem: let f and g be continuous functions at $x = c$. Then the following functions are also continuous at $x = c$:

1. $f + g$ 2. $f - g$ 3. fg

4. $\dfrac{f}{g}$ $(g \neq 0)$ 5. $f \circ g$ 6. $g \circ f$

(4.4) Asymptotes: let $f(x) = P(x)/D(x)$, where $P(x)$ and $D(x)$ have no common factors. Then

1. the *vertical line* given by $x = r$ is as asymptote if $D(r) = 0$;
2. the *horizontal line* given by $y = k$ is an asymptote if $\lim\limits_{|x|\to\infty} f(x) = k$;

3. the *slant line* given by $y = mx + b$ is an asymptote if the degree of $P(x)$ is one more than the degree of $D(x)$ and

$$\frac{P(x)}{D(x)} = mx + b + \frac{R(x)}{D(x)}.$$

(4.4) Graphing rational functions
(4.5) Decomposing into partial fractions

REVIEW PROBLEMS

1. a. (4.1) Simplify

$$\frac{x + 2}{x} + \frac{3 - x}{x^2} + \frac{x - 1}{x^3}.$$

b. (4.5) Decompose

$$\frac{x^3 + x^2 + 4x - 1}{x^3}$$

into the sum of partial fractions.
c. (4.1) What is the domain for the expression in part b?
d. (4.1) Is

$$\frac{x^5 + x^4 + 5x^3 + 4x - 1}{x^5 + x^3}$$

equal to the rational expression given in part b?

2. (4.1) Factor the given expression by first factoring out all fractions.
a. $x^{-1}(x + 2) + x^{-2}(3 - x) + x^{-3}(x - 1)$
b. $x^{-1} + y^{-1} + 1$
c. $x(x^2 + 5)^{-4}(2x)(-3) + (x^2 + 5)^{-3}$

3. (4.1) Simplify the given expressions.

a. $\dfrac{2x + y}{(x + y)^2} - \dfrac{x^2 - y^2}{(x + y)^3}$ b. $\dfrac{x^3 - y^3}{x^2 - y^2} \div \dfrac{x^2 + xy + y^2}{x + y}$

c. $\dfrac{(x/y) - (x^2/y^2)}{x - (1/y) + (x/y^2)}$

4. (4.2) Evaluate the given limits.

a. $\lim\limits_{x \to 2} (x^2 - 3x + 4)$ b. $\lim\limits_{x \to 2} \dfrac{x^3 - 5x^2 + 10x - 8}{x - 2}$

c. $\lim\limits_{x \to 0} \dfrac{x + 2}{x^2 - 4}$ d. $\lim\limits_{|x| \to \infty} \dfrac{3x^2 - 4x - 7}{2x^2 - 9x + 4}$

e. $\lim\limits_{|x| \to \infty} \dfrac{(x - 2)(x + 1)(2x - 1)}{x^3}$

5. (4.5) Decompose the given rational functions into sums of partial fractions.

a. $\dfrac{x + 2}{x^2 - 2x + 1}$ b. $\dfrac{5x^2 - 19x + 17}{(x - 1)(x - 2)^2}$

c. $\dfrac{5x^4 - 2x^3 + 22x^2 - 9x + 16}{(x - 1)(x^2 + 3)^2}$

(4.3, 4.4) *Graph the rational expressions given in Problems 6–10 by first finding all asymptotes and points of discontinuity, if any.*

6. $f(x) = \dfrac{1}{x - 2} + 2$ 7. $f(x) = \dfrac{3x^2 + 2x - 5}{x - 1}$

8. $f(x) = \dfrac{2x^2 - 3x - 1}{x^2 - x - 2}$ 9. $f(x) = \dfrac{x^3 - x - 6}{x - 2}$

10. $f(x) = \dfrac{2x^2 - 3x - 1}{x - 2}$

Exponential and Logarithmic Functions

5

5.1 EXPONENTIAL FUNCTIONS

ALGEBRAIC AND
TRANSCENDENTAL
FUNCTIONS
By algebraic operations we
mean addition, subtraction,
multiplication, division, or
extraction of roots.

EXPONENTIAL
FUNCTION

Recall that x is called the
exponent and b is called the
base.

The linear, quadratic, polynomial, and rational functions considered in the first part of this book are all called *algebraic functions*. In general, an algebraic function is a function that can be expressed in terms of algebraic operations alone. If a function is not algebraic, then it is called a *transcendental function*. In this chapter, two transcendental functions, *exponential* and *logarithmic* functions, are considered. In the next chapter, several other transcendental functions will be considered.

The function $f(x) = b^x$, where b is a positive constant, is called an *exponential function*. But in Chapter 3, we defined b^n only for integers n. It is possible to extend the domain to include all real numbers, as shown below.

Represent $\sqrt{2}$ using exponential notation by finding an x so that

$$\sqrt{2} = 2^x.$$

From the definition of square root,

Recall, $\sqrt{a} = b$ only if $a = b^2$.

$$2 = (2^x)^2,$$

so if the properties of exponents are to hold,

$$2^1 = 2^{2x}.$$

This is true only when

$$1 = 2x$$

Notice that $2^{1/2} \cdot 2^{1/2} = 2^1 = 2$ and $\sqrt{2}\sqrt{2} = 2$. This indicates $2^{1/2} = \sqrt{2}$.

$$\frac{1}{2} = x.$$

Hence, the use of a rational number as an exponent will preserve the previous laws of exponents and give us an alternative choice of notation for roots if we adopt the following definition:

$\sqrt[n]{b}$ is called the *principal nth root of b.*

DEFINITION: For $b > 0$ and m, n positive integers,

$$b^{1/n} = \sqrt[n]{b} \quad \text{and} \quad b^{m/n} = (b^{1/n})^m = \sqrt[n]{b^m} = (\sqrt[n]{b})^m$$

EXAMPLES

1. $16^{1/2} = \sqrt{16} = 4$

2. $-16^{1/2} = -\sqrt{16} = -4$

3. $(-16)^{1/2} = \sqrt{-16}$ is not defined (b must be greater than zero, by definition)

4. $(343)^{2/3} = (7^3)^{2/3} = 7^2 = 49$

5. $(100)^{3/2} = (10^2)^{3/2} = 10^3 = 1000$

Recall, $(x^p)^q = x^{pq}$, so

$(7^3)^{2/3} = 7^{3(2/3)} = 7^2$.

6. $25^{-3/2} = \dfrac{1}{25^{3/2}} = \dfrac{1}{(5^2)^{3/2}} = \dfrac{1}{5^3} = \dfrac{1}{125}$

First use the definition

$b^{-p} = \dfrac{1}{b^p}$.

7. $x(x^{2/3} + x^{1/2}) = x^1 x^{2/3} + x^1 x^{1/2}$
$= x^{1 + \frac{2}{3}} + x^{1 + \frac{1}{2}}$
$= x^{5/3} + x^{3/2}$

8. $(x^{1/2} + y^{1/2})(x^{1/2} - y^{1/2}) = x^{1/2} x^{1/2} - x^{1/2} y^{1/2} + x^{1/2} y^{1/2} - y^{1/2} y^{1/2}$
$= x - y$

Recall, $x^p x^q = x^{p+q}$, so

$x^1 x^{2/3} = x^{1 + (2/3)}$.

The next step in enlarging the domain for x in $f(x) = b^x$ requires the following property:

For any real number x, positive real b, $b \neq 1$:

1. b^x is a unique real number; and
2. if $h < x < k$ for rational numbers h and k, then

$$b^h < b^x < b^k.$$

We want to give meaning to expressions such as

$2^{\sqrt{3}}$.

Consider the graph of the function $f(x) = 2^x$ by plotting the points shown in the table, as in Figure 5.1.

Notice that f is continuous for all real numbers.

x	$y = f(x)$
-3	$2^{-3} = \frac{1}{8}$
-2	$2^{-2} = \frac{1}{4}$
-1	$2^{-1} = \frac{1}{2}$
0	$2^0 = 1$
1	$2^1 = 2$
2	$2^2 = 4$
3	$2^3 = 8$

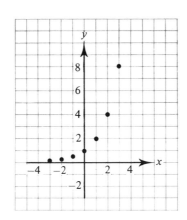

Figure 5.1   Selected points that satisfy $f(x) = 2^x$

If these points are connected with a smooth curve, as shown in Figure 5.2, you can see that $2^{\sqrt{3}}$ is defined and is between 2^1 and 2^2.

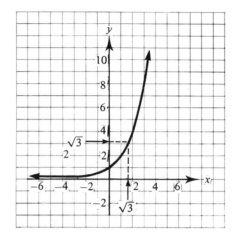

Figure 5.2   Graph of $f(x) = 2^x$

$\lim\limits_{|x|\to\infty} f(x) = 0$, so $y = 0$ is the equation of a horizontal asymptote.

The number $2^{\sqrt{3}}$ can be approximated to any desired degree of accuracy. Since

$$1 < \sqrt{3} < 2$$
$$1.7 < \sqrt{3} < 1.8$$
$$1.73 < \sqrt{3} < 1.74$$
$$1.732 < \sqrt{3} < 1.733$$
$$\vdots$$

we have

$$2^1 < 2^{\sqrt{3}} < 2^2$$
$$2^{1.7} < 2^{\sqrt{3}} < 2^{1.8}$$
$$2^{1.73} < 2^{\sqrt{3}} < 2^{1.74}$$
$$2^{1.732} < 2^{\sqrt{3}} < 2^{1.733}$$
$$\vdots$$

This can be visualized by looking at portions of Figure 5.2, as shown in Figure 5.3.

Although the function $f(x) = 2^x$ is a special case of the general exponential function $f(x) = b^x$, some observations can be made about the graph of $f(x) = b^x$.

Historical Note

Nicole Oredme (1323–1382), a university teacher and churchman, tried to find a notation for $x^{\sqrt{2}}$. This was the first hint in the history of mathematics of a higher transcendental function. Unfortunately, Oredme's ideas were ahead of his time, and he did not have adequate terminology or notation to develop his ideas on irrational powers.

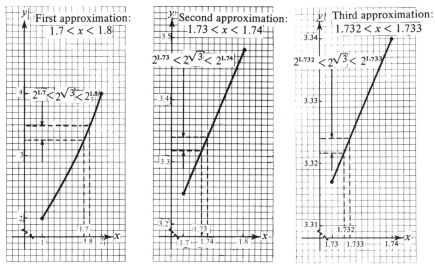

Figure 5.3   Successive approximations for locating $2^{\sqrt{3}}$

1. It passes through the point $(0, 1)$.
2. $f(x) > 0$ for all x.
3. If $b > 1$, f is an increasing function, and since

$$\lim_{x \to -\infty} b^x = 0,$$

then $y = 0$ is the equation of a horizontal asymptote.
4. If $b < 1$, f is a decreasing function and since

$$\lim_{x \to \infty} b^x = 0,$$

then $y = 0$ is the equation of a horizontal asymptote.

For any b, $b^0 = 1$.

A function f is *increasing* on an interval I if

$$f(x) < f(y)$$

and *decreasing* on I if

$$f(x) > f(y)$$

for all real numbers x and y of I with $x < y$.

Even though it is beyond the scope of this book to prove that the usual laws of exponents hold for all real exponents, we will accept them as axioms and state them here for easy reference.

Let a and b be any nonzero real numbers and let m and n be any real numbers, except that the form $0/0$ and division by zero are excluded.

First law: $b^m \cdot b^n = b^{m+n}$

Second law: $\dfrac{b^m}{b^n} = b^{m-n}$

Third law: $(b^n)^m = b^{mn}$

LAWS OF EXPONENTS

These are the same as those stated in Section 3.1, except that we are now allowing the exponents to be any real numbers.

Fourth law: $(ab)^m = a^m b^m$

Fifth law: $\left(\dfrac{a}{b}\right)^m = \dfrac{a^m}{b^m}$

Let's sketch several exponential functions and observe their behavior.

EXAMPLES: Sketch the given function.

9. $f(x) = \left(\tfrac{1}{2}\right)^x$

x	$f(x) = \left(\tfrac{1}{2}\right)^x$
-3	$(2^{-1})^{-3} = 8$
-2	$(2^{-1})^{-2} = 4$
-1	$(2^{-1})^{-1} = 2$
0	$(2^{-1})^{0} = 1$
1	$(1/2)^{1} = 1/2$
2	$(1/2)^{2} = 1/4$
3	$(1/2)^{3} = 1/8$

Notice

$y = \left(\tfrac{1}{2}\right)^x$
$\quad = (2^{-1})^x$

or

$y = 2^{-x}.$

We use whichever form is most convenient.

f is a continuous, decreasing function with a horizontal asymptote at $y = 0$. The points are plotted in Figure 5.4 and connected by a smooth curve.

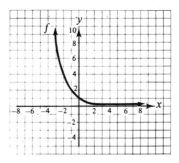

Figure 5.4   Graph of $f(x) = \left(\tfrac{1}{2}\right)^x$

10. $f(x) = 10^x$

x	$f(x) = 10^x$
-3	$10^{-3} = 1/1000$
-2	$10^{-2} = 1/100$
-1	$10^{-1} = 1/10$
0	$10^{0} = 1$
1	$10^{1} = 10$
2	$10^{2} = 100$
3	$10^{3} = 1000$

Notice that it is often necessary to alter the scale for exponential functions.

f is a continuous, increasing function with a horizontal asymptote at $y = 0$.

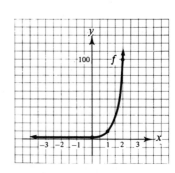

Figure 5.5   Graph of $f(x) = 10^x$

11. $f(x) = 2^{|x|}$

x	$f(x) = 2^{|x|}$
-3	$2^3 = 8$
-2	$2^2 = 4$
-1	$2^1 = 2$
0	$2^0 = 1$
1	$2^1 = 2$
2	$2^2 = 4$
3	$2^3 = 8$

f is continuous for all real numbers and has no asymptotes.

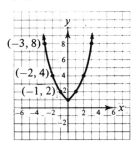

Figure 5.6   Graph of $= 2^{|x|}$

Notice that

$$y = 2^x$$

is $y = 2^x$ for $x \geq 0$ and $y = 2^{-x}$ for $x < 0$.

12. $f(x) = 2^{-x^2}$

x	$f(x) = 2^{-x^2}$
-3	$2^{-9} = 1/512$
-2	$2^{-4} = 1/16$
-1	$2^{-1} = 1/2$
0	$2^0 = 1$
1	$2^{-1} = 1/2$
2	$2^{-4} = 1/16$
3	$2^{-9} = 1/512$

f is continuous for all real numbers. Also,

$$\lim_{|x| \to \infty} 2^{-x^2} = 0$$

so $y = 0$ is the equation of a horizontal asymptote.

A calculator could be used to estimate additional points.

x	$y = 2^{-x^2}$
-1.5	0.21
-0.5	0.84
0.5	0.84
1.5	0.21

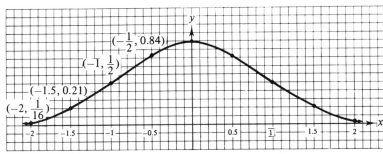

Figure 5.7   Graph of $f(x) = 2^{-x^2}$

PROBLEM SET 5.1

A Problems

Simplify each expression in Problems 1–19 to find the value of the given letter.

1. $A = 25^{1/2}$
2. $B = -25^{1/2}$
3. $C = -27^{1/3}$
4. $D = 216^{1/3}$
5. $E = -216^{1/3}$
6. $F = 729^{1/6}$
7. $G = 64^{2/3}$
8. $H = 8^{2/3}$
9. $I = -64^{2/3}$
10. $J = (-64)^{3/2}$
11. $L = 64^{3/2}$
12. $M = -64^{3/2}$

13. $N = 7^{1/3} \cdot 7^{2/3}$
14. $R = 8^{4/3} \cdot 8^{-1/3}$
15. $S = \dfrac{10^{4/3}}{10^{1/3}}$

16. $T = (2^{1/3} \cdot 3^{1/2})^6$
17. $U = (2^6 \cdot 3^{12})^{1/6}$
18. $Y = 1000^{-2/3}$
19. $Z = 0.001^{-2/3}$ $(10^{-3})^{-2/3} = 10^2 = 100$

20. Replace the value within each set of parentheses with the value of the corresponding variable from Problems 1–19. The letter "oh" has been filled in for you.

(512)(O)(16)   (5)   (8)(−16)(108)(4)(−512)(10)
(6)(O)(18)(−5)(512)(−6)(6)   (108)(4)(−6)
(512)(−16)(3)(−6)   (O)(3)   (108)(4)(−6)
(5)(10)(108)(8)(O)(7)(O)(−512)(−6)(8)
(−5)(.01)   (10)(4)(O)(8)(108)(−16)(7)(16)
(108)(4)(−6)   (512)(5)(−5)(O)(8)(10).

Sketch the graph of each function given in Problems 21–33. (Hint for Problems 27–33: shift the coordinate axes.)

21. $y = 3^x$
22. $y = (\frac{1}{3})^x$
23. $y = 3^{-x}$
24. $y = 4^x$
25. $y = 4^{-x}$
26. $y = (\frac{1}{4})^x$
27. $y = 2^{x-1}$
28. $y = 2^{x-2}$
29. $y - 2 = 2^x$
30. $y - 5 = 2^{x+4}$
31. $y - 3 = 2^x$
32. $y = 2^{x-1} + 3$
33. $y - 4 = 2^{x-3}$

B Problems

Simplify the expressions in Problems 34–40.

34. $x^{2/3}(x^{-2/3} + x^{1/3})$
35. $(x^{2/3}y^{-1/3})^3$
36. $(x^{2^2+1}x^5)^{1/10}$
37. $(x^{1/2} - y^{1/2})^2$
38. $(x^{1/2} + y^{1/2})^2$
39. $(x^{1/3} + y^{1/3})(x^{2/3} - x^{1/3}y^{1/3} + y^{2/3})$
40. $(x^{1/3} - y^{1/3})(x^{2/3} + x^{1/3}y^{1/3} + y^{2/3})$

Sketch the graph of each function given in Problems 41–46.

41. $y = 3^{|x|}$
42. $y = 2^{-|x|}$
43. $y = 2^{x^2}$
44. $y = 3^{x^2}$
45. $y = 3^{-x^2}$
46. $y = 10^{-x^2}$

Historical Quote

In Section 5.3, logarithms will be defined and discussed. Logarithms were invented by John Napier (see the historical note on page 235). Napier's invention was a tremendous labor-saving device and was enthusiastically adopted throughout Europe. The quotation given here is attributed to Pierre-Simon Laplace (1749–1827), who is famous for his work in celestial mechanics, probability, differential equations, and geodesy. Laplace considered beginners in mathematics as his stepchildren, and several instances occurred where he withheld publication of one of his own discoveries to allow a beginner the opportunity to publish first.

47. In the definition of the exponential function $f(x) = b^x$, we require that b is a positive constant.
    a. What happens if $b = 1$? Draw the graph of $f(x) = b^x$ where $b = 1$. Is this an algebraic or a transcendental function?
    b. What happens if $b = 0$? Draw the graph of $f(x) = b^x$ where $b = 0$. Is this an algebraic or a transcendental function?

48. In the definition of the exponential function $f(x) = b^x$, we require that b is a positive constant. What happens if $b < 0$, say $b = -2$? For what values of x is f defined? Describe the graph of $f(x)$ in this case.

49. Use graphical methods to estimate the value of $2^{\sqrt{2}}$.

50. Use graphical methods to estimate the value of $3^{\sqrt{2}}$.

51. Use graphical methods to estimate the value of $10^{\sqrt{2}}$.

52. Graph $y = 10^x$, $-1 \le x \le 1$ and approximate x if:
    a. $10^x = 5$ b. $10^x = 0.5$ c. $10^x = 2$ d. $10^x = 8.4$
    e. $10^x = -1$

53. *Calculator Problem.* Graph $y = \sqrt{2}^x$. Estimate the value of $\sqrt{2}^{\sqrt{3}}$.

54. *Calculator Problem.* Graph $y = \sqrt{3}^x$. Estimate the value of $\sqrt{3}^{\sqrt{2}}$.

55. Radioactive argon-39 has a half-life of four minutes. This means that the time required for half of the argon-39 to decompose is four minutes. If we start with 100 milligrams (mg) of argon-39, the amount (A) left after t minutes is given by

    $$A = 100\left(\frac{1}{2}\right)^{t/4}.$$

    Graph this function.

56. Carbon-14, used for archaeological dating, has a half-life of 5700 years. This means that the time required for half of the carbon-14 to decompose is 5700 years. If we start with 100 milligrams (mg) of carbon-14, the amount (A) left after t years is given by

    $$A = 100\left(\frac{1}{2}\right)^{t/5700}.$$

    Graph this function.

57. *Calculator Problem.* The population of Fort Lauderdale, Florida, in 1960 was about 84,000 and in 1970 was about 140,000. If we assume a constant rate of growth for the period 1960–1970, this would be a growth rate of about 5.2%. The formula expressing the population of Fort Lauderdale for t years after 1960 is given by

    $$P = 84(1 + 0.052)^t$$
    $$= 84(1.052)^t,$$

    where P is the population in thousands. Graph this function for 1960–1980.

58. *Calculator Problem.* In 1974, the world population was about 3.89 billion. If we assume a growth rate of 2%, the formula expressing the world population for t years after 1974 is given by

    $$P = 3.89(1 + 0.02)^t$$
    $$= 3.89(1.02)^t,$$

    where P is the population in billions. Graph this function for 1974–1984.

Mind Boggler

59. *Calculator Problem.* Use graphical methods to determine which is larger.
   a. $(\sqrt{3})^{\pi}$ or $\pi^{\sqrt{3}}$ b. $(\sqrt{5})^{\pi}$ or $\pi^{\sqrt{5}}$ c. $(\sqrt{6})^{\pi}$ or $\pi^{\sqrt{6}}$
   d. Consider $(\sqrt{N})^{\pi} = \pi^{\sqrt{N}}$, where N is a positive real number. From parts a–c, notice that $(\sqrt{N})^{\pi}$ is larger for some values of N and $\pi^{\sqrt{N}}$ is larger for others. For $N = \pi^2$,

$$(\sqrt{N})^{\pi} = \pi^{\sqrt{N}}$$

   is obviously true. Using a graphical method, find another value (approximately) for which the given statement is true.

5.2 EXPONENTIAL EQUATIONS

Historical Note

The carbon-14 method for dating artifacts was introduced in 1946 by the Nobel-Prize-winning chemist Willard Libby. All living tissues absorb carbon-14 through carbon dioxide, and these tissues maintain it at a constant level until they die. This method of dating is reasonably accurate up to about 40,000 or 50,000 years.

In Problem 56 of the previous section, we mentioned the archaeological method for determining the age of various artifacts. The decomposition of carbon-14 is used because most artifacts contain carbon and also because carbon decays very slowly. The half-life of carbon-14 is 5700 years, so if there are A_0 milligrams of carbon-14 present at a particular time, then the amount, A, present after t years is given by

$$A = A_0\left(\frac{1}{2}\right)^{t/5700}$$

This means that if an artifact is found with 25% of its original amount of carbon-14 present, the age of the object can be determined by solving for t:

$$\frac{A}{A_0} = \left(\frac{1}{2}\right)^{t/5700}$$

$$\frac{1}{4} = \left(\frac{1}{2}\right)^{t/5700} \qquad \left(25\% = \frac{1}{4}\right)$$

$$\left(\frac{1}{2}\right)^2 = \left(\frac{1}{2}\right)^{t/5700}$$

$$2 = \frac{t}{5700}$$

$$11{,}400 = t.$$

The estimated age of the artifact is 11,400 years.

The equation just solved is different from the previous equations solved because the unknown appeared as an exponent. Such an equation is called an *exponential equation*. The method of solution illustrated by this example is typical of many types of exponential equations in which the bases of the numbers involved are the same. It seems reasonable that if equal bases are raised to some power, and the results are equal, then the exponents must also be equal. This is summarized by the following result:

EXPONENTIAL EQUATION

For positive real b, $b \neq 1$,

$$\text{if} \quad b^x = b^y, \quad \text{then} \quad x = y.$$

EXPONENTIAL PROPERTY OF EQUALITY

EXAMPLES: Solve for x.

1. $7^x = 343$
   $7^x = 7^3$
   $x = 3$

2. $25^x = 125$
   $(5^2)^x = 5^3$
   $5^{2x} = 5^3$
   $2x = 3$
   $x = \frac{3}{2}$

3. $36^x = \frac{1}{6}$
   $(6^2)^x = 6^{-1}$
   $6^{2x} = 6^{-1}$
   $2x = -1$
   $x = \frac{-1}{2}$

Not all exponential equations are as easy to solve as those in Examples 1–3. Suppose Blondie saves 20¢ per day; in one year, she saves $73. She now puts the money into the bank at 6% interest compounded annually. One year later she has in the bank

$$73 \quad + 0.06(73)$$
$$\text{principal} + \text{interest}$$
$$= P \qquad + \quad rP$$
$$= P(1 + r).$$

Principal is the amount invested; denote this by P.
Interest is the rate denoted by r.

Two years later, her bank balance is

$$\text{amount from first year} + \text{interest for second year}$$
$$= \quad P(1 + r) \qquad\qquad + \qquad rP(1 + r)$$
$$= \quad P(1 + r)(1 + r)$$
$$= \quad P(1 + r)^2.$$

Common factoring

If we continue this process, the amount P invested for t years and compounded annually at the rate r is given by

$$A = P(1 + r)^t.$$

How long does it take for Blondie's $73 to double after she deposits it in the bank?

$$146 = 73(1 + 0.06)^n$$
$$2 = (1.06)^n.$$
The amount doubles⟶

Notice that we don't have to know the amount originally invested in order to answer this question, since $A/P = 2$ regardless of the original value of P.

This is an exponential equation, but it cannot be solved by the technique used in Examples 1–3. What can we do? By calculator, we can find

$$(1.06)^2 \approx 1.12$$
$$(1.06)^4 \approx 1.26$$
$$(1.06)^8 \approx 1.59$$
$$(1.06)^{12} \approx 2.01.$$

The symbol $\approx$ is used because the answer is rounded. We could write

$$(1.06)^2 = 1.1236$$
$$(1.06)^4 = 1.26247696.$$

But even with a ten-place calculator, we cannot use equality for $(1.06)^8$ since we only have an approximate answer,

$$(1.06)^8 \approx 1.593848075.$$

So it would take 12 years for the money to double. But a calculator solution using this method is not very satisfactory. Even before the development of calculators, interest tables were developed for various interest rates, as listed in Table 5.1. From the table, we see that Blondie's deposit of $73 doubles in 12 years and triples in 19 years. (Notice that at 8% it triples in 15 years.)

The disadvantages of working this problem by using tables are: (1) a different table is needed for each different interest rate, and (2) the tables are not always available. It would seem desirable to approach the problem by solving the equation for the exponent directly. We'll do this in the next section, but first let's consider the question of compound interest. If Blondie's money is deposited in a 6% account that compounds semiannually, then the period is one-half year and each period pays 3%. The advantage to Blondie is that she begins earning interest on the interest after one-half year instead of after one year. In general, if the money is compounded n times per year, then the rate per period is r/n and the interest formula becomes

$$A = P\left(1 + \frac{r}{n}\right)^{nt}.$$

What difference does it make to Blondie how often the interest is compounded? The amounts are shown in the following table below.

Length of time (years)	Compounded annually	Compounded semiannually
0	$73.00	$73.00
1	77.38	77.45
5	97.69	98.11
12	146.89	148.39
50	1344.67	1402.96

The differences in these amounts seem insignificant, but if the amount invested were $73,000, the difference between them would be $70 for one year with accounts paying the *same* interest rate!

Many savings institutions compound interest quarterly ($n = 4$), monthly ($n = 12$), or daily ($n = 360$). (Note: most savings institutions assume a 360-day year.) Amounts after one year are compared below for a deposit of $73 at 6%.

Method	Formula	Amount
Annually, $n = 1$	$73(1 + 0.06)^1$	$77.38
Semiannually, $n = 2$	$73\left(1 + \dfrac{0.06}{2}\right)^2$	77.45
Quarterly, $n = 4$	$73\left(1 + \dfrac{0.06}{4}\right)^4$	77.48
Monthly, $n = 12$	$73\left(1 + \dfrac{0.06}{12}\right)^{12}$	77.50
Daily, $n = 360$	$73\left(1 + \dfrac{0.06}{360}\right)^{360}$	77.51

TABLE 5.1 Amount at Compound Interest $(1 + r)^n$

Periods	Rate r				
n	.06 (6%)	.065 (6½%)	.07 (7%)	.075 (7½%)	.08 (8%)
1	1.0600 0000	1.0650 0000	1.0700 0000	1.0750 0000	1.0800 0000
2	1.1236 0000	1.1342 2500	1.1449 0000	1.1556 2500	1.1664 0000
3	1.1910 1600	1.2079 4963	1.2250 4300	1.2422 9688	1.2597 1200
4	1.2624 7696	1.2864 6635	1.3107 9601	1.3354 6914	1.3604 8896
5	1.3382 2558	1.3700 8666	1.4025 5173	1.4356 2933	1.4693 2808
6	1.4185 1911	1.4591 4230	1.5007 3035	1.5433 0153	1.5868 7432
7	1.5036 3026	1.5539 8655	1.6057 8148	1.6590 4914	1.7138 2427
8	1.5938 4807	1.6549 9567	1.7181 8618	1.7834 7783	1.8509 3021
9	1.6894 7896	1.7625 7039	1.8384 5921	1.9172 3866	1.9990 0463
10	1.7908 4770	1.8771 3747	1.9671 5136	2.0610 3156	2.1589 2500
11	1.8982 9856	1.9991 5140	2.1048 5195	2.2156 0893	2.3316 3900
12	2.0121 9647	2.1290 9624	2.2521 9159	2.3817 7960	2.5181 7012
13	2.1329 2826	2.2674 8750	2.4098 4500	2.5604 1307	2.7196 2373
14	2.2609 0396	2.4148 7418	2.5785 3415	2.7524 4405	2.9371 9362
15	2.3965 5819	2.5718 4101	2.7590 3154	2.9588 7735	3.1721 6911
16	2.5403 5168	2.7390 1067	2.9521 6375	3.1807 9315	3.4259 4264
17	2.6927 7279	2.9170 4637	3.1588 1521	3.4193 5264	3.7000 1805
18	2.8543 3915	3.1066 5438	3.3799 3228	3.6758 0409	3.9960 1950
19	3.0255 9950	3.3085 8691	3.6165 2754	3.9514 8940	4.3157 0106
20	3.2071 3547	3.5236 4506	3.8696 8446	4.2478 5110	4.6609 5714
21	3.3995 6360	3.7526 8199	4.1405 6237	4.5664 3993	5.0338 3372
22	3.6035 3742	3.9966 0632	4.4304 0174	4.9089 2293	5.4365 4041
23	3.8197 4966	4.2563 8573	4.7405 2986	5.2770 9215	5.8714 6365
24	4.0489 3464	4.5330 5081	5.0723 6695	5.6728 7406	6.3411 8074
25	4.2918 7072	4.8276 9911	5.4274 3264	6.0983 3961	6.8484 7520
26	4.5493 8296	5.1414 9955	5.8073 5292	6.5557 1508	7.3963 5321
27	4.8223 4594	5.4756 9702	6.2138 6763	7.0473 9371	7.9880 6147
28	5.1116 8670	5.8316 1733	6.6488 3836	7.5759 4824	8.6271 0639
29	5.4183 8790	6.2106 7245	7.1142 5705	8.1441 4436	9.3172 7490
30	5.7434 9117	6.6143 6616	7.6122 5504	8.7549 5519	10.0626 5689
31	6.0881 0064	7.0442 9996	8.1451 1290	9.4115 7683	10.8676 6944
32	6.4533 8668	7.5021 7946	8.7152 7080	10.1174 4509	11.7370 8300
33	6.8405 8988	7.9898 2113	9.3253 3975	10.8762 5347	12.6760 4964
34	7.2510 2528	8.5091 5950	9.9781 1354	11.6919 7248	13.6901 3361
35	7.6860 8679	9.0622 5487	10.6765 8148	12.5688 7042	14.7853 4429
36	8.1472 5200	9.6513 0143	11.4239 4219	13.5115 3570	15.9681 7184
37	8.6360 8712	10.2786 3603	12.2236 1814	14.5249 0088	17.2456 2558
38	9.1542 5235	10.9467 4737	13.0792 7141	15.6142 6844	18.6252 7563
39	9.7035 0749	11.6582 8595	13.9948 2041	16.7853 3858	20.1152 9768
40	10.2857 1794	12.4160 7453	14.9744 5784	18.0442 3897	21.7245 2150
41	10.9028 6101	13.2231 1938	16.0226 6989	19.3975 5689	23.4624 8322
42	11.5570 3267	14.0826 2214	17.1442 5678	20.8523 7366	25.3394 8187
43	12.2504 5463	14.9979 9258	18.3443 5475	22.4163 0168	27.3666 4042
44	12.9854 8191	15.9728 6209	19.6284 5959	24.0975 2431	29.5559 7166
45	13.7646 1083	17.0110 9813	21.0024 5176	25.9048 3863	31.9204 4939
46	14.5904 8748	18.1168 1951	22.4726 2338	27.8477 0153	34.4740 8534
47	15.4659 1673	19.2944 1278	24.0457 0702	29.9362 7915	37.2320 1217
48	16.3938 7173	20.5485 4961	25.7289 0651	32.1815 0008	40.2105 7314
49	17.3775 0403	21.8842 0533	27.5299 2997	34.5951 1259	43.4274 1899
50	18.4201 5427	23.3066 7868	29.4570 2506	37.1897 4603	46.9016 1251

Even though there is very little difference in the amounts, can we continue to increase the amount by compounding more frequently? Is there a limit to the amount received? Consider $1 deposited at 100% interest for one year.

Number of periods	Formula	Amount
Annually, $n = 1$	$\left(1 + \dfrac{1}{1}\right)^{1}$	$2.00
Semiannually, $n = 2$	$\left(1 + \dfrac{1}{2}\right)^{2}$	2.25
Quarterly, $n = 4$	$\left(1 + \dfrac{1}{4}\right)^{4}$	2.44
Monthly, $n = 12$	$\left(1 + \dfrac{1}{12}\right)^{12}$	2.61
Daily, $n = 360$	$\left(1 + \dfrac{1}{360}\right)^{360}$	2.715
Hourly, $n = 8640$	$\left(1 + \dfrac{1}{8640}\right)^{8640}$	2.7181

Remember, the general formula is

$$\left(1 + \frac{1}{n}\right)^{n}.$$

These calculations are done using a calculator.

It looks like

$$\lim_{n \to \infty} \left(1 + \frac{1}{n}\right)^{n}$$

exists. Indeed, if we continue our calculations for

$n = 10,000$ the formula yields	2.718145926
$n = 100,000$	2.718268237
$n = 1,000,000$	2.718280469
$n = 10,000,000$	2.718281828
$n = 100,000,000$	2.718281828.

The calculator can no longer distinguish the values of $(1 + 1/n)^{n}$ for very large n. We denote the number this sequence is approaching by e and define it as follows:

$$e = \lim_{n \to \infty} \left(1 + \frac{1}{n}\right)^{n}$$

DEFINITION OF e

e is an irrational number approximately equal to

2.718281828459045235

Historical Note

The number e is sometimes called *Euler's number* in honor of Leonhard Euler, the famous Swiss mathematician who was the first to publish a paper on e, in 1748.

The number e has many applications because it is used to describe the process of continuous growth or decay. For interest compounded continuously, we use the formula

$$A = P e^{rt}.$$

Most natural phenomena grow or decay continuously and are characterized by the formula

$$y = y_0 e^{kt}$$

where y is the amount of something present after t time intervals and the amount present at the start is y_0. The constant k depends on the particular phenomenon under investigation.

EXAMPLES

4. Suppose Blondie deposits her $73 at 6% interest at a bank that offers continuous compounding. How long will it take for her money to double?
Solution:

$$A = P e^{rt}$$
$$2 = e^{0.06t}$$

We are faced with the previous problem of solving for t. However, tables for powers of e are readily available (Table 2 at the back of this book, for example). From Table 2, we locate the number in a column headed e^x that is as close to 2 as possible. We find it in the middle column (about halfway down):

A process called *linear interpolation* can be used for more accurate results. We'll discuss this later. Here, we assume that we need x only to the nearest hundredth, as given by the table.

$$e^x = 2.014 \quad \text{if} \quad x = 0.70$$

Thus,

$$e^{0.70} \approx e^{0.06t}$$

so

$$0.70 \approx 0.06t$$
$$t \approx 11.67 \text{ years.}$$

5. Human population growth is described by

$$P = P_0 e^{rt},$$

where a population of P_0 grows to a population P after t years at an annual rate of r. In the years 1960–1970, the population of Juneau, Alaska, increased from 6797 to 13,556. What was Juneau's growth rate for this period?
Solution: $P = 13,556, \quad P_0 = 6797, \quad t = 10$

$$13{,}556 = 6797e^{10r}$$
$$1.9944 = e^{10r}$$
$$e^{0.69} \approx e^{10r}$$
$$0.69 \approx 10r$$
$$0.069 \approx r$$

From Table 2,

$$1.9944 \approx e^{0.69}.$$

The growth rate was about 6.9%.

6. The radioactive decay of a substance is described by

$$A = A_0\, e^{-kt},$$

where an initial amount of A_0 decays to an amount A after t years. The value of the constant k depends on the type of substance. Find the half-life of radium if $k = 0.000425$.

Solution: $\frac{1}{2} = e^{-0.000425t}$

In Table 2, we look for $e^{-x} = 0.5$ and find $x \approx 0.69$:

The half-life of a substance is the time it takes for half of the original amount to decay.

$$e^{-0.69} \approx e^{-0.000425t}$$

$$-0.69 \approx -0.000425t$$

$$t \approx \frac{0.69}{0.000425}$$

$$t \approx 1624 \quad \text{(by calculator)}$$

Therefore, it takes about 1600 years for any amount of radium to decay to half of its initial quantity.

PROBLEM SET 5.2

A Problems

Solve the exponential equations in Problems 1–15.

1. $2^x = 128$
2. $3^x = 243$
3. $8^x = 32$
4. $9^x = 27$
5. $125^x = 25$
6. $4^x = \frac{1}{16}$
7. $27^x = \frac{1}{81}$
8. $(\frac{1}{2})^x = 8$
9. $(\frac{1}{2})^x = \frac{1}{8}$
10. $(\frac{2}{3})^x = \frac{9}{4}$
11. $(\frac{3}{4})^x = \frac{16}{9}$
12. $2^{3x+1} = \frac{1}{2}$
13. $3^{4x-3} = \frac{1}{9}$
14. $27^{2x+1} = 3$
15. $8^{5x+2} = 16$

Use Table 5.1, on page 216, to answer Problems 16–19.

$ 16. If $1000 is invested at 7% interest compounded annually, how much money will there be in 25 years?

$ 17. If $1000 is invested at 12% interest compounded semiannually, how much money will there be in ten years?

$ 18. If $1000 is invested at 12% interest compounded semiannually, how long will it take (to the nearest year) for the money to double?

$ 19. If $1000 is invested at 6% interest compounded annually, how long will it take (to the nearest year) for the money to quadruple?

B Problems

Use Table 2 at the back of the book to answer Problems 20–30.

$ 20. If $1000 is invested at 6% interest compounded continuously, how much money will there be in 25 years?

$ 21. If $1000 is invested at 8% interest compounded continuously, how much money will there be in ten years?

$ 22. If $1000 is invested at 8% interest compounded continuously, how long will it take (to the nearest year) for the money to triple?

23. In the years 1960–1970, the population of Honolulu, Hawaii, increased from 294,000 to 325,000. What was Honolulu's growth rate for this period?

24. In the years 1960–1970, the population of Salt Lake City, Utah, decreased from 190,000 to 176,000. What was Salt Lake City's growth rate for this period?

25. *Calculator Problem.* If the half-life of nickel is 92 years, find the constant k for which

$$A = A_0 e^{-kt}.$$

26. *Calculator Problem.* If the half-life of cesium-137 is 30 years, find the constant k for which

$$A = A_0 e^{-kt}.$$

27. *Calculator Problem.* Find the half-life of strontium-90 if $k = 0.0246$.

28. *Calculator Problem.* Find the half-life of krypton if $k = 0.0641$.

29. The atmospheric pressure P in pounds per square inch (psi) is given by

$$P = 14.7\, e^{-0.21a},$$

where a is the altitude above sea level (in miles). What is the pressure in Denver, Colorado, the "Mile-High City"?

30. If the atmospheric pressure, as given by the formula in Problem 29, is 1.79 psi, what is the altitude?

31. *Calculator Problem.* In 1975 ($t = 0$), the world use of petroleum, P_0, was 19,473 million barrels of oil. If the world reserves are 584,600 million barrels and the growth rate for the use of oil is k, then the total amount, A, used during a time interval $t > 0$ is given by

$$A = \frac{P_0}{k}(e^{kt} - 1).$$

How long will it be before the world reserves are depleted if
a. $k = 0.08$ (8%)?   b. $k = 0.052$ (5.2%)?

$ 32. *Calculator Problem.* The problem we solved concerning the Blondie cartoon was simplified for ease of calculation. We assumed that she made a single $73 deposit at the end of a year of saving in a cookie jar. The cartoon, however, implies she will make a $73 deposit *every* year. That is, Blondie will make a series of equal payments over equal time intervals. This is called an *annuity*, *A*, where *P* dollars per year are invested at interest rate *r* compounded continuously over *t* years. The total amount at the end of *t* years is approximated by the formula

ANNUITY

$$A = \frac{P}{r}(e^{rt} - 1).$$

For example, if Blondie deposits $73 every year for ten years at 8%, the amount is found by

$$A = \frac{73}{0.08}(e^{0.08(10)} - 1)$$

$$\approx \$1118.31.$$

For how many years will she have to deposit $73 a year in order to save $3650 at 8% interest compounded continuously?

Graph the functions given in Problems 33–40.

33. $f(x) = e^x$
35. $f(x) = -e^x$

34. $f(x) = e^{-x}$
36. $f(x) = -e^{-x}$

37. $f(x) = \left(1 + \frac{1}{x}\right)^x, x > 0$

38. $f(x) = e^{-x^2}$

39. $c(x) = \dfrac{e^x + e^{-x}}{2}$

40. $s(x) = \dfrac{e^x - e^{-x}}{2}$

Mind Bogglers

41. The functions *c* and *s* of Problems 39 and 40 are called the *hyperbolic cosine* and *hyperbolic sine* functions. These are defined by

HYPERBOLIC
FUNCTIONS

$$\cosh x = \frac{e^x + e^{-x}}{2} \quad \text{and} \quad \sinh x = \frac{e^x - e^{-x}}{2}.$$

a. Show that $\cosh^2 x - \sinh^2 x = 1$.
b. Show that the hyperbolic cosine is an even function and that the hyperbolic sine is an odd function.
c. Show that $\sinh 2x = 2 \sinh x \cosh x$.
d. Graph $y = \cosh x + \sinh x$.
e. Define

Recall, a function *f* is even if $f(-x) = f(x)$ and is odd if $f(-x) = -f(x)$.

$$\tanh x = \frac{\sinh x}{\cosh x}.$$

Graph tanh *x*.

42. The function given in Problem 38 is very closely related to the normal curve of probability, statistics, and psychology. The function

$$\phi(x) = \frac{1}{\sqrt{2\pi}} e^{-(1/2)x^2}$$

is called the normal density function. Graph $\phi(x)$.

43. *Calculator Problem.* In calculus, it is shown that

$$e^x = 1 + x + \frac{x^2}{2} + \frac{x^3}{2 \cdot 3} + \frac{x^4}{2 \cdot 3 \cdot 4} + \cdots.$$

a. What are the next two terms?
b. What is the *r*th term?
c. Calculate *e* correct to the nearest thousandth using the above equation.
d. Calculate $\sqrt{e} = e^{0.5}$ correct to the nearest thousandth using the above equation.

5.3 LOGARITHMIC FUNCTIONS

In the last section, an analysis of Blondie's bank deposit required solving an exponential equation. If the exponential equation is simple or if we have tables available, we can solve the problem. But, in general, if we are given the exponential equation

$$A = b^x$$

Remember, b is called the base.

$(b > 1, b \neq 1)$, how can we solve this equation for x?
We see that

x is the exponent of b that yields A.

This can be rewritten

x = exponent of b to get A.

It appears that the equation is now solved for x, but this is simply a notational change. The expression "exponent of b to get A" is called, for historical reasons, "the log of A to the base b." That is,

The term log is an abbreviation for logarithm.

x = log A to the base b.

But this phrase is shortened to the notation

$x = \log_b A$.

This form still forces us to use tables or calculators to find the value of x. However, certain properties of logarithms make tables for them much more readily available than tables for the exponential equations of the last section. Let's investigate some of these properties.

DEFINITION OF
LOGARITHMIC
FUNCTION

DEFINITION: For $b > 0$, $b \neq 1$,

$$x = b^y \qquad \text{is equivalent to} \qquad y = \log_b x.$$

The function $f(x) = \log_b x$ is called the *logarithmic function.*

It is important to recognize this as a notational change only:

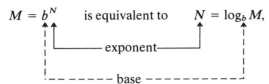

$M = b^N$ is equivalent to $N = \log_b M,$

or

$Q = \log_b S$ is equivalent to $S = b^Q.$

EXAMPLES: Change from exponential form to logarithmic form.

Remember, the log form solves for the exponent.

1. $5^2 = 25 \Leftrightarrow \log_5 25 = 2$

We use the symbol $\Leftrightarrow$ to mean "is equivalent to."

Remember, this is the base.

2. $3^2 = 9 \Leftrightarrow \log_3 9 = 2$

base   exponent

3. $\frac{1}{8} = 2^{-3} \Leftrightarrow \log_2 \frac{1}{8} = -3$

4. $\sqrt{16} = 4 \Leftrightarrow \log_{16} 4 = \frac{1}{2}$

Remember, $\sqrt{16} = 16^{1/2}$.

EXAMPLES: Change from logarithmic form to exponential form.

5. $\log_{10} 100 = 2 \Leftrightarrow 10^2 = 100$

base   exponent

6. $\log_{10} \frac{1}{1000} = -3 \Leftrightarrow 10^{-3} = \frac{1}{1000}$

7. $\log_e 1 = 0 \Leftrightarrow e^0 = 1$

By relating logarithms back to exponential form, we find that we already know a great deal about the logarithmic form. For example, the graph of $y = \log_2 x$ is the same as the graph of $2^y = x$ (see Figure 5.8).

y	x
-3	$2^{-3} = \frac{1}{8}$
-2	$2^{-2} = \frac{1}{4}$
-1	$2^{-1} = \frac{1}{2}$
0	$2^0 = 1$
1	$2^1 = 2$
2	$2^2 = 4$
3	$2^3 = 8$

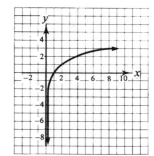

Figure 5.8   Graph of $y = \log_2 x$

Compare Figure 5.8 with Figure 5.2, on page 206. These figures are similar but not the same. They are, however, related. Let $f(x) = \log_2 x$ and $g(x) = 2^x$. Consider $f \circ g$.

$$(f \circ g)(x) = f[g(x)] = \log_2 2^x.$$

But by the definition of logarithm, if

$$N = \log_2 2^x,$$

$$\uparrow \qquad \uparrow$$
$$\text{exponent} \quad \text{base}$$

then

$$\downarrow \qquad \downarrow$$
$$2^N = 2^x$$

and

$$N = x.$$

Therefore,

$$x = \log_2 2^x \qquad \text{or} \qquad (f \circ g)(x) = x,$$

which proves that f and g are inverse functions.

We can also find certain logarithms by using the techniques of the last section, as illustrated by Examples 8–11.

EXAMPLES: Evaluate the given logarithms.

8. $\log_2 64$

It is usually necessary to supply a variable to convert to exponential form. We'll use N in these examples.

$$\log_2 64 = N \qquad \text{or} \qquad \begin{aligned} 2^N &= 64 \\ 2^N &= 2^6 \\ N &= 6 \end{aligned}$$

Thus, $\log_2 64 = 6$.

9. $\log_3 \frac{1}{9}$

$$\begin{aligned} 3^N &= \tfrac{1}{9} \\ 3^N &= 3^{-2} \\ N &= -2 \end{aligned}$$

Thus, $\log_3 \frac{1}{9} = -2$.

10. $\log_{10} 0.1$

$$\begin{aligned} 10^N &= 0.1 \\ 10^N &= 10^{-1} \\ N &= -1 \end{aligned}$$

Thus, $\log_{10} 0.1 = -1$.

11. $\log_9 27$

$$\begin{aligned} 9^N &= 27 \\ 3^{2N} &= 3^3 \\ 2N &= 3 \\ N &= \tfrac{3}{2} \end{aligned}$$

Thus, $\log_9 27 = \frac{3}{2}$.

The fact most commonly forgotten by students working with logarithms is that logarithmic equations are equivalent to exponential equations and that the properties of exponents can be applied. For example, the first law of exponents,

$$b^x b^y = b^{x+y},$$

can be restated in logarithmic form. Let

$$A = b^x \qquad \text{and} \qquad B = b^y;$$

then

$$\log_b A = x \qquad \text{and} \qquad \log_b B = y$$

so

$$AB = b^x b^y$$
$$= b^{x+y}.$$

But

$$AB = b^{x+y} \Leftrightarrow \log_b AB = x + y;$$

therefore,

$$\log_b AB = x + y$$
$$= \log_b A + \log_b B.$$

Since $x = \log_b A$ and $y = \log_b B$

The other laws for exponents also lead to logarithmic forms, as summarized below.

If A and B are positive real numbers and b is a positive real number other than 1, then:

FIRST LAW OF LOGARITHMS
The log of the product of two numbers is the sum of the logs of those numbers.

1. $\log_b AB = \log_b A + \log_b B$

    This follows from the First Law of Exponents,

    $$b^x b^y = b^{x+y}.$$

SECOND LAW OF LOGARITHMS
The log of the quotient of two numbers is the log of the numerator minus the log of the denominator.

2. $\log_b \dfrac{A}{B} = \log_b A - \log_b B$

    This follows from the Second Law of Exponents,

    $$\frac{b^x}{b^y} = b^{x-y},$$

    where $A = b^x$ and $B = b^y$.

THIRD LAW OF LOGARITHMS
The log of the pth power of a number is p times the log of that number.

3. $\log_b A^p = p \log_b A$

    This follows from the Third Law of Exponents,

    $$(b^x)^y = b^{xy},$$

    where $A = b^x$ and $y = p$.

We can use these properties to solve logarithmic equations by noting that

$$b^{\log_b x} = x,$$

Do you see how this follows from the definition?

which follows directly from the definition of logarithm. Then, if we know that the logarithms of two numbers are equal, the numbers must be equal since

$$\log_b A = \log_b B \Leftrightarrow b^{\log_b B} = A$$

by the definition of logarithm. But

$$b^{\log_b B} = B \quad \text{so} \quad B = A,$$

which results in the following theorem:

LOG OF BOTH SIDES THEOREM: If A, B, and b are positive reals with $b \neq 1$, then

$$\log_b A = \log_b B \Leftrightarrow A = B.$$

We now use these laws of logarithms to solve logarithmic equations.

EXAMPLES: Solve for x.

12.　　$\log_b 3 + \frac{1}{2}\log_b 25 = \log_b x$
　　　$\log_b 3 + \log_b 25^{1/2} = \log_b x$
　　　　$\log_b 3 + \log_b 5 = \log_b x$

　　　　　$\log_b(3 \cdot 5) = \log_b x$

　　　　　　　$15 = x$
　　　　　　　$\{15\}$

13.　$\log_b 2x + \frac{1}{3}\log_b 125 = 2(\log_b 10 - \log_b 2)$
　　　$\log_b 2x + \log_b 5 = 2(\log_b 10 - \log_b 2)$
　　　　　$\log_b 10x = 2\log_b 5$
　　　　　$\log_b 10x = \log_b 25$
　　　　　　$10x = 25$
　　　　　　　$x = 2.5$
　　　　　　　$\left\{\frac{5}{2}\right\}$

14.　　　　$5\log_x 2 - \frac{1}{2}\log_x 8 = 2 - \frac{1}{2}\log_x 2$

　　　　$\log_x 2^5 - \log_x \sqrt{8} = 2 - \log_x \sqrt{2}$

　　$\log_x 32 - \log_x 2\sqrt{2} + \log_x \sqrt{2} = 2$

　　　　　$\log_x\left(\frac{32}{2\sqrt{2}} \cdot \sqrt{2}\right) = 2$

　　　　　　　$\log_x 16 = 2$

　　　　　　　　$x^2 = 16$

　　　　　　　　$x = \pm 4$

By the definition of a logarithm, x must be positive, so the solution set is $\{4\}$.

Definition:

$$\log_b x = y \Leftrightarrow b^y = x$$

and obviously

$$\log_b x = \log_b x.$$

Apply the definition where $y = \log_b x$:

$$\log_b x = \log_b x \Leftrightarrow b^{\log_b x} = x.$$

Third Law of Logarithms

$$25^{1/2} = \sqrt{25} = 5$$

First Law of Logarithms

Log of Both Sides Theorem

Third Law of Logarithms,

$$125^{1/3} = (5^3)^{1/3} = 5$$

Definition of Logarithm

15. $\log_b x - \frac{1}{2}\log_b 2 = \frac{1}{2}\log_b(x + 4)$

$$\log_b x - \log_b \sqrt{2} = \log_b \sqrt{x + 4}$$

$$\log_b\left(\frac{x}{\sqrt{2}}\right) = \log_b \sqrt{x + 4}$$

$$\frac{x}{\sqrt{2}} = \sqrt{x + 4}$$

$$\frac{x^2}{2} = x + 4$$

$$x^2 - 2x - 8 = 0$$

$$(x - 4)(x + 2) = 0$$

$$x = 4, -2$$

Notice that $\log_b(-2)$ is not defined, so $x = -2$ is an extraneous root. Therefore, the solution set is $\{4\}$.

PROBLEM SET 5.3

A Problems

Write Problems 1–12 in logarithmic form.

1. $64 = 2^6$
2. $100 = 10^2$
3. $1000 = 10^3$
4. $64 = 8^2$
5. $125 = 5^3$
6. $a = b^c$
7. $m = n^p$
8. $1 = e^0$
9. $9 = \left(\frac{1}{3}\right)^{-2}$
10. $8 = \left(\frac{1}{2}\right)^{-3}$
11. $\frac{1}{3} = 9^{-1/2}$
12. $\frac{1}{2} = 4^{-1/2}$

Write Problems 13–24 in exponential form.

13. $\log_{10} 10,000 = 4$
14. $\log_{10} 0.01 = -2$
15. $\log_{10} 1 = 0$
16. $\log_2 32 = 5$
17. $\log_2 \frac{1}{8} = -3$
18. $\log_4 2 = \frac{1}{2}$
19. $\log_e e^2 = 2$
20. $\log_a b = c$
21. $\log_m n = p$
22. $\log_{1/3} 81 = -4$
23. $\log_{1/2} 16 = -4$
24. $\log_4 \sqrt[3]{4} = \frac{1}{3}$

Evaluate the expressions in Problems 25–36.

25. $\log_b b^2$
26. $\log_t t^3$
27. $\log_e e^4$
28. $\log_\pi \sqrt{\pi}$
29. $\log_\pi \dfrac{1}{\pi}$
30. $\log_2 8$
31. $\log_3 9$
32. $\log_{10} 0.0001$
33. $\log_{10} 1,000,000$
34. $\log_{19} 1$
35. $\log_8 32$
36. $\log_9 3$

B Problems

Solve for x in Problems 37–49.

37. $\log_2 8\sqrt{2} = x$

38. $\log_3 27\sqrt{3} = x$

39. $\log_{10} x = 5$

40. $\log_2 x = 5$

41. $\log_x 1 = 0$

42. $\log_x 10 = 0$

43. $\log_b 5 + \frac{1}{2}\log_b 9 = \log_b x$

44. $\log_b 10 - \frac{1}{2}\log_b 25 = \log_b x$

45. $\frac{1}{2}\log_b x = 3\log_b 5 - \log_b x$

46. $\log_b x - \frac{1}{2}\log_b 4 = \frac{1}{2}\log_b(2x - 3)$

47. $\log_b x - \frac{1}{2}\log_b 3 = \frac{1}{2}\log_b(x + 6)$

48. $\log_b x - \frac{1}{2}\log_b 9 = \frac{1}{2}\log_b(x - 2)$

49. $3\log_e \dfrac{e}{\sqrt[3]{5}} = 3 - \log_e x$

Graph the functions given in Problems 50–55.

50. $y = \log_3 x$

51. $y = \log_{1/2} x$

52. $y = \log_{1/3} x$

53. $y = \log_{10} x$

54. $y = \log_e x$

55. $y = \log_\pi x$

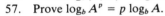

56. Prove $\log_b(A/B) = \log_b A - \log_b B$.

57. Prove $\log_b A^p = p\log_b A$.

58. Prove $\log_b b^p = p$.

Mind Boggler

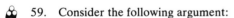

59. Consider the following argument:

$$4 > 3 \quad \text{is obviously true}$$
$$4\log_{10}\tfrac{1}{3} > 3\log_{10}\tfrac{1}{3} \quad \text{Multiply both sides by } \log_{10}\tfrac{1}{3}$$
$$\log_{10}(\tfrac{1}{3})^4 > \log_{10}(\tfrac{1}{3})^3 \quad \text{Property of exponents}$$
$$(\tfrac{1}{3})^4 > (\tfrac{1}{3})^3 \quad \text{Theorem of logarithms}$$
$$\tfrac{1}{81} > \tfrac{1}{27} \quad \text{is obviously false}$$

Can you find the error?

5.4 ELECTRONIC CALCULATORS VERSUS LOGARITHMIC TABLES

In the last section, we solved

$$A = b^x \qquad x = \log_b A.$$

As we pointed out, this is a notational change. Suppose we want to know how long it would take for $73 to double if invested at 3% interest. To answer this question, we must solve

$$2 = (1.03)^x.$$

Writing this as a logarithm,

$$\log_{1.03} 2 = x,$$

doesn't seem to help much unless we have an appropriate interest table. However, tables for base ten are readily available, so we can solve the problem by using the Log of Both Sides Theorem.

$$\log_{10} 2 = \log_{10} 1.03^x$$

$$\log_{10} 2 = x \log_{10} 1.03$$

$$x = \frac{\log_{10} 2}{\log_{10} 1.03}$$

Table 3 in the back of the book gives us these values. A portion of this table is reproduced here. The numbers in the column N are assumed to have a decimal point after the first digit. The values across the top indicate the digit following the two digits listed under N. Thus, $\log_{10} 1.03$ is found by locating the first two digits in column N and then reading over to the column headed 3 (see circles). A decimal point is then affixed in front of this number to give the result

$$\log_{10} 1.03 \approx 0.0128.$$

Remember what this means:

$$10^{0.0128} \approx 1.03$$

$$10^{0.3010} \approx 2$$

For $\log_{10} 2$, find the value 2.00 in the table (see boxes):

$$\log_{10} 2 \approx 0.3010$$

N	⓪	1	2	③	4	5	6	7	8	9
⑩	0000	0043	0086	**0128**	0170	0212	0253	0294	0334	0374
11	0414	0453	0492	0531	0569	0607	0645	0682	0719	0755
12	0792	0828	0864	0899	0934	0969	1004	1038	1072	1106
13	1139	1173	1206	1239	1271	1303	1335	1367	1399	1430
14	1461	1492	1523	1553	1584	1614	1644	1673	1703	1732
15	1761	1790	1818	1847	1875	1903	1931	1959	1987	2014
16	2041	2068	2095	2122	2148	2175	2201	2227	2253	2279
17	2304	2330	2355	2380	2405	2430	2455	2480	2504	2529
18	2553	2577	2601	2625	2648	2672	2695	2718	2742	2765
19	2788	2810	2833	2856	2878	2900	2923	2945	2967	2989
⑳	**3010**	3032	3054	3075	3096	3118	3139	3160	3181	3201

The result is

$$x = \frac{\log_{10} 2}{\log_{10} 1.03} \approx \frac{0.3010}{0.0128}$$

$$\approx 23.5 \text{ years.}$$

It will take about 23.5 years for the money to double.

The use of a calculator can take the drudgery out of working problems like the one above. Calculators (especially just a four-function calculator) are readily available for under $10. To find x on a calculator that has keys for logarithms, you need to keep in mind the following points.

1. If you have a calculator with algebraic logic, push the keys in this order:

$$\boxed{2}\quad\boxed{\log}\quad\boxed{\div}\quad\boxed{1.03}\quad\boxed{\log}\quad\boxed{=}$$

The answer is 23.44977225. While this is still approximate, it is more accurate than using tables.

2. If you have a calculator with RPN logic:

$$\boxed{2}\quad\boxed{\log}\quad\boxed{1.03}\quad\boxed{\log}\quad\boxed{\div}$$

The answer is 23.44977227.

If you don't have a calculator available, a long division problem like the one above is not very pleasant. For many years complicated calculations were accomplished with the aid of logarithms. We'll see how this is done later in this section.

This problem suggests another useful result, whether calculators or tables are used. Only two bases are ordinarily used. The first is base ten; logarithms to base ten are called *common logarithms*. The second is base e; logarithms to base e are called *natural logarithms* because of their relationship to natural phenomena such as growth and decay. We write $\log N$ instead of $\log_{10} N$ and assume ten is the base when no base is written. When e is the base, we write

$$\ell n\ N \qquad \text{instead of} \qquad \log_e N.$$

Tables for natural logarithms (Table 2) and common logarithms (Table 3) are readily available, and these functions are also found on many calculators. Since we occasionally encounter a problem with some other base, the first example of this section suggests a method for changing from one base to another. Recall that

$$x = b^y \Leftrightarrow \log_b x = y,$$

which is in terms of base b. Suppose we want to find y in terms of base a (ten or e, for example). We can take the logarithm of both sides of the original equation:

$$x = b^y$$

$$\log_a x = \log_a b^y$$

By long division, or if you have a four-function calculator, you can carry out this division as follows:

ALGEBRAIC LOGIC	RPN LOGIC
$\boxed{0.3010}$	$\boxed{0.3010}$
$\boxed{\div}$	$\boxed{\text{ENTER}}$
$\boxed{0.0128}$	$\boxed{0.0128}$
$\boxed{=}$	$\boxed{\div}$

The answer is approximately 23.515625.

Notice that the four answers we've given for this problem all vary slightly depending on the particular calculator or method used to solve the problem. For our purposes in this chapter, either table or calculator accuracy is acceptable. Significant digits are discussed in Appendix C. However, for this problem, since the bank compounds annually, the length of time is 24 years.

ℓn is read "lon."

$$\log_a x = y \log_a b$$

$$y = \frac{\log_a x}{\log_a b}.$$

But if $x = b^y$, then $y = \log_b x$; so by substitution, we obtain the following result:

$$\log_b x = \frac{\log_a x}{\log_a b}$$

CHANGE OF BASE THEOREM
To change from base b to a base a, change the base b to base a and divide by the logarithm to the base $\dot{a}$ of the old base b.

EXAMPLES: Find the logarithms correct to three significant digits using a table or calculator.

1. $\log 7.68 = x$
    Solution:
    a. From Table 3:
    $$\log 7.68 \approx 0.8854$$
    b. By calculator:
    $$\boxed{7.68} \quad \boxed{\log} \approx 0.88536122$$
    To three significant digits, $x = 0.885$.

2. $\ln 4.015 = x$
    Solution:
    a. $e^x = 4.015$ so from Table 2:
    $$x \approx 1.39$$
    b. By calculator:
    $$\boxed{4.015} \; \boxed{\ell n} \approx 1.390037347$$
    To three significant digits, $x = 1.39$.

3. $\log_7 4.28$
    Solution: We must change to base ten or base e.
    $$\log_7 4.28 = \frac{\log 4.28}{\log 7}$$

    a. From Table 3:
    $$\log_7 4.28 \approx \frac{0.6314}{0.8451} \approx 0.7471$$

    b. By calculator with algebraic logic:
    $$\boxed{4.28} \; \boxed{\log} \; \boxed{\div} \; \boxed{7} \; \boxed{\log} \; \boxed{=} \approx 0.7471840415$$

c. By calculator with RPN logic:

$$\boxed{4.28} \;\; \boxed{\log} \;\; \boxed{7} \;\; \boxed{\log} \;\; \boxed{\div} \;\; \approx 0.7471840417$$

To three significant digits, $x = 0.747$.

Notice that there are limitations to Tables 2 and 3. Common logarithms are given for numbers between 1 and 10 and natural logarithms are given for numbers between 0 and 3. But we can use properties of logarithms to find other values not contained in the tables. For example, the logarithm of 852 can be found by writing 852 in scientific notation.

$$852 = 8.52 \cdot 10^2$$

Thus,

$$\begin{aligned}
\log 852 &= \log(8.52 \cdot 10^2) \\
&= \log 8.52 + \log 10^2 \\
&= \log 8.52 + 2 \log 10 \\
&\approx 0.9304 + 2 \\
&= 2.9304 \\
&\approx 2.93.
\end{aligned}$$

We see that if n is written in scientific notation as

$$n = M \cdot 10^C \quad (1 \le M < 10),$$

then

$$\begin{aligned}
\log n &= \log (M \cdot 10^C) \\
&= \log M + C \log 10 \\
&= C + \log M.
\end{aligned}$$

Since $1 \le M < 10$, $\log M$ can be found in Table 3. Log M is called the *mantissa* and C is called the *characteristic*; C can usually be found by inspection.

Recall, a number is in *scientific notation* if it is a power of ten or written as a number between 1 and 10 times an integral power of ten.

The value of log 852 is found in Table 3 and log 10 = 1.

If n is a power of ten, then $M = 1$

EXAMPLES: Find each logarithm to three significant digits using a table or calculator.

4. $\log 2420 = x$
    Solution:
    a. By table: the mantissa is $\log 2.420 = 0.3838$ and the characteristic is 3, so

    $$x = \log 2420 \approx 3.3838.$$

b.   By calculator:

$$x = \boxed{2420} \ \boxed{\log} \approx 3.383815366$$

To three significant digits, $x = 3.38$.

5. $\log 2426 = x$
Solution:
a.   By table: the mantissa is log 2.426 and the characteristic is 3. But 2.426 is not listed in the table so we proceed as follows:

This procedure is called *linear interpolation.* In practice, linear interpolation for logarithms is no longer necessary. Table 3 is a four-place table, and if greater accuracy is needed, a five-place or six-place table can be used. However, because of the availability of calculators, they would be used in any application requiring a greater accuracy than given by the four-place table.

$$6\begin{bmatrix} 2.420 & 0.3838 \\ 2.426 & ? \\ 2.430 & 0.3856 \end{bmatrix} \text{ difference is } 18$$

The result is 6/10 of 18 or 10.8. Thus, log 2426 $\approx$ 3.3849. However, in this book, we will only require the accuracy found in a four-place table; since 2.426 is closer to 2.430 than 2.420, the answer

$$x = \log 2430 \approx 3.3856$$

is acceptable.
b.   By calculator:

$$x = \boxed{2426} \ \boxed{\log} \approx 3.384890797$$

To three significant digits, $x = 3.38$.

6. $\log 0.00728 = x$
Solution:
a.   By table: the mantissa is log 7.28 = 0.8621 and the characteristic is -3, so

$$x = \log 0.00728 \approx 0.8621 - 3$$
$$= -2.1379$$

Since mantissas are always positive but characteristics are negative for any number less than one, the answer is sometimes left in the form

$$0.8621 - 3$$

or

$$0.8621 - 3$$
$$= (0.8621 + 10) - 10 - 3$$

$$= 10.8621 - 13.$$

Remember, log $e \approx$ log 2.72.

b.   By calculator:

$$\boxed{0.00728} \ \boxed{\log} \approx -2.137868621$$

To three significant digits, $x = -2.14$.

7. $\ell \text{n} \, 426 = x$
Solution:
a.   By table: rewrite as $e^x = 426$. Since this is not contained in Table 2 we change to base ten:

$$\ln 426 = \frac{\log 426}{\log e}$$

$$\approx \frac{2.6294}{0.4346}$$

$$\approx 6.050$$

b.   By calculator:

$$x = \boxed{426} \quad \boxed{\ell n} \approx 6.054439346$$

To three significant digits, $x = 6.05$.

CALCULATIONS

For many years, logarithms were taught in elementary mathematics primarily as an aid to computation. As we've seen in this chapter, logarithms are useful as functions in mathematics, but with the widespread use of calculators, they are no longer necessary for complicated calculations. In the remaining part of this section we'll consider some problems and give their logarithmic and calculator solutions so you can compare the different methods. If you do not have a calculator available, you can work the problems using the four-place tables in the back of the book.

EXAMPLES

8.   In October 1974, the inflation rate was 12%. If a person at that time was earning a $10,000 salary, how much salary will be necessary to equal this salary in 15 years, assuming this same rate of inflation? Assume that the formula is

$$A = P(1 + r)^n,$$

or, for this problem,

$$A = 10,000(1 + 0.12)^{15}$$
$$= 10,000(1.12)^{15}.$$

Solution:

a.   By calculator with algebraic logic:

$$\boxed{1.12} \quad \boxed{y^x} \quad \boxed{15} \quad \boxed{\times} \quad \boxed{10,000} \quad \boxed{=}$$

$$\approx 54,736$$

b.   By calculator with RPN logic:

$$\boxed{15} \quad \boxed{\text{ENTER}} \quad \boxed{1.12} \quad \boxed{x^y} \quad \boxed{10,000} \quad \boxed{\times}$$

$$\approx 54,736$$

c.   By logarithms:

$$A = 10,000(1.12)^{15}$$
$$\log A = \log[10,000(1.12)^{15}]$$
$$= \log 10,000 + 15 \log 1.12$$
$$\approx 4 + 15(0.0492)$$
$$= 4 + 0.7380$$
$$= 4.7380$$

Historical Note

John Napier (1550–1617) was the Isaac Asimov of his day, having envisioned the tank, the machine gun, and the submarine. He also predicted that the end of the world would occur between 1688 and 1700. He is best known today as the inventor of logarithms, which up until the advent of the low-cost electronic calculator were used extensively in trigonometry to aid in the calculations necessary for the solution of triangles. Curiously enough, Napier believed that his reputation with posterity would rest on his predictions about the end of the world and on the fact that he thought he had proved, in 1593, that the Pope was the Anti-Christ. He considered logarithms merely an interesting recreational diversion.

The problem now is to use Table 3 in reverse by finding the mantissa 0.7380 in the table; it is log 5.47. Thus, if

$$\log A = 4.7380,$$

then

$$A \approx 5.47 \cdot 10^4$$
$$= 54{,}700.$$

To two significant digits the answer is $55,000.

9. In Section 5.2, we found that the population in Juneau, Alaska, increased at the rate of 6.9% and that the population in 1970 was 13,556. Predict the population in 1980. The growth formula is

$$P = P_0 \, e^{rt},$$

or, for this problem,

$$P = 13{,}556e^{0.069(10)}$$
$$= 13{,}556e^{0.69}$$

Solution:

a. By calculator with algebraic logic:

$$\boxed{0.69} \; \boxed{e^x} \; \boxed{x} \; \boxed{13{,}556} \; \boxed{=} \; \approx 27{,}027$$

b. By calculator with RPN logic:

$$\boxed{0.69} \; \boxed{e^x} \; \boxed{13{,}556} \; \boxed{x} \; \approx 27{,}027$$

c. By logarithms:

$$P = 13{,}556e^{0.69}$$
$$\log P = \log (13{,}556e^{0.69})$$
$$= \log 13{,}556 + 0.69 \log e$$
$$\approx 4.1335 + 0.69(0.4346)$$
$$= 4.433374$$
$$P \approx 2.71 \cdot 10^4$$
$$= 27{,}100$$

Approximate value:

$$13{,}556 \approx 13{,}600$$

$$e \approx 2.72$$

Then use Table 2. A better approximation could be found by using linear interpolation or by using five- or six-place logarithm tables.

To three significant digits, the answer is 27,000.

10. Calculate $x = 0.55^{4.9}$.
    Solution:
    a. By calculator with algebraic logic:

$$\boxed{0.55} \; \boxed{y^x} \; \boxed{4.9} \; \boxed{=} \; \approx 0.0534290163$$

b. By calculator with RPN logic:

$$\boxed{4.9} \; \boxed{\text{ENTER}} \; \boxed{0.55} \; \boxed{x^y} \; \approx 0.0534290163$$

   c.   By logarithms:

$$x = 0.55^{4.9}$$
$$\log x = \log(0.55^{4.9})$$
$$= 4.9 \log 0.55$$
$$\approx 4.9(0.7404 - 1)$$
$$= 4.9(-0.2596)$$
$$= -1.27204$$
$$= 0.72796 - 2$$
$$x \approx 5.34 \cdot 10^{-2}$$
$$= 0.0534$$

To two significant digits, the answer is 0.053.

We have to do this because all mantissas in Table 3 are positive.

PROBLEM SET 5.4

A Problems

Use a calculator, Table 2, or Table 3 to evaluate the expressions in Problems 1–18 to three significant digits.

1. $\log 4.27$	2. $\log 1.08$	3. $\log 8.43$
4. $\ell n\, 2.27$	5. $\ell n\, 16.77$	6. $\ell n\, 2$
7. $\log_3 4.41$	8. $\log_6 9.54$	9. $\log_2 7.57$
10. $\log 9760$	11. $\log 71{,}600$	12. $\log 37{,}000{,}000$
13. $\log 0.042$	14. $\log 0.321$	15. $\log 0.0532$
16. $\ell n\, 521$	17. $\ell n\, 1000$	18. $\ell n\, 360$

Use a calculator or logarithms to evaluate the expressions in Problems 19–31. Be sure to round off your answers to the appropriate number of significant digits. (See Appendix C for a discussion of significant digits.)

19. $(14)(351)$

20. $(218)(263)$

21. $(2.00)^4(1245)(277)$

22. $(3.00)^3(182)$

23. $\dfrac{(1979)(1356)}{452}$

24. $\dfrac{(515)(20{,}600)}{200}$

25. $(990)(1117)(342) - 89$

26. $[0.14 + (197)(25.08)](19)$

27. $\dfrac{1.00}{0.005 + 0.02}$

28. $3478.06 + 2256.028 + 1979.919 + 0.00091$

29. $57{,}300 + 0.094 + 32.3 + 2.09 + 0.0074$

30. $\dfrac{(2.51)^2 + (5.48)^2 - (3.72)^2}{2(2.51)(5.48)}$

31. $\dfrac{241^2 + 568^2 - 351^2}{2(241)(568)}$

For a quick check if you are using a calculator to do problems 19–29, answer the following questions or clues by turning over your calculator and reading the word formed before rounding your answer. (A calculator with 10 decimal places is necessary.) For example, if an answer is

531907018

and you turn over the calculator, it will spell

BIOLOGIES

19. What is opposite of low?
20. You step on them.
21. How do you like your calculator?
22. How are you feeling?
23. What is above your feet?
24. What is below your feet?
25. This problem is illegible.
26. How will you raise the money?
27. This is fun!
28. Where do you live on the pipeline?
29. How do you look taller?

THE PRESS DEMOCRAT

The Redwood Empire's Leading Newspaper   **25** cents

SANTA ROSA, CALIF., SUNDAY, MARCH 28, 1976

World population 4 billion tonight

By EDWARD K. DeLONG

WASHINGTON (UPI) — By midnight tonight, the Earth's population will reach the 4 billion mark, twice the number of people living on the planet just 46 years ago, the Population Reference Bureau said Saturday.

The bureau expressed no joy at the new milestone.

It said global birth rates are too high, placing serious pressures on all aspects of future life and causing "major concern" in the world scientific community, and more than one-third of the present population has yet to reach child-bearing age.

The PRB found cause for optimism, however, in that some governments are stressing birth control to blunt the impact of "explosive growth" and the population growth rate dropped slightly in the past year.

"In 1976, each new dawn brings a formidable increase of approximately 195,000 newborn infants to share the resources of our finite world," it said.

One expert warned that a lack of jobs, rather than too little food, may be the "ultimate threat" facing society as the planet becomes more and more crowded.

It took between two and three million years for the human race to hit the one billion mark in 1850, the PRB said. By 1930, 80 years later, the population stood at 2 billion. A mere 31 years after that, in 1961, it was 3 billion. The growth from 3 to the present 4 billion took just 16 years.

The world could find it has 5 billion people by 1989 — just 13 years from now — if population growth continues at the present rate of 1.8 per cent a year, said Dr. Leon F. Bouvier, vice president of the private, nonprofit PRB.

Bouvier said the newly calculated growth rate is a little lower than the 1.9 per cent estimated last year. Thanks to that slowdown, the passing of the 4 billion milestone came a year later than some demographers had predicted.

"I really think the rate of growth is going to start declining ever so slightly because of declining fertility," Bouvier said. "I think there is some evidence of progress — ever so slow, much too slow."

The new PRB figures show there were 3,982,815,000 people on Earth on Jan. 1. By March 1 the number had grown to 3,994,812,000, the organization said, and by April 1 the total will be 4,000,824,000.

The bureau said its calculations are based on estimates of 328,000 live births per day minus 133,000 deaths.

A growing number of governments are taking steps to slow growth rates, the PRB said.

Singapore appears likely to meet the goal of the two-child family "well before the target date of 1980," it said, and several states in India, which yearly adds the equivalent of the population of Australia, are considering financial incentives to birth control and mandatory sterilization after the birth of two children.

Dr. Paul Ehrlich of Stanford University, one of several population experts contacted by PRB, said he was sad to realize at the age of 44 he had lived through a doubling of Earth's population. He expressed fear the next 44 years could see population growth halted "by a horrifying increase in death rates."

"At this point, hunger does not seem the greatest issue presented by the ever growing number of people," said Dr. Louis M. Hellman, chief of population staff at the Health, Education and Welfare Department.

"Rather, the threat appears to lie in the increasing numbers who can find no work. As these masses of unemployed migrate toward the cities, they create a growing impetus toward political unrest and instability."

Use the formula

$$\dot{P} \doteq P_0\, e^{rt}$$

for Problems 32–36.

 32. On March 28, 1976, the world population reached 4 billion. According to the news article, it took 80 years to increase from 1 to 2 billion people. What is the growth rate for this period?

33. According to the news article, it took 31 years to increase from 2 to 3 billion people. What is the growth rate for this period?

34. According to the news article, it took 16 years to increase from 3 to 4 billion people. What is the growth rate for this period?

35. According to the news article, the current growth rate is 1.8%. Is the article correct when it predicts 5 billion people in 13 years?

36. How long will it take the population to double from 4 to 8 billion if we use the current growth rate of 1.8%?

37. The formula used for carbon-14 dating in archaeology is

$$A = A_0\left(\frac{1}{2}\right)^{t/5700}$$

or

$$P = \left(\frac{1}{2}\right)^{t/5700},$$

where P is the percentage of carbon-14 present after t years. Solve for t.

 38. Some bone artifacts were found at the Lindenmeier site in northeastern Colorado and were tested for their carbon-14 content. If 25% of the original carbon-14 was still present, what is the probable age of the artifacts?

 39. An artifact was discovered at the Debert site in Nova Scotia. Tests showed that 28% of the original carbon-14 was still present. What is the probable age of the artifact?

See Problem 37 for the appropriate formula.

 40. The pH (hydrogen potential) of a solution is given by

$$pH = \log\frac{1}{[H^+]},$$

where $[H^+]$ is the concentration of hydrogen ions in a water solution given in moles per liter. Find the pH for the substances with the $[H^+]$ given.

a. Lemon juice, $5.01 \cdot 10^{-3}$ b. Milk, $3.98 \cdot 10^{-7}$
c. Vinegar, $7.94 \cdot 10^{-4}$ d. Rainwater, $5.01 \cdot 10^{-7}$
e. Seawater, $4.35 \cdot 10^{-9}$

A solution is considered neutral if its pH is 7, acid if the $pH < 7$, and alkaline if the $pH > 7$.

41. A satellite has an initial radioisotope power supply of 50 watts (W). The power output in watts is given by the equation

$$P = 50\, e^{-t/250},$$

where t is the time in days.

a. How much power will be available at the end of one year?
b. What is the half-life of the power supply?
c. The equipment aboard the satellite requires 10 W of power to operate properly. What is the operational life of the satellite?

If your calculator has scientific notation, you should use it for Problem 40. For example, for part a:

$ 42. *Calculator Problem.* The monthly payment, m, for a loan of A dollars at an interest rate of r for t months is

$$m = \frac{A(r/12)[1 + (r/12)]^t}{[1 + (r/12)]^t - 1}.$$

a. What is the monthly payment for a $30,000 loan at 9.75% for 30 years?
b. What is the total interest paid on this loan?

Explanation	Algebraic	RPN
number	5.01	5.01
exponent on power of 10	EE	EEx
negative exponent	+/−	CHS
evaluation of the proper logarithm	1/x LOG	1/x LOG

43. The "learning curve" describes the rate at which a person learns certain tasks. If a person sets a goal of typing N words per minute (wpm) the length of time, t days to achieve this goal is given by

$$t = -62.5\,\ell n\left(1 - \frac{N}{80}\right).$$

a. How long would it take to learn to type 30 wpm?
b. If we accept this formula, is it possible to learn to type 80 wpm?
c. Solve for N.

44. In Problem 43, an equation for learning was given. Psychologists are also concerned with forgetting. In a certain experiment, students were asked to remember a set of "nonsense syllables," such as "htm." The students then had to recall the syllables after t seconds. The equation that was found to describe forgetting was

$$R = 80 - 27\ell n\, t \quad (t \geq 1),$$

where R is the percentage of students who remember the syllables after t seconds.

a. What percentage of the students remembered the syllables after 3 sec?
b. In how many seconds would only 10% of the students remember?
c. Solve for t.

45. Newton's law of cooling states that an object at temperature B surrounded by air temperature A will cool to a temperature T after t minutes according to the equation

$$\cdot T = A + (B - A)\, e^{-kt},$$

where k is a constant depending on the item being cooled.

a. If you draw a tub of 120°F water for a bath and let it stand in a 75°F room, what is the temperature of the water after 30 minutes if $k = 0.01$?
b. What is k for an apple pie taken from a 375°F oven and cooled to 75°F after it is left in a 72°F room for one hour?
c. Solve the given equation for t.

46. The police coroner can use the formula given in Problem 45 to find the approximate time a murder was committed. Suppose the coroner arrives at the scene of a crime at 4:00 P.M. and finds the body temperature of the victim to be 93.2°F. The coroner also notes that the room temperature is 68°F and then waits one hour and takes the body temperature again. This time it is 92.5°F. When was the murder committed, if you assume that the victim's temperature was normal (98.6°F) when death occurred?

47. An advertising agency conducted a survey and found that the number of units sold, N, is related to the amount, a, spent on advertising (in dollars) by the following formula:

$$N = 1500 + 300\ell n\, a \quad (a \geq 1).$$

a. How many units are sold after spending $1000?
b. How many units are sold after spending $50,000?
c. If the company wants to sell 5000 units, how much money will it have to spend?

Historical Note

The *decibel* is named after Alexander Graham Bell, the inventor of the telephone.

48. The intensity of sound is measured in decibels, D, and is given by

$$D = \log\left(\frac{I}{I_0}\right)^{10},$$

where I is the power of the sound in watts per cubic centimeter (W/cm³) and $I_0 = 10^{-16}$ W/cm³ (the power of sound just below the threshold of hearing). Find the number of decibels of the given sound.

a. A whisper, 10^{-13} W/cm³
b. Normal conversation, $3.16 \cdot 10^{-10}$ W/cm³
c. The world's loudest shout by Skipper Kenny Leader, 10^{-5} W/cm³
d. A rock concert, $5.23 \cdot 10^{-6}$ W/cm³

 49. A test of a rocket engine for a certain spacecraft on a launch pad, shows the noise level to be 100 decibels outside the spacecraft and 45 decibels inside. How many times greater is the noise intensity outside the spacecraft than inside?

 50. The Richter scale for measuring earthquakes was developed by Gutenberg and Richter. It relates the energy, E, (in ergs), to the magnitude of the earthquake, M, by the formula

$$M = \frac{\log E - 11.8}{1.5}.$$

a. Solve for E.
b. What was the energy of the 1974 Alaska earthquake, which measured 8.5 on the Richter scale?
c. A small earthquake is one that releases 15^{15} ergs of energy. What is the magnitude of such an earthquake on the Richter scale?
d. A large earthquake is one that releases 10^{25} ergs of energy. What is the magnitude of such an earthquake on the Richter scale.

See Problem 48, for the appropriate equation to use for Problem 49.

The symbol M is the Richter number for the magnitude of an earthquake. From a scientific point of view, the magnitude scale has serious limitations as a means for determining the energy of an earthquake. It is, however, sufficiently accurate for our purposes.

RECENT MAJOR
EARTHQUAKES

India, 1897	8.7
California, 1906	8.3
Japan, 1923	8.3
Chile, 1960	8.5
Alaska, 1964	8.5
Peru, 1970	7.7

Mind Bogglers

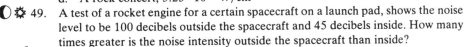

 51. From your local Chamber of Commerce, obtain the population figures for your city for 1950, 1960, and 1970.
a. Find the rate of growth for each period.
b. Forecast the population of your city for the year 2000. Which of the rates you obtained in part a is most accurate for this purpose?

52. *Calculator Problem.* Use the table of oil consumption given here for this problem.
a. Establish that the use of petroleum has grown exponentially.
b. Estimate the rate of growth of the worldwide use of petroleum.
c. Forecast figures for 1975, 1980, 1990, ...
d. Verify forecast figures for years for which data are available.
e. How do these figures compare to estimated world petroleum reserves? Graph use and remaining reserves on one grid. What are your conclusions?

5.5 SUMMARY AND REVIEW

TERMS

Algebraic function (5.1)
Base (5.1)
Characteristic (5.4)
Common logarithm (5.4)

Exponential function (5.1)
Logarithmic function (5.3)
Mantissa (5.4)
Natural logarithm (5.4)

World Consumption of
Petroleum Products

Year	Millions of barrels
1915	43
1920	68
1925	110
1930	150
1935	170
1940	220
1945	260
1950	340
1955	480
1960	780
1965	1130
1970	1670

e (5.2)   Principal nth root (5.1)
Exponent (5.1)   Scientific notation (5.4)
Exponential equation (5.2)   Transcendental function (5.1)

IMPORTANT IDEAS

(5.1) The laws of exponents hold for any real exponents.
(5.1) Sketching exponential functions
(5.2) Solving exponential equations: if $b^x = b^y$, then $x = y$.
(5.2) e is an irrational number approximately equal to 2.72.
(5.3) $y = \log_b x \Leftrightarrow x = b^y$ $(b > 0, b \neq 1)$
(5.3) Laws of logarithms (A and B positive):

   1. $\log_b AB = \log_b A + \log_b B$

   2. $\log_b \dfrac{A}{B} = \log_b A - \log_b B$

   3. $\log_b A^p = p \log_b A$

(5.3) $b^{\log_b x} = x$
(5.3) Sketching logarithmic functions
(5.3) Solving logarithmic equations: $\log_b A = \log_b B \Leftrightarrow A = B$
(5.4) Change of base of a logarithm:

$$\log_b x = \frac{\log_a x}{\log_a b}$$

(5.4) Evaluating logarithms using tables or a calculator
(5.4) Solving applied problems involving exponential and logarithmic functions

REVIEW PROBLEMS

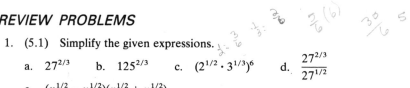

1. (5.1) Simplify the given expressions.

   a. $27^{2/3}$ b. $125^{2/3}$ c. $(2^{1/2} \cdot 3^{1/3})^6$ d. $\dfrac{27^{2/3}}{27^{1/2}}$

   e. $(x^{1/2} - y^{1/2})(x^{1/2} + y^{1/2})$

2. (5.1, 5.3) Sketch the graph of the given functions.

   a. $y = (\frac{1}{2})^x$ b. $y = -2^x$ c. $y = 2^{-x}$ d. $y = \log x$

   e. $y = \ell n\, x$

3. (5.2, 5.3) Solve for x.

   a. $5^x = 125$ b. $25^x = \sqrt{5}$ c. $5^{2x+1} = 0.2$

   d. $\log_5 25 = x$ e. $\log_x(x + 6) = 2$

4. (5.3) Evaluate the given logarithms without a calculator or tables.

   a. $\log_3 27$ b. $\log_2 0.125$ c. $\log 0.01$ d. $\ell n\, e^5$ e. $\log_8 16$

5. (5.3) Solve for x.

   a. $3 \log 3 - \frac{1}{2} \log 3 = \log \sqrt{x}$ b. $2\ell n \dfrac{e}{\sqrt{7}} = 2 - \ell n\, x$

6. (5.4) Evaluate the given expression to three significant digits using tables or a calculator.

   a. $\log 729,000$ b. $\log 0.00216$ c. $\log_5 4.51$ d. $\log_2 818$

   e. $\ell n\, 624$

7. (5.2) The half-life of arsenic-76 is 26.5 hours. Find the constant k for which

$$A = A_0 e^{-kt}$$

8. (5.4) In the years 1960–1970, the population of Key West, Florida, decreased from 34,000 to 29,000 people. What was the rate of decline for this period?

Use the formula $P = P_0 e^{rt}$ for Problem 8.

9. (5.4) If a person's present salary is $20,000 per year, use the formula

$$A = P(1 + r)^n$$

to determine the salary necessary to equal this salary in 15 years if you assume the given inflation rate remains constant.

a. 5.5% (United States inflation rate for 1976)
b. 4.5% (West Germany inflation rate for 1976)
c. 16.0% (Great Britain inflation rate for 1976)

10. (5.4) *Calculator Problem.* The formula for the present value of an annuity is

$$A = P\left[\frac{1 - (1 + r)^{-n}}{r}\right].$$

a. This formula may be used for determining the amount, A, you can borrow if the interest rate, r, and monthly payment, P, are known. Suppose you can obtain a $9\frac{1}{4}$% loan for 30 years and can afford $260 per month for repayment. How much can you borrow?

b. Solve for n.

Note: the interest rate stated is usually a yearly rate, but in this problem you are using a period of one month. Therefore,

$$r = \frac{\text{yearly rate}}{12} = \frac{0.0925}{12}.$$

Also, n is the number of months so

$$n = 12(\text{number of years})$$
$$= 12(30)$$
$$= 360.$$

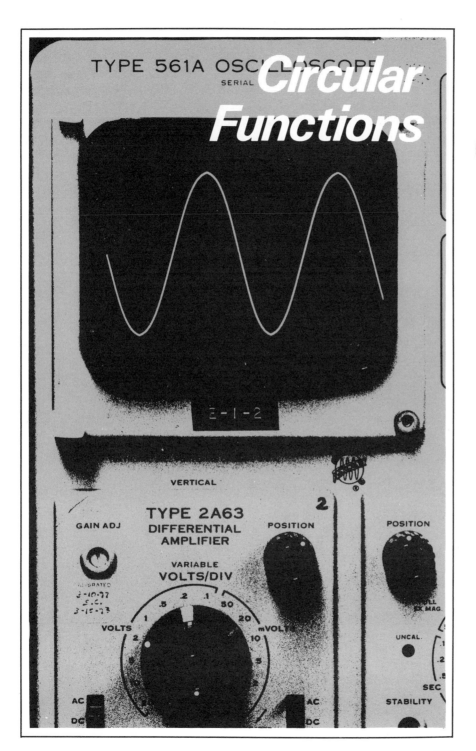

Circular Functions

6

6.1 INTRODUCTION—ANGLES

In mathematics, it is often useful to consider functions of angles; however, the definition of an angle depends on the context in which it is being used. In geometry, an angle is usually defined as the union of two rays with a common endpoint. In more advanced mathematics courses, a more general definition is used.

ANGLE

> DEFINITION OF AN ANGLE: An *angle* is formed by rotating a ray about its endpoint (called the *vertex*) from some initial position (called the *initial side*) to some terminal position (called the *terminal side*). The measure of an angle is the amount of rotation. An angle is also formed if a line segment is rotated about one of its endpoints.

If the rotation of the ray is in a counterclockwise direction, the measure of the angle is called *positive*. If the rotation is in a clockwise direction, the measure is called *negative*. The notation ∠*ABC* means the measure of an angle with vertex *B* and points *A* and *C* (different from *B*) on the sides; ∠*B* denotes the measure of an angle with vertex at *B*, and a curved arrow is used

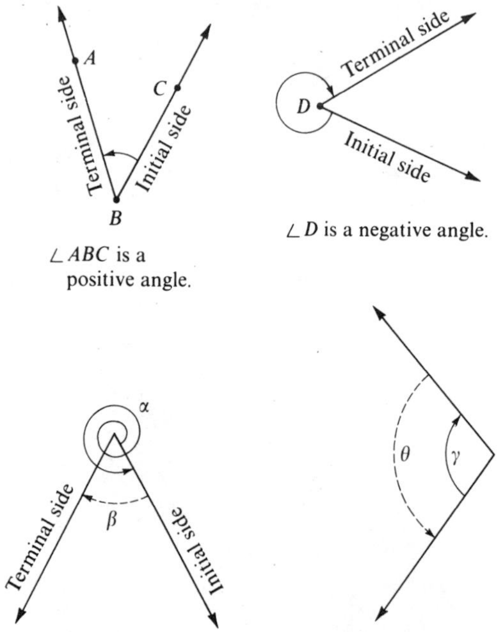

Table of Commonly Used
Greek Letters

Symbol	Name
α	alpha
β	beta
γ	gamma
θ	theta
λ	lambda
ϕ or φ	phi
ω	omega

π (pi) is a lowercase Greek letter that will not be used to represent an angle. It denotes an irrational number approximately equal to 3.141592654.

∠*ABC* is a positive angle.

∠*D* is a negative angle.

α is a positive angle.
β is a negative angle.

θ is a positive angle.
γ is a negative angle.

Figure 6.1   Examples of angles

to denote the direction and amount of rotation, as shown in Figure 6.1. If no arrow is shown, the measure of the angle is considered to be the smallest positive rotation. Lowercase Greek letters are also used to denote the measure of angles. Some examples are shown in Figure 6.1.

A Cartesian coordinate system may be superimposed on an angle so that the vertex is at the origin and the initial side is along the positive *x*-axis. In this case, the angle is in *standard position*. Angles in standard position having the same terminal sides are *coterminal angles*. Given any angle α, there is an unlimited number of coterminal angles (some positive and some negative). In Figure 6.2, β is coterminal with α. Can you find other angles coterminal with α?

STANDARD-POSITION ANGLES
COTERMINAL ANGLES

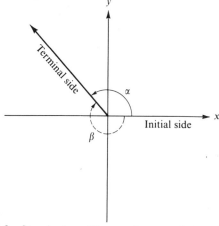

Figure 6.2 Standard-position angles α and β; α is a positive angle, and β is a negative angle; α and β are coterminal angles

Several methods are used for measuring angles. Let α be an angle in standard position with a point P not the vertex but on the terminal side. As this side is rotated through one revolution, the trace of the point P forms a circle. The measure of the angle is one revolution, but, since much of our work will be with amounts less than one revolution, we need to define measures of smaller angles. Historically, the most common scheme divides one revolution into 360 equal parts with each part called a *degree*. Sometimes even finer divisions are necessary, so a degree is divided into 60 equal parts, each called a *minute* (thus, $1° = 60'$). Furthermore, a minute can be divided into 60 equal parts, each called a *second* (thus, $1' = 60''$). For most applications, we will write decimal parts of degrees instead of minutes and seconds. That is, 32.5° is preferred over 32° 30'.

DEGREE MEASURE FOR ANGLES

In calculus and in scientific work, another measure for angles is defined. This method uses real numbers to measure angles. Let's draw a circle with any nonzero radius *r*. Next, we measure out an arc with length *r*. Figure 6.3a shows the case in which $r = 1$ and Figure 6.3b shows $r = 2$. Regardless of our choice for r, the angle determined by this arc of length r is the same (it is labeled θ in the figure). This angle is used as a basic unit of measurement and is called a *radian*. Notice that the circumference, C, generates an angle of one revolution. Since $C = 2\pi r$, and since the basic unit of measurement on the circle is r, we see that

$$\text{one revolution} = \frac{C}{r}$$

$$= \frac{2\pi r}{r}$$

$$= 2\pi.$$

Thus, $\frac{1}{2}$ revolution is $\frac{1}{2}(2\pi) = \pi$ radians; $\frac{1}{4}$ revolution is $\frac{1}{4}(2\pi) = \pi/2$ radians.

RADIAN MEASURE FOR ANGLES

𝕳istorical 𝕹ote

The term *radian* was introduced in 1873 by the physicist James T. Thompson to simplify certain formulas in mathematics and physics.

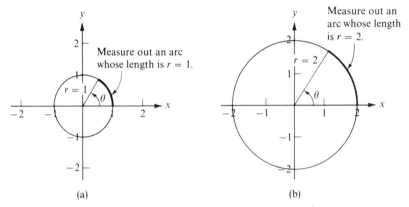

Figure 6.3   Radian measure

Do not attempt to change these problems to degrees. Work them directly as shown in the examples. You will have to work at thinking in terms of radian measure. You should memorize the approximate size of an angle of measure 1 radian in much the same way you have memorized the approximate size of an angle of measure 45°.

EXAMPLES: Let $r = 1$. Draw the angles with the given measures.

1. $\theta = 2$ radians

2. $\theta = 3$ radians

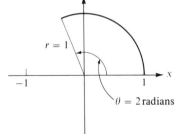

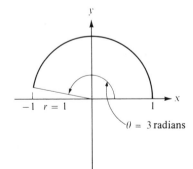

3. $\theta = 0.5$ radian

4. $\theta = -\dfrac{\pi}{4}$ radian

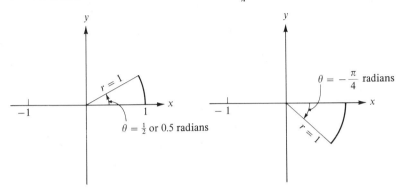

We are led to the following relationships.

one revolution is measured by 360°
one revolution is measured by 2π radians

If θ is the measure of any angle, then

$$\text{number of revolutions} = \frac{\theta \text{ in degrees}}{360}$$

$$\text{number of revolutions} = \frac{\theta \text{ in radians}}{2\pi}$$

Therefore:

$$\frac{\theta \text{ in degrees}}{360} \quad \Leftrightarrow \quad \frac{\theta \text{ in radians}}{2\pi}$$

RELATIONSHIP
BETWEEN DEGREE AND
RADIAN MEASURE

EXAMPLES

5. Change 45° to radians.

$$\frac{45}{360} = \frac{\theta}{2\pi}$$

$$\frac{90\pi}{360} = \theta$$

$$\frac{\pi}{4} = \theta$$

An alternative method is to remember that π radians is 180°. That is, since 45° is $\frac{1}{4}$ of 180°, you know that the radian measure is $\pi/4$.

6. Change 1° to radians.

$$\frac{1}{360} = \frac{\theta}{2\pi}$$

$$\frac{\pi}{180} = \theta$$

Even though this solution is in the desired form, you might be interested in performing the above division on a calculator: $1° \approx 0.0174532925$ radian.

7. Change 1 to degrees.

$$\frac{\theta}{360} = \frac{1}{2\pi}$$

$$\theta = \frac{360}{2\pi}$$

Note that if units are not specified, radians are understood.

On a calculator, we find $\theta \approx 57.29577951°$, or 57° 17′ 45″.

Decimal degrees is the preferred form.

For the more common measures of angles, it is a good idea to memorize the equivalent degree and radian measures. If you keep in mind that 180° in radian measure is π, the rest of the values will be easy to remember.

This relationship between degree measure and radian measure should be memorized.

DEGREES	RADIANS
0	0
30	$\pi/6$
45	$\pi/4$
60	$\pi/3$
90	$\pi/2$
180	π
270	$3\pi/2$
360	2π

An *arc* is part of a circle; thus, *arc length* is the distance around part of a circle. The arc length corresponding to one revolution is the circumference.

We now relate the radian measure of an angle to a circle to find the *arc length*. Let s be the length of an arc and let θ be the angle measured in radians. Then

$$\text{angle in revolutions} = \frac{s}{2\pi r},$$

since one revolution has an arc length (circumference of the circle) of $2\pi r$. Also,

$$\text{angle in radians} = (\text{angle in revolutions})(2\pi).$$

Substituting, we get

$$\theta = \frac{s}{2\pi r}(2\pi)$$

$$= \frac{s}{r}.$$

This result says that the arc length, s, cut by a central angle θ, *measured in radians*, of a circle of radius r is given by

$$s = r\theta.$$

EXAMPLE 8: The length of the arc subtended (cut off) by a central angle of 36° in a circle with a radius of 20 centimeters (cm) is found by first changing 36° to radians so that we can use the formula given above.

$$\frac{36}{360} = \frac{\theta}{2\pi}$$

Solving for θ, we find

$$\frac{\pi}{5} = \theta.$$

Thus,

$$s = 20\left(\frac{\pi}{5}\right)$$

$$= 4\pi.$$

The length of the arc is 4π cm. This is about 12.6 cm.

THINK METRIC

There is an important relationship between angles and circles. Draw a unit circle with an angle θ in standard position, as shown in Figure 6.4.

The *unit circle* is the circle centered at the origin with radius $r = 1$. The equation of the unit circle is

$$x^2 + y^2 = 1.$$

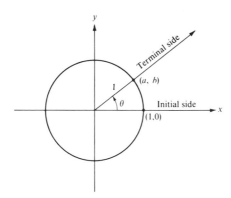

Figure 6.4   The unit circle

The initial side intersects the unit circle at $(1, 0)$, and the terminal side intersects the unit circle at (a, b). Define functions of θ as follows:

$$c(\theta) = a \quad \text{and} \quad s(\theta) = b.$$

We could have used f for the first component and s for the second component, but we've already used f for general functions, so we use c for first component and s for second component. We'll use this later in the book, so remember this notation.

EXAMPLES: Evaluate the given functions.

9. $s(90°)$; this is the second component of the ordered pair (a, b), where (a, b) is the point of intersection of the terminal side of a 90° angle and the unit circle. By inspection, it is 0. Thus, $s(90°) = 1$.

10. $c(90°) = 0$.

11. $s(-270°) = 1$; this is the same as Example 9, since the angle $-270°$ is coterminal with 90°.

We'll continue our discussion of the functions $c(\theta)$ and $s(\theta)$ in the next section.

PROBLEM SET 6.1

A Problems

1. From memory, give the radian measure for each angle with degree measure stated below.

    a. 30°   b. 90°   c. 270°   d. 45°   e. 360°   f. 60°   g. 180°

2. From memory, give the degree measure for each angle with radian measure stated below.

    a. π b. $\pi/4$ c. $\pi/3$ d. 2π e. $\pi/2$ f. $3\pi/2$ g. $\pi/6$

3. *Without* first changing to degrees, sketch the angle with the measure given (remember that π measures a straight angle).

    a. $\pi/2$ b. $2\pi/3$ c. $-\pi/4$ d. 2.5   e. -1

4. *Without* first changing to degrees, sketch the angle with the measure given (remember that π measures a straight angle).

    a. $-3\pi/2$ b. $\pi/6$ c. $3\pi/4$ d. 6   e. -1.2365

Find a positive angle less than one revolution that is coterminal with each angle in Problems 5–18.

5. a. 400°   b. 540°   6. a. 750°   b. 1050°
7. a. 3π b. $13\pi/6$ 8. a. $11\pi/3$ b. $17\pi/4$
9. a. $-30°$ b. $-200°$ 10. a. $-55°$ b. $-320°$

11. a. $-\pi/4$	b. $-\pi$		12. a. 7	b. -2	
13. a. 9	b. -5		14. a. $\sqrt{50}$	b. -6	
15. a. 6.2832	b. -3.1416		16. a. 9.4247	b. 6.8068	
17. a. -0.7854	b. $3\sqrt{5}$		18. a. 150	b. 30	

It would be helpful to have a calculator for Problems 12–18.

Evaluate each of the functions given in Problems 19–22.

19. a. $c(-270°)$ b. $s(-3\pi/2)$ c. $s(270°)$ d. $c(\pi)$ e. $s(2\pi)$
20. a. $s(90°)$ b. $c(\pi/2)$ c. $c(270°)$ d. $s(\pi)$ e. $c(2\pi)$
21. a. $s(3\pi)$ b. $c(-\pi)$ c. $s(630°)$ d. $c(-270°)$ e. $s(-450°)$
22. a. $c(3\pi)$ b. $s(-\pi)$ c. $c(630°)$ d. $s(-270°)$ e. $c(-450°)$
23. Let $[c(\theta)]^2$ and $[s(\theta)]^2$ be written as $c^2(\theta)$ and $s^2(\theta)$.
   a. Find $c^2(0°) + s^2(0°)$.
   b. Find $c^2(270°) + s^2(270°)$.
   c. Find $c^2(\pi/2) + s^2(\pi/2)$.
   d. Make a conjecture about $c^2(\theta) + s^2(\theta)$ for any angle θ based on your answers to parts a–c.
   e. Prove $c^2(\theta) + s^2(\theta) = 1$.

B Problems

Change the given measures of the angles in Problems 24–31 to decimal degrees to the nearest hundredth.

EXAMPLE 12: 14° 20′

The problem is to change 20′ to a decimal number. This means that we have 20/60, which by division gives 0.333 . . . . To the nearest hundredth, 14° 20′ = 14.33°.

---

24. 52° 30′	25. 65° 40′	26. 146° 50′	27. $-85°$ 20′
28. $-127°$ 10′	29. 315° 25′	30. 16° 42′	31. 29° 17′

Change the given measures of the angles in Problems 32–36 to decimal degrees to the nearest thousandth.

EXAMPLE 13: 42° 13′ 40″

Since $1″ = 1/3600$ of a degree, we have

$$13′\,40″ = \frac{13}{60} + \frac{40}{3600}$$

$$= \frac{820}{3600}.$$

By division, $13′\,40″ = 0.22777\ldots°$. To the nearest thousandth, 42° 13′ 40″ = 42.228°. As you can see, it is quite appropriate to use a calculator to work these problems. If you have a four-function (+ , − , × , ÷) calculator, you can do the problem as follows:

For a calculator with algebraic logic:

$\boxed{\text{number of minutes}}$ $\boxed{\times}$ $\boxed{60}$ $\boxed{+}$ $\boxed{\text{number of seconds}}$ $\boxed{=}$

$\boxed{\div}$ $\boxed{3600}$ $\boxed{=}$ $\boxed{+}$ $\boxed{\text{number of degrees}}$ $\boxed{=}$

The desired answer is now displayed.

For a calculator with RPN logic:

$\boxed{\text{number of minutes}}$ $\boxed{\text{ENTER}}$ $\boxed{60}$ $\boxed{\times}$ $\boxed{\text{number of seconds}}$

$\boxed{+}$ $\boxed{3600}$ $\boxed{\div}$ $\boxed{\text{number of degrees}}$ $\boxed{+}$

Many calculators that are more sophisticated have a single key conversion for this problem. Check the operating guide for your calculator.

---

32. 14° 30′ 50″   33. 48° 28′ 10″   34. 12′ 24″
35. −94° 21′ 31″   36. 281° 31′ 36″

Change the angles in Problems 37–42 to degrees.

EXAMPLE 14: $\pi/9$

Since

$$\frac{\text{(angle in degrees)}}{360} = \frac{\text{(angle in radians)}}{2\pi},$$

we have

$$\theta = \frac{360}{2\pi} \text{ (angle in radians)}$$

$$= \frac{180}{\pi}\left(\frac{\pi}{9}\right)$$

$$= 20.$$

Thus, $\pi/9$ is an angle with a measure of 20°.

---

37. $2\pi/9$ 38. $\pi/10$ 39. $\pi/30$ 40. $5\pi/3$ 41. $-11\pi/12$ 42. $3\pi/18$

Change the angles in Problems 43–48 to degrees correct to the nearest hundredth.

EXAMPLE 15: 2.3

$$\theta = \frac{180}{\pi}(2.3)$$

$$\approx (57.296)(2.3)$$

$$\approx 131.7808$$

To the nearest hundredth, the angle is 131.78°.

If you have a calculator, you can obtain a much more accurate answer by using a better approximation for $180/\pi$. For example:

Algebraic logic	RPN logic
180	180
÷	π
π	÷
×	2.3
2.3	×
=	

The result is 131.7802929.

---

43. 2   44. −3   45. −0.25   46. −2.5   47. 0.4   48. 0.51

Change the angles in Problems 49–54 to radians.

EXAMPLE 16: 125°

$$\frac{\text{angle in degrees}}{360} = \frac{\text{angle in radians}}{2\pi}$$

$$\theta = \frac{2\pi}{360}(\text{angle in degrees})$$

$$= \frac{\pi}{180}(125)$$

$$= \frac{25\pi}{36}$$

---

49. 40°   50. 20°   51. −64°   52. −220°   53. 254°   54. 85°

Change the angles in Problems 55–58 to radians correct to the nearest hundredth.

EXAMPLE 17: 43°

$$\theta = \frac{\pi}{180}(43)$$

$$\approx (0.0175)(43)$$

$$= 0.7525$$

To the nearest hundredth, the angle is 0.75.

If you have a calculator, you can obtain a much more accurate answer by using a better approximation for $\pi/180$.

---

55. 112°   56. 314°   57. −62.8°   58. 350°

In Problems 59–66, find the intercepted arc to the nearest hundredth if the central angle and radius are as given.

59. Angle 1, radius 1 m
60. Angle 2.34, radius 6 cm
61. Angle 3.14, radius 10 m
62. Angle $\pi/3$, radius 4 m
63. Angle $3\pi/2$, radius 15 cm
64. Angle 40°, radius 7 ft
65. Angle 72°, radius 10 ft
66. Angle 112°, radius 7.2 cm
67. How far does the tip of an hour hand on a clock move in three hours if the hour hand is 2 cm long?

 68. A 50-cm pendulum on a clock swings through an angle of 100°. How far does the tip travel in one arc?

 69. In about 230 B.C., a mathematician named Eratosthenes estimated the radius of the earth using the following information: Syene and Alexandria in Egypt are on the same line of longitude. They are also 500 miles apart. At noon on the longest day of the year, when the sun was directly overhead in Syene, Eratosthenes measured the sun to be 7.2° from the vertical in Alexandria. Because of the distance of the earth from the sun, he assumed that the rays were parallel. Thus, he concluded that the central angle subtending rays from the center of the earth to Syene and Alexandria was also 7.2°. Using this information, find the approximate radius of the earth.

 70. *Calculator Problem.* Omaha, Nebraska, is located at approximately 97° west longitude, 41° north latitude; Wichita, Kansas, is located at approximately 97° west longitude, 37° north latitude. Notice that these two cities have about the same longitude. If we know that the radius of the earth is about 6370 kilometers (km), what is the distance between these cities to the nearest 10 km?

 71. *Calculator Problem.* Entebbe, Uganda, is located at approximately 33° east longitude, and Stanley Falls in Zaire is located at 25° east longitude. Also, both these cities lie approximately on the equator. If we know that the radius of the earth is about 6370 km, what is the distance between the cities to the nearest 10 km?

Mind Bogglers

 72. *Calculator Problem.* Suppose it is known that the moon subtends an angle of 45.75′ at the center of the earth. It is also known that the center of the moon is 238,866 miles from the surface of the earth. What is the diameter of the moon to the nearest 10 miles?

73. One side of a triangle is 20 cm longer than another, and the angle between them is 60°. If two circles are drawn with these sides as diameters, one of the points of intersection of the two circles is the common vertex. How far from the third side is the other point of intersection?

6.2 SINE AND COSINE FUNCTIONS

In the last section, we considered the problem of drawing a unit circle with an angle θ in standard position and also the intersection of the angle's sides and the circle at the points $(1, 0)$ and (a, b), as shown in Figure 6.5. Then we defined functions c and s by $c(\theta) = a$ and $s(\theta) = b$. In this section, this idea is generalized. Suppose that $P(x, y)$ is a point on the terminal side at a distance r from the origin ($r \neq 0$). If points A, B, and Q are labeled as shown in Figure 6.5, then $\triangle AOB \sim \triangle POQ$, and therefore the sides are proportional.

$$a = \frac{a}{1} = \frac{x}{r} \quad \text{and} \quad b = \frac{b}{1} = \frac{y}{r}$$

The notation $\triangle AOB \sim \triangle POQ$ means that triangles AOB and POQ are *similar*. Recall that two triangles are similar if two angles of one are congruent to two angles of the other. For these triangles, $\angle OBA$ is congruent to $\angle OQP$ since they are both right angles; also, $\angle O$ is congruent to $\angle O$ since equal angles are congruent. The important property of similar triangles is that corresponding parts of similar triangles are proportional.

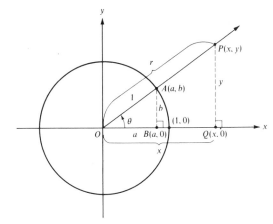

Figure 6.5   A unit circle with similar triangles AOB and POQ

Notice that, since $\triangle POQ$ is a right triangle,

$$x^2 + y^2 = r^2$$

or

$$r = \sqrt{x^2 + y^2}$$

by the Pythagorean Theorem. Thus,

$$c(\theta) = a = \frac{x}{r} \quad \text{and} \quad s(\theta) = b = \frac{y}{r}.$$

This generalization allows us to find additional values for the functions c and s. For example, if you want to find $c(45°)$, you can find $c(45°) = a$, which is the first component of the point (a, b) on the unit circle that is intersected by the terminal side of a 45° angle. (See Figure 6.6.) But instead of finding a, pick any point on the terminal side different from $(0, 0)$—say $(2, 2)$. Notice that an angle of 45° determines a ray that bisects the first quadrant. Since (x, y) represents a point on the terminal side (x and y are both positive in the first quadrant), $x = y$.

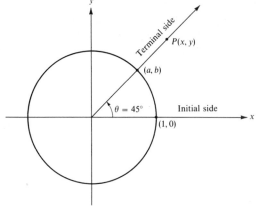

Figure 6.6   A unit circle with any point $P(x, y)$ on the terminal side of a 45° angle

$$c(45°) = \frac{x}{r}$$

$$= \frac{x}{\sqrt{x^2 + y^2}}$$

$$= \frac{2}{\sqrt{4 + 4}}$$

$$= \frac{2}{2\sqrt{2}}$$

$$= \frac{1}{\sqrt{2}} \quad \text{or} \quad \frac{1}{2}\sqrt{2}$$

What if you choose $(1, 1)$ on the terminal side? Then

$$c(45°) = \frac{1}{\sqrt{1 + 1}}$$

$$= \frac{1}{2}\sqrt{2}.$$

Any point different from $(0, 0)$ on the terminal side gives the same result.

EXAMPLES

1. Find $s(45°)$. By definition, $s(\theta) = b = y/r$, so choose *any* point except $(0, 0)$ on the terminal side—say $(1, 1)$—and find

$$s(45°) = \frac{y}{\sqrt{x^2 + y^2}}$$

$$= \frac{1}{\sqrt{1 + 1}}$$

$$= \frac{1}{\sqrt{2}} \quad \text{or} \quad \frac{1}{2}\sqrt{2}.$$

Notice that for $\theta = 45°$, $s(\theta) = c(\theta)$, as expected.

2. Find $c(5\pi/4)$. Find a point on the terminal side, as shown in Figure 6.7. Notice that the terminal side bisects the third quadrant and that x and y are both negative. Thus, $x = y$, so choose *any* point except $(0, 0)$ on the terminal side—say $(-1, -1)$.

$$c\left(\frac{5\pi}{4}\right) = \frac{x}{r}$$

$$= \frac{-1}{\sqrt{(-1)^2 + (-1)^2}}$$

$$= \frac{-1}{\sqrt{2}}$$

$$= -\frac{1}{2}\sqrt{2}$$

Notice from Figure 6.7 that $(1, 1)$ is not on the terminal side of the angle $5\pi/4$.

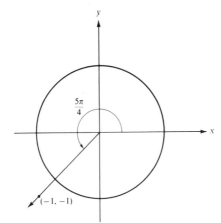

Figure 6.7   A unit circle with an angle of $5\pi/4$

These functions c and s are called the *cosine* and *sine functions* in trigonometry and are usually written $c(\theta) = \cos\theta$ and $s(\theta) = \sin\theta$. There are many properties associated with these functions. For example, $\cos\theta$ will have the same value as $\cos\theta'$ if θ and θ' are coterminal, since the definition of cosine requires that we choose any point different from the vertex and on the terminal side and, by definition, coterminal angles have the same terminal

COSINE AND SINE FUNCTIONS

side. This point can be summarized by saying

$$\cos{(\theta + 2\pi)} = \cos{\theta} \quad \text{and} \quad \sin{(\theta + 2\pi)} = \sin{\theta}$$

since $\theta + 2\pi$ and θ are coterminal angles. The following relationships are also true:

$$\cos{(\theta - 2\pi)} = \cos{\theta} \quad \sin{(\theta - 2\pi)} = \sin{\theta}$$
$$\cos{(\theta + 4\pi)} = \cos{\theta} \quad \sin{(\theta - 6\pi)} = \sin{\theta}$$

In general:

Cosine and sine are periodic with period 2π.

$$\cos{(\theta \pm 2n\pi)} = \cos{\theta} \quad \text{and} \quad \sin{(\theta \pm 2n\pi)} = \sin{\theta} \quad \text{for any integer } n$$

In other words, the functional values of the cosine and the sine functions repeat themselves after $\pm 2\pi$ ($\pm 360°$) radians. We say that these functions are *periodic* with period 2π.

EXAMPLES

3. Find $\cos 405°$. Since $405°$ is coterminal with $45°$ ($405° = 360° + 45°$), we have

$$\cos 405° = \cos 45° = \frac{1}{2}\sqrt{2}.$$

4. Find $\sin{(-7\pi/4)}$. Since $-7\pi/4$ is coterminal with $\pi/4$, we have

$$\sin\left(\frac{-7\pi}{4}\right) = \sin\frac{\pi}{4} = \frac{1}{2}\sqrt{2}.$$

5. Find $\cos 60°$. Find a point (x, y) on the terminal side of this angle, as shown in Figure 6.8. In geometry, it is shown that for a $30°$–$60°$ right triangle, the length of the hypotenuse is twice the length of the shorter leg and the longer leg is $\sqrt{3}$ times the length of the shorter leg. Thus, $r = 2x$ and

$$\cos 60° = \frac{x}{r}$$

$$= \frac{x}{2x}$$

$$= \frac{1}{2}.$$

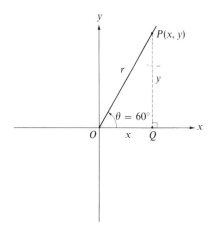

Figure 6.8   A point $P(x, y)$ on the terminal side of an angle of 60°

PROBLEM SET 6.2

A Problems

Each of the angles in Problems 1–6 is coterminal with an angle θ so that $0 \leq \theta \leq 90°$. Find θ using the unit of measurement (radians or degrees) given in the problem.

1. $-300°$ 2. $-330°$ 3. $17\pi/4$ 4. -6π 5. $9\pi/2$ 6. $-675°$

Find the functional values in Problems 7–26.

7. $\sin 0°$ 8. $\sin 30°$ 9. $\sin (\pi/4)$ 10. $\sin 60°$
11. $\sin (\pi/2)$ 12. $\sin 180°$ 13. $\sin 270°$ 14. $\cos 0$
15. $\cos (\pi/6)$ 16. $\cos 45°$ 17. $\cos (\pi/2)$ 18. $\cos 90°$
19. $\sin \pi$ 20. $\cos (3\pi/2)$ 21. $\cos (-300°)$ 22. $\sin 390°$
23. $\sin (17\pi/4)$ 24. $\cos (-6\pi)$ 25. $\cos (9\pi/2)$ 26. $\sin (-675°)$

Hint for Problem 8: use the result from geometry given in Example 5, $r = 2y$, $x = \sqrt{3}y$.

Simplify the expressions in Problems 27–41.

27. $\sqrt{2x^2}$

28. $\sqrt{2x^2}$ if x is positive

29. $\sqrt{2x^2}$ if x is negative

30. $\dfrac{x}{\sqrt{2x^2}}$ if x is negative

31. $\dfrac{x}{\sqrt{4x^2}}$ if x is positive

32. $\sin 30° + \cos 0$

33. $2 \cos (\pi/2)$

34. $\cos 2(\pi/2)$

Hint for Problems 27–30: recall that $\sqrt{x^2} = |x|$. This means that $\sqrt{x^2} = x$ if x is positive and $-x$ if x is negative.

Hint for Problem 38:
remember that $\cos^2(\pi/4)$
means $[\cos(\pi/4)]^2$. We write
$\cos^2\theta = (\cos\theta)^2$ and $\sin^2\theta$
$= (\sin\theta)^2$. Notice, $\cos^2 x$
$\neq \cos(\cos x)$ and $\cos^2 x$
$= (\cos x)^2 \neq \cos x^2$.

35. $\sin 2(\pi/4)$

36. $2\sin(\pi/4)$

37. $\sin(\pi/2) + 3\cos(\pi/2)$

38. $\cos^2(\pi/4)$

39. $\sin^2 60°$

40. $\sin^2(\pi/2) + \cos^2(\pi/2)$

41. $\sin^2(\pi/6) + \cos^2(\pi/2)$

B Problems

42. In this section, $\cos 45°$ was found by choosing *particular points* other than the vertex on the terminal side. If we let $x = y$, you can find $\cos 45°$ by working with an *arbitrary point* (x, y) on the terminal side.

$$\cos 45° = \frac{x}{\sqrt{x^2 + y^2}}$$

$$= \frac{x}{\sqrt{x^2 + x^2}}$$

$$= \frac{x}{\sqrt{2x^2}}$$

$$= \frac{x}{x\sqrt{2}}$$

$$= \frac{1}{\sqrt{2}} \quad \text{or} \quad \frac{1}{2}\sqrt{2}$$

Now, from Problem 27, $\sqrt{2x^2} \neq x\sqrt{2}$. Why is it that we can say $\sqrt{2x^2} = x\sqrt{2}$ for this problem?

43. In this section, $\cos(5\pi/4)$ was found by choosing a particular point on the terminal side. Notice that the terminal side bisects the third quadrant and that x and y are both negative. Show that $\cos(5\pi/4) = -\frac{1}{2}\sqrt{2}$ by working with an arbitrary point (x, y)—not $(0, 0)$—as shown in Problem 42.

Find the values in Problems 44–47 by working with an arbitrary point (x, y)—not $(0, 0)$—as shown in Problem 42.

44. $\cos 135°$

45. $\sin(-\pi/4)$

46. $\sin 210°$

47. $\cos 210°$

Mind Bogglers

48. What is the smaller angle between the hands of a clock at 12:25 P.M.?

49. a. Let $P(x, y)$ be any point in the plane. Show that $P(r\cos\theta, r\sin\theta)$ is a representation for P where θ is the standard-position angle formed by drawing ray $\overrightarrow{OP}$.

   b. Let $A(\cos\alpha, \sin\alpha)$ and $B(\cos\beta, \sin\beta)$ be any two points on a unit circle. Use the distance formula to show that

$$|AB| = \sqrt{2 - 2(\cos\alpha\cos\beta + \sin\alpha\sin\beta)}.$$

6.3 TANGENT AND COFUNCTIONS

In our discussion of sines and cosines in the last section, we considered certain ratios formed by choosing a point (x, y) a distance $r > 0$ units from the origin on the terminal side of an angle θ. That is, $\cos \theta = x/r$ and $\sin \theta = y/r$ were defined. There are four other ratios to consider—namely, r/x, r/y, y/x, and x/y. In trigonometry, each of these ratios has a name. As you saw in the last section, these ratios are functions of the angle θ and do not depend on the choice of the point P on the terminal side.

DEFINITION OF THE CIRCULAR FUNCTIONS: Let θ be an angle in standard position with a point $P(x, y)$ on the terminal side a distance $r = \sqrt{x^2 + y^2}$ from the origin ($r \neq 0$). Then the six circular functions are defined as follows:

cosine function: $\cos \theta = \dfrac{x}{r}$ **secant** function: $\sec \theta = \dfrac{r}{x}$ $(x \neq 0)$

sine function: $\sin \theta = \dfrac{y}{r}$ **cosecant** function: $\csc \theta = \dfrac{r}{y}$ $(y \neq 0)$

tangent function: $\tan \theta = \dfrac{y}{x}$ $(x \neq 0)$ **cotangent** function: $\cot \theta = \dfrac{x}{y}$ $(y \neq 0)$

Using these definitions, various functional values for θ can now be found.

CIRCULAR FUNCTIONS

Some books make a distinction between the terms *circular* functions and *trigonometric* functions. We will use these terms synonymously.

Also, notice that these functions are functions of an angle θ and cannot be expressed in terms of algebraic operations alone; thus, they are transcendental functions.

EXAMPLES

1. Find $\tan 45°$. For a $45°$ angle, you know that $x = y$, so by the definition of tangent, $\tan 45° = y/x = x/x = 1$.

2. Find $\sec 60°$. For a $60°$ angle the length of the hypotenuse, r, is twice the length of the shorter leg. Thus, $r = 2x$ and

$$\sec 60° = \frac{r}{x} = \frac{2x}{x} = 2.$$

3. Find $\tan 90°$. For a $90°$ angle, $x = 0$, and from the definition, $\tan \theta = y/x$ $(x \neq 0)$. So the tangent is *not defined* for $90°$. Notice also from the definition that $\sec 90°$ is not defined. For what angles are the cosecant and cotangent not defined?

By considering the examples of Sections 6.1 and 6.2, you should be able to verify the entries in Table 6.1. This table of values is used extensively, and you should memorize it as you did multiplication tables in elementary school. There are, however, some hints that will make it easier to learn.

This table of *exact values* is important, and you should understand how to obtain each entry in this table.

TABLE 6.1

angle θ / function	0	$\dfrac{\pi}{6}$	$\dfrac{\pi}{4}$	$\dfrac{\pi}{3}$	$\dfrac{\pi}{2}$	π	$\dfrac{3\pi}{2}$
$\cos\theta$	1	$\dfrac{\sqrt{3}}{2}$	$\dfrac{\sqrt{2}}{2}$	$\dfrac{1}{2}$	0	-1	0
$\sin\theta$	0	$\dfrac{1}{2}$	$\dfrac{\sqrt{2}}{2}$	$\dfrac{\sqrt{3}}{2}$	1	0	-1
$\tan\theta$	0	$\dfrac{\sqrt{3}}{3}$	1	$\sqrt{3}$	undef.	0	undef.
$\sec\theta$	1	$\dfrac{2}{\sqrt{3}}$	$\dfrac{2}{\sqrt{2}}$	2	undef.	-1	undef.
$\csc\theta$	undef.	2	$\dfrac{2}{\sqrt{2}}$	$\dfrac{2}{\sqrt{3}}$	1	undef.	-1
$\cot\theta$	undef.	$\dfrac{3}{\sqrt{3}}$	1	$\dfrac{\sqrt{3}}{3}$	0	undef.	0

From the definition, $\cos\theta = x/r$ and $\sin\theta = y/r$, so

$$\cos^2\theta + \sin^2\theta = \frac{x^2}{r^2} + \frac{y^2}{r^2}$$

$$= \frac{x^2 + y^2}{r^2}.$$

But $r^2 = x^2 + y^2$, so

$$\cos^2\theta + \sin^2\theta = \frac{r^2}{r^2}.$$

Compare this with Problem 23 of Section 6.1. It is the first of several properties involving the circular functions. We will consider this property in detail in Chapter 7.

$$\cos^2\theta + \sin^2\theta = 1$$

This equation restricts the values of $\cos\theta$ and $\sin\theta$: $\cos^2\theta \le 1$ and $\sin^2\theta \le 1$ (why?). Thus, *both* sine and cosine have functional values between -1 and $+1$, inclusive:

$$-1 \le \cos\theta \le 1 \qquad \text{and} \qquad -1 \le \sin\theta \le 1.$$

Also, notice the following pattern from the table:

	0	$\dfrac{\pi}{6}$	$\dfrac{\pi}{4}$	$\dfrac{\pi}{3}$	$\dfrac{\pi}{2}$	
$\cos \theta$	1	$\dfrac{\sqrt{3}}{2}$	$\dfrac{\sqrt{2}}{2}$	$\dfrac{1}{2}$	0	⟵ simplified form
$\cos \theta$	$\dfrac{\sqrt{4}}{2}$	$\dfrac{\sqrt{3}}{2}$	$\dfrac{\sqrt{2}}{2}$	$\dfrac{\sqrt{1}}{2}$	$\dfrac{\sqrt{0}}{2}$	⟵ easy-to-remember form

equal values

And the sine is simply the reversal of this pattern:

	0	$\dfrac{\pi}{6}$	$\dfrac{\pi}{4}$	$\dfrac{\pi}{3}$	$\dfrac{\pi}{2}$	
$\sin \theta$	0	$\dfrac{1}{2}$	$\dfrac{\sqrt{2}}{2}$	$\dfrac{\sqrt{3}}{2}$	1	⟵ simplified form
$\sin \theta$	$\dfrac{\sqrt{0}}{2}$	$\dfrac{\sqrt{1}}{2}$	$\dfrac{\sqrt{2}}{2}$	$\dfrac{\sqrt{3}}{2}$	$\dfrac{\sqrt{4}}{2}$	⟵ easy-to-remember form

For the tangent, use the following pattern:

	0	$\dfrac{\pi}{6}$	$\dfrac{\pi}{4}$	$\dfrac{\pi}{3}$	$\dfrac{\pi}{2}$	
$\tan \theta$	0	$\dfrac{\sqrt{3}}{3}$	1	$\sqrt{3}$	undef.	⟵ simplified form
$\tan \theta$	$\sqrt{\dfrac{0}{4}}$	$\sqrt{\dfrac{1}{3}}$	$\sqrt{\dfrac{2}{2}}$	$\sqrt{\dfrac{3}{1}}$	$\sqrt{\dfrac{4}{0}}$	⟵ easy-to-remember form (division by 0 means undefined)

Another way to remember the values of the tangent is to notice that

$$\frac{\sin \theta}{\cos \theta} = \frac{\dfrac{y}{r}}{\dfrac{x}{r}} \quad (\cos \theta \neq 0)$$

$$= \frac{y}{r} \cdot \frac{r}{x} \quad (x \neq 0)$$

$$= \frac{y}{x} \quad (x \neq 0),$$

which is the definition of tangent. Therefore:

This is the second property involving the circular functions.

$$\tan \theta = \frac{\sin \theta}{\cos \theta} \quad (\cos \theta \neq 0)$$

The rest of the table is easy to remember; the definition of the secant says that it is the reciprocal of the cosine. This means that $\sec \theta = 1/\cos \theta$ since

$$\frac{1}{\cos \theta} = \frac{1}{x/r} \qquad \text{(from the definition of cosine)}$$

$$= 1 \cdot \frac{r}{x} \quad (x \neq 0)$$

$$= \frac{r}{x} \qquad \text{(the definition of secant)}.$$

Similar results hold for the cosecant and cotangent.

These properties of the circular functions are called the *reciprocal* relationships.

Don't confuse these reciprocal relationships with the idea of cofunctions. The sine and cosine are cofunctions, as are the secant and cosecant, and the tangent and cotangent.

$$\sec \theta = \frac{1}{\cos \theta} \quad (\cos \theta \neq 0)$$

$$\csc \theta = \frac{1}{\sin \theta} \quad (\sin \theta \neq 0)$$

$$\cot \theta = \frac{1}{\tan \theta} \quad (\tan \theta \neq 0)$$

When you use these definitions to remember the table, keep in mind that the reciprocal of zero is undefined.

Consider four examples of an angle, one in each of the four quadrants, as shown in Figure 6.9.

REFERENCE ANGLE

Given an angle θ, the *reference angle* θ' is defined by the smallest positive acute angle the terminal side makes with the x-axis. Notice that the reference angle θ' and angle θ are the same in the first quadrant and that the reference angles for all the examples in the figure are the same. The only change in the values of the functions is in the sign of the result. In other words, if f represents a circular function, then:

This relationship is called the *reduction principle*.

$$f(\theta) = \pm f(\theta')$$

⌐ This sign depends on the
quadrant and the function.

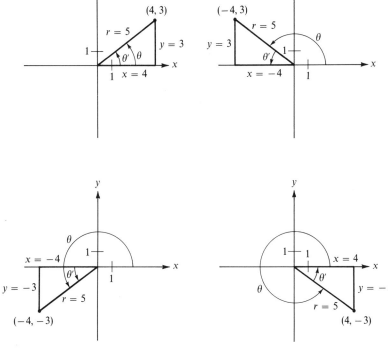

Figure 6.9   Reference angles

We determine the signs from the definition of the functions, as indicated in Table 6.2. The values of sec θ, csc θ, and cot θ are the reciprocals of the values given in the table.

Table 6.2 can be summarized by remembering the following:

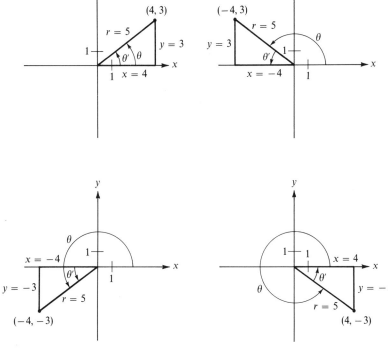

Easy-to-remember   form: *A Smart Trig Class*

TABLE 6.2

	Quadrant I *x pos* *y pos* *r pos*	*Quadrant II* *x neg* *y pos* *r pos*	*Quadrant III* *x neg* *y neg* *r pos*	*Quadrant IV* *x pos* *y neg* *r pos*
$\cos \theta = x/r$	pos	neg	neg	pos
$\sin \theta = y/r$	pos	pos	neg	neg
$\tan \theta = y/x$	pos	neg	pos	neg
Summary	*A*ll pos	*S*ine pos	*T*angent pos	*C*osine pos

EXAMPLES

4. $\tan 210° = \quad + \quad \tan 30° \quad = \quad \dfrac{\sqrt{3}}{3}$

           ↑   ↑   ↑

      Quadrant III:

      tangent pos,

      secant neg   reference angle   from memorized table

           ↓

5. $\sec 210° = \quad - \quad \sec 30° \quad = -\dfrac{2}{\sqrt{3}} \text{ or } -\dfrac{2}{3}\sqrt{3}$

6. $\csc \dfrac{3\pi}{2} = -1$

If it is not in any quadrant, then is comes directly from the memorized table.

7. $\sec\left(-\dfrac{5\pi}{3}\right) = +\sec \dfrac{\pi}{3} = 2$

Sketch the angle if necessary to find the quadrant and the reference angle.

8. $\cot\left(-\dfrac{7\pi}{6}\right) = -\cot \dfrac{\pi}{6} = -\sqrt{3}$

PROBLEM SET 6.3

Historical Note

The origin of the words we use for the circular functions is rather interesting. Consider the figure given below:

tan θ and sec θ are the same

A Problems

In Problems 1–6, give the functional value from memory.

1. a. $\tan(\pi/4)$ b. $\cos 0$ c. $\sin 60°$ d. $\sec(\pi/6)$ e. $\csc 0$
    f. $\tan(\pi/6)$

2. a. $\cos 270°$ b. $\cos 30°$ c. $\sec(\pi/4)$ d. $\tan 180°$ e. $\sin 45°$
    f. $\sin \pi$

3. a. $\sin(\pi/2)$ b. $\sec 0°$ c. $\tan 0$ d. $\sec(\pi/6)$ e. $\cos(\pi/4)$
    f. $\cot 90°$

4. a. $\cot \pi$ b. $\sec(\pi/4)$ c. $\tan 90°$ d. $\tan 60°$ e. $\cos(\pi/3)$
    f. $\sec(\pi/3)$

5. a. $\cot 45°$ b. $\cos \pi$ c. $\sin(3\pi/2)$ d. $\sin 0°$ e. $\sec(\pi/2)$
    f. $\sec 270°$

6. a. $\csc(\pi/2)$ b. $\cos 90°$ c. $\sec \pi$ d. $\tan 270°$ e. $\sin(\pi/6)$
    f. $\csc(3\pi/2)$

In Problems 7–16, write the answer in simplest form.

7. $\sin(\pi/6)\csc(\pi/6)$ 8. $\csc(\pi/2)\sin(\pi/2)$

9. $\sin^2(\pi/3) + \cos^2(\pi/3)$ 10. $\sin^2(\pi/6) + \cos^2(\pi/3)$

11. a. $\cos(\pi/4 - \pi/2)$ b. $\cos(\pi/4) - \cos(\pi/2)$
12. a. $\tan(2 \cdot 30°)$ b. $2\tan 30°$

13. a. $\csc\left(\dfrac{1}{2} \cdot 60°\right)$ b. $\dfrac{\csc 60°}{2}$

14. a. $\cos(\pi/2 - \pi/6)$ b. $\cos(\pi/2)\cos(\pi/6) + \sin(\pi/2)\sin(\pi/6)$

15. a. $\tan(2 \cdot 60°)$ b. $\dfrac{2\tan 60°}{1 - \tan^2 60°}$

16. . a. $\cos\left(\dfrac{1}{2} \cdot 60°\right)$ b. $\sqrt{\dfrac{1 + \cos 60°}{2}}$

B Problems

Find the values of the six circular functions for an angle θ in standard position with terminal side passing through the points given in Problems 17–28. Draw a picture showing θ and the reference angle θ'.

17. $(3, 4)$ 18. $(-3, 4)$ 19. $(-3, -4)$ 20. $(3, -4)$
21. $(5, 12)$ 22. $(-5, -12)$ 23. $(5, -12)$ 24. $(-5, 12)$
25. $(2, -5)$ 26. $(-6, 1)$ 27. $(-4, -5)$ 28. $(-5, 2)$

Find the values of the six circular functions for each of the angles given in Problems 29–40.

29. $450°$ 30. $-5\pi/4$ 31. $-\pi/6$ 32. $120°$
33. -2π 34. $7\pi/6$ 35. $390°$ 36. $-120°$
37. $-135°$ 38. $135°$ 39. 3π 40. $8\pi/3$

VALUES OF CIRCULAR FUNCTIONS BY APPROXIMATION

Although we've been limiting ourselves to angles with reference angles that are $0°$, $30°$, $45°$, $60°$, or $90°$, we can estimate the functional values for other angles. From the definitions of the circular functions, estimate to one decimal place the numbers in Problems 41–51.

EXAMPLE 9: For $\tan 110°$ we use a unit circle, as shown in Figure 6.10. Draw the terminal side at $110°$. Estimate $y \approx 0.92$ and $x \approx -0.35$. Then, by the definition, $\tan 110° \approx 0.93/(-0.35) \approx -2.7$ to the nearest tenth.

---

as the lengths of the tangent and secant lines drawn on a unit circle:

$$\tan\theta = \frac{y}{x} = \frac{y}{1} = |\overrightarrow{PA}|$$

$\overrightarrow{PA}$ is tangent to the circle.

$$\sec\theta = \frac{r}{x} = r = |\overleftrightarrow{PO}|$$

$\overleftrightarrow{PO}$ is a secant to the circle. However, the sine has a more interesting origin according to Howard Eves in his book *In Mathematical Circles* (Boston: Prindle, Weber, & Schmidt, 1969). Āryabhata called it *jyā-adhā* (chord half) and abbreviated it *jyā*, which the Arabs wrote as *jîba* but shortened to *jb*. Later writers saw *jb*, and since *jîba* was meaningless to them, they substituted *jaib* (cove or bay). The Latin equivalent for *jaib* is *sinus*, from which our present word *sine* is derived. The other cofunctions are derived from these, with *co*sine meaning *complement's sine*.

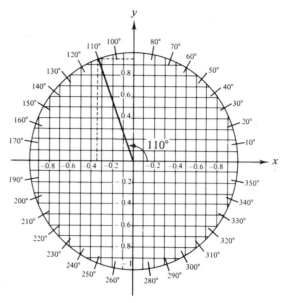

Figure 6.10   Approximate values of circular functions

---

41. cos 50°	42. sec 70°	43. csc 150°	44. cot 250°
45. tan (−20°)	46. csc 190°	47. sec (−190°)	48. sin 20°
49. sin 320°	50. tan 80°	51. cot (−100°)	

VALUES OF CIRCULAR FUNCTIONS BY CALCULATOR

Many calculators have built-in trigonometric functions. However, attention must be given to the unit of measure used. For a calculator with a degree-radian switch, simply depress the given angle (in degrees or radians) and then the appropriate trig-function key.

EXAMPLES

10.   sin 50°: switch key to degrees; depress the following keys:

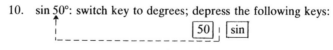

The answer 0.7660444431 is now displayed.

11.   sin (π/12): switch key to radians; depress the following keys:

algebraic logic: $\boxed{\pi}$ $\boxed{\div}$ $\boxed{12}$ $\boxed{=}$ $\boxed{\text{SIN}}$

RPN logic: $\boxed{\pi}$ $\boxed{12}$ $\boxed{\div}$ $\boxed{\text{SIN}}$

The answer 0.2588190451 is now displayed.

12. tan 70° 23′ 40″: switch key to degrees; this measure must first be changed to a decimal degree measure, as discussed earlier:

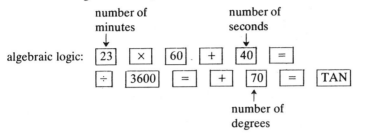

The answer 2.807464818 is displayed.

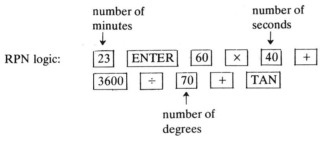

The answer 2.807464819 is displayed.

___

52. *Calculator Problem.* Find the values of the numbers in Problems 41–46 correct to six decimal places.
53. *Calculator Problem.* Find the values of the numbers in Problems 47–51 correct to six decimal places.

VALUES OF CIRCULAR FUNCTIONS BY TABLE

Sometimes we require more accuracy than can be obtained by the methods of Problems 41–51 but have no access to a calculator. Mathematicians have developed tables to find very precise trigonometric values. Table 4 in the back of this book gives the values of the circular functions of angles from 0° to 90°, and Table 5 gives the values for the sine, tangent, cotangent, and cosine measured in radians from 0 to 2.

For angles between 0° and 45°, find the angle to the nearest tenth in the *left-hand* column headed *Deg.* Next, read across that row to find the value of the desired function named at the *top.* For example, cos 22.1° is found as follows:

If your calculator does not work with radians, use 180/12 since $\pi = 180°$. In general, you can convert radians to degrees (as shown in Section 6.1) by using the formula

$$\left(\begin{array}{c}\text{angle in}\\ \text{degrees}\end{array}\right) = \frac{180}{\pi}\left(\begin{array}{c}\text{angle in}\\ \text{radians}\end{array}\right)$$

𝕳istorical 𝕹ote

During the second half of second century B.C., the astronomer Hipparchus of Nicaea (about 180–125 B.C.) tabulated the first trigonometric tables in 12 books. It was he who used the 360° circle of the Babylonians and thus introduced trigonometry with the angle measure we still use today. Hipparchus' work formed the foundation for Ptolemy's *Mathematical Syntaxis*, the most significant early work in trigonometry. Ptolemy acknowledged the earlier contributions of Hipparchus, whom he described as "a labor-loving and truth-loving man." It should be pointed out that these early works do not use the ideas of trigonometric ratios introduced in this section; rather, they use trigonometric lines that take the form of chords of a circle. For a discussion of this method, see Carl Boyer's *A History of Mathematics*, Chap. 10 (New York: Wiley, 1968).

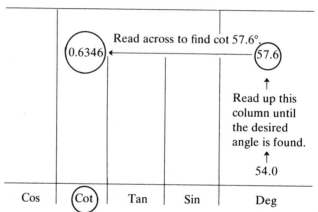

For angles between 45° and 90°, find the desired angle in the *right-hand* column. Then read across that row to find the value of the desired function named at the *bottom*. For example, cot 57.6° is found as follows:

Since all reference angles are between 0° and 90°, Table 4 can be used to approximate any of the trigonometric values.

EXAMPLES: Verify the following from Table 4:

13. $\sin 34.4° = 0.5650$ 　　　　14. $\cos 54.2° = 0.5850$
15. $\tan 70.2° = 2.778$ 　　　　16. $\cot 46.7° = 0.9424$

---

54. Find the values of the numbers in Problems 41–46 correct to four decimal places by using Table 4.
55. Find the values of the numbers in Problems 47–51 correct to four decimal places by using Table 4.

In Problems 56–64, use Table 4, Table 5, or a calculator to find the functional value. Remember that if degrees are not stated, the units are in radians.

56. tan 56.2°   57. cot 78.4°   58. sin 1°

59. sin 1   60. tan (−1.5)   61. cos (−0.48)

62. cos 48.3°   63. sin 21.48°   64. cos (−65.21°)

Note: for Problems 63 and 64, approximate the angle to the nearest tenth and use Table 4, interpolation, or a calculator.

 65. a. If θ is in Quadrant I, then $\theta + \pi$ is in Quadrant III with a reference angle θ. Use this fact and the reduction principle to show that $\sin (\theta + \pi) = -\sin \theta$ if θ is in Quadrant I.

    b. Show that $\sin (\theta + \pi) = -\sin \theta$ if θ is in Quadrant II.

    c. Show that $\sin (\theta + \pi) = -\sin \theta$ if θ is in Quadrant III.

    d. Show that $\sin (\theta + \pi) = -\sin \theta$ if θ is in Quadrant IV.

    e. By considering parts a–d, show that $\sin (\theta + \pi) = -\sin \theta$ for any angle θ.

 66. Show that $\cos (\theta + \pi) = -\cos \theta$ for any angle θ. (Hint: see Problem 65.)

Mind Bogglers

 67. *Calculator Problem.* In more advanced mathematics courses, it is shown that

$$\sin x = x - \frac{x^3}{3!} + \frac{x^5}{5!} - \frac{x^7}{7!} + \cdots,$$

where $n! = n(n - 1)(n - 2) \cdots 3 \cdot 2 \cdot 1$. Find sin 1 correct to four decimal places by using this equation. (Remember that the 1 in sin 1 refers to radian measure.)

 68. *Calculator Problem.* It is known that

$$\cos x = 1 - \frac{x^2}{2!} + \frac{x^4}{4!} - \frac{x^6}{6!} + \cdots.$$

Use this equation to find cos 1 correct to four decimal places.

69. *Computer Problem.* If you have access to a computer, write a program that will output a table of trigonometric values for the sine, cosine, and tangent for every degree from 0° to 45°.

6.4 GRAPHS OF SINE, COSINE, AND TANGENT FUNCTIONS

As with the polynomial and rational functions, we are interested in the graphs of the circular functions. First, we will determine the general shape of the circular functions by plotting points and then generalize so we can graph the functions without too many calculations concerning points.

SINE FUNCTION

To graph $y = \sin x$, we can begin by plotting familiar values for the sine:

We are using exact values here, but you could also use Table 4, Table 5, or a calculator to generate these values. See Problems 18 and 19 in the problem set.

x = real number	0	$\pi/6$	$\pi/4$	$\pi/3$	$\pi/2$	π	$3\pi/2$
$y = \sin x$	0	1/2	$\sqrt{2}/2$	$\sqrt{3}/2$	1	0	−1
y (approximate)	0	0.5	0.71	0.87	1	0	−1

The difficulty with this method is that approximate values must be plotted. We can help matters a little by setting up a scale on the x-axis that is in units of π (we chose 12 squares = π units in Figure 6.11). Plot additional values using the reduction principle of the previous section.

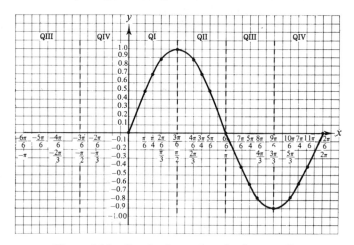

Figure 6.11   Graph of $y = \sin x$ for $0 \le x \le 2\pi$

x = real number	$2\pi/3$	$3\pi/4$	$5\pi/6$	$7\pi/6$	$5\pi/4$	$4\pi/3$	$5\pi/3$	$7\pi/4$	$11\pi/6$
Quadrant; sign of $\sin x$	II; +	II; +	II; +	III; −	III; −	III; −	IV; −	IV; −	IV; −
Reference angle	$\pi/3$	$\pi/4$	$\pi/6$	$\pi/6$	$\pi/4$	$\pi/3$	$\pi/3$	$\pi/4$	$\pi/6$
$y = \sin x$	$\sqrt{3}/2$	$\sqrt{2}/2$	1/2	−1/2	$-\sqrt{2}/2$	$-\sqrt{3}/2$	$-\sqrt{3}/2$	$-\sqrt{2}/2$	−1/2
y (approximate)	0.87	0.71	0.5	−0.5	−0.71	−0.87	−0.87	−0.71	−0.5

Notice that, when x is in Quadrant I, then $0 < x < \pi/2$, which does *not* correspond to the first quadrant of the graph $y = \sin x$. In Figure 6.11, we've shown the intervals corresponding to the quadrants of the angle x. The points from the preceding table are plotted in Figure 6.11 and are connected by a smooth curve called the *sine curve*.

What about values other than $0 \le x \le 2\pi$? Since the sine function is periodic, with period 2π, we see that angles greater than 2π or less than 0 simply repeat the values already plotted. The entire sine curve continues as indicated in Figure 6.12. Notice that for the base period shown, the sine curve

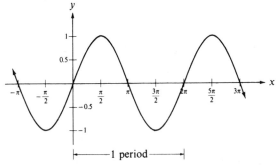

Figure 6.12   Graph of $y = \sin x$

starts at $(0, 0)$, goes *up* to $(\pi/2, 1)$, then *down* to $(3\pi/2, -1)$ passing through $(\pi, 0)$, and then goes back *up* to $(2\pi, 0)$, which completes one period. We can summarize a technique for sketching the sine curve (see Figure 6.13):

1. Plot the endpoints of the base period—namely $(0, 0)$ and $(2\pi, 0)$.
2. Plot the midpoint $(\pi, 0)$.
3. Halfway between these points, plot the highest (up) point $(\pi/2, 1)$, and the lowest (down) point $(3\pi/2, -1)$.
4. Now the sine curve is "framed"; draw the curve.

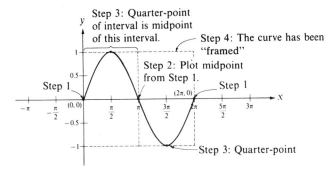

Figure 6.13   "Framing" the sine curve

COSINE FUNCTION

The cosine curve can be graphed by plotting points, as with the sine curve. We'll leave the details of plotting these points as an exercise and summarize the results by "framing" the cosine curve (see Figure 6.14):

1. Plot the endpoints of the base period—namely $(0, 1)$ and $(2\pi, 1)$.
2. Plot the midpoint $(\pi, -1)$.
3. Halfway between these points, plot the points $(\pi/2, 0)$ and $(3\pi/2, 0)$.
4. Now the cosine curve is "framed"; draw the curve.

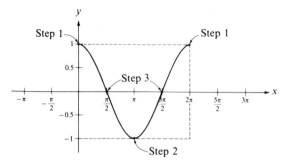

Figure 6.14 "Framing" the cosine curve

Since values for x greater than 2π or less than 0 are coterminal with those already considered, we indicate the entire cosine curve in Figure 6.15.

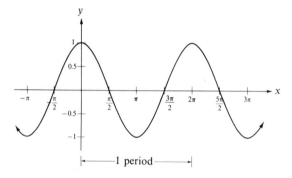

Figure 6.15 Graph of $y = \cos x$

TANGENT FUNCTION

By setting up a table of values and plotting points (the details are left as an exercise), we notice that $y = \tan x$ doesn't exist at $\pi/2, 3\pi/2$, or $\pi/2 \pm n\pi$ for any integer n. The lines $x = \pi/2, x = 3\pi/2, \ldots, x = \pi/2 \pm n\pi$ for which the tangent is not defined are vertical asymptotes.

We now "frame" the tangent curve as follows (see Figure 6.16):

1. Plot the center point $(0, 0)$.
2. Draw a pair of adjacent asymptotes—say, $-\pi/2$ and $\pi/2$.
3. Halfway between the center point and the asymptotes, plot the points $(\pi/4, 1)$ and $(-\pi/4, -1)$.
4. The tangent curve is now "framed"; draw the curve.

The entire tangent curve is indicated in Figure 6.17. Even though the curve repeats for values of x greater than 2π or less than 0, it also repeats after it has passed through an interval with length π. This result can be shown algebraically if we use the answers to Problems 65 and 66 of Section 6.3. Since $\sin(\theta + \pi) = -\sin\theta$ and $\cos(\theta + \pi) = -\cos\theta$, we have

$$\tan(\theta + \pi) = \frac{\sin(\theta + \pi)}{\cos(\theta + \pi)} = \frac{-\sin\theta}{-\cos\theta} = \frac{\sin\theta}{\cos\theta} = \tan\theta.$$

Since $\tan(\theta + \pi) = \tan\theta$, then $\tan(\theta + n\pi) = \tan\theta$ for any integer n, and we see that the tangent has a period of π.

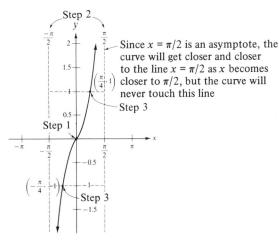

Since $x = \pi/2$ is an asymptote, the curve will get closer and closer to the line $x = \pi/2$ as x becomes closer to $\pi/2$, but the curve will never touch this line

Figure 6.16 "Framing" the tangent curve

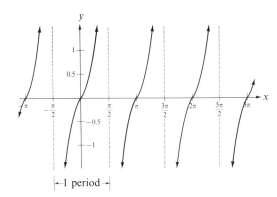

Figure 6.17 Graph of $y = \tan x$

PROBLEM SET 6.4

A Problems

In Problems 1–6, give the functional value from memory.

1. a. $\sec 270°$ b. $\cot(3\pi/2)$ c. $\sin 0$ d. $\cos(\pi/6)$ e. $\csc 0$ f. $\tan 0°$

2. a. csc 30°   b. tan π c. sec $(\pi/2)$ d. sin $(\pi/6)$ e. cot 180° f. sin 45°
3. a. tan 90°   b. sec π c. cot 90°   cos 0   e. csc 90°   f. cos $(\pi/3)$
4. a. çot $(\pi/6)$ b. sin 60°   c. cos 45°   d. sec 0°   e. tan $(\pi/6)$ f. csc π
5. a. sin $(\pi/2)$ b. csc 45°   c. tan 60°   d. cot $(\pi/4)$ e. cos $(\pi/2)$ f. sin 270°
6. a. csc $(3\pi/2)$ b. sin π c. csc 60°   d. tan 45°   e. cos 90°   f. cot $(\pi/6)$
7. Complete the following table of values for $y = \cos x$:

x = angle	$2\pi/3$	$3\pi/4$	$5\pi/6$	$7\pi/6$	$5\pi/4$	$4\pi/3$	$7\pi/4$	$11\pi/6$
Quadrant; sign of cos x								
$y = \cos x$								
y (approximate)								

8. Use the table in Problem 7, along with other values if necessary, to plot $y = \cos x$.
9. Complete a table of values like the one in Problem 7 for $y = \tan x$.
10. Use the table in Problem 9, along with other values if necessary, to plot $y = \tan x$.
11. Draw a quick sketch of $y = \cos x$ from memory by framing the curve. (Note: this is *not* the same thing you did in Problem 8.)
12. Draw a quick sketch of $y = \sin x$ from memory by framing the curve.
13. Draw a quick sketch of $y = \tan x$ from memory by framing the curve.

B Problems

There are other methods for sketching the sine, cosine, and tangent curves besides plotting points or framing the curve. Some of these methods are developed in Problems 14–17.

14. Draw a unit circle, as shown in Figure 6.18. By definition, $\sin \theta = y/r = y/1 = y$. Notice that, as we choose different values for θ, the y value is the height of the point P above the x-axis. For example, when $\theta = \pi/4$, the point to plot is the

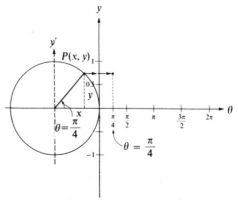

Figure 6.18   Graphing a sine curve using a unit circle

intersection of the line $\theta = \pi/4$ and the horizontal line passing through P. As P makes one revolution on the unit circle and the values are plotted in this fashion, one period of the sine curve results. Plot $y = \sin \theta$ in this fashion.

15. Draw a unit circle, as shown in Figure 6.19. By definition, $\cos \theta = x/r = x/1 = x$. Notice the orientation of the axes for the unit circle. Repeat the procedure outlined in Problem 14 to graph $y = \cos \theta$.

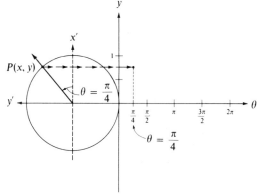

Figure 6.19 Graphing a cosine curve using a unit circle

16. Draw a unit circle with $P(x, y)$ the intersection of the terminal side of an angle θ and the unit circle. Let $|PQ|$ be perpendicular to the x-axis. Draw a line through $S(1, 0)$ perpendicular to the x-axis. Let R be the point of intersection of the line and the terminal side of angle θ. In Problem 14, we showed that $\sin \theta = |PQ|$, and in Problem 15, that $\cos \theta = |OQ|$. Show that $\tan \theta = |RS|$.

17. With the information from Problem 16, graph $y = \tan \theta$ using a technique similar to the one outlined in Problem 14.

18. *Calculator Problem.* We've emphasized the fact that the sine function is a function of a real number, x. Instead of using units of π, graph the sine function by finding additional values for the table shown here.

x = real number	0	1	2	3	4	5	6
$y = \sin x$	0	0.84	0.91	0.14	−0.76	−0.96	−0.27

Plot the ordered pairs—namely $(0, 0)$, $(1, 0.84)$, $(2, 0.91)$, . . .—and complete the graph of $y = \sin x$.

19. *Calculator Problem.* We've emphasized the fact that the cosine function is a function of a real number, x. Instead of using units of π, graph the cosine function by finding additional values for the table shown here.

x = real number	0	1	2	3	4	5	6
$y = \cos x$	1	0.54	−0.42	−0.99	−0.65	0.28	0.96

Plot the ordered pairs—namely, $(0, 1)$, $(1, 0.54)$, $(2, −0.42)$, . . .—and complete the graph of $y = \cos x$.

Use the technique of plotting points to graph the functions in Problems 20–29.

20. $y = \sec x$ 21. $y = \csc x$ 22. $y = \cot x$
23. $y = 2 \sin x$ 24. $y = \sin 2x$ 25. $y = 2 \sin x + 1$
26. $y = 2 \tan x$ 27. $y = \tan 2x$ 28. $y = \sin x + \cos x$
29. $y = x + \sin x$

The graphs of the other circular functions can be found by using the following relationships:

$$\sec \theta = \frac{1}{\cos \theta} \qquad \csc \theta = \frac{1}{\sin \theta} \qquad \cot \theta = \frac{1}{\tan \theta}$$

EXAMPLE 1: Sketch $y = \sec x$.

Solution: Begin by sketching the reciprocal, $y = \cos x$ (dotted curve). Wherever $\cos x = 0$, $\sec x$ is undefined. Draw asymptotes at these places, as shown in Figure 6.20. Now plot points by finding the reciprocals of already plotted points. For example, when $y = \cos x = \frac{1}{2}$, the reciprocal is

$$y = \sec x = \frac{1}{\cos x} = \frac{1}{\frac{1}{2}} = 2.$$

The completed graph is shown in Figure 6.20.

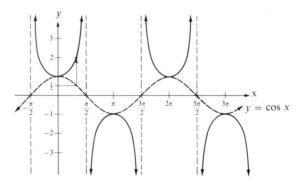

Figure 6.20   Graph of $y = \sec x$

Graph the curves in Problems 30 and 31 using this technique.

30. $y = \csc x$ 31. $y = \cot x$

Mind Bogglers

32. Plot points to graph $y = 2 \sin [3x + (\pi/4)] + 1$.
33. Plot points to graph $y = (\sin x)/x$.

34. Table 6.3, below, gives the times of sunrise and sunset for various latitudes. Plot the sunrise and sunset for the latitude nearest your home. Can either of these curves be easily represented as a circular function?

35. Plot the sunrise and sunset for your own latitude by interpolating as in the following example: the latitude for Santa Rosa, California, is about 38°. This falls between 35° and 40°. The difference is 5°, while the difference between 35° and 38° is 3°, so we add 3/5 of the difference of times for a particular date to the time shown for 35°. For January 1, the difference of time is 14 min so

$$\frac{3}{5}(14) = 8.4 \quad \text{or} \quad 8 \text{ (to the nearest minute).}$$

Your local Chamber of Commerce will likely be able to tell you your town's latitude.

Thus, sunrise in Santa Rosa would be about 7:16 A.M. (7:08 + 8).

36. Plot the length of daylight for the curves you drew in Problems 34 and 35. Can the curves be represented as a circular function?

TABLE 6.3 Sunrise and Sunset

| | Time of Sunrise | | | | | | Time of Sunset | | | | | |
| | 20° N. Latitude (Hawaii) | 30° N. Latitude (New Orleans) | 35° N. Latitude (Albuquerque) | 40° N. Latitude (Philadelphia) | 45° N. Latitude (Minneapolis) | 60° N. Latitude (Alaska) | 20° N. Latitude (Hawaii) | 30° N. Latitude (New Orleans) | 35° N. Latitude (Albuquerque) | 40° N. Latitude (Philadelphia) | 45° N. Latitude (Minneapolis) | 60° N. Latitude (Alaska) |
Date	h m	h m	h m	h m	h m	h m	h m	h m	h m	h m	h m	h m
Jan. 1	6 35	6 56	7 08	7 22	7 38	9 03	17 31	17 10	16 58	16 44	16 28	15 03
Jan. 15	6 38	6 57	7 08	7 20	7 35	8 48	17 41	17 22	17 12	16 59	16 44	15 31
Jan. 30	6 36	6 52	7 01	7 11	7 23	8 19	17 51	17 35	17 27	17 17	17 05	16 09
Feb. 14	6 30	6 41	6 48	6 55	7 03	7 42	17 59	17 48	17 42	17 34	17 26	16 48
Mar. 1	6 20	6 26	6 30	6 34	6 39	6 59	18 05	17 59	17 56	17 52	17 47	17 27
Mar. 16	6 08	6 09	6 10	6 11	6 11	6 15	18 10	18 09	18 08	18 08	18 07	18 04
Mar. 31	5 55	5 51	5 49	5 46	5 43	5 29	18 14	18 18	18 20	18 23	18 26	18 41
Apr. 15	5 42	5 34	5 28	5 23	5 16	4 44	18 18	18 27	18 32	18 38	18 45	19 18
Apr. 30	5 32	5 18	5 11	5 02	4 51	4 01	18 23	18 37	18 44	18 53	19 04	19 55
May 15	5 24	5 07	4 57	4 45	4 31	3 23	18 29	18 46	18 56	19 08	19 22	20 31
May 30	5 20	5 00	4 48	4 34	4 18	2 53	18 35	18 55	19 07	19 21	19 38	21 04
June 14	5 20	4 58	4 45	4 30	4 13	2 37	18 40	19 02	19 15	19 30	19 48	21 24
June 29	5 23	5 02	4 49	4 34	4 16	2 40	18 43	19 05	19 18	19 33	19 51	21 26
July 14	5 29	5 08	4 56	4 43	4 26	3 01	18 43	19 03	19 15	19 29	19 45	21 09
July 29	5 34	5 17	5 07	4 55	4 41	3 33	18 39	18 56	19 06	19 17	19 31	20 38
Aug. 13	5 39	5 26	5 18	5 09	4 59	4 09	18 30	18 43	18 51	19 00	19 10	19 59
Aug. 28	5 43	5 35	5 29	5 24	5 17	4 45	18 19	18 27	18 33	18 38	18 45	19 16
Sept. 12	5 47	5 43	5 40	5 38	5 35	5 20	18 06	18 10	18 12	18 14	18 17	18 31
Sept. 27	5 50	5 51	5 51	5 52	5 53	5 55	17 52	17 51	17 50	17 49	17 49	17 45
Oct. 12	5 54	6 00	6 03	6 07	6 11	6 31	17 39	17 33	17 29	17 26	17 21	17 01
Oct. 22	5 57	6 06	6 12	6 18	6 25	6 56	17 32	17 22	17 17	17 11	17 04	16 32
Nov. 6	6 04	6 18	6 26	6 35	6 45	7 35	17 23	17 10	17 02	16 52	16 42	15 52
Nov. 21	6 12	6 30	6 40	6 52	7 05	8 12	17 19	17 02	16 51	16 40	16 26	15 19
Dec. 6	6 22	6 42	6 54	7 07	7 23	8 44	17 20	17 00	16 48	16 35	16 19	14 58
Dec. 21	6 30	6 52	7 04	7 18	7 35	9 02	17 26	17 05	16 52	16 38	16 21	14 54

Table courtesy of U.S. Naval Observatory. This table of sunrise and sunset may be used in any year of the 20th century with an error not exceeding two minutes and generally less than one minute. It may also be used anywhere in the vicinity of the stated latitude with an additional error of less than one minute for each 9 miles.

Note: a shift of negative units to the right is a shift to the left.

6.5 GENERAL SINE, COSINE, AND TANGENT CURVES

In Section 1.5, we showed that $y - k = f(x - h)$ can be sketched by translating the coordinate axes to a point (h, k) and then graphing the related function $y = f(x)$ on this new coordinate system. Thus, if $f(x) = \sin x$, then $f(x - h) = \sin (x - h)$ is a sine curve shifted h units to the right.

EXAMPLE 1: Graph $y = \sin (x + \pi/2)$.

Solution:

Step 1. Frame the curve, as shown in Figure 6.21.
  a. Plot $(h, k) = (-\pi/2, 0)$.
  b. The period of the sine curve is 2π, and it has a high point up one unit and a low point down one unit.

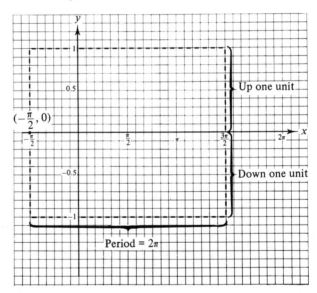

Figure 6.21   Framing the curve: this step is the same regardless of whether you are graphing a sine or a cosine

Step 2. Plot the five critical points (two endpoints, the midpoint, and two quarter-points). For the sine curve, plot the endpoint (h, k) and use the frame to plot the other endpoint and the midpoint. For the quarter-points, remember that the sine curve is "up–down"; use the frame to plot the quarter-points, as shown in Figure 6.22.

Step 3. Remembering the shape of the sine curve, sketch one period of $y = \sin [x + (\pi/2)]$ using the frame and the five critical points. If you want to show more than one period, just repeat the same pattern.

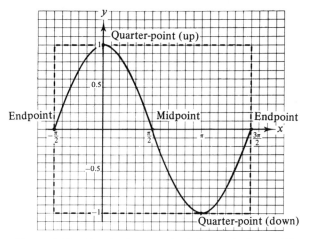

Figure 6.22   Graph of one period of $y = \sin(x + \pi/2)$

Notice from Figure 6.22 that the graph of $y = \sin(x + \pi/2)$ is the same as the graph of $y = \cos x$. Thus,

$$\sin\left(x + \frac{\pi}{2}\right) = \cos x.$$

EXAMPLES

2. Graph $y - 2 = \cos(x - \pi/6)$.
    Solution:
    Step 1.   Frame the curve, as shown in Figure 6.23. Notice that (h, k) = $(\pi/6, 2)$.
    Step 2.   Plot the five critical points. For the cosine curve the left and right endpoints are at the upper corners of the frame; the midpoint is at the bottom of the frame; the quarter-points are on a line through the middle of the frame.
    Step 3.   Draw one period of the curve, as shown in Figure 6.23.

3. Sketch $y + 3 = \tan(x + \pi/3)$.
    Solution:
    Step 1.   Frame the curve, as shown in Figure 6.24. Notice that (h, k) = $(-\pi/3, -3)$, and remember that the period of the tangent is π.

    Step 2.   For the tangent curve, (h, k) is the midpoint of the frame. The endpoints, which are each a distance of $\frac{1}{2}$ the period from the midpoint, determine the location of the asymptotes. The top and bottom of the frame are one unit from (h, k). Locate the quarter-points at the top and bottom of the frame, as shown in Figure 6.24.

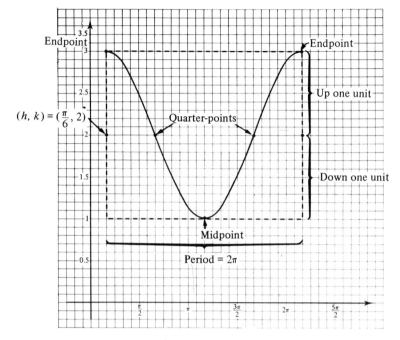

Figure 6.23 Graph of one period of $y - 2 = \cos (x - \pi/6)$

Step 3. Sketch one period of the curve, as shown in Figure 6.24. Remember that the tangent curve is not contained within the frame.

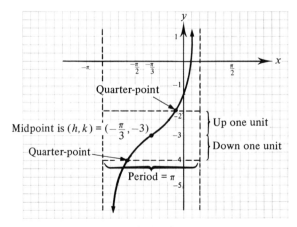

Figure 6.24 Graph of one period of $y + 3 = \tan (x + \pi/3)$

We will now discuss two additional changes for the function defined by $y = f(x)$. The first, $y = af(x)$, changes the scale on the y-axis; the second, $y = f(bx)$, changes the scale on the x-axis.

For a function $y = af(x)$, it is clear that the y value is a times the corresponding value of $f(x)$, which means that $f(x)$ is stretched or shrunk in the y direction by the multiple of a. For example, if $y = f(x) = \cos x$, then $y = 3f(x) = 3 \cos x$ is the graph of $\cos x$ that has been stretched so that the high point is at three units and the low point is at negative three units. In general, given

$$y = af(x),$$

where f represents a trigonometric function, $|a|$ gives the height of the frame for f. To graph $y = 3 \cos x$, we frame the cosine curve using an amplitude of 3 rather than 1 (see Figure 6.25).

For the sine and cosine curves, $|a|$ is called the *amplitude* of the function. When $a = 1$, the amplitude is 1, so $y = \sin x$ and $y = \cos x$ are said to have amplitude 1.

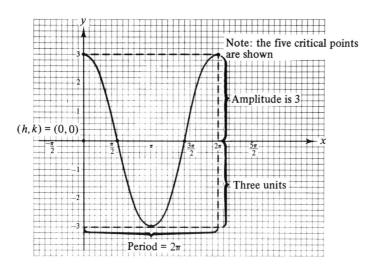

Figure 6.25   Graph of one period of $y = 3 \cos x$

For a function $y = f(bx)$, $b > 0$, b affects the scale on the x-axis. Recall that $y = \sin x$ has a period of 2π $(f(x) = \sin x$, so $b = 1)$. A function $y = \sin 2x$ $(f(x) = \sin x$ and $f(2x) = \sin 2x)$ must complete one period as $2x$ varies from 0 to 2π. This means that one period is completed as x varies from 0 to π. (Remember that for each value of x the result is doubled *before* we find the sine of that number.) In general, the period of $y = \sin bx$ is $2\pi/b$, and the period of $y = \cos bx$ is $2\pi/b$. However, since the period of $y = \tan x$ is π, we can see that $y = \tan bx$ has a period of π/b. Therefore, when framing the curve, we use $2\pi/b$ for the sine and cosine and π/b for the tangent.

The period of sine and cosine is $2\pi/b$, and the period of tangent is π/b.

EXAMPLE 4: Graph one period of $y = \sin 2x$.

Solution: The period is $2\pi/2 = \pi$; thus, the endpoints of the frame are $(0, 0)$ and $(\pi, 0)$, as shown in Figure 6.26.

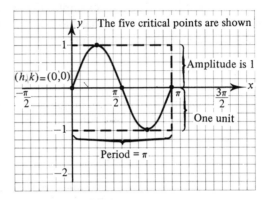

Figure 6.26   Graph of one period of $y = \sin 2x$

Summarizing all the above results, we have the *general* sine, cosine, and tangent curves:

We call these the *translated forms* of the general sine, cosine, and tangent functions.

$$y - k = a \sin b(x - h);$$
$$y - k = a \cos b(x - h);$$
$$y - k = a \tan b(x - h).$$

1. The origin has been translated or shifted to the point (h, k).
2. The height of the frame is $|a|$.
3. The length of the frame is $2\pi/b$ for the sine and cosine curves and π/b for the tangent curve.
4. The curves are sketched by translating the origin to the point (h, k) and then framing the curve to complete the graph.

EXAMPLES: Graph the curves.

5. $y + 1 = 2 \sin \dfrac{2}{3}\left(x - \dfrac{\pi}{2}\right)$

Notice that $(h, k) = (\pi/2, -1)$ and that the amplitude is 2; the period is $2\pi/(2/3) = 3\pi$. Now plot (h, k) and frame the curve. Then plot the five critical points (two endpoints, the midpoint, and two quarter-points).

Finally, after sketching one period, draw the other periods, as shown in Figure 6.27.

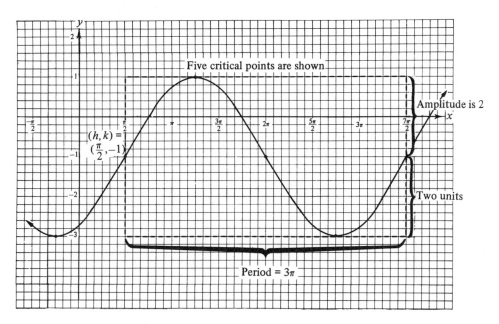

Figure 6.27 Graph of $y + 1 = 2 \sin \dfrac{2}{3}\left(x - \dfrac{\pi}{2}\right)$; one period inside the frame is drawn first, and then the curve is extended outside the frame.

6. $y = 3 \cos\left(2x + \dfrac{\pi}{2}\right) - 2$

Rewrite in standard form to obtain

$$y + 2 = 3 \cos 2\left(x + \dfrac{\pi}{4}\right).$$

Notice that $(h, k) = (-\pi/4, -2)$; the amplitude is 3, and the period is $2\pi/2 = \pi$. Plot (h, k) and frame the curve, as shown in Figure 6.28.

7. $y - 2 = 3 \tan \dfrac{1}{2}\left(x - \dfrac{\pi}{3}\right)$

Notice that $(h, k) = (\pi/3, 2)$, $a = 3$, and the period is $\pi/(1/2) = 2\pi$. Plot (h, k) and frame the curve, as shown in Figure 6.29.

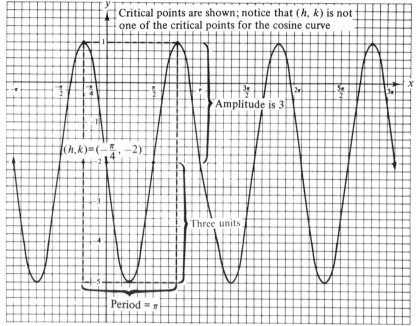

Figure 6.28   Graph of $y + 2 = 3 \cos 2\left(x + \dfrac{\pi}{4}\right)$

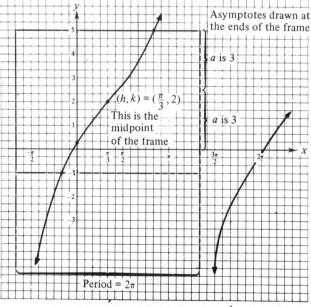

Figure 6.29   Graph of $y - 2 = 3 \tan \dfrac{1}{2}\left(x - \dfrac{\pi}{3}\right)$

PROBLEM SET 6.5

A Problems

In Problems 1–6, give the functional value from memory.

1. a. $\cos 0$ b. $\sec(3\pi/2)$ c. $\tan \pi$ d. $\cot \pi$ e. $\csc(3\pi/2)$
    f. $\sin 0$

2. a. $\cot(\pi/2)$ b. $\cos(\pi/6)$ c. $\sec \pi$ d. $\tan(\pi/2)$ e. $\sin(\pi/6)$
    f. $\csc \pi$

3. a. $\cot(\pi/3)$ b. $\csc(\pi/6)$ c. $\cos(\pi/4)$ d. $\sin(\pi/4)$ e. $\tan(\pi/3)$
    f. $\sec(\pi/2)$

4. a. $\csc(\pi/4)$ b. $\cot(\pi/4)$ c. $\sin(\pi/2)$ d. $\cos(\pi/2)$ e. $\sec(\pi/3)$
    f. $\tan(\pi/4)$

5. a. $\cot(\pi/6)$ b. $\sin \pi$ c. $\csc(\pi/3)$ d. $\sec(\pi/4)$ e. $\cos \pi$
    f. $\tan(\pi/6)$

6. a. $\sin(3\pi/2)$ b. $\csc(\pi/4)$ c. $\sec(\pi/6)$ d. $\tan 2\pi$ e. $\tan 0$
    f. $\cos(3\pi/2)$

Graph one period of each function given in Problems 7–20.

7. $y = \sin(x + \pi)$ 8. $y = \cos(x + \pi/2)$
9. $y = \cos(x - 3\pi/2)$ 10. $y = \sin(x - \pi/3)$
11. $y = 3 \sin x$ 12. $y = 2 \cos x$
13. $y = \sin 3x$ 14. $y = \cos 2x$
15. $y = \tan(x - 3\pi/2)$ 16. $y = \tan(x + \pi/6)$
17. $y = \frac{1}{2} \sin x$ 18. $y = \frac{1}{3} \tan x$
19. $y = 4 \tan x$ 20. $y = 5 \sin x$

B Problems

Graph one period of each function given in Problems 21–30.

21. $y - 2 = \sin(x - \pi/2)$ 22. $y + 1 = \cos(x + \pi/3)$
23. $y - 3 = \tan(x + \pi/6)$ 24. $y - \frac{1}{2} = \frac{1}{2} \cos x$
25. $y - 1 = 2 \cos(x - \pi/4)$ 26. $y - 1 = \cos 2(x - \pi/4)$
27. $y + 2 = 3 \sin(x + \pi/6)$ 28. $y + 2 = \sin 3(x + \pi/6)$
29. $y + 2 = \tan(x - \pi/4)$ 30. $y = 1 + \tan 2(x - \pi/4)$

Graph the curves given in Problems 31–40.

31. $y = \sin(4x + \pi)$ 32. $y = \sin(3x + \pi)$
33. $y = \tan(2x - \pi/2)$ 34. $y = \tan(x/2 + \pi/3)$
35. $y = \frac{1}{2} \cos(x + \pi/6)$ 36. $y = \cos(\frac{1}{2}x + \pi/12)$
37. $y = 3 \cos(3x + 2\pi) - 2$ 38. $y = 4 \sin(\frac{1}{2}x + 2)$
39. $y = \sqrt{2} \cos(x - \sqrt{2}) - 1$ 40. $y = \sqrt{3} \sin(\frac{1}{3}x - \sqrt{\frac{1}{3}})$

So far we have limited ourselves to $a > 0$. If $a < 0$, the curve is reflected through the x-axis. Graph the curves in Problems 41–46.

41. $y = -\sin x$ 42. $y = -\cos x$

43. $y = -\tan x$ 44. $y = -2 \sin 3x$
45. $y = -3 \cos 2x$ 46. $y = -2 \tan (x - \pi/3)$

In Section 1.3, parabolas were graphed by considering the sum of two functions. The same procedure can be used to graph the sum of two trigonometric functions. For example, if

$$y = 3 \sin 2x + 2 \cos 3x,$$

we can consider

$$y = f_1 + f_2,$$

where $f_1(x) = 3 \sin 2x$ and $f_2(x) = 2 \cos 3x$. We first graph f_1 and f_2 on the same axes, as shown in Figure 6.30. Next, we select a particular x value—say x_1—and graphically find $y_1 = f_1(x_1)$ and $y_2 = f_2(x_1)$. Then we plot $y = y_1 + y_2$. Do this for several points, as shown, and draw a smooth curve to represent $y = f_1(x) + f_2(x)$.

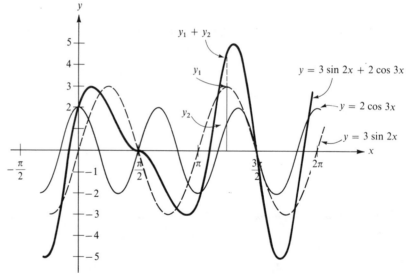

Figure 6.30   Graph of $y = 3 \sin 2x + 2 \cos 3x$

Graph the curves given in Problems 47–52.

47. $y = \sin x + \cos x$ 48. $y = \sin 2x + \cos x$
49. $y = \sin 2x + 2 \sin x$ 50. $y = 2 \cos x + \cos 2x$
51. $y = 3 \sin x + \cos 3x$ 52. $y = 2 \cos x + \sin 2x$

Mind Bogglers

 53. *What is your biorhythm?* Some people believe that one's entire life can be charted into cycles of good days and bad days. These cycles begin at the moment of birth and continue throughout life. There are three distinct biorhythm charts:

Physical—a sine curve with a period of 23 days; it shows a cycle of strength, endurance, and energy.

Sensitivity—a sine curve with a period of 28 days; it shows a cycle of feelings, intuition, moodiness, and creativity.

Mental—a sine curve with a period of 33 days; it shows a cycle of intelligence, memory, reasoning, power, and reaction.

You can easily chart your own biorhythm by following this step-by-step procedure.

(1) Find the number of days since your birth.

    a. Multiply your present age by 365.

    b. Add one day for each leap year lived (1936, 1940, 1944, 1948, 1952, 1956, 1960, 1964, 1968, 1972, 1976, 1980, ...). If you were born between March and December of a leap year, do not count that year. Add this number to the answer from part a.

    c. Count the number of days from your last birthday to the first day of the chart you are preparing (include your birthday and the first day of the chart). Add that answer to the answers from parts a and b.

(2) Divide the number of days since your birth (the total from part 1) by 23, 28, and 33, keeping track of the remainders.

(3) Plot sine curves with periods 23, 28, and 33 (the amplitudes are not relevant). The remainders found in step 2 tell what day of each cycle you are in for the first day of the chart.

For example, let's use Elvis Presley's birthday, 1/8/35, and calculate his biorhythm chart for the month he died, August 1977.

(1) a. $365 \times 42 = 15{,}330$

    b. 11 leap
years   =   11
Total   15,341

    c.

Jan.	24
Feb.	28
Mar.	31
Apr.	30
May	31
June	30
July	31
Aug.	1
Total	15,547

(2)

Divide by	Remainder
23	22
28	7
33	4

(3) On August 1, Elvis was in the 22nd day of a physical cycle, the 7th day of a sensitivity cycle, and the 4th day of a mental cycle.

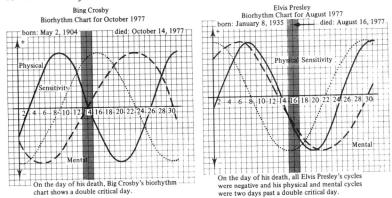

Bing Crosby
Biorhythm Chart for October 1977

born: May 2, 1904 — died: October 14, 1977

Physical

Sensitivity

Mental

On the day of his death, Big Crosby's biorhythm chart shows a double critical day.

Elvis Presley
Biorhythm Chart for August 1977

born: January 8, 1935 — died: August 16, 1977

Physical Sensitivity

Mental

On the day of his death, all Elvis Presley's cycles were negative and his physical and mental cycles were two days past a double critical day.

Figure 6.31   Biorhythm charts for Elvis Presley and Bing Crosby

Write equations for Bing Crosby's or Elvis Presley's physical, sensitivity, and mental cycles assuming that each cycle has an amplitude of 1.

54. Calculate your own biorhythm chart using the information given in Problem 53.
55. In all honesty, I calculated many, many charts before finding the two examples used in Figure 6.31, and many did not seem interesting from the biorhythm-theory standpoint. Investigate this matter on your own by selecting several personalities and calculating their biorhythms.

56. Any ordinary nonperiodic curve of finite length can also be analyzed by *harmonic analysis*. For example, the profile of a human face is analyzed by D. C. Miller in his book *The Science of Musical Sounds* and has the equation

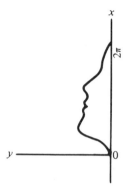

$$y = 49.6 \sin (\theta + 302°) + 17.4 \sin (2\theta + 298°) + 13.8 \sin (3\theta + 195°)$$
$$+ 7.1 \sin (4\theta + 215°) + 4.5 \sin (5\theta + 80°) + 0.6 \sin (6\theta + 171°)$$
$$+ 2.7 \sin (7\theta + 34°) + 0.6 \sin (8\theta + 242°) + 1.6 \sin (9\theta + 331°)$$
$$+ 1.3 \sin (10\theta + 208°) + 0.3 \sin (11\theta + 89°) + 0.5 \sin (12\theta + 229°)$$
$$+ 0.7 \sin (13\theta + 103°) + 0.3 \sin (14\theta + 305°) + 0.4 \sin (15\theta + 169°)$$
$$+ 0.5 \sin (16\theta + 230°) + 0.5 \sin (17\theta + 207°) + 0.4 \sin (18\theta + 64°).$$

Do some research and write a report or make a presentation to the class on the subject of harmonic analysis.

6.6 INVERSE CIRCULAR FUNCTIONS

In Section 1.4, the notion of the inverse of a function was introduced. The ideas presented there apply to circular as well as algebraic functions. Recall that, if a function f is given, then its inverse is the set of ordered pairs with the x and y values interchanged. Thus, if

$$y = \sin x$$

is a given function, then its *inverse* is

$$x = \sin y.$$

It is also customary to solve the resulting equation for y. To do this, we define

$$y = \sin^{-1} x \qquad \text{to mean} \qquad x = \sin y.$$

That is, *y is the angle with sine equal to x*, which is sometimes also written $y = \arcsin x$. It is important to remember that $y = \sin^{-1} x$ is simply a notational change for $x = \sin y$.

In each of the problem sets of this chapter there have been problems in which you were to find the sine of some number, say sin 30°, and you should know that $\sin 30° = \frac{1}{2}$. Now, we can reverse the procedure and ask "For which angle θ does $\sin \theta = \frac{1}{2}$?" Using the above notation, we look for $\sin^{-1}(\frac{1}{2})$. We see that $\sin^{-1}(\frac{1}{2}) = 30°$. But is this answer unique? Since $\sin 150° = \frac{1}{2}$, we see that $\sin^{-1}(\frac{1}{2}) = 150°$. In fact, there are many values of θ for which $\sin \theta = \frac{1}{2}$ (see Figure 6.32). Thus, the inverse of sin x—namely

$\sin^{-1} x$—is not a function, as we can see by examining the graphs of $y = \sin x$ and $y = \sin^{-1} x$ in Figure 6.33.

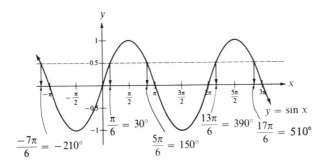

Figure 6.32 Graph of $y = \sin x$

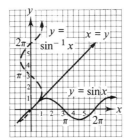

Figure 6.33 Graphs of $y = \sin x$ and $y = \sin^{-1} x$

However, we can define a function $y = \text{Sin } x$ that is identical to $y = \sin x$ except that its domain is limited so that its inverse is a function. From Figure 6.33, notice that, if $y = \text{Sin } x$, where $-\pi/2 \le x \le \pi/2$, then the inverse

$$y = \text{Sin}^{-1} x, \quad \text{where } -\pi/2 \le y \le \pi/2,$$

is a function. Thus, $\text{Sin}^{-1}(\frac{1}{2}) = \pi/6$. The way we have restricted the function is somewhat arbitrary at this point, and other choices might have been made. However, this method is the commonly accepted one in mathematics and is used to simplify later work in this book and in calculus.

Remember that our goal was simply to restrict the original function in such a way that its inverse would also be a function that includes all values of the range.

EXAMPLES: Find the values.

1. $\text{Sin}^{-1}(\frac{1}{2}\sqrt{3}) = \theta$
   Solution: Find the angle θ with sine equal to $\frac{1}{2}\sqrt{3}$. Also, since the arcsine is capitalized, we want $-\pi/2 \le \theta \le \pi/2$. From the memorized table of values, we find $\sin(\pi/3) = \frac{1}{2}\sqrt{3}$, and since $\pi/3$ is between $-\pi/2$ and $\pi/2$, we have $\text{Sin}^{-1}(\frac{1}{2}\sqrt{3}) = \pi/3$.

2. $\mathrm{Sin}^{-1}(-\frac{1}{2}\sqrt{3}) = \theta$

    Solution: Find θ where $-\pi/2 \leq \theta \leq \pi/2$ so that $\sin \theta = -\frac{1}{2}\sqrt{3}$. The sine is negative in both the third and fourth quadrants, but we choose the fourth-quadrant value since $-\pi/2 \leq \theta \leq \pi/2$. Thus, $\mathrm{Sin}^{-1}(-\frac{1}{2}\sqrt{3}) = -\pi/3$.

3. $\sin^{-1}(\frac{1}{2}\sqrt{3}) = \theta$

    Solution: Find θ so that $\sin \theta = \frac{1}{2}\sqrt{3}$. There are many values satisfying this relationship. The sine is positive in the first and second quadrants, so we seek angles in those quadrants with reference angles that are $\pi/3$. Thus, $\sin^{-1}(\frac{1}{2}\sqrt{3}) = \pi/3$ and $\sin^{-1}(\frac{1}{2}\sqrt{3}) = 2\pi/3$. Also, since the period of the sine is 2π, we have

$$\sin^{-1}(\tfrac{1}{2}\sqrt{3}) = \begin{cases} \dfrac{\pi}{3} + 2n\pi \\[2ex] \dfrac{2\pi}{3} + 2n\pi \end{cases} \qquad \text{where } n \text{ is an integer.}$$

The other circular functions are handled similarly.

<table>
<tr><td></td><td>GIVEN
FUNCTION</td><td>INVERSE</td><td colspan="2">OTHER NOTATIONS
FOR INVERSE</td></tr>
<tr><td rowspan="6">*Notation for the* inverse
circular functions</td><td>$y = \cos x$</td><td>$x = \cos y$</td><td>$y = \cos^{-1} x$</td><td>$y = \arccos x$</td></tr>
<tr><td>$y = \sin x$</td><td>$x = \sin y$</td><td>$y = \sin^{-1} x$</td><td>$y = \arcsin x$</td></tr>
<tr><td>$y = \tan x$</td><td>$x = \tan y$</td><td>$y = \tan^{-1} x$</td><td>$y = \arctan x$</td></tr>
<tr><td>$y = \sec x$</td><td>$x = \sec y$</td><td>$y = \sec^{-1} x$</td><td>$y = \operatorname{arcsec} x$</td></tr>
<tr><td>$y = \csc x$</td><td>$x = \csc y$</td><td>$y = \csc^{-1} x$</td><td>$y = \operatorname{arccsc} x$</td></tr>
<tr><td>$y = \cot x$</td><td>$x = \cot y$</td><td>$y = \cot^{-1} x$</td><td>$y = \operatorname{arccot} x$</td></tr>
</table>

These are the same.

Because the inverse secant and inverse cosecant are rarely used, we will limit our study to the other four inverse circular functions. Consider the graphs in Figure 6.34. We have to restrict each circular function so that the inverse is also a function, but we also want to include all possible values in the range of the original function. For the sine curve, x was restricted so that $-\pi/2 \leq x \leq \pi/2$. Then the inverse is the function

$$y = \mathrm{Sin}^{-1} x, \quad \text{where } -\pi/2 \leq y \leq \pi/2.$$

The same restrictions (leaving out the values $-\pi/2$ and $\pi/2$) will apply for the tangent and arctangent curves. However, for the cosine function notice that, by restricting x to the same interval, you obtain only positive values for

$y = \cos x$. Thus, to include the entire range of the cosine curve, x can be restricted so that $0 \le x \le \pi$. Then the inverse is the function

$$y = \text{Cos}^{-1} x, \quad \text{where } 0 \le y \le \pi.$$

The cotangent function is restricted in almost the same way, and the results are summarized in the definitions below. Notice that the inverse functions are indicated in Figure 6.34 by the darker parts of the curves.

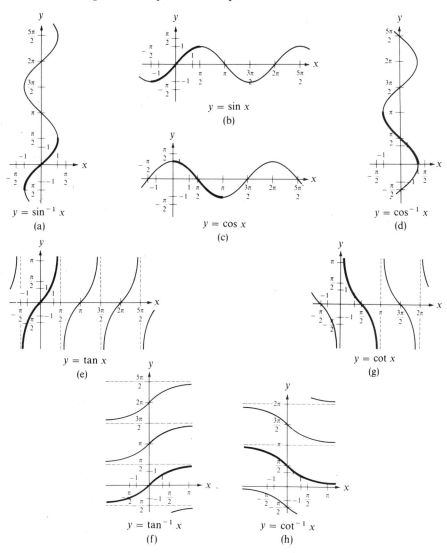

Figure 6.34   Inverse circular functions

INVERSE CIRCULAR
FUNCTIONS

> DEFINITION
>
> | INVERSE FUNCTION | | | DOMAIN | RANGE |
> |---|---|---|---|---|
> | $y = \text{Arccos } x$ | or | $y = \text{Cos}^{-1} x$ | $-1 \le x \le 1$ | $0 \le y \le \pi$ |
> | $y = \text{Arcsin } x$ | or | $y = \text{Sin}^{-1} x$ | $-1 \le x \le 1$ | $-\pi/2 \le y \le \pi/2$ |
> | $y = \text{Arctan } x$ | or | $y = \text{Tan}^{-1} x$ | All reals | $-\pi/2 < y < \pi/2$ |
> | $y = \text{Arccot } x$ | or | $y = \text{Cot}^{-1} x$ | All reals | $0 < y < \pi$ |

EXAMPLES: Find the given angle.

4. Arctan $1 = \theta$
    Solution: We are looking for an angle θ with tangent equal to 1. Since this is an exact value, we know that $\theta = \pi/4$ or 45°.

5. Arccot $(-\sqrt{3}) = \theta$
    Solution: Find θ so that cot $\theta = -\sqrt{3}$; the reference angle is 30°, and the cotangent is negative in Quadrants II and IV. Since Arccot x is defined in Quadrant II but not in Quadrant IV, we see that Arccot $(-\sqrt{3}) = 150°$.

6. Arcsin $(-0.4695) = \theta$
    Solution: Find θ so that sin $\theta = -0.4695$; from Table 4 in the back of the book we find that the reference angle is 28°. Since Arcsin x is defined for values between $-90°$ and 90°, we see that Arcsin $(-0.4695) = -28°$.

    Alternate Solution by Calculator: Check the algorithm used in the operation manual of your own calculator.

    | ENTER VALUE | INV | TRIG FUNCTION |

    The INV and TRIG FUNCTION keys pushed in succession give the inverse trigonometric functions. For this example, push the following keys:

    | 4695 | +/− | INV | SIN |

    The angle is now displayed: -28.00184535, which is the decimal representation of the angle in degrees.

Some calculators have an ARC button instead of an INV button and a CHS button instead of a +/− button.

Another way of looking at this is to consider the keys pressed on a calculator to find cotangent. For example, cot 30°:

| 30 | TAN | 1/x |

The answer is 1.73205080. Thus, to find arccot

7. Arccot $(-2.747) = \theta$
    Solution: We can use Table 4 to find the reference angle 20°. Since $0 < \text{cot}^{-1} x < \pi$, we must place this angle in Quadrant II, so $\theta = 160°$. Since calculators don't have a cotangent function, we note that if cot $y = x$, then tan $y = 1/x$. Thus,

    $$y = \text{cot}^{-1} x \quad \text{and} \quad y = \tan^{-1}(1/x) \quad (x > 0),$$
    $$\text{cot}^{-1} x = \tan^{-1}(1/x).$$

    This tells us that to find the inverse cotangent on a calculator we must first take the reciprocal of the given value and then complete the problem.

Another difficulty is the negative value, since the range for the tangent and cotangent is not the same. Proceed as follows if using a calculator:

```
┌─────────────────────────────┐
│  ENTER ABSOLUTE VALUE       │
│       OF NUMBER             │
└─────────────────────────────┘
```

$\boxed{2.747}$ $\boxed{1/x}$ $\boxed{INV}$ $\boxed{TAN}$

The result, 20.00320032°, gives the reference angle. Thus, the result in Quadrant II is 160° (159.9967997°).

8. Arccos (cos 2)

Solution: Let $f(x) = \cos x$. Then $f^{-1}(x) = \arccos x$. From Section 1.4, $(f^{-1} \circ f)(x) = x$. Thus, Arccos (cos 2) = 2. This example can be seen even more clearly on a calculator.

BUTTONS PRESSED	DISPLAY
$\boxed{2}$	2
$\boxed{COS}$	0.4161468365
$\boxed{INV}$ $\boxed{COS}$	2

Try to find Arccos (cos θ) for other values of θ, where $0 \le \theta < \pi$, on your calculator.

9. tan (Arctan 2.463) = 2.463

1.73205080, just reverse the steps:

$\boxed{1/x}$ $\boxed{INV}$ $\boxed{TAN}$

The answer is 30.

Check Example 9 on a calculator:

BUTTONS	DISPLAY
$\boxed{2.463}$	2.463
$\boxed{TAN}$	1.181306656
$\boxed{INV}$$\boxed{TAN}$	2.463

PROBLEM SET 6.6

A Problems

In Problems 1–7, obtain the given angle from memory.

1. a. Arcsin 0   b. $\text{Tan}^{-1}(\sqrt{3}/3)$ c. Arccot $\sqrt{3}$ d. Arccos 1
2. a. $\text{Cos}^{-1}(\sqrt{3}/2)$ b. Arcsin $\frac{1}{2}$ c. $\text{Tan}^{-1} 1$ d. $\text{Sin}^{-1} 1$
3. a. Arctan $\sqrt{3}$ b. $\text{Cos}^{-1}(\sqrt{2}/2)$ c. Arcsin $(\frac{1}{2}\sqrt{2})$ d. Arccot 1
4. a. Arcsin (-1) b. $\text{Cot}^{-1}(-1)$ c. Arcsin $(-\sqrt{3}/2)$ d. $\text{Cos}^{-1}(-1)$
5. a. $\text{Cot}^{-1}(-\sqrt{3})$ b. Arctan (-1) c. $\text{Sin}^{-1}(-\frac{1}{2}\sqrt{2})$ d. $\text{Cos}^{-1}(-\frac{1}{2})$
6. a. Arccos $(-\sqrt{2}/2)$ b. $\text{Cot}^{-1}(-\sqrt{3}/3)$ c. $\text{Sin}^{-1}(-\frac{1}{2})$ d. Arctan $(-\sqrt{3}/3)$
7. a. $\text{Tan}^{-1} 0$ b. Arccot $(\sqrt{3}/3)$ c. Arccos $\frac{1}{2}$ d. $\text{Sin}^{-1}(\sqrt{3}/2)$

Use Table 4 in the back of the book or a calculator to find the values (in degrees) given in Problems 8–22.

8. Arcsin 0.2079
9. $\text{Cos}^{-1} 0.8387$
10. Arctan 1.171
11. $\text{Cot}^{-1} 0.0875$
12. $\text{Tan}^{-1}(-3.732)$
13. Arccos 0.9455
14. $\text{Sin}^{-1}(-0.4695)$
15. Arccot 0.7265
16. Arctan 1.036
17. $\text{Cot}^{-1}(-0.3249)$
18. $\text{Sin}^{-1} 0.3584$
19. $\text{Cos}^{-1} 0.3584$
20. $\text{Tan}^{-1} 2.050$
21. Arcsin (-0.9135)
22. Arccot (-1.235)

Simplify the expressions in Problems 23–30.

23. cot (Arccot 1)
24. Arccos [cos $(\pi/6)$]
25. sin (Arcsin $\frac{1}{3}$)
26. Tan^{-1} [tan $(\pi/15)$]
27. Arcsin [sin $(2\pi/15)$]
28. cos (Arccos $\frac{2}{3}$)
29. tan (Arctan 0.4163)
30. Arccot (cot 35°)

B Problems

Find the values given in Problems 31–39.

31. $\sin^{-1}\frac{1}{2}$
32. arccos $\frac{1}{2}$
33. $\tan^{-1}\sqrt{3}$
34. $\cot^{-1}(-\sqrt{3})$
35. $\cos^{-1}(-\sqrt{3}/2)$
36. arcsin $(-\sqrt{2}/2)$
37. arcsin 0.3907
38. arccos 0.2924
39. arctan 1.376

In Problems 40–45, graph the given pair of curves on the same axes.

40. $y = \sin x; x = \sin y$
41. $y = \cos x; x = \cos y$
42. $y = \tan x; x = \tan y$
43. $y = \cot x; x = \cot y$
44. $y = \sec x; x = \sec y$
45. $y = \csc x; x = \csc y$

Mind Boggler

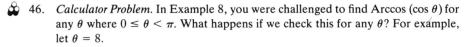

46. *Calculator Problem.* In Example 8, you were challenged to find Arccos (cos θ) for any θ where $0 \le \theta < \pi$. What happens if we check this for any θ? For example, let $\theta = 8$.

CALCULATOR	DISPLAY
8	8
COS	0.145500338
INV	0.145500338
COS	1.716814693

But 8 ≠ 1.716814693. The difficulty here is that $\theta > \pi$, but, in Section 1.4, we stated that

$$f^{-1} \circ f = x$$

if f and f^{-1} are inverse functions. Reconcile these seemingly contradictory results.

6.7 TRIGONOMETRIC FORM OF COMPLEX NUMBERS

One important use of the sine and cosine functions is in their relationship to complex numbers. The set of complex numbers is the set of all numbers of the form

$$a + bi,$$

where a and b are real numbers and $i = \sqrt{-1}$. The operations on complex numbers are reviewed in Appendix A, but their relationship to the sine and cosine functions is found by considering the graphical representation of a complex number. To give a graphical representation of complex numbers, such as

$$2 + 3i, \quad -i, \quad -3 - 4i, \quad 3i, \quad -2 + \sqrt{2}\, i, \quad \frac{3}{2} - \frac{5}{2}i,$$

a two-dimensional coordinate system is used: the horizontal axis represents the *real axis* and the vertical axis is the *imaginary axis*, so that $a + bi$ is represented by the ordered pair (a, b). Remember that a and b represent real numbers, so we plot (a, b) in the usual manner, as shown in Figure 6.35.

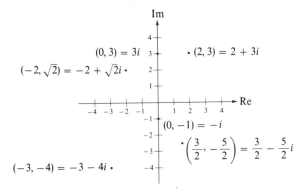

Figure 6.35   Complex plane

The coordinate system in Figure 6.35 is called the *complex plane* or the *Gaussian plane*, in honor of Carl Friedrich Gauss (1777–1855).

The *absolute value* of a complex number z is, graphically, the distance between z and the origin (just as it is for real numbers). The absolute value of a complex number is also called the *modulus*. The distance formula leads to the following definition:

COMPLEX OR GAUSSIAN PLANE

DEFINITION: If $z = a + bi$, then the *absolute value*, or *modulus*, of z is denoted by $|z|$ and defined by

$$|z| = \sqrt{a^2 + b^2}.$$

ABSOLUTE VALUE, OR MODULUS, OF A COMPLEX NUMBER

EXAMPLES: Find the absolute value.

1. $3 + 4i$

$$\text{absolute value: } |3 + 4i| = \sqrt{3^2 + 4^2}$$
$$= \sqrt{25}$$
$$= 5$$

2. $-2 + \sqrt{2}\,i$

$$\text{absolute value: } |-2 + \sqrt{2}\,i| = \sqrt{4 + 2}$$
$$= \sqrt{6}$$

3. -3

$$\text{absolute value: } |-3 + 0i| = \sqrt{9 + 0}$$
$$= 3$$

Notice from Example 3 that the definition we give here is consistent with the definition of absolute value given for real numbers.

The form $a + bi$ is called the *rectangular form*, but another very useful representation uses trigonometry. Consider the graphical representation of a complex number $a + bi$, as shown in Figure 6.36. Let r be the distance from the origin to (a, b) and let θ be the angle the segment makes with the real axis. Then

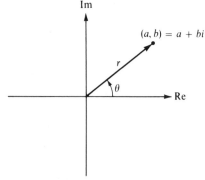

Figure 6.36   Trigonometric form and rectangular form of a complex number

$$r = \sqrt{a^2 + b^2}$$

ARGUMENT OF A COMPLEX NUMBER

and θ, called the *argument*, is chosen so that it is the smallest nonnegative angle the terminal side makes with the real axis. From the definition of the trigonometric functions,

$$\cos \theta = \frac{a}{r}, \qquad \sin \theta = \frac{b}{r}, \qquad \tan \theta = \frac{b}{a},$$

$$a = r \cos \theta, \qquad b = r \sin \theta.$$

Therefore,

$$a + bi = r \cos \theta + ir \sin \theta$$
$$= r(\cos \theta + i \sin \theta).$$

Sometimes, $r(\cos \theta + i \sin \theta)$ is abbreviated by

$$r(\cos \theta + i \sin \theta)$$
$$\downarrow\downarrow \qquad \downarrow\downarrow \quad \downarrow$$
$$r(c \qquad i\,s \quad \theta) \qquad \text{or } r \text{ cis } \theta.$$

> **DEFINITION:** The *trigonometric*, or *polar form*, of a complex number $z = a + bi$ is
>
> $$r(\cos \theta + i \sin \theta) = r \text{ cis } \theta,$$
>
> where $r = \sqrt{a^2 + b^2}$; $\tan \theta = b/a$ if $a \neq 0$; $a = r \cos \theta$; $b = r \sin \theta$. This representation is unique for $0 \leq \theta < 360°$ for all z except $0 + 0i$.

TRIGONOMETRIC, OR
POLAR, FORM OF A
COMPLEX NUMBER

The placement of θ in the proper quadrant is an important consideration because there are two values of $0 \leq \theta < 360°$ that will satisfy the relationship

$$\tan \theta = \frac{b}{a}.$$

For example, compare the following.

$$(1) \quad -1 + i \qquad a = -1, b = 1 \qquad \tan \theta = \frac{1}{-1} \quad \text{or} \quad \tan \theta = -1$$

$$(2) \quad 1 - i \qquad a = 1, b = -1 \qquad \tan \theta = \frac{-1}{1} \quad \text{or} \quad \tan \theta = -1$$

Notice the same trigonometric equation for both complex numbers, even though $-1 + i$ is in Quadrant II and $1 - i$ is in Quadrant IV. This consideration of quadrants is even more important when we're doing the problem on a calculator, since the proper sequence of steps for this example is

$$\boxed{1} \quad \boxed{+/-} \quad \boxed{\text{INV}} \quad \boxed{\text{TAN}},$$

giving the result $-45°$, which is not true for either example since $0 \leq \theta < 360°$. The entire process can be taken care of quite simply if we let θ' be the reference angle for θ. Then we find the reference angle

$$\theta' = \tan^{-1} \left| \frac{b}{a} \right|.$$

After you know the reference angle and the quadrant, it is easy to find θ. For these examples,

$$\theta' = \tan^{-1}|-1|$$
$$= 45°.$$

On a calculator:

$\boxed{1}$ $\boxed{\text{INV}}$ $\boxed{\text{TAN}}$

For Quadrant II, $\theta = 135°$; for Quadrant IV, $\theta = 315°$.

EXAMPLES: Change the complex numbers to trigonometric form.

4. $1 - \sqrt{3}\,i$
    Solution: $a = 1$ and $b = -\sqrt{3}$; the number is in Quadrant IV.

The reference angle is 60°; in Quadrant IV: $\theta = 300°$.

$$r = \sqrt{1^2 + (-\sqrt{3})^2} \qquad \theta' = \tan^{-1}\left|\frac{-\sqrt{3}}{1}\right|$$
$$= \sqrt{4} \qquad\qquad\qquad = 60°$$
$$= 2 \qquad\qquad\qquad \theta = 300° \text{ (Quad IV)}$$

On a calculator:

$\boxed{3}$ $\boxed{\sqrt{}}$ $\boxed{\text{INV}}$ $\boxed{\text{TAN}}$

Thus, $1 - \sqrt{3}\,i = 2 \text{ cis } 300°$.

5. $-3 - 5i$
    Solution: $a = -3$ and $b = -5$; the number is in Quadrant III.

The reference angle is 59°; in Quadrant III: $\theta = 239°$.

$$r = \sqrt{(-3)^2 + (-5)^2} \qquad \theta' = \tan^{-1}\left|\frac{5}{3}\right|$$
$$= \sqrt{9 + 25} \qquad\qquad\quad \approx 59°$$
$$= \sqrt{34} \qquad\qquad\quad \theta \approx 239° \text{ (Quad III)}$$

On a calculator with

Algebraic logic:	RPN logic:
$\boxed{5}$	$\boxed{5}$
$\boxed{\div}$	$\boxed{\text{ENTER}}$
$\boxed{3}$	$\boxed{3}$
$\boxed{=}$	$\boxed{\div}$
$\boxed{\text{INV}}$	$\boxed{\text{ARC}}$
$\boxed{\text{TAN}}$	$\boxed{\text{TAN}}$

Notice that $\tan \theta$ is not defined for $\theta = 90°$.

Thus, $-3 - 5i \approx \sqrt{34} \text{ cis } 239°$.

6. $4.310 + 5.516i$
    Solution: $a = 4.310$ and $b = 5.516$; the number is in Quadrant I.

$$r = \sqrt{(4.310)^2 + (5.516)^2} \qquad \theta = \tan^{-1}\left|\frac{5.516}{4.310}\right|$$
$$= \sqrt{49} \qquad\qquad\qquad \approx \tan^{-1}(1.2798)$$
$$= 7 \qquad\qquad\qquad \approx 52°$$

Thus, $4.310 + 5.516i \approx 7 \text{ cis } 52°$.

7. $6i$
    Solution: $a = 0$ and $b = 6$; by inspection, $6i = 6 \text{ cis } 90°$.

EXAMPLES: Change the complex number to rectangular form.

8. $4 \text{ cis } 330°$
    Solution: $r = 4$ and $\theta = 330°$

$$a = 4 \cos 330° \qquad b = 4 \sin 330°$$

$$= 4\left(\frac{\sqrt{3}}{2}\right) \qquad = 4\left(-\frac{1}{2}\right)$$

$$= 2\sqrt{3} \qquad = -2$$

Thus, $4 \text{ cis } 330° = 2\sqrt{3} - 2i$.

9. $5(\cos 38° + i \sin 38°)$
    Solution: $r = 5$ and $\theta = 38°$

$$a = 5 \cos 38° \qquad b = 5 \sin 38°$$
$$\approx 3.94 \qquad \approx 3.08$$

Thus, $5(\cos 38° + i \sin 38°) \approx 3.94 + 3.08i$.

In the next section we'll consider some of the advantages of writing a complex number in trigonometric form.

PROBLEM SET 6.7

A Problems

Plot the complex numbers given in Problems 1–12.

1. $3 + i$
2. $7 - i$
3. $3 + 2i$
4. $-3 - 2i$
5. $-1 + 3i$
6. $2 + 4i$
7. $5 + 6i$
8. $2 - 5i$
9. $-2 + 5i$
10. $-5 + 4i$
11. $4 - i$
12. $-1 + i$

B Problems

Plot and then change each of the numbers in Problems 13–28 to trigonometric form. Use exact values whenever possible.

13. $1 + i$
14. $1 - i$
15. $\sqrt{3} - i$
16. $\sqrt{3} + i$
17. $1 - \sqrt{3}\,i$
18. $-1 - \sqrt{3}\,i$
19. $\sqrt{12} - 2i$
20. $-\sqrt{6} + \sqrt{2}\,i$
21. 1
22. 5
23. $-i$
24. $-4i$
25. $1.7207 + 2.4575i$
26. $5.7956 - 1.5529i$
27. $-0.6946 + 3.9392i$
28. $1.5321 - 1.2856i$

Plot and then change each of the numbers in Problems 29–44 to rectangular form. Use exact values whenever possible.

29. $2(\cos 45° + i \sin 45°)$
30. $3(\cos 60° + i \sin 60°)$

31. $4(\cos 315° + i \sin 315°)$
32. $5(\cos 4\pi/3 + i \sin 4\pi/3)$
33. $\cos (5\pi/6) + i \sin (5\pi/6)$
34. $5 \operatorname{cis} (3\pi/2)$
35. $\operatorname{cis} 0$
36. $4 \operatorname{cis} 30°$
37. $3 \operatorname{cis} (2\pi/3)$
38. $2 \operatorname{cis} \pi$
39. $10 \operatorname{cis} 65°$
40. $8 \operatorname{cis} 24°$
41. $7 \operatorname{cis} 135°$
42. $6 \operatorname{cis} 247°$
43. $9 \operatorname{cis} 190°$
44. $10 \operatorname{cis} 371°$

Mind Bogglers

45. If $z = a + bi$ and $\bar{z} = a - bi$, show that

$$|z| = \sqrt{z \cdot \bar{z}}.$$

This relationship is called the *triangle inequality*.

46. If $z_1 = a + bi$ and $z_2 = c + di$, show that

$$|z_1 + z_2| \leq |z_1| + |z_2|.$$

6.8 OPERATIONS IN TRIGONOMETRIC FORM

In algebra, the operations involving complex numbers were fairly simple, so you encountered little difficulty when carrying them out. For example,

Remember, $i^2 = -1$.

$$(1 + 2i)(2 + 3i) = 2 + 7i + 6i^2$$
$$= 2 + 7i + 6(-1)$$
$$= -4 + 7i.$$

But consider

$$(1.9319 + 0.5176i)(2.5981 + 1.5i) = 5.0191 + 2.8979i + 1.3448i + 0.7764i^2$$
$$= 4.2427 + 4.2427i.$$

You almost need a calculator to carry out the operations. If these numbers are changed to trigonometric form, are the calculations easier?

$$1.9319 + 0.5176i = 2 \operatorname{cis} 15°$$
$$2.5981 + 1.5i \quad = 3 \operatorname{cis} 30°$$

The result, changed to trigonometric form, is

Do you notice any relationship between the problem and the answer in polar form?

$$4.2427 + 4.2427i = 6 \operatorname{cis} 45°.$$

Let's consider another example:

$$-1 + \sqrt{3}\, i = 2 \operatorname{cis} 120°$$
$$-\sqrt{3} - i = 2 \operatorname{cis} 210°.$$

If the pattern illustrated by the first example holds in general, it would indicate that the product is

$$2 \cdot 2 \text{ cis } (120° + 210°) = 4 \text{ cis } 330°.$$

Checking this result by direct calculation, we get

$$(-1 + \sqrt{3}\, i)(-\sqrt{3} - i) = \sqrt{3} + i - 3i - \sqrt{3}\, i^2$$
$$= 2\sqrt{3} - 2i.$$

This can be changed to trigonometric form:

$$2\sqrt{3} - 2i = 4 \text{ cis } 330°.$$

Since the results agree, a general result is suggested.

Let $z_1 = r_1 \text{ cis } \theta_1$ and $z_2 = r_2 \text{ cis } \theta_2$ be complex numbers. Then

$$z_1 z_2 = r_1 r_2 \text{ cis } (\theta_1 + \theta_2).$$

Multiplication of complex numbers in polar form

This theorem for the multiplication of complex numbers in trigonometric form is extremely useful, as Examples 1–4 illustrate.

EXAMPLES: Simplify.

1. $5 \text{ cis } 38° \cdot 4 \text{ cis } 75° = 5 \cdot 4 \text{ cis } (38° + 75°)$
   $$= 20 \text{ cis } 113°$$

2. $\sqrt{2} \text{ cis } 188° \cdot 2\sqrt{2} \text{ cis } 310° = 4 \text{ cis } 498°$
   $$= 4 \text{ cis } 138°$$

3. $(2 \text{ cis } 48°)^3 = (2 \text{ cis } 48°)(2 \text{ cis } 48°)^2$
   $$= (2 \text{ cis } 48°)(4 \text{ cis } 96°)$$
   $$= 8 \text{ cis } 144°$$

   Notice that this result is the same as

   $$(2 \text{ cis } 48°)^3 = 2^3 \text{ cis } (3 \cdot 48°)$$
   $$= 8 \text{ cis } 144°.$$

4. $(1 - \sqrt{3}\, i)^5$

   We first change to trigonometric form.

   $$a = 1; b = -\sqrt{3}; \text{ Quadrant IV}$$
   $$r = \sqrt{1 + 3} \qquad \theta' = \tan^{-1}|-\sqrt{3}/1|$$
   $$= 2 \qquad\qquad = 60°$$
   $$\theta = 300° \text{ (Quadrant IV)}$$

   $$(1 - \sqrt{3}\, i)^5 = (2 \text{ cis } 300°)^5$$
   $$= 2^5 \text{ cis } (5 \cdot 300°)$$

$$= 32 \text{ cis } 1500°$$
$$= 32 \text{ cis } 60°$$

If we want the answer in rectangular form, we can now change back.

$$a = 32 \cos 60° \qquad b = 32 \sin 60°$$
$$= 32(\tfrac{1}{2}) \qquad\qquad = 32(\tfrac{1}{2}\sqrt{3})$$
$$= 16 \qquad\qquad\quad = 16\sqrt{3}$$

Thus, $(1 - \sqrt{3}\, i)^5 = 16 + 16\sqrt{3}\, i$.

As you can see from Example 4, multiplication in polar form extends quite nicely for any positive integral power.

Raising a number in polar form to a power—we will generalize this result in the next section.

> If n is a natural number, then
> $$(r \text{ cis } \theta)^n = r^n \text{ cis } n\theta$$
> for a complex number $r \text{ cis } \theta = r(\cos \theta + i \sin \theta)$.

For division, consider the number

$$10(\cos 150° + i \sin 150°) \div 2(\cos 120° + i \sin 120°)$$
$$= (-5\sqrt{3} + 5i) \div (-1 + \sqrt{3}\, i)$$

Algebraic (rectangular form):

$$\frac{-5\sqrt{3} + 5i}{-1 + \sqrt{3}\, i} = \frac{-5\sqrt{3} + 5i}{-1 + \sqrt{3}\, i} \cdot \frac{-1 - \sqrt{3}\, i}{-1 - \sqrt{3}\, i}$$

$$= \frac{5\sqrt{3} + 15i - 5i - 5\sqrt{3}\, i^2}{1 - 3i^2}$$

$$= \frac{10\sqrt{3} + 10i}{4}$$

$$= \frac{5}{2}\sqrt{3} + \frac{5}{2}i$$

$$= 5 \text{ cis } 30°.$$

But from the trigonometric forms,

$$\frac{10 \text{ cis } 150°}{2 \text{ cis } 120°} = 5 \text{ cis } 30°$$

would seem to indicate that

$$\frac{10 \text{ cis } 150°}{2 \text{ cis } 120°} = \frac{10}{2} \text{ cis } (150° - 120°),$$

which is true not only for this example but also in general.

Let $z_1 = r_1 \text{ cis } \theta_1$ and $z_2 = r_2 \text{ cis } \theta_2$ be nonzero complex numbers. Then

$$\frac{z_1}{z_2} = \frac{r_1}{r_2} \text{ cis } (\theta_1 - \theta_2).$$

DIVISION OF
COMPLEX NUMBERS
IN POLAR FORM

EXAMPLE 5: Find

$$\frac{15(\cos 48° + i \sin 48°)}{5(\cos 125° + i \sin 125°)}.$$

Solution: $\dfrac{15 \text{ cis } 48°}{5 \text{ cis } 125°} = 3 \text{ cis } (48° - 125°)$

$$= 3 \text{ cis } (-77°)$$

$$= 3 \text{ cis } 283°$$

PROBLEM SET 6.8

A Problems

Perform the indicated operations in Problems 1–19. You may leave your answers in trigonometric or rectangular form.

1. $2 \text{ cis } 60° \cdot 3 \text{ cis } 150°$ 2. $3 \text{ cis } 48° \cdot 5 \text{ cis } 92°$
3. $4(\cos 65° + i \sin 65°) \cdot 12(\cos 87° + i \sin 87°)$
4. $7(\cos 83° + i \sin 83°) \cdot 8(\cos 12° + i \sin 12°)$

5. $\dfrac{8 \text{ cis } 30°}{4 \text{ cis } 15°}$ 6. $\dfrac{12 \text{ cis } 250°}{4 \text{ cis } 120°}$

7. $\dfrac{5(\cos 315° + i \sin 315°)}{2(\cos 48° + i \sin 48°)}$ 8. $\dfrac{9(\cos 87° + i \sin 87°)}{5(\cos 28° + i \sin 28°)}$

9. $\dfrac{20(\cos 40° + i \sin 40°)}{10(\cos 210° + i \sin 210°)}$ 10. $\dfrac{18(\cos 25° + i \sin 25°)}{9(\cos 135° + i \sin 135°)}$

11. $6(\cos 215° + i \sin 215°) \cdot 3(\cos 312° + i \sin 312°)$
12. $\sqrt{5}(\cos 125° + i \sin 125°) \cdot \sqrt{45}(\cos 312° + i \sin 312°)$
13. $(2 \text{ cis } 50°)^3$ 14. $(3 \text{ cis } 60°)^4$
15. $(\cos 210° + i \sin 210°)^5$ 16. $(\cos 85° + i \sin 85°)^9$

17. $(2 - 2i)^4$ 18. $(\sqrt{3} - i)^8$

19. $(1 + i)^6$

B Problems

Perform the indicated operations in Problems 20–29. Leave your answers in trigonometric form.

20. $(1 + i)(\sqrt{3} - i)$ 21. $\dfrac{1 + i}{\sqrt{3} - i}$

22. $\dfrac{1 - \sqrt{3}\, i}{\sqrt{3} + i}$ 23. $(1 - \sqrt{3}\, i)(\sqrt{3} + i)$

24. $(2 + 2i)(1 + i)^2$ 25. $3i(4 - 4i)$

26. $4i(5 - 5i)(4 - 4i)(3 - 3i)$ 27. $(1 + i)^5$

28. $(1 + i\sqrt{3})^4$ 29. $\dfrac{(1 + i)^2}{1 + \sqrt{3}\, i}$

Work each of Problems 30–45 using whichever form you wish, but leave your answer in the same form that is given in the problem.

30. $2(\cos 60° + i \sin 60°) \cdot 3(\cos 210° + i \sin 210°)$
31. $3(\cos 300° + i \sin 300°) \cdot 4(\cos 30° + i \sin 30°)$
32. $8(\cos 45° + i \sin 45°) \cdot 2(\cos 150° + i \sin 150°)$
33. $(3\sqrt{3} - 3i)(1 + i)$ 34. $(2 + 2i)(1 - \sqrt{3}\, i)$

35. $3i(\sqrt{3} + i)$ 36. $\dfrac{4(\cos 60° + i \sin 60°)}{2(\cos 210° + i \sin 210°)}$

37. $\dfrac{12(\cos 300° + i \sin 300°)}{4(\cos 30° + i \sin 30°)}$ 38. $\dfrac{8(\cos 45° + i \sin 45°)}{2(\cos 150° + i \sin 150°)}$

39. $\dfrac{3\sqrt{3} - 3i}{1 + i}$ 40. $\dfrac{2 + 2i}{1 - \sqrt{3}\, i}$

41. $\dfrac{3i}{\sqrt{3} + i}$ 42. $(5 \text{ cis } 48°)(3 \text{ cis } 25°)$

43. $(-3.2253 + 8.4022i)(3.4985 + 1.9392i)$

44. $\dfrac{7 \text{ cis } 135°}{5 \text{ cis } 53°}$ 45. $\dfrac{-3.2253 + 8.4022i}{3.4985 + 1.9392i}$

Mind Bogglers

46. If

$$\cos \theta = 1 - \frac{\theta^2}{2!} + \frac{\theta^4}{4!} - \frac{\theta^6}{6!} + \cdots + \frac{(-1)^n \theta^{2n}}{(2n)!} + \cdots ,$$

$$\sin \theta = \theta - \frac{\theta^3}{3!} + \frac{\theta^5}{5!} - \frac{\theta^7}{7!} + \cdots + \frac{(-1)^n \theta^{2n+1}}{(2n + 1)!} + \cdots ,$$

and

$$e^{i\theta} = 1 + (i\theta) + \frac{(i\theta)^2}{2!} + \frac{(i\theta)^3}{3!} + \frac{(i\theta)^4}{4!} + \cdots + \frac{(i\theta)^n}{n!} + \cdots,$$

show that $e^{i\theta} = \cos\theta + i\sin\theta$. This equation is called *Euler's Formula*.

47. Using Problem 46, show that $e^{i\pi} = -1$.

6.9 DE MOIVRE'S THEOREM

In the last section, we used

$$(r\ \text{cis}\ \theta)^n = r^n\ \text{cis}\ n\theta$$

for n a natural number. However, we also noted that this result is true for any multiple of 360°, so that we could write expressions such as

$$\text{cis}\ 1500° = \text{cis}\ 60°.$$

That is, we could write

$$(r\ \text{cis}\ \theta)^n = r^n\ \text{cis}\ n(\theta + 360°k),$$

where k is any integer. As long as we restrict n to a natural number, this second form merely gives the same values as the formula of the last section. However, our intent is to generalize the result by allowing n to be any real number. Making this generalization gives a result called *De Moivre's Theorem*, which is stated here without proof.

DE MOIVRE'S THEOREM: If n is any real number,

$$(r\ \text{cis}\ \theta)^n = r^n\ \text{cis}\ n(\theta + 360°k) \quad \text{for any integer } k.$$

EXAMPLES

1. Find the square roots of

$$-\frac{9}{2} + \frac{9}{2}\sqrt{3}\ i.$$

Solution: First change to polar form.

$$r = \sqrt{\left(-\frac{9}{2}\right)^2 + \left(\frac{9}{2}\sqrt{3}\right)^2}$$

$$= \sqrt{\frac{81}{4} + \frac{81 \cdot 3}{4}}$$

$$\theta' = \tan^{-1}\left|\frac{\frac{9}{2}\sqrt{3}}{-\frac{9}{2}}\right|$$

DE MOIVRE'S THEOREM

Historical Note

The naming of results in the history of mathematics is a strange phenomenon. De Moivre's Theorem was named after Abraham de Moivre (1667–1754), who used his famous formula in *Philosophical Transactions* (1707) and in *Miscellanea analytica* (1730). However, De Moivre's Theorem was really first used by Roger Cotes (1682–1716). On the other hand, De Moivre was the inventor of a result called *Stirling's Formula*. De Moivre was unable to receive a professorship, so he supported himself by making a London coffeehouse his headquarters and solving problems for anyone who came to him.

$$= \sqrt{81\left(\frac{1}{4} + \frac{3}{4}\right)}$$

$$= 9$$

$$= \tan^{-1}(\sqrt{3})$$

$$= 60°$$

$$\theta = 120° \text{ (Quadrant II)}$$

$$\sqrt{-\frac{9}{2} + \frac{9}{2}\sqrt{3}\,i} = \left(-\frac{9}{2} + \frac{9}{2}\sqrt{3}\,i\right)^{1/2}$$

$$= (9 \text{ cis } 120°)^{1/2}$$

$$= 9^{1/2} \text{ cis}\left(\frac{120° + 360°k}{2}\right) \quad \text{(by De Moivre's Theorem)}$$

$$= 3 \text{ cis } (60° + 180°k)$$

$$k = 0: \quad 3 \text{ cis } 60° = \frac{3}{2} + \frac{3}{2}\sqrt{3}\,i$$

Notice that we found two square roots, as expected. In fact, a complex number will have *n* *n*th roots.

$$k = 1: \quad 3 \text{ cis } 240° = -\frac{3}{2} - \frac{3}{2}\sqrt{3}\,i$$

All other integral values of k repeat one of the previously found roots. For example,

$$k = 2: \quad 3 \text{ cis } 420° = \frac{3}{2} + \frac{3}{2}\sqrt{3}\,i.$$

Check:

$$\left(\frac{3}{2} + \frac{3}{2}\sqrt{3}\,i\right)^2 = \frac{9}{4} + \frac{9}{2}\sqrt{3}\,i + \frac{9}{4} \cdot 3i^2$$

$$= -\frac{9}{2} + \frac{9}{2}\sqrt{3}\,i$$

$$\left(-\frac{3}{2} - \frac{3}{2}\sqrt{3}\,i\right)^2 = \frac{9}{4} + \frac{9}{2}\sqrt{3}\,i + \frac{9}{4} \cdot 3i^2$$

$$= -\frac{9}{2} + \frac{9}{2}\sqrt{3}\,i$$

You might ask "Why not do Example 2 on a calculator?" The reason you can't is that it will give you only one of the roots—not all five. A calculator will, however, help you find the real number *r*.

2. Find the five fifth roots of 32.
   Solution: $32 = 32 \text{ cis } 0°$; thus:

$$\sqrt[5]{32} = (32)^{1/5} \qquad = 32^{1/5} \text{ cis}\left(\frac{0° + 360°k}{5}\right)$$

$\uparrow$ This represents all roots of 32.

$\uparrow$ This is *r*, which represents a length and therefore is the positive real root.

$\downarrow$

$$= 2 \text{ cis } 72°k$$

$$k = 0: \quad 2 \text{ cis } 0° \quad = 2$$
$$k = 1: \quad 2 \text{ cis } 72° \quad = 0.6180 + 1.9021i$$
$$k = 2: \quad 2 \text{ cis } 144° = -1.6180 + 1.1756i$$
$$k = 3: \quad 2 \text{ cis } 216° = -1.1680 - 1.1756i$$
$$k = 4: \quad 2 \text{ cis } 288° = 0.6180 - 1.9021i$$

All other integral values for k repeat those listed here.

If we represent the fifth roots of 32 graphically, as shown in Figure 6.37, we notice that they all lie on a circle of radius 2 and are equally spaced.

If n is a positive integer, then

$$(a + bi)^{1/n} = (r \text{ cis } \theta)^{1/n},$$

and the roots are equally spaced on the circle of radius r centered at the origin.

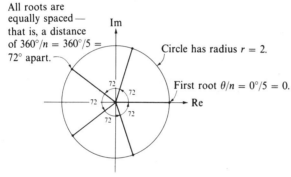

Figure 6.37   Graphical representation of the fifth roots of 32

The first root, which is located so that its argument is θ/n, is called the *principal nth root*.

PROBLEM SET 6.9

A Problems

Find the indicated roots of the numbers in Problems 1–20. Leave your answers in trigonometric form.

1. Square roots of 9 cis 240°
2. Square roots of 16 cis 100°
3. Square roots of 25 cis 300°
4. Cube roots of 8 cis 240°
5. Cube roots of 27 cis 300°
6. Cube roots of 64 cis 216°
7. Fourth roots of 16 cis 352°
8. Fourth roots of 81 cis 88°
9. Fourth roots of 256 cis 224°
10. Fifth roots of 32 cis 200°
11. Fifth roots of 32 cis 160°
12. Fifth roots of 243 cis 60°
13. Cube roots of −1
14. Cube roots of 8
15. Cube roots of 27
16. Fourth roots of i
17. Fourth roots of $-1 - i$
18. Fourth roots of $1 + i$
19. Fifth roots of $-i$
20. Fifth roots of 1

𝕳istorical 𝕹ote

A story about De Moivre's death is reported in Howard Eves' *In Mathematical Circles* (Boston: Prindle, Weber, & Schmidt, 1969). According to the story, De Moivre noticed that each day he required a quarter of an hour more sleep than on the preceding day. De Moivre said he would die when the arithmetic progression reached 24 hours. Finally, after sleeping longer and longer every night, he went to bed and slept for more than 24 hours and then died in his sleep.

B Problems

Find the indicated roots of the numbers in Problems 21–26. Leave your answers in rectangular form. Also, show the roots graphically.

21. Cube roots of 1

22. Square roots of $\dfrac{25}{2} - \dfrac{25\sqrt{3}}{2}i$

23. Cube roots of -8

24. Fourth roots of 1

25. Cube roots of $4\sqrt{3} - 4i$

26. Fourth roots of $12.2567 + 10.2846i$

Find the indicated roots of the numbers in Problems 27–34.

27. Sixth roots of -64

28. Sixth roots of $64i$

29. Ninth roots of 1

30. Ninth roots of $-1 + i$

31. Tenth roots of i

32. Tenth roots of 1

33. $(4\sqrt{2} + 4\sqrt{2}\,i)^{2/3}$

34. $(-16 + 16\sqrt{3}\,i)^{3/5}$

Mind Bogglers

35. Solve $x^5 - 1 = 0$.

36. Solve $x^4 + x^3 + x^2 + x + 1 = 0$.

6.10 SUMMARY AND REVIEW

TERMS

Absolute value (6.7)
Amplitude (6.5)
Angle (6.1)
Arc (6.1)
Arccosecant (6.6)
Arccosine (6.6)
Arccotangent (6.6)
Arcsecant (6.6)
Arcsine (6.6)
Arctangent (6.6)
Argument (6.7)
Complex number (6.7)
Complex plane (6.7)
Cosecant (6.3)
Cosine (6.2)
Cotangent (6.3)
Coterminal angles (6.1)
Degree (6.1)
De Moivre's Theorem (6.9)
Frame (6.4, 6.5)
Gaussian plane (6.7)

Imaginary axis (6.7)
Inverse (6.6)
Minute (6.1)
Modulus (6.7)
Negative angle (6.1)
Period (6.2)
Polar form of a complex number (6.7)
Positive angle (6.1)
Radian (6.1)
r cis θ (6.7)
Real axis (6.7)
Rectangular form of a complex number (6.7)
Reduction principle (6.3)
Reference angle (6.3)
Secant (6.3)
Second (6.1)
Sine (6.2)
Tangent (6.3)
Trigonometric form of a complex number (6.7)
Unit circle (6.1)

IMPORTANT IDEAS

(6.1) Measure of angles:

$$\frac{\theta \text{ in degrees}}{360} \Leftrightarrow \frac{\theta \text{ in radians}}{2\pi}$$

(6.1) Equivalencies of angles:

Degrees	0°	30°	45°	60°	90°	180°	270°	360°
Radians	0	$\pi/6$	$\pi/4$	$\pi/3$	$\pi/2$	π	$3\pi/2$	2π

(6.1) The arc length, s, cut by a central angle θ, measured in radians, of a circle of radius r is given by $s = r\theta$.

(6.2) Let θ be an angle in standard position with a point $P(x, y)$ on the terminal side a distance of $r = \sqrt{x^2 + y^2}$ from the origin ($r \neq 0$). Then we define the six circular functions as follows:

$$\cos \theta = \frac{x}{r} \qquad \tan \theta = \frac{y}{x} \ (x \neq 0) \qquad \csc \theta = \frac{r}{y} \ (y \neq 0)$$

$$\sin \theta = \frac{y}{r} \qquad \sec \theta = \frac{r}{x} \ (x \neq 0) \qquad \cot \theta = \frac{x}{y} \ (y \neq 0)$$

(6.2) $\cos (\theta \pm 2n\pi) = \cos \theta$, $\sin (\theta \pm 2n\pi) = \sin \theta$, and $\tan (\theta + n\pi) = \tan \theta$; that is, the cosine and sine have period 2π and the tangent has period π.

(6.3) Exact values

function \\ angle θ	0	$\dfrac{\pi}{6}$	$\dfrac{\pi}{4}$	$\dfrac{\pi}{3}$	$\dfrac{\pi}{2}$	π	$\dfrac{3\pi}{2}$
$\cos \theta$	1	$\dfrac{\sqrt{3}}{2}$	$\dfrac{\sqrt{2}}{2}$	$\dfrac{1}{2}$	0	-1	0
$\sin \theta$	0	$\dfrac{1}{2}$	$\dfrac{\sqrt{2}}{2}$	$\dfrac{\sqrt{3}}{2}$	1	0	-1
$\tan \theta$	0	$\dfrac{\sqrt{3}}{3}$	1	$\sqrt{3}$	undef.	0	undef.
$\sec \theta$	1	$\dfrac{2}{\sqrt{3}}$	$\dfrac{2}{\sqrt{2}}$	2	undef.	-1	undef.
$\csc \theta$	undef.	2	$\dfrac{2}{\sqrt{2}}$	$\dfrac{2}{\sqrt{3}}$	1	undef.	-1
$\cot \theta$	undef.	$\dfrac{3}{\sqrt{3}}$	1	$\dfrac{\sqrt{3}}{3}$	0	undef.	0

(6.3) $-1 \le \cos\theta \le 1$, $-1 \le \sin\theta \le 1$, $\cos^2\theta + \sin^2\theta = 1$, and $\tan\theta = \sin\theta/\cos\theta$

(6.3) $\sec\theta = 1/\cos\theta$, $\csc\theta = 1/\sin\theta$, and $\cot\theta = 1/\tan\theta$

(6.3)   Reduction principle: if f represents a trigonometric function, then $f(\theta)$ $= \pm f(\theta')$, where θ' is the reference angle for θ and the sign depends on the quadrant and the function:

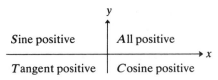

(6.4)   Framing the cosine, sine, and tangent curves

(6.5)   To graph the general cosine, sine, and tangent curves:
$$y - k = a \cos b(x - h); \qquad y - k = a \sin b(x - h);$$
$$y - k = a \tan b(x - h).$$

a is called the *amplitude* for the sine and cosine curves; the length of the frame is the *period*.

Translate the origin to the point (h, k). The height of the frame is $|a|$, and the length of the frame is $2\pi/b$ for the cosine and sine curves and π/b for the tangent curve. Next, frame the curve. Then plot the critical values and sketch the curve.

(6.6)   The inverse functions are defined as follows:

INVERSE FUNCTION		DOMAIN	RANGE
$y = \text{Arccos } x$	or $\quad y = \text{Cos}^{-1} x$	$-1 \le x \le 1$	$0 \le y \le \pi$
$y = \text{Arcsin } x$	or $\quad y = \text{Sin}^{-1} x$	$-1 \le x \le 1$	$-\dfrac{\pi}{2} \le y \le \dfrac{\pi}{2}$
$y = \text{Arctan } x$	or $\quad y = \text{Tan}^{-1} x$	All reals	$-\dfrac{\pi}{2} < y < \dfrac{\pi}{2}$
$y = \text{Arccot } x$	or $\quad y = \text{Cot}^{-1} x$	All reals	$0 < y < \pi$

(6.7)   To change from the rectangular form of a complex number $a + bi$ to trigonometric form, use

$$r = \sqrt{a^2 + b^2}, \qquad \theta' = \tan^{-1}\left|\frac{b}{a}\right|,$$

where θ' is the reference angle for θ. Place θ in the proper quadrant by noting the signs of a and b.

(6.7)   To change from the trigonometric form of a complex number r cis θ to rectangular form, use

$$a = r \cos\theta, \qquad b = r \sin\theta.$$

(6.8)   If r_1 cis θ_1 and r_2 cis θ_2 are complex numbers, then their product is

$$r_1 r_2 \text{ cis } (\theta_1 + \theta_2),$$

and their quotient is

$$\frac{r_1}{r_2} \text{ cis } (\theta_1 - \theta_2).$$

(6.9) De Moivre's Theorem: if n is any real number, then

$$(r \text{ cis } \theta)^n = r^n \text{ cis } n(\theta + 360°k) \quad \text{for any integer } k.$$

REVIEW PROBLEMS

1. (6.3) Give the functional value from memory.
   a. $\cos (\pi/3)$ b. $\sin (\pi/6)$ c. $\tan (\pi/4)$ d. $\sec 0$
   e. $\csc \pi$ f. $\cot (\pi/3)$ g. $\sin (\pi/4)$ h. $\tan (\pi/3)$
2. (6.6) Give the angle from memory.
   a. $\text{Arcsin} (1/2)$ b. $\text{Arccos} (\sqrt{3}/2)$ c. $\text{Tan}^{-1}(-\sqrt{3})$ d. $\text{Arccot} 1$
   e. $\text{Cos}^{-1} 0$ f. $\text{Sin}^{-1}(-\sqrt{2}/2)$ g. $\text{Cot}^{-1}(\frac{1}{3}\sqrt{3})$ h. $\text{Arctan} 0$
3. (6.3, 6.6) Give the values of the functions by using Table 4, Table 5, or a calculator.
   a. $\sec 23.4°$ b. $\cot 2.5$ c. $\csc 43.28°$
   d. $\text{Arcsin} 0.3140$ e. $\text{Arccos} (-0.6494)$ f. $\text{Arctan} 3.271$
4. (6.4, 6.6) From memory, draw a quick sketch of each curve.
   a. $y = \cos x$ b. $y = \sin x$ c. $y = \tan x$
   d. $y = \arccos x$ e. $y = \arcsin x$ f. $y = \arctan x$
5. (6.5) Graph each curve
   a. $y - 2 = \sin [x - (\pi/6)]$ b. $y = \cos [x + (\pi/4)]$
   c. $y = \tan [x - (\pi/3)] - 2$ d. $y = 2 \cos \frac{2}{3}x$
   e. $y = \frac{1}{3} \sin 2x$ f. $y = \frac{1}{2} \tan \frac{1}{2}x$
6. (6.8) Perform the operations and leave your answer in rectangular form.

   a. $(\sqrt{12} - 2i)^4$ b. $\dfrac{(3 + 3i)(\sqrt{3} - i)}{1 + i}$

7. (6.8) Perform the operations and leave your answer in trigonometric form.

   a. $2i(-1 + i)(-2 + 2i)$ b. $\dfrac{2 \text{ cis } 158° \cdot 4 \text{ cis } 212°}{(2 \text{ cis } 312°)^3}$

(6.9) Find the roots indicated in Problems 8 and 9 and represent them in rectangular and trigonometric form.

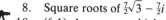

8. Square roots of $\frac{7}{2}\sqrt{3} - \frac{7}{2}i$ 9. Fourth roots of 1
10. (6.1) A curve on a highway is laid out as the arc of a circle of radius 500 m. If the curve subtends a central angle of 18°, what is the distance around this section of road? Give the exact answer and an answer rounded off to the nearest meter.

THINK METRIC

Trigonometric Identities and Applications

7

7.1 EQUATIONS AND IDENTITIES

In mathematics, we consider several types of equations; they are summarized as follows:

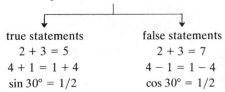

Equations without a Variable

true statements	false statements
$2 + 3 = 5$	$2 + 3 = 7$
$4 + 1 = 1 + 4$	$4 - 1 = 1 - 4$
$\sin 30° = 1/2$	$\cos 30° = 1/2$

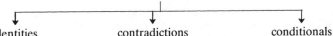

Variable Equations

identities (true for all replacements of the variable)	contradictions (false for all replacements of the variable)	conditionals (true for some replacements of the variable and false for others)
$2x + 3x = 5x$	$2x + 3 = 2x + 7$	$2x + 3x = 7$
$a + b = b + a$	$a + 6 = 5 + a$	$a + 6 = 2a$
$\cos^2 \theta + \sin^2 \theta = 1$	$\sin \theta = 2$	$\sin \theta = 1/2$

In arithmetic, attention is focused primarily on true equations, whereas in algebra, the main concentration is on conditional equations. However, in order to solve equations, you first have to learn some identities (sometimes called *laws* or *properties*), such as the following:

commutativity: $a + b = b + a$; $\quad ab = ba$
associativity: $a + (b + c) = (a + b) + c$; $\quad a(bc) = (ab)c$
distributivity: $a(b + c) = ab + ac$

In this chapter, we'll focus on trigonometric identities, and trigonometric equations applying these results to the solution of triangles. Recall the following definitions:

EQUATION

> An *equation* consists of two members connected by an equals sign (=). We call these members the left-hand side (L) and the right-hand side (R), so that the equation has the form
>
> $$L = R.$$

VARIABLE EQUATION

> If the equation contains at least one variable, it is called a *variable equation*.

CONDITIONAL EQUATION

> A *conditional equation* is a variable equation that is true for some (at least one), but not all, replacements of the variable for which the members of the equation are defined.

IDENTITY

> An *identity* is a variable equation that is true for all replacements of the variable for which the members of the equation are defined.

All our work with trigonometric identities is ultimately based on eight basic identities, called the *fundamental identities*. Notice that these identities are classified into three categories and are numbered for later reference.

FUNDAMENTAL IDENTITIES

Reciprocal Identities:

(1) $\qquad \sec \theta = \dfrac{1}{\cos \theta}, \quad \cos \theta \neq 0$

(2) $\qquad \csc \theta = \dfrac{1}{\sin \theta}, \quad \sin \theta \neq 0$

(3) $\qquad \cot \theta = \dfrac{1}{\tan \theta}, \quad \tan \theta \neq 0$

Ratio Identities:

(4) $\qquad \tan \theta = \dfrac{\sin \theta}{\cos \theta}, \quad \cos \theta \neq 0$

(5) $\qquad \cot \theta = \dfrac{\cos \theta}{\sin \theta}, \quad \sin \theta \neq 0$

Pythagorean Identities:

(6) $\qquad \sin^2 \theta + \cos^2 \theta = 1$
(7) $\qquad 1 + \tan^2 \theta = \sec^2 \theta$
(8) $\qquad 1 + \cot^2 \theta = \csc^2 \theta$

These eight identities are *indeed* fundamental for your study—not only for this chapter, but also for future mathematics courses. For this reason, they should be memorized.

The proofs of these identities follow directly from the definitions of the circular functions. Some proofs were given in the last chapter, but they are repeated here and in Problem Set 7.1.

PROOFS OF THE FUNDAMENTAL IDENTITIES

Let θ be an angle in standard position with point $P(x, y)$ on the terminal side a distance of r from the origin, with $r \neq 0$.

(1) $\sec \theta = 1/\cos \theta$
By definition, $\cos \theta = x/r$; thus,

$$\frac{1}{\cos \theta} = \frac{1}{x/r}$$

$$= 1 \cdot \frac{r}{x} \qquad \text{division of fractions}$$

$$= \frac{r}{x} \qquad \text{multiplication of fractions}$$

$$= \sec \theta \qquad \text{by definition of } \sec \theta$$

Identities (2) and (3) are proved in precisely the same way and are left as problems.

(4) $\tan \theta = \sin \theta / \cos \theta$

$$\frac{\sin \theta}{\cos \theta} = \frac{y/r}{x/r} \qquad \text{by definition of } \sin \theta \text{ and } \cos \theta$$

$$= \frac{y}{r} \cdot \frac{r}{x} \qquad \text{division of fractions}$$

$$= \frac{y}{x} \qquad \text{multiplication and simplification of fractions}$$

$$= \tan \theta \qquad \text{by definition of } \tan \theta$$

This identity can be proved by using the definition of the circular function in the same manner as for Identity (4). However, we can also use previously proved identities as illustrated here.

(5) $\cot \theta = \cos \theta / \sin \theta$

$$\cot \theta = \frac{1}{\tan \theta} \qquad \text{by identity (3)}$$

$$= \frac{1}{\sin \theta / \cos \theta} \qquad \text{identity (4)}$$

$$= 1 \cdot \frac{\cos \theta}{\sin \theta} \qquad \text{division of fractions}$$

$$= \frac{\cos \theta}{\sin \theta} \qquad \text{multiplication of fractions}$$

In proving identities, it is important not to start with the given identity, because we don't know if it is true until we prove it true! Notice that in all the above examples we *began* with something we knew was true and *ended* with the identity we were trying to prove.

(6) $\sin^2 \theta + \cos^2 \theta = 1$

For identities (6), (7), and (8), we begin with the Pythagorean Theorem (which is why these are called the *Pythagorean identities*).

$$x^2 + y^2 = r^2 \quad \text{by the Pythagorean Theorem}$$

To prove (6), we divide both sides by r^2; for (7), we divide by x^2; and for (8), we divide by y^2. We will show the details for (6) and leave (7) and (8) as problems.

$$\frac{x^2}{r^2} + \frac{y^2}{r^2} = \frac{r^2}{r^2} \qquad \text{dividing both sides by } r^2 \ (r \neq 0)$$

$$\left(\frac{x}{r}\right)^2 + \left(\frac{y}{r}\right)^2 = 1 \qquad \text{properties of exponents}$$

$$(\cos \theta)^2 + (\sin \theta)^2 = 1 \quad \text{definition of } \cos \theta \text{ and } \sin \theta$$

$$\sin^2 \theta + \cos^2 \theta = 1 \qquad \text{commutative property}$$

EXAMPLE 1: Write all the trigonometric functions in terms of $\sin \theta$.

Solution: a. $\sin \theta = \sin \theta$

b. $\cos \theta = \pm\sqrt{1 - \sin^2 \theta}$ from (6)

c. $\tan \theta = \dfrac{\sin \theta}{\cos \theta}$ from (4)

      $= \dfrac{\sin \theta}{\pm\sqrt{1 - \sin^2 \theta}}$ from part b

d. $\cot \theta = \dfrac{1}{\tan \theta}$ from (3)

      $= \dfrac{\pm\sqrt{1 - \sin^2 \theta}}{\sin \theta}$ from part c

e. $\csc \theta = \dfrac{1}{\sin \theta}$ from (2)

f. $\sec \theta = \dfrac{1}{\cos \theta}$ from (1)

      $= \dfrac{1}{\pm\sqrt{1 - \sin^2 \theta}}$ from part b

The $\pm$ sign we've been using, as in

$$\cos \theta = \pm\sqrt{1 - \sin^2 \theta},$$

means that $\cos \theta$ is positive for some values of θ and negative for other values of θ. The $+$ or the $-$ sign is chosen by determining the proper quadrant, as shown in Example 2.

EXAMPLE 2: Given $\sin \theta = 3/5$ and $\tan \theta < 0$, find the other functions of θ.

Solution: Since the tangent is negative and the sine is positive, we see that the proper quadrant is II. Thus,

$$\cos \theta = -\sqrt{1 - \sin^2 \theta} \quad \text{(since the cosine is negative in Quadrant II)}$$
$$= -\sqrt{1 - (3/5)^2}$$
$$= -\sqrt{1 - 9/25}$$
$$= -\sqrt{16/25}$$
$$= -4/5.$$

Also,

$$\tan \theta = \frac{\sin \theta}{\cos \theta} = \frac{3/5}{-4/5} = -\frac{3}{4}.$$

Using the reciprocal identities, we find $\cot \theta = -4/3$, $\sec \theta = -5/4$, and $\csc \theta = 5/3$.

PROBLEM SET 7.1

A Problems

1. State from memory the eight fundamental identities.

In Problems 2–9, state the quadrant in which θ may lie to make the expression true.

2. $\sin \theta = \sqrt{1 - \cos^2 \theta}$

3. $\sin \theta = -\sqrt{1 - \cos^2 \theta}$

4. $\sec \theta = -\sqrt{1 + \tan^2 \theta}$

5. $\sec \theta = \sqrt{1 + \tan^2 \theta}$

6. $\csc \theta = \sqrt{1 + \cot^2 \theta}$; $\tan \theta < 0$

7. $\cos \theta = -\sqrt{1 - \sin^2 \theta}$; $\sin \theta > 0$

8. $\tan \theta = \sqrt{\sec^2 \theta - 1}$; $\cos \theta < 0$

9. $\cot \theta = \sqrt{1 + \cot^2 \theta}$; $\sin \theta > 0$

Write each of the expressions in Problems 10–19 as a single trigonometric function of some angle by using one of the eight fundamental identities.

10. $\dfrac{\sin 50°}{\cos 50°}$

11. $\dfrac{\cos (A + B)}{\sin (A + B)}$

12. $\dfrac{1}{\sec 75°}$

13. $\dfrac{1}{\cot (\pi/15)}$

14. $\tan 42° \cos 42°$

15. $\cot (\pi/8) \sin (\pi/8)$

16. $1 - \cos^2 18°$

17. $-\sqrt{1 - \sin^2 127°}$

18. $\sec^2(\pi/6) - 1$

19. $1 + \tan^2(\pi/5)$

Evaluate the expressions in Problems 20–25 by using one of the eight fundamental identities.

20. $\cos 128° \sec 128°$

21. $\sin^2(\pi/3) + \cos^2(\pi/3)$

22. $\sec^2(\pi/6) - \tan^2(\pi/6)$

23. $\cot^2 45° - \csc^2 45°$

24. $\tan^2 135° - \sec^2 135°$

25. $\csc 85° \sin 85°$

26. Prove that $\csc \theta = 1/\sin \theta$.

27. Prove that $\cot \theta = 1/\tan \theta$.

28. Prove that $1 + \tan^2 \theta = \sec^2 \theta$.

29. Prove that $1 + \cot^2 \theta = \csc^2 \theta$.

B Problems

In Problems 30–34, write all the trigonometric functions in terms of the given function.

30. $\cos \theta$ 31. $\tan \theta$ 32. $\cot \theta$ 33. $\sec \theta$ 34. $\csc \theta$

In Problems 35–44, find the other functions of θ using the given information.

35. $\cos \theta = 5/13; \tan \theta < 0$
36. $\cos \theta = 5/13; \tan \theta > 0$
37. $\tan \theta = 5/12; \sin \theta > 0$
38. $\tan \theta = 5/12; \sin \theta < 0$
39. $\sin \theta = 2/3; \sec \theta > 0$
40. $\sin \theta = 2/3; \sec \theta < 0$
41. $\sec \theta = \sqrt{34}/5; \tan \theta < 0$
42. $\sec \theta = \sqrt{34}/5; \tan \theta > 0$
43. $\csc \theta = -\sqrt{10}/3; \cos \theta > 0$
44. $\csc \theta = -\sqrt{10}/3; \cos \theta < 0$

Simplify the expressions in Problems 45–50 using only sines, cosines, and the fundamental identities.

45. $\dfrac{1 - \cos^2 \theta}{\sin \theta}$

46. $\dfrac{1 - \sin^2 \theta}{\cos \theta}$

47. $\dfrac{\dfrac{\sin \theta}{\cos \theta} + \dfrac{\cos \theta}{\sin \theta}}{\dfrac{1}{\sin \theta \cos \theta}}$

48. $\sin \theta + \dfrac{\cos^2 \theta}{\sin \theta}$

49. $\dfrac{\cos \theta + \dfrac{\sin^2 \theta}{\cos \theta}}{\sin \theta}$

50. $\dfrac{\dfrac{\cos^4 \theta}{\sin^2 \theta} + \cos^2 \theta}{\dfrac{\cos^2 \theta}{\sin^2 \theta}}$

Reduce the expressions in Problems 51–58 so that they involve only sines and cosines, and then simplify.

51. $\sin \theta + \cot \theta$
52. $\sec \theta + \tan \theta$

53. $\dfrac{\tan \theta + \cot \theta}{\sec \theta \csc \theta}$

54. $\dfrac{\sec \theta + \csc \theta}{\tan \theta \cot \theta}$

55. $\sec^2 \theta + \tan^2 \theta$
56. $\csc^2 \theta + \cot^2 \theta$
57. $(\cot \theta - \sec \theta)(\sin \theta \cos \theta)$
58. $(\tan \theta - \csc \theta)(\cos \theta \sin \theta)$

Mind Bogglers

59. a. Explain why this number trick works: Pick any number. Square the number. Add 25. Add 10 times the original number. Divide by 5 more than the original number. Subtract your original number. Your answer is 5.
    b. What term introduced in this section can be applied to your explanation of part a?
    c. Will this trick always work?

60. Consider this problem: An elephant and a bird want to play together on a teeter-totter. The bird says that it is an impossible idea, but the elephant assures the little bird that it will work out and that he will prove it, since he has had a little algebra. He presents the following argument to the little bird: let

E = the weight of the elephant,
b = the weight of the bird.

Now there must be some weight, w (probably very large), so that

$$E = b + w.$$

Multiply both sides by $E - b$.

$$\begin{aligned}
E(E - b) &= (b + w)(E - b) \\
E^2 - Eb &= bE + wE - b^2 - wb \\
E^2 - Eb - wE &= bE - b^2 - wb \qquad \text{(subtracting } wE \text{ from both sides)} \\
E(E - b - w) &= b(E - b - w) \\
E &= b \qquad \text{(dividing both sides by } E - b - w)
\end{aligned}$$

This last equation says that the weight of the elephant is the same as the weight of the bird. "Now," says the elephant, "since our weights are the same, we'll have no problem on the teeter-totter."

"Wait!" hollers the bird. "Obviously this is false." But where is the error in the reasoning?

7.2 PROVING IDENTITIES

In the last section, we considered eight fundamental identities, which are used to simplify and change the form of a variety of trigonometric expressions. Suppose we're given a trigonometric equation such as

$$\tan \theta + \cot \theta = \sec \theta \csc \theta$$

and are asked to show that it is an identity. We must be careful not to treat this problem as though it is an algebraic equation. When asked to prove an identity, we do *not* start with the given expression, since we cannot assume it is true. What, then, can we do if we cannot begin with the given equation? We can *begin* with what we know is true and *end* with the given identity. This means that, when we are given an identity to prove such as

$$L = R,$$

there are three ways to proceed:

1. Reduce the left-hand side to the right-hand side by using algebra and the fundamental identities. Thus,

$$\begin{aligned}
L &= L \\
&= L_1 \\
&= L_2 \\
&\;\;\vdots \\
&= L_n,
\end{aligned}$$

so that $L_n = R$ after n steps.

2. Reduce the right-hand side to the left-hand side. Thus,

$$R = R$$
$$= R_1$$
$$= R_2$$
$$\vdots$$
$$= R_m,$$

so that $R_m = L$.

3. Reduce both sides independently to the same expression.

$L = L$	$R = R$
$= L_1$	$= R_1$
$= L_2$	$= R_2$
$\vdots$	$\vdots$
$= L_p$	$= R_q$

If these two are identical, then $L = R$.

We now illustrate these techniques with several examples.

EXAMPLE 1: Prove that

$$\tan \theta + \cot \theta = \sec \theta \csc \theta.$$

Solution, Method I: Reduce the left-hand side to the right-hand side.

$$\tan \theta + \cot \theta = \frac{\sin \theta}{\cos \theta} + \frac{\cos \theta}{\sin \theta}$$

$$= \frac{\sin^2 \theta}{\cos \theta \sin \theta} + \frac{\cos^2 \theta}{\cos \theta \sin \theta}$$

$$= \frac{\sin^2 \theta + \cos^2 \theta}{\cos \theta \sin \theta}$$

$$= \frac{1}{\cos \theta \sin \theta}$$

$$= \frac{1}{\cos \theta} \cdot \frac{1}{\sin \theta}$$

$$= \sec \theta \csc \theta$$

Thus, $\tan \theta + \cot \theta = \sec \theta \csc \theta.$

Note: in proving identities, it is often advantageous to change all the trigonometric functions involved to sines and cosines.

Note: when we arrive at the given identity, we are finished with the problem.

Solution, Method II: Reduce the right-hand side to the left-hand side.

$$\sec \theta \csc \theta = \frac{1}{\cos \theta} \cdot \frac{1}{\sin \theta}$$

$$= \frac{1}{\cos \theta \sin \theta}$$

$$= \frac{\sin^2 \theta + \cos^2 \theta}{\cos \theta \sin \theta}$$

$$= \frac{\sin^2 \theta}{\cos \theta \sin \theta} + \frac{\cos^2 \theta}{\cos \theta \sin \theta}$$

$$= \frac{\sin \theta}{\cos \theta} + \frac{\cos \theta}{\sin \theta}$$

$$= \tan \theta + \cot \theta$$

Thus, $\tan \theta + \cot \theta = \sec \theta \csc \theta$.

Solution, Method III: Reduce both sides to a form that is the same.

$$\tan \theta + \cot \theta = \frac{\sin \theta}{\cos \theta} + \frac{\cos \theta}{\sin \theta} \qquad \bigg| \qquad \sec \theta \csc \theta = \frac{1}{\cos \theta} \cdot \frac{1}{\sin \theta}$$

$$= \frac{\sin^2 \theta + \cos^2 \theta}{\cos \theta \sin \theta} \qquad \bigg| \qquad = \frac{1}{\cos \theta \sin \theta}$$

$$= \frac{1}{\cos \theta \sin \theta}$$

These are identical.

Therefore, $\tan \theta + \cot \theta = \sec \theta \csc \theta$.

Of course, we would pick only one of these methods when proving a particular identity. Usually, it is easier to begin with the more complicated side and try to reduce it to the simpler side. If both sides seem equally complicated, you might change all the functions to sines and cosines and then simplify.

EXAMPLE 2: Prove that

$$\frac{1}{1 + \cos \theta} + \frac{1}{1 - \cos \theta} = 2 \csc^2 \theta.$$

Solution: $\dfrac{1}{1 + \cos\theta} + \dfrac{1}{1 - \cos\theta} = \dfrac{(1 - \cos\theta) + (1 + \cos\theta)}{(1 + \cos\theta)(1 - \cos\theta)}$

We begin with the more complicated side.

$$= \dfrac{2}{1 - \cos^2\theta}$$

$$= \dfrac{2}{\sin^2\theta}$$

$$= 2\csc^2\theta$$

EXAMPLE 3: Prove that

$$\sec 4\theta + \cos 4\theta = 2\sec 4\theta - \tan 4\theta \sin 4\theta.$$

Solution: $\sec 4\theta + \cos 4\theta = \dfrac{1}{\cos 4\theta} + \cos 4\theta$

When both sides seem equally simple, a good procedure is to change everything to sines and cosines and then simplify.

$$= \dfrac{1 + \cos^2 4\theta}{\cos 4\theta}$$

$$= \dfrac{1 + (1 - \sin^2 4\theta)}{\cos 4\theta}$$

$$= \dfrac{2 - \sin^2 4\theta}{\cos 4\theta}$$

$$= \dfrac{2}{\cos 4\theta} - \dfrac{\sin^2 4\theta}{\cos 4\theta}$$

$$= 2 \cdot \dfrac{1}{\cos 4\theta} - \dfrac{\sin 4\theta}{\cos 4\theta} \cdot \dfrac{\sin 4\theta}{1}$$

$$= 2\sec 4\theta - \tan 4\theta \sin 4\theta$$

EXAMPLE 4: Prove that

$$\dfrac{\sec 2\lambda + \cot 2\lambda}{\sec 2\lambda} = 1 + \csc 2\lambda - \sin 2\lambda.$$

Solution: We begin with the left-hand side. When we are working with a fraction consisting of a single function as a denominator, it is often helpful to separate the fraction into the sum of several fractions.

$$\dfrac{\sec 2\lambda + \cot 2\lambda}{\sec 2\lambda} = \dfrac{\sec 2\lambda}{\sec 2\lambda} + \dfrac{\cot 2\lambda}{\sec 2\lambda}$$

$$= 1 + \cot 2\lambda \cdot \dfrac{1}{\sec 2\lambda}$$

$$= 1 + \dfrac{\cos 2\lambda}{\sin 2\lambda} \cdot \cos 2\lambda$$

$$= 1 + \frac{\cos^2 2\lambda}{\sin 2\lambda}$$

$$= 1 + \frac{1 - \sin^2 2\lambda}{\sin 2\lambda}$$

$$= 1 + \frac{1}{\sin 2\lambda} - \frac{\sin^2 2\lambda}{\sin 2\lambda}$$

$$= 1 + \csc 2\lambda - \sin 2\lambda$$

There are many "tricks of the trade" that can be used in proving identities. Some of these are developed in Examples 5–7.

EXAMPLE 5: Prove that

$$\frac{\cos \theta}{1 - \sin \theta} = \frac{1 + \sin \theta}{\cos \theta}.$$

Since $a + b$ is called the *conjugate* of $a - b$, the procedure illustrated in Example 5 is sometimes called *multiplying by the conjugate.*

Solution: Sometimes, when there is a binomial in the numerator or denominator, the identity can be proved by multiplying one side by 1, where 1 is written in the form of the conjugate of the binomial. When changing one side, keep a sharp eye on the other side, since it often gives a clue about what to do. Thus, in this example we can multiply the numerator and denominator of the left-hand side by $1 + \sin \theta$:

We could also prove this identity by multiplying the numerator and denominator of the right-hand side by $1 - \sin \theta$.

$$\frac{\cos \theta}{1 - \sin \theta} = \frac{\cos \theta}{1 - \sin \theta} \cdot \frac{1 + \sin \theta}{1 + \sin \theta}$$

$$= \frac{\cos \theta (1 + \sin \theta)}{1 - \sin^2 \theta}$$

$$= \frac{\cos \theta (1 + \sin \theta)}{\cos^2 \theta}$$

$$= \frac{1 + \sin \theta}{\cos \theta}.$$

EXAMPLE 6: Prove that

$$\frac{\sec^2 2\theta - \tan^2 2\theta}{\tan 2\theta + \sec 2\theta} = \frac{\cos 2\theta}{1 + \sin 2\theta}.$$

Solution: Sometimes the identity can be proved by factoring.

$$\frac{\sec^2 2\theta - \tan^2 2\theta}{\tan 2\theta + \sec 2\theta} = \frac{(\sec 2\theta + \tan 2\theta)(\sec 2\theta - \tan 2\theta)}{\tan 2\theta + \sec 2\theta}$$

$$= \sec 2\theta - \tan 2\theta$$

$$= \frac{1}{\cos 2\theta} - \frac{\sin 2\theta}{\cos 2\theta}$$

$$= \frac{1 - \sin 2\theta}{\cos 2\theta}$$

$$= \frac{1 - \sin 2\theta}{\cos 2\theta} \cdot \frac{1 + \sin 2\theta}{1 + \sin 2\theta}$$

$$= \frac{1 - \sin^2 2\theta}{\cos 2\theta(1 + \sin 2\theta)}$$

$$= \frac{\cos^2 2\theta}{\cos 2\theta(1 + \sin 2\theta)}$$

$$= \frac{\cos 2\theta}{1 + \sin 2\theta}$$

EXAMPLE 7: Prove that

$$\frac{-2 \sin \theta \cos \theta}{1 - \sin \theta - \cos \theta} = 1 + \sin \theta + \cos \theta.$$

Solution: Sometimes, when there is a fraction on one side, the identity can be proved by multiplying the other side by 1 written so that the desired denominator is obtained. Thus, for this example,

$1 + \sin \theta + \cos \theta$

$$= (1 + \sin \theta + \cos \theta) \cdot \frac{1 - \sin \theta - \cos \theta}{1 - \sin \theta - \cos \theta}$$

$$= \frac{(1 + \sin \theta + \cos \theta)(1 - \sin \theta - \cos \theta)}{1 - \sin \theta - \cos \theta}$$

$$= \frac{1 - \sin \theta - \cos \theta + \sin \theta - \sin^2 \theta - \sin \theta \cos \theta + \cos \theta - \cos \theta \sin \theta - \cos^2 \theta}{1 - \sin \theta - \cos \theta}$$

$$= \frac{1 - (\sin^2 \theta + \cos^2 \theta) - 2 \sin \theta \cos \theta}{1 - \sin \theta - \cos \theta}$$

$$= \frac{-2 \sin \theta \cos \theta}{1 - \sin \theta - \cos \theta}.$$

Remember,

$\sin^2 \theta + \cos^2 \theta = 1,$

so

$1 - (\sin^2 \theta + \cos^2 \theta) = 0.$

In summary, there is no one best way to proceed in proving identities. However, the following hints should help:

TRICKS OF THE TRADE

1. If one side contains one function only, write all the trigonometric functions on the other side in terms of that function.
2. If the denominator of a fraction consists of only one function, break up the fraction.
3. Simplify by combining fractions.
4. Factoring is sometimes helpful.
5. Change all trigonometric functions to sines and cosines and simplify.
6. Multiply by the conjugate of either the numerator or the denominator.
7. Avoid the introduction of radicals.

PROBLEM SET 7.2

A Problems

Prove the identities given in Problems 1–30.

1. $\sin \theta = \sin^3 \theta + \cos^2 \theta \sin \theta$

2. $\sec \theta = \sec \theta \sin^2 \theta + \cos \theta$

3. $\tan \theta = \cot \theta \tan^2 \theta$

4. $\dfrac{\sin \theta \cos \theta + \sin^2 \theta}{\sin \theta} = \cos \theta + \sin \theta$

5. $\tan^2 \theta - \sin^2 \theta = \tan^2 \theta \sin^2 \theta$

6. $\cot^2 \theta \cos^2 \theta = \cot^2 \theta - \cos^2 \theta$

7. $\tan A + \cot A = \sec A \csc A$

8. $\cot A = \csc A \sec A - \tan A$

9. $\sin x + \cos x = \dfrac{\sec x + \csc x}{\csc x \sec x}$

10. $\dfrac{\cos \gamma + \tan \gamma \sin \gamma}{\sec \gamma} = 1$

11. $\dfrac{1 - \sec^2 t}{\sec^2 t} = -\sin^2 t$

12. $\dfrac{1 + \cot^2 t}{\cot^2 t} = \sec^2 t$

13. $(\sec \theta - \cos \theta)^2 = \tan^2 \theta - \sin^2 \theta$

14. $\dfrac{\sin \theta}{\csc \theta} + \dfrac{\cos \theta}{\sec \theta} = 1$

15. $1 - \sin 2\theta = \dfrac{1 - \sin^2 2\theta}{1 + \sin 2\theta}$

16. $\dfrac{1 - \tan^2 3\theta}{1 - \tan 3\theta} = 1 + \tan 3\theta$

17. $\sin \lambda = \dfrac{\sin^2 \lambda + \sin \lambda \cos \lambda + \sin \lambda}{\sin \lambda + \cos \lambda + 1}$

18. $\dfrac{1 + \cot 2\lambda \sec 2\lambda}{\tan 2\lambda + \sec 2\lambda} = \cot 2\lambda$

19. $\sin 2\alpha \cos 2\alpha (\tan 2\alpha + \cot 2\alpha) = 1$

20. $(\sin \beta - \cos \beta)^2 + (\sin \beta + \cos \beta)^2 = 2$

21. $\csc 3\beta - \cos 3\beta \cot 3\beta = \sin 3\beta$

22. $\dfrac{1 + \cot^2 A}{1 + \tan^2 A} = \cot^2 A$

23. $\dfrac{\sin^2 B - \cos^2 B}{\sin B + \cos B} = \sin B - \cos B$ 24. $\dfrac{\tan^2 \gamma - \cot^2 \gamma}{\tan \gamma + \cot \gamma} = \tan \gamma - \cot \gamma$

25. $\tan^2 2\gamma + \sin^2 2\gamma + \cos^2 2\gamma = \sec^2 2\gamma$ 26. $\cot^2 C + \cos^2 C + \sin^2 C = \csc^2 C$

27. $\dfrac{\tan \theta + \cot \theta}{\sec \theta \csc \theta} = 1$ 28. $\dfrac{\tan \theta - \cot \theta}{\sec \theta \csc \theta} = \sin^2 \theta - \cos^2 \theta$

29. $\dfrac{1}{\sin \theta + \cos \theta} + \dfrac{1}{\sin \theta - \cos \theta} = \dfrac{\sin \theta}{\sin^2 \theta - (1/2)}$

30. $\dfrac{1}{\sec \theta + \tan \theta} + \dfrac{1}{\sec \theta - \tan \theta} = 2 \sec \theta$

B Problems

Prove the identities given in Problems 31–57.

31. $\dfrac{1 + \tan C}{1 - \tan C} = \dfrac{\sec^2 C + 2 \tan C}{2 - \sec^2 C}$

32. $(\cot x + \csc x)^2 = \dfrac{\sec x + 1}{\sec x - 1}$

33. $\dfrac{\sin^3 x - \cos^3 x}{\sin x - \cos x} = 1 + \sin x \cos x$

34. $\dfrac{\tan^3 t - \cot^3 t}{\tan t - \cot t} = \sec^2 t + \cot^2 t$

35. $\sqrt{(3 \cos \theta - 4 \sin \theta)^2 + (3 \sin \theta + 4 \cos \theta)^2} = 5$

36. $\dfrac{1 - \cos \theta}{1 + \cos \theta} = \left(\dfrac{1 - \cos \theta}{\sin \theta}\right)^2$

37. $\dfrac{(\sec^2 \gamma + \tan^2 \gamma)^2}{\sec^4 \gamma - \tan^4 \gamma} = 1 + 2 \tan^2 \gamma$

38. $\dfrac{(\cos^2 \gamma - \sin^2 \gamma)^2}{\cos^4 \gamma - \sin^4 \gamma} = 2 \cos^2 \gamma - 1$

39. $(\sec 2\theta + \csc 2\theta)^2 = \dfrac{1 + 2 \sin 2\theta \cos 2\theta}{\cos^2 2\theta \sin^2 2\theta}$

40. $\dfrac{1}{\sec \theta + \tan \theta} = \sec \theta - \tan \theta$

41. $\csc \theta + \cot \theta = \dfrac{1}{\csc \theta - \cot \theta}$

42. $\sec^2 \lambda - \csc^2 \lambda = (2 \sin^2 \lambda - 1)(\sec^2 \lambda + \csc^2 \lambda)$

43. $2 \csc A = 2 \csc A - \cot A \cos A + \cos^2 A \csc A$

44. $\sec^2 2\lambda + \csc^2 2\lambda = \csc^2 2\lambda \sec^2 2\lambda$

45. $\dfrac{\tan \theta}{\cot \theta} - \dfrac{\cot \theta}{\tan \theta} = \sec^2 \theta - \csc^2 \theta$

46. $\dfrac{\cos^4 \theta - \sin^4 \theta}{(\cos^2 \theta - \sin^2 \theta)^2} = \dfrac{\cos \theta}{\cos \theta + \sin \theta} + \dfrac{\sin \theta}{\cos \theta - \sin \theta}$

47. $\dfrac{1 + \tan^3 \theta}{1 + \tan \theta} = \sec^2 \theta - \tan \theta$

48. $\dfrac{1 - \sec^3 \theta}{1 - \sec \theta} = \tan^2 \theta + \sec \theta + 2$

49. $\dfrac{\cos^2 \theta - \cos \theta \csc \theta}{\cos^2 \theta \csc \theta - \cos \theta \csc^2 \theta} = \sin \theta$

50. $\dfrac{\tan^2 \theta - 2 \tan \theta}{2 \tan \theta - 4} = \dfrac{1}{2} \tan \theta$

51. $\dfrac{\cos \theta + \cos^2 \theta}{\cos \theta + 1} = \dfrac{\cos \theta \sin \theta + \cos^2 \theta}{\sin \theta + \cos \theta}$

52. $\dfrac{\csc \theta + 1}{\cot^2 \theta + \csc \theta + 1} = \dfrac{\sin^2 \theta + \sin \theta \cos \theta}{\sin \theta + \cos \theta}$

53. $\sin \theta + \cos \theta + 1 = \dfrac{2 \sin \theta \cos \theta}{\sin \theta + \cos \theta - 1}$

54. $\dfrac{2 \tan^2 \theta + 2 \tan \theta \sec \theta}{\tan \theta + \sec \theta - 1} = \tan \theta + \sec \theta + 1$

55. $\dfrac{\csc \theta + 1}{\csc \theta - 1} - \dfrac{\sec \theta - \tan \theta}{\sec \theta + \tan \theta} = 4 \tan \theta \sec \theta$

56. $\dfrac{\cos \theta + \sin \theta}{\cos \theta - \sin \theta} + \dfrac{\cot \theta - 1}{\cot \theta + 1} = \dfrac{-2}{\sin^2 \theta - \cos^2 \theta}$

57. $\dfrac{\cos \theta + 1}{\cos \theta - 1} + \dfrac{1 - \sec \theta}{1 + \sec \theta} = -2 \cot^2 \theta - 2 \csc^2 \theta$

Mind Bogglers

58. a. What is an identity?
    b. Are the Fundamental Identities always true? If not, give examples of angles for which they are not true.
    c. Are the identities in Problems 1, 21, 33, and 37 always true? If not, give angles for which they are not true.

59. Prove that

$$\frac{\sin\theta}{1-\cos\theta} + \frac{\cos\theta}{1-\sin\theta} = (\sin\theta + \cos\theta + 1)(\sec\theta\csc\theta).$$

7.3 SPECIAL IDENTITIES

ADDITION LAWS

When proving identities, it is sometimes necessary to simplify the functional value of the sum or difference of two angles. That is, if α and β represent any two angles, we know that in general

$$\cos(\alpha - \beta) \neq \cos\alpha - \cos\beta.$$

In fact, we will now show that

$$\cos(\alpha - \beta) = \cos\alpha\cos\beta + \sin\alpha\sin\beta.$$

This result not only will simplify the cosine of the difference of two angles but also will provide the cornerstone upon which we'll build a great many additional identities that are essential in calculus.

We begin by finding the length of any chord (in a unit circle) with a corresponding arc intercepted by the central angle θ, where θ is in standard position. Let A be the point $(1, 0)$ and P the point on the intersection of the terminal side of angle θ and the unit circle. This means that the coordinates of P are $(\cos\theta, \sin\theta)$. Now, find the length of the chord AP (see Figure 7.1) by using the distance formula.

For example, if $\alpha = 60°$ and $\beta = 30°$, then

$$\cos(60° - 30°) = \cos 30°$$
$$= \frac{\sqrt{3}}{2},$$

$$\cos 60° = \frac{1}{2},$$

and

$$\cos 30° = \sqrt{3}/2,$$

so

$$\cos(60° - 30°)$$
$$\neq \cos 60° - \cos 30°.$$

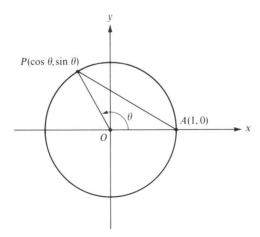

Figure 7.1 Length of a chord determined by an angle θ

$$AP = \sqrt{(1 - \cos \theta)^2 + (0 - \sin \theta)^2}$$
$$= \sqrt{1 - 2 \cos \theta + \cos^2 \theta + \sin^2 \theta}$$
$$= \sqrt{1 - 2 \cos \theta + 1}$$
$$= \sqrt{2 - 2 \cos \theta}$$

Next, apply this result to a chord determined by any two angles α and β, as shown in Figure 7.2. Let P_α and P_β be the points on the unit circle

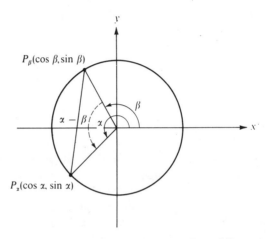

Figure 7.2   Distance between P_α and P_β

determined by the angles α and β, respectively. By the previous result, we see that

$$P_\alpha P_\beta = \sqrt{2 - 2 \cos (\alpha - \beta)}.$$

But we could also have found this distance directly via the distance formula.

$$P_\alpha P_\beta = \sqrt{(\cos \beta - \cos \alpha)^2 + (\sin \beta - \sin \alpha)^2}$$
$$= \sqrt{\cos^2 \beta - 2 \cos \alpha \cos \beta + \cos^2 \alpha + \sin^2 \beta - 2 \sin \alpha \sin \beta + \sin^2 \alpha}$$
$$= \sqrt{(\cos^2 \beta + \sin^2 \beta) + (\cos^2 \alpha + \sin^2 \alpha) - 2(\cos \alpha \cos \beta + \sin \alpha \sin \beta)}$$
$$= \sqrt{2 - 2(\cos \alpha \cos \beta + \sin \alpha \sin \beta)}$$

Finally, we equate these quantities, since they both represent the distance between P_α and P_β.

$$\sqrt{2 - 2 \cos (\alpha - \beta)} = \sqrt{2 - 2(\cos \alpha \cos \beta + \sin \alpha \sin \beta)}$$
$$2 - 2 \cos (\alpha - \beta) = 2 - 2(\cos \alpha \cos \beta + \sin \alpha \sin \beta)$$
$$-2 \cos (\alpha - \beta) = -2(\cos \alpha \cos \beta + \sin \alpha \sin \beta)$$
$$\cos (\alpha - \beta) = \cos \alpha \cos \beta + \sin \alpha \sin \beta$$

We can use this identity as follows to find the exact value of certain angles:

EXAMPLE 1: $\cos 345° = \cos 15°$

$$= \cos (45° - 30°)$$

$$= \cos 45° \cos 30° + \sin 45° \sin 30°$$

$$= \frac{1}{2}\sqrt{2} \cdot \frac{1}{2}\sqrt{3} + \frac{1}{2}\sqrt{2} \cdot \frac{1}{2}$$

$$= \frac{\sqrt{6}}{4} + \frac{\sqrt{2}}{4}$$

$$= \frac{\sqrt{6} + \sqrt{2}}{4}$$

Reduction principle
since $45° - 30° = 15°$

Applying our new identity

Using exact values

Even though this identity is helpful for making evaluations (as in the preceding example), its real value lies in the fact that it is true for *any* choice of α and β. By making some particular choices for α and β, we find several useful special cases of this identity. We begin numbering with 9 since we considered the eight fundamental identities in Section 7.1. Identities (9)–(11) tell us that the cofunctions of complementary angles are equal.

(9) $\cos \left(\dfrac{\pi}{2} - \theta \right) = \sin \theta$

To prove this, let $\alpha = \pi/2$ and $\beta = \theta$.

$$\cos \left(\frac{\pi}{2} - \theta \right) = \cos \frac{\pi}{2} \cos \theta + \sin \frac{\pi}{2} \sin \theta$$

$$= 0 \cdot \cos \theta + 1 \cdot \sin \theta$$

$$= \sin \theta$$

(10) $\sin \left(\dfrac{\pi}{2} - \theta \right) = \cos \theta$

(11) $\tan \left(\dfrac{\pi}{2} - \theta \right) = \cot \theta$

The next three identities tell us how to deal with a circular function of the opposite of an angle.

All the identities in this section are based on the identity

$$\cos (\alpha - \beta)$$
$$= \cos \alpha \cos \beta + \sin \alpha \sin \beta.$$

METHOD OF PROOF
(10) Let $\alpha = \pi/2$ and
        $\beta = \pi/2 - \theta$.

(11) Write

$$\tan \left(\frac{\pi}{2} - \theta \right)$$

$$= \frac{\sin [(\pi/2) - \theta]}{\cos [(\pi/2) - \theta]}.$$

(12)   Let $\alpha = 0$ and $\beta = \theta$.

(13)   Let $\alpha = \pi/2$ and $\beta = -\theta$.

(14)   Write
$$\tan(-\theta) = \frac{\sin(-\theta)}{\cos(-\theta)}.$$

(12) $\cos(-\theta) = \cos\theta$

(13) $\sin(-\theta) = -\sin\theta$

(14) $\tan(-\theta) = -\tan\theta$

The next identities are analogous to the main identity of this section.

(15)   Write $\cos(\alpha + \beta)$ $= \cos[\alpha - (-\beta)]$.

(16)   Stated here again for completeness.

(17)   Write $\sin(\alpha + \beta)$ $= \cos[\pi/2 - (\alpha + \beta)]$.

(18)   Write $\sin(\alpha - \beta)$ $= \sin[\alpha + (-\beta)]$.

(19), (20)   Write the tangent in terms of sine and cosine.

(15) $\cos(\alpha + \beta) = \cos\alpha\cos\beta - \sin\alpha\sin\beta$

(16) $\cos(\alpha - \beta) = \cos\alpha\cos\beta + \sin\alpha\sin\beta$

(17) $\sin(\alpha + \beta) = \sin\alpha\cos\beta + \cos\alpha\sin\beta$

(18) $\sin(\alpha - \beta) = \sin\alpha\cos\beta - \cos\alpha\sin\beta$

(19) $\tan(\alpha + \beta) = \dfrac{\tan\alpha + \tan\beta}{1 - \tan\alpha\tan\beta}$

(20) $\tan(\alpha - \beta) = \dfrac{\tan\alpha - \tan\beta}{1 + \tan\alpha\tan\beta}$

DOUBLE-ANGLE IDENTITIES

There are two additional special cases of the sum and difference identities of the previous section, and they will now be considered. The first is that of the *double-angle identities*. For this case, we let $\alpha = \theta$ and $\beta = \theta$.

DOUBLE-ANGLE IDENTITIES
(21) $\cos 2\theta = \cos^2\theta - \sin^2\theta$ $= 2\cos^2\theta - 1$ $= 1 - 2\sin^2\theta$
(22) $\sin 2\theta = 2\sin\theta\cos\theta$
(23) $\tan 2\theta = \dfrac{2\tan\theta}{1 - \tan^2\theta}$

(21)
$$\cos 2\theta = \cos(\theta + \theta) = \cos\theta\cos\theta - \sin\theta\sin\theta$$
$$= \cos^2\theta - \sin^2\theta$$

(22)
$$\sin 2\theta = \sin(\theta + \theta) = \sin\theta\cos\theta + \cos\theta\sin\theta$$
$$= 2\sin\theta\cos\theta$$

(23)
$$\tan 2\theta = \tan(\theta + \theta) = \frac{\tan\theta + \tan\theta}{1 - \tan\theta\tan\theta}$$
$$= \frac{2\tan\theta}{1 - \tan^2\theta}$$

EXAMPLES

2. $\tan\dfrac{\pi}{8} = \tan\left(2 \cdot \dfrac{\pi}{16}\right) = \dfrac{2\tan(\pi/16)}{1 - \tan^2(\pi/16)}$

3. $\cos 100x = \cos(2 \cdot 50x) = \cos^2 50x - \sin^2 50x$

4. $\sin \alpha = 2 \sin \dfrac{1}{2}\alpha \cos \dfrac{1}{2}\alpha$

5. $\cos 3\theta = \cos(2\theta + \theta) = \cos 2\theta \cos \theta - \sin 2\theta \sin \theta$
$$= (\cos^2 \theta - \sin^2 \theta)\cos \theta - (2 \sin \theta \cos \theta)\sin \theta$$
$$= \cos^3 \theta - \sin^2 \theta \cos \theta - 2 \sin^2 \theta \cos \theta$$
$$= \cos^3 \theta - (1 - \cos^2 \theta)\cos \theta - 2(1 - \cos^2 \theta)\cos \theta$$
$$= \cos^3 \theta - \cos \theta + \cos^3 \theta - 2 \cos \theta + 2 \cos^3 \theta$$
$$= 4 \cos^3 \theta - 3 \cos \theta$$

HALF-ANGLE IDENTITIES

Sometimes, as in Example 5, we want to write $\cos 2\theta$ in terms of cosines, and at other times we want to write it in terms of sines. Thus, we have

$$\cos 2\theta = \cos^2 \theta - \sin^2 \theta$$
$$= \cos^2 \theta - (1 - \cos^2 \theta)$$
$$= 2 \cos^2 \theta - 1,$$

and

$$\cos 2\theta = \cos^2 \theta - \sin^2 \theta$$
$$= (1 - \sin^2 \theta) - \sin^2 \theta$$
$$= 1 - 2 \sin^2 \theta.$$

These last two identities lead us to the second important special case of the sum and difference identities, called the *half-angle identities*. Solve $\cos 2\alpha = 2 \cos^2 \alpha - 1$ for $\cos^2 \alpha$.

$$2 \cos^2 \alpha - 1 = \cos 2\alpha$$

$$2 \cos^2 \alpha = 1 + \cos 2\alpha$$

$$\cos^2 \alpha = \frac{1 + \cos 2\alpha}{2}$$

Now, if $\alpha = \frac{1}{2}\theta$, then $2\alpha = \theta$ and

$$\cos^2 \frac{1}{2}\theta = \frac{1 + \cos \theta}{2}.$$

If $\frac{1}{2}\theta$ is in Quadrant I or IV, then

$$\cos \frac{1}{2}\theta = \sqrt{\frac{1 + \cos \theta}{2}};$$

if $\frac{1}{2}\theta$ is in Quadrant II or III, then

$$\cos \frac{1}{2}\theta = -\sqrt{\frac{1 + \cos \theta}{2}}.$$

These results are summarized by writing

(24) $\qquad \cos\dfrac{1}{2}\theta = \pm\sqrt{\dfrac{1 + \cos\theta}{2}}.$

This situation is not consistent with the use of $\pm$ from algebra. For example, when using $\pm$ in the quadratic formula, we mean to indicate possibly *two* correct roots. In this trigonometric identity, we will obtain *one* correct value depending on the quadrant of $\frac{1}{2}\theta$.

However, *you must be careful.* The sign $+$ or $-$ is chosen according to the quadrant of $\frac{1}{2}\theta$. The formula requires either $+$ or $-$, but not both.

For the sine, solve $\cos 2\alpha = 1 - 2\sin^2\alpha$ for $\sin^2\alpha$.

$$\cos 2\alpha = 1 - 2\sin^2\alpha$$

$$2\sin^2\alpha = 1 - \cos 2\alpha$$

$$\sin^2\alpha = \frac{1 - \cos 2\alpha}{2}$$

Replace α with $\frac{1}{2}\theta$:

$$\sin^2\frac{1}{2}\theta = \frac{1 - \cos\theta}{2}$$

or

(25) $\qquad \sin\dfrac{1}{2}\theta = \pm\sqrt{\dfrac{1 - \cos\theta}{2}},$

where the sign depends on the quadrant of $\frac{1}{2}\theta$. If $\frac{1}{2}\theta$ is in Quadrant I or II, we use $+$; if it is in Quadrant III or IV, we use $-$.

Finally, to find the half-angle identity for the tangent, we write

$$\tan^2\frac{1}{2}\theta = \frac{\sin^2\frac{1}{2}\theta}{\cos^2\frac{1}{2}\theta} = \frac{\dfrac{1 - \cos\theta}{2}}{\dfrac{1 + \cos\theta}{2}} = \frac{1 - \cos\theta}{1 + \cos\theta}.$$

Thus,

$$\tan\frac{1}{2}\theta = \pm\sqrt{\frac{1 - \cos\theta}{1 + \cos\theta}}$$

$$= \pm\sqrt{\frac{1 - \cos\theta}{1 + \cos\theta} \cdot \frac{1 - \cos\theta}{1 - \cos\theta}}$$

$$= \pm\sqrt{\frac{(1 - \cos\theta)^2}{\sin^2\theta}}$$

(26) $\qquad = \dfrac{1 - \cos\theta}{\sin\theta}.$

HALF-ANGLE IDENTITIES

(24) $\cos\dfrac{1}{2}\theta = \pm\sqrt{\dfrac{1 + \cos\theta}{2}}$

(25) $\sin\dfrac{1}{2}\theta = \pm\sqrt{\dfrac{1 - \cos\theta}{2}}$

(26) $\tan\dfrac{1}{2}\theta = \dfrac{1 - \cos\theta}{\sin\theta}$

$\qquad\quad = \dfrac{\sin\theta}{1 + \cos\theta}$

Remember that $1 - \cos^2\theta = \sin^2\theta$

You can also show that

$$\tan \frac{1}{2}\theta = \frac{\sin \theta}{1 + \cos \theta}.$$

EXAMPLES

6. Find $\cos (9\pi/8)$.
   Solution:

$$\cos \frac{9\pi}{8} = \cos \left(\frac{1}{2} \cdot \frac{9\pi}{4} \right) = -\sqrt{\frac{1 + \cos (9\pi/4)}{2}}$$

$\uparrow$

Choose a negative sign, since $9\pi/8$
is in Quadrant III and the cosine is
negative in this quadrant.

$$= -\sqrt{\frac{1 + \cos (\pi/4)}{2}}$$

Note: $\cos (9\pi/4) = \cos (\pi/4)$
(coterminal angles).

$$= -\sqrt{\frac{1 + \sqrt{2}/2}{2}}$$

$$= -\sqrt{\frac{2 + \sqrt{2}}{4}}$$

$$= -\frac{1}{2}\sqrt{2 + \sqrt{2}}$$

7. If $\cot 2\theta = 3/4$, find $\cos \theta$, $\sin \theta$, and $\tan \theta$ where 2θ is in Quadrant I.
   Solution: Choose the point $(3, 4)$ on the terminal side of angle 2θ. Then, from the definition of the trigonometric function,

$$\cot 2\theta = \frac{x}{y} \quad \text{and} \quad \cos 2\theta = \frac{x}{\sqrt{x^2 + y^2}}.$$

Example 7 illustrates a
process you will need in
analytic geometry when
rotating certain types of
curves.

Thus,

$$\cos 2\theta = \frac{3}{\sqrt{9 + 16}} = \frac{3}{5}$$

$$\cos \theta = +\sqrt{\frac{1 + 3/5}{2}} \qquad \sin \theta = +\sqrt{\frac{1 - 3/5}{2}}$$

$\uparrow$ $\uparrow$

positive value chosen because θ is in Quadrant I

$$= \frac{2}{\sqrt{5}} \quad \text{or} \quad \frac{2}{5}\sqrt{5} \qquad = \frac{1}{\sqrt{5}} \quad \text{or} \quad \frac{1}{5}\sqrt{5}$$

$$\tan \theta = \frac{\sin \theta}{\cos \theta} = \frac{1/\sqrt{5}}{2/\sqrt{5}} = \frac{1}{2}$$

8. Prove that

$$\sin \theta = \frac{2 \tan \frac{1}{2}\theta}{1 + \tan^2 \frac{1}{2}\theta}.$$

Solution: When proving identities involving functions of different angles, we should write all the trigonometric functions in the problems as functions of a single angle. As before, we begin with the more complicated side and change to sines and cosines.

$$\frac{2 \tan \frac{1}{2}\theta}{1 + \tan^2 \frac{1}{2}\theta} = \frac{2 \dfrac{\sin \frac{1}{2}\theta}{\cos \frac{1}{2}\theta}}{1 + \dfrac{\sin^2 \frac{1}{2}\theta}{\cos^2 \frac{1}{2}\theta}} = \frac{\dfrac{2 \sin \frac{1}{2}\theta}{\cos \frac{1}{2}\theta}}{\dfrac{\cos^2 \frac{1}{2}\theta + \sin^2 \frac{1}{2}\theta}{\cos^2 \frac{1}{2}\theta}} = \frac{2 \sin \frac{1}{2}\theta \cos \frac{1}{2}\theta}{\cos^2 \frac{1}{2}\theta + \sin^2 \frac{1}{2}\theta}$$

$$= 2 \sin \tfrac{1}{2}\theta \cos \tfrac{1}{2}\theta$$

$$= \sin \theta$$

PRODUCT AND SUM IDENTITIES

It is sometimes convenient, or even necessary, to write a trigonometric sum as a product or a product as a sum. To do so, we again turn to the identities for the sum and difference of two angles. Add and subtract the following pair of identities:

$$\cos \alpha \cos \beta + \sin \alpha \sin \beta = \cos (\alpha - \beta)$$
$$\cos \alpha \cos \beta - \sin \alpha \sin \beta = \cos (\alpha + \beta)$$

(27)   Adding: $2 \cos \alpha \cos \beta = \cos (\alpha - \beta) + \cos (\alpha + \beta)$
(28)   Subtracting: $2 \sin \alpha \sin \beta = \cos (\alpha - \beta) - \cos (\alpha + \beta)$

Also:

$$\sin \alpha \cos \beta + \cos \alpha \sin \beta = \sin (\alpha + \beta)$$
$$\sin \alpha \cos \beta - \cos \alpha \sin \beta = \sin (\alpha - \beta)$$

(29)   Adding: $2 \sin \alpha \cos \beta = \sin (\alpha + \beta) + \sin (\alpha - \beta)$
(30)   Subtracting: $2 \cos \alpha \sin \beta = \sin (\alpha + \beta) - \sin (\alpha - \beta)$

These identities are called the *product identities*.

PRODUCT IDENTITIES

> **PRODUCT IDENTITIES**
>
> (27) $2 \cos \alpha \cos \beta = \cos (\alpha - \beta) + \cos (\alpha + \beta)$
> (28) $2 \sin \alpha \sin \beta = \cos (\alpha - \beta) - \cos (\alpha + \beta)$
> (29) $2 \sin \alpha \cos \beta = \sin (\alpha + \beta) + \sin (\alpha - \beta)$
> (30) $2 \cos \alpha \sin \beta = \sin (\alpha + \beta) - \sin (\alpha - \beta)$

To return to the original formulas for a sum, let $x = \alpha + \beta$ and $y = \alpha - \beta$. Solving for α and β, $\alpha = \frac{1}{2}(x + y)$ and $\beta = \frac{1}{2}(x - y)$. Then, substitute into the product formulas to obtain the *sum identities*.

SUM IDENTITIES

(31) $\qquad \cos x + \cos y = 2 \cos \left(\dfrac{x + y}{2} \right) \cos \left(\dfrac{x - y}{2} \right)$

(32) $\qquad \cos x - \cos y = -2 \sin \left(\dfrac{x + y}{2} \right) \sin \left(\dfrac{x - y}{2} \right)$

(33) $\qquad \sin x + \sin y = 2 \sin \left(\dfrac{x + y}{2} \right) \cos \left(\dfrac{x - y}{2} \right)$

(34) $\qquad \sin x - \sin y = 2 \sin \left(\dfrac{x - y}{2} \right) \cos \left(\dfrac{x + y}{2} \right)$

SUM IDENTITIES

The derivations of these last identities are left as problems.

EXAMPLES

9. Write $\sin 40° \cos 12°$ as the sum of two functions.
   Solution: $2 \sin 40° \cos 12° = \sin (40° + 12°) + \sin (40° - 12°)$
   Therefore, $\sin 40° \cos 12° = \frac{1}{2} (\sin 52° + \sin 28°)$.

10. Write $\sin 35° + \sin 27°$ as a product.
    Solution: $x = 35°$, $y = 27°$, and $(x + y)/2 = (35° + 27°)/2 = 31°$;
    $(x - y)/2 = 4°$.
    Therefore, $\sin 35° + \sin 27° = 2 \sin 31° \cos 4°$.

PROBLEM SET 7.3

A Problems

Using the identities of this section, find the exact values of the sine, cosine, and tangent of the angles given in Problems 1–6.

1. $-15°$ 2. $195°$ 3. $75°$ 4. $165°$ 5. $345°$ 6. $105°$

In Problems 7–12, find the exact values of the sine, cosine, and tangent of $\frac{1}{2}\theta$ and 2θ.

7. $\sin \theta = 3/5$; θ in Quadrant I
8. $\sin \theta = 5/13$; θ in Quadrant II
9. $\tan \theta = -5/12$; θ in Quadrant IV
10. $\tan \theta = -3/4$; θ in Quadrant II
11. $\cos \theta = 5/9$; θ in Quadrant I
12. $\cos \theta = -5/13$; θ in Quadrant III

Change each of the expressions in Problems 13–18 to a function of θ only.

13. $\cos (30° + \theta)$
14. $\sin (\theta - 45°)$
15. $\tan (\pi/4 - \theta)$
16. $\cos (\theta - 45°)$
17. $\sin (120° + \theta)$
18. $\tan (\theta - 225°)$

Write each of the expressions in Problems 19–24 as the sum of two functions.

19. $2 \cos 75° \cos 35°$ 20. $2 \cos 46° \cos 18°$
21. $2 \sin 35° \sin 24°$ 22. $2 \sin 53° \cos 24°$
23. $\sin 2\theta \sin 5\theta$ 24. $\cos \theta \cos 3\theta$

Write each of the expressions in Problems 25–30 as a product.

25. $\sin 43° + \sin 63°$ 26. $\sin 22° - \sin 6°$
27. $\cos 81° - \cos 79°$ 28. $\cos 78° + \cos 25°$
29. $\sin x - \sin 2x$ 30. $\cos 3\theta + \cos 2\theta$

Use the double-angle or half-angle identities to evaluate Problems 31–40 using exact values.

31. $2 \cos^2 22.5° - 1$ 32. $\dfrac{2 \tan (\pi/8)}{1 - \tan^2 (\pi/8)}$ 33. $\sqrt{\dfrac{1 - \cos 60°}{2}}$

34. $\cos^2 15° - \sin^2 15°$ 35. $1 - 2 \sin^2 90°$ 36. $-\sqrt{\dfrac{1 - \cos 420°}{2}}$

37. $\sin 22.5°$ 38. $\cos (\pi/8)$ 39. $\tan 22.5°$ 40. $\sin 105°$

Evaluate the expressions in Problems 41–46. Use Table 4 or a calculator.

41. $\sin 158° \cos 92° - \cos 158° \sin 92°$ 42. $\cos 114° \cos 85° + \sin 114° \sin 85°$
43. $\cos 30° \cos 48° - \sin 30° \sin 48°$ 44. $\sin 18° \cos 23° + \cos 18° \sin 23°$

45. $\dfrac{\tan 32° + \tan 18°}{1 - \tan 32° \tan 18°}$ 46. $\dfrac{\tan 59° - \tan 25°}{1 + \tan 59° \tan 25°}$

B Problems

47. Prove that

$$\tan (\alpha + \beta) = \frac{\tan \alpha + \tan \beta}{1 - \tan \alpha \tan \beta}.$$

48. Prove that

$$\tan (\alpha - \beta) = \frac{\tan \alpha - \tan \beta}{1 + \tan \alpha \tan \beta}.$$

49. Derive a formula for $\cot (\alpha + \beta)$ in terms of $\cot \alpha$ and $\cot \beta$.
50. Derive a formula for $\cot (\alpha - \beta)$ in terms of $\cot \alpha$ and $\cot \beta$.

In Problems 51–54, let $\sin \alpha = 3/5$ and $\sin \beta = 5/13$, where α and β are both acute. Find the given value.

EXAMPLE 11: $\sin (\alpha - \beta)$

Solution: $\sin (\alpha - \beta) = \sin \alpha \cos \beta - \cos \alpha \sin \beta$

$$= \left(\frac{3}{5}\right) \cos \beta - \cos \alpha \left(\frac{5}{13}\right) \qquad \begin{array}{l}\text{(since } \sin \alpha \text{ and } \sin \beta \\ \text{are given)}\end{array}$$

We find the other values by using Fundamental Identities.

$$\cos \beta = \pm\sqrt{1 - \sin^2 \beta} \qquad \cos \alpha = \pm\sqrt{1 - \sin^2 \alpha}$$
$$= \pm\sqrt{1 - (5/13)^2} \qquad = \pm\sqrt{1 - (3/5)^2}$$
$$= \pm\sqrt{144/169} \qquad = \pm\sqrt{16/25}$$
$$= 12/13 \qquad = 4/5$$

The positive values are chosen since it is given that α and β are both acute. We can now finish the problem.

$$\sin (\alpha - \beta) = \left(\frac{3}{5}\right)\left(\frac{12}{13}\right) - \left(\frac{4}{5}\right)\left(\frac{5}{13}\right)$$

$$= \frac{36}{65} - \frac{20}{65}$$

$$= \frac{16}{65}$$

---

51. $\cos (\alpha + \beta)$ 52. $\cos (\alpha - \beta)$ 53. $\sin (\alpha + \beta)$ 54. $\tan (\alpha + \beta)$

In Problems 55–60, find $\cos \theta$, $\sin \theta$, *and* $\tan \theta$ *when* θ *is in Quadrant I and* $\cot 2\theta$ *is given.*

55. $\cot 2\theta = -3/4$ 56. $\cot 2\theta = 0$ 57. $\cot 2\theta = 1/\sqrt{3}$
58. $\cot 2\theta = -1/\sqrt{3}$ 59. $\cot 2\theta = -4/3$ 60. $\cot 2\theta = 4/3$

Prove the identities in Problems 61–75.

61. $\dfrac{\cos 5\theta}{\sin \theta} - \dfrac{\sin 5\theta}{\cos \theta} = \dfrac{\cos 6\theta}{\sin \theta \cos \theta}$

62. $\sin (\alpha + \beta) \cos \beta - \cos (\alpha + \beta) \sin \beta = \sin \alpha$

63. $\dfrac{\sin (\theta + h) - \sin \theta}{h} = \cos \theta \left(\dfrac{\sin h}{h}\right) - \sin \theta \left(\dfrac{1 - \cos h}{h}\right)$

64. $\dfrac{\cos (\theta + h) - \cos \theta}{h} = -\sin \theta \left(\dfrac{\sin h}{h}\right) - \cos \theta \left(\dfrac{1 - \cos h}{h}\right)$

Problems 63 and 64 are problems you will encounter in calculus in a process called differentiation.

65. $\sin (\alpha + \beta + \gamma) = \sin \alpha \cos \beta \cos \gamma + \cos \alpha \sin \beta \cos \gamma + \cos \alpha \cos \beta \sin \gamma$
$- \sin \alpha \sin \beta \sin \gamma$

66. $\tan \dfrac{1}{2}\theta = \dfrac{1 - \cos \theta}{\sin \theta}$

67. $\sin x - \sin y = 2 \sin \left(\dfrac{x - y}{2}\right) \cos \left(\dfrac{x + y}{2}\right)$

68. $\sin \alpha = 2 \sin (\alpha/2) \cos (\alpha/2)$

69. $\cos 4\theta = \sin 2\theta$

70. $\dfrac{\sin 5\theta + \sin 3\theta}{\cos 5\theta + \cos 3\theta} = \tan 4\theta$

71. $\dfrac{\cos 3\theta - \cos \theta}{\sin \theta - \sin 3\theta} = \tan 2\theta$

72. $\cos 2y = 2\cos^2 y - 1$
73. $\tan (B/2) = \csc B - \cot B$
74. $\sin 4\theta = 4\sin \theta \cos \theta - 8\sin^3 \theta \cos \theta$
75. $\sin 3\theta = 3\sin \theta - 4\sin^3 \theta$

Mind Bogglers

76. Here is an alternate proof for the formula
    $\cos (\alpha + \beta) = \cos \alpha \cos \beta - \sin \alpha \sin \beta$.
    a. Case i: let $\alpha = 0$ and $\beta = 0$. Prove the given identity.
    b. Case ii: let $\alpha \neq 0$ and $\beta \neq 0$. Let α be in standard position and β be any angle drawn so that its initial side is along the terminal side of α (see Figure 7.3).

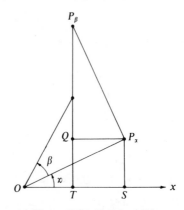

Figure 7.3   Problem 76

Let P_α be an arbitrary point on the terminal side of α. Let P_β be the point on the terminal side of β so that $P_\alpha P_\beta$ is perpendicular to OP_α. Draw perpendiculars $P_\alpha S$ and $P_\beta T$ to the x-axis. Draw $P_\alpha Q$ perpendicular to $P_\beta T$. Thus,

$$\cos (\alpha + \beta) = \frac{OT}{OP_\beta}.$$

Now show that

$$\cos (\alpha + \beta) = \frac{OS}{OP_\alpha} \cdot \frac{OP_\alpha}{OP_\beta} - \frac{QP_\alpha}{P_\alpha P_\beta} \cdot \frac{P_\alpha P_\beta}{OP_\beta},$$

and therefore

$$\cos (\alpha + \beta) = \cos \alpha \cos \beta - \sin \alpha \sin \beta.$$

77. Prove that

$$\cos^4 \theta = \frac{1}{8}(3 + 4\cos 2\theta + \cos 4\theta).$$

78. Prove that

$$\frac{\sqrt{6} + \sqrt{2}}{2} = \sqrt{2 + \sqrt{3}}.$$

7.4   USING IDENTITIES TO SOLVE EQUATIONS

Up to this point we have been working with identities alone. Sometimes, though, we will have to solve trigonometric equations. To do this, we use the techniques of algebra coupled with the trigonometric identities. Our goal when solving a trigonometric equation is to solve first for $F(x)$, where F is one of the six circular functions. Next, we solve for x by using the inverse relations presented in Section 6.6 or by using the table of exact values. However, for convenience we'll seek those solutions in the interval $0 \leq x < 2\pi$. This restriction is somewhat subtle in work with multiple angles, as shown by Examples 1 and 2.

EXAMPLES

1. $\sin x = \frac{1}{2}$
   From the table of exact values, the reference angle is $\pi/6$. Since the sine is positive in Quadrants I and II, $x = \pi/6$ and $5\pi/6$.

2. $\sin 2x = \frac{1}{2}$
   Since $0 \leq x < 2\pi$, we solve for $2x$ such that $0 \leq 2x < 4\pi$. Thus, $2x = \pi/6$, $2x = 5\pi/6$, $2x = 13\pi/6$, and $2x = 17\pi/6$; $x = \pi/12, 5\pi/12, 13\pi/12$, and $17\pi/12$.

If you are using a calculator, find $x = \text{Sin}^{-1}(\frac{1}{2})$.

$$\boxed{0.5}$$

$$\boxed{\text{ARC}}$$

$$\boxed{\text{SIN}}$$

The display now shows 30 (in degrees) or 0.5235987756 (in radians). Notice that the calculator gives $\text{Sin}^{-1}(\frac{1}{2})$ and not $\sin^{-1}(\frac{1}{2})$. This means that even using a calculator, you will have to consider quadrants to obtain all solutions between 0 and 2π.

Notice from Example 2 that calculating $F(nx)$ requires that we find the values $0 \leq nx < 2n\pi$ for a given trigonometric function F. This means that $0 \leq x < 2\pi$. You may have to solve for $F(x)$ algebraically by factoring or by using the Quadratic Formula, as shown by Examples 3 and 4.

EXAMPLES

3. $2\cos\theta\sin\theta = \sin\theta$
   Solution: $2\cos\theta\sin\theta - \sin\theta = 0$

$$\sin \theta (2 \cos \theta - 1) = 0$$

$$\sin \theta = 0 \qquad 2 \cos \theta - 1 = 0$$

$$\cos \theta = \frac{1}{2}$$

$$\theta = 0, \pi \qquad\qquad \theta = \frac{\pi}{3}, \frac{5\pi}{3}$$

$$\left\{ 0, \pi, \frac{\pi}{3}, \frac{5\pi}{3} \right\}$$

4. $2 \sin^2 \theta = 1 + 2 \sin \theta$

Solution: $2 \sin^2 \theta - 2 \sin \theta - 1 = 0$

Since we can't factor the left side, we use the Quadratic Formula to solve for $\sin \theta$.

$$\sin \theta = \frac{2 \pm \sqrt{4 - 4(2)(-1)}}{2(2)}$$

$$= \frac{1 \pm \sqrt{3}}{2}$$

$$\approx 1.366, -0.366025$$

$$\underset{\uparrow}{\quad} \text{Reject, since } 0 \leq \sin \theta \leq 1.$$

Solve $\sin \theta = -0.366025$ by using Table 5 or a calculator to find a reference angle of 0.3747. Since the sine is negative in Quadrants III and IV, the solutions are $\pi + 0.3747 \approx 3.5163$ and $2\pi - 0.3747 \approx 5.9084$.

In degrees, $\theta = 201.47°$ and 338.53°.

Finally, it may be necessary to apply some trigonometric identities before solving. If different functions appear in the equation, use the Fundamental Identities to express the equation in terms of a single function— or at least in terms of factors that contain only a single function. If the *angles* are different (as in θ, 2θ, or 3θ), then use identities to write the equation in terms of a single angle. If it is necessary to multiply both sides by a trigonometric function or to square both sides, be sure to check for extraneous roots.

EXAMPLES

5. $\sin^2 3\theta + \sin 3\theta = \cos^2 3\theta - 1$

Solution: $\sin^2 3\theta - \cos^2 3\theta + \sin 3\theta + 1 = 0$

Changing to a single function

$$\sin^2 3\theta - (1 - \sin^2 3\theta) + \sin 3\theta + 1 = 0$$

$$2 \sin^2 3\theta + \sin 3\theta = 0$$

Factoring

$$\sin 3\theta (2 \sin 3\theta + 1) = 0$$

$$\sin 3\theta = 0, \ -\frac{1}{2}$$

$$3\theta = 0, \ \pi, \ 2\pi, \ 3\pi, \ 4\pi, \ 5\pi$$

From the first factor,
$\sin 3\theta = 0$.

$$3\theta = \frac{7\pi}{6}, \frac{11\pi}{6}, \frac{19\pi}{6}, \frac{23\pi}{6}, \frac{31\pi}{6}, \frac{35\pi}{6}$$

$$\left\{ 0, \frac{\pi}{3}, \frac{2\pi}{3}, \pi, \frac{4\pi}{3}, \frac{5\pi}{3}, \frac{7\pi}{18}, \frac{11\pi}{18}, \frac{19\pi}{18}, \frac{23\pi}{18}, \frac{31\pi}{18}, \frac{35\pi}{18} \right\}$$

From the second factor,
$\sin 3\theta = -\frac{1}{2}$ (the reference
angle is $\pi/6$ in Quadrants III
and IV), since

6. $\cos 6\theta + \sin 3\theta + 1 = 0$

   Solution: Notice that the angles 6θ and 3θ are different. We could use a half-angle identity to change $\sin 3\theta$ to a function of 6θ, but this process would introduce a radical. So instead we'll use a double-angle identity to change $\cos 6\theta$ to a function of 3θ as follows:

$$\cos^2 3\theta - \sin^2 3\theta + \sin 3\theta + 1 = 0.$$

$$0 \le \theta < 2\pi,$$
$$0 \le 3\theta < 6\pi.$$

The problem is now similar to the one solved in Example 5.

$$(1 - \sin^2 3\theta) - \sin^2 3\theta + \sin 3\theta + 1 = 0$$

$$1 - 2\sin^2 3\theta + \sin 3\theta + 1 = 0$$

$$2\sin^2 3\theta - \sin 3\theta - 2 = 0$$

$$\sin 3\theta = \frac{1 \pm \sqrt{1 - 4(2)(-2)}}{2(2)}$$

$$= \frac{1 \pm \sqrt{17}}{4}$$

$$\approx 1.2807, \ -0.7808$$

$$3\theta \approx 4.038, \ 5.3872, \ 10.3207, \ 11.6705, \ 16.6039, \ 17.9536$$

$$\theta \approx 1.3458, \ 1.7958, \ 3.4402, \ 3.8902, \ 5.5346, \ 5.9845$$

Since
$$0 \le \theta < 2\pi,$$
$$0 \le 3\theta < 6\pi \approx 18.8496.$$

In degrees, the answer to
Example 6 is {42.89°,
102.89°, 162.89°, 222.89°,
282.89°, 342.89°}.

To summarize the procedure of solving a trigonometric equation, we must carry out three steps.

Steps in solving a
trigonometric equation

1.   Solve for a trigonometric function. (Algebraic solution, although some trigonometric identities may be required.)
2.   Solve for the angle. (Trigonometric solution; may use exact values or inverse functions on a calculator.)
3.   Solve for the unknown. (Algebraic solution.) The unknown may or may not be the same as the angle. For example, in Example 6, the angle is 3θ and the unknown is θ.

PROBLEM SET 7.4

A Problems

Solve the equations in Problems 1–10 for $0 \le x < 2\pi$.

1. $\cos x = 1/2$
2. $\cos 2x = 1/2$
3. $\sin 2x = \sqrt{2}/2$
4. $\sin 2x = -\sqrt{3}/2$
5. $\tan 3x = 1$
6. $\sec 2x = -2\sqrt{3}/3$
7. $(\sin x)\cos x = 0$
8. $(\sec 2x)\tan x = 0$
9. $(\sec x - 2)(\sin x - 1) = 0$
10. $(\csc x - 2)(2 \cos x - 1) = 0$

B Problems

Solve the equations in Problems 11–30 for $0 \le x < 2\pi$.

11. $\tan^2 x = \sqrt{3} \tan x$
12. $\tan^2 x = \tan x$
13. $\sin x(3 \cos x - 1) = 0$
14. $\sin^2 x = 1$
15. $4 \sin^2 x = 1$
16. $2 \cos 2x \sin 2x = \sin 2x$
17. $\sin x = \cos x$
18. $\sin^2 x = 1/2$
19. $\cos^2 x = 1/2$
20. $\sin 2x = \cos 2x$
21. $\sin^2 x - \sin x - 2 = 0$
22. $\sin^2 x + \cos x = 0$
23. $4 \cos^2 x - 8 \cos x + 3 = 0$
24. $\tan^2 x - 3 \tan x + 1 = 0$
25. $\sec^2 x - \sec x - 1 = 0$
26. $2 \sin^2 x - \cos 2x = 0$
27. $\cos 2x = 3 \sin x$
28. $\sin x = \cos 2x$
29. $4 \sin^3 x + \sin 3x - 3 \sin x + \sqrt{3} = 2 \sin 2x$
30. $4 \cos^3 x - 3 \cos x - \cos 3x + 2 \cos x = 1$

Mind Boggler

31. For $0 \le \theta \le \pi/2$, solve

$$\left(\frac{16}{81}\right)^{\sin^2 \theta} + \left(\frac{16}{81}\right)^{\cos^2 \theta} = \frac{26}{27}.$$

7.5   RIGHT TRIANGLES

One of the most important uses of trigonometry is in solving triangles. Recall from geometry that every triangle has three sides and three angles, which are called the six *parts* of the triangle. We say that a *triangle is solved* if all six parts are known. Typically, three parts will be given, or known, and we will want to find the other three parts. We will usually label a triangle as shown in Figure 7.4. The vertices are labeled A, B, and C, with the sides opposite those vertices a, b, and c, respectively. The angles are labeled α, β, and γ, respectively. In this section, we'll limit our examples to right triangles, and we'll further agree that γ denotes the right angle and c the hypotenuse.

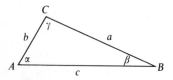

Figure 7.4   Correctly labeled triangle

According to the definition of the circular functions, the angle under consideration must be in standard position. This requirement is sometimes inconvenient, so we use that definition to create a special casé, which applies to any acute angle θ of a right triangle. Notice that in Figure 7.4, θ might be α or β, but it would not be γ since γ is not an acute angle. Also notice that the hypotenuse is one of the sides of both acute angles. The other side making up the angle is called the *adjacent side*. Thus, side a is adjacent to β and side b is adjacent to α. The third side of the triangle (the one not making up the angle) is called the *side opposite* the angle. Thus, side a is opposite α and side b is opposite β.

If θ is an acute angle in a right triangle, then

$$\cos \theta = \frac{\text{adjacent side}}{\text{hypotenuse}},$$

$$\sin \theta = \frac{\text{opposite side}}{\text{hypotenuse}},$$

$$\tan \theta = \frac{\text{opposite side}}{\text{adjacent side}}.$$

The other trigonometric functions are the reciprocals of these relationships.

RIGHT TRIANGLE DEFINITION OF THE CIRCULAR FUNCTIONS

We can now use this definition to solve some given triangles.

EXAMPLES: Solve the right triangles ($\gamma = 90°$) with the given information.

1. $a = 50; \alpha = 35°$

When we write $\alpha = 35°$, we mean the measure of angle α is 35°.

> *Solution:* $\alpha = 35°$ (given)
> $\beta = 55°$ (since $\alpha + \beta = 90°$ for any right
>   triangle with right angle at C)
> $\gamma = 90°$ (given)
> $a = 50$ (given)
>
> $b:\ \tan 35° = \dfrac{50}{b}$
>
> $b = \dfrac{50}{\tan 35°}$

From Table 4, tan 35°
≈ 0.7002.
By division on a four-
function calculator

This avoids division.
From Table 4

(1) By table:

$$b = \frac{50}{\tan 35°} \approx \frac{50}{0.7002}$$

$$\approx 71.4082$$

or

$b = 50 \cot 35°$
   $\approx 50(1.4281)$
   $= 71.405$

(2) By calculator with algebraic logic:

| 50 | | ÷ | | 35 | | TAN | | = | ≈ 71.40740034

(3) By calculator with RPN logic:

| 50 | | ENTER | | 35 | | TAN | | ÷ |

   ≈ 71.40740034

Notice that some of the answers above
differ. However, to two significant
digits, $b = 71$.

$$c: \ \sin 35° = \frac{50}{c}$$

$$c = \frac{50}{\sin 35°}$$

See Appendix C for a
discussion on how to
determine the correct number
of significant digits.
The intermediate steps for
finding c by table or by
calculator are left to you.

$$\approx 87 \quad \text{(to two significant digits)}$$

2. $a = 32; b = 58$

Solution: α: $\tan \alpha = \dfrac{32}{58}$ or $\alpha = \tan^{-1}\left(\dfrac{32}{58}\right)$

(1) By table:

$$\frac{32}{58} \approx 0.5517,$$

so from Table 4,

$\alpha \approx 28.9°$.

(2) By calculator with algebraic logic:

| 32 | | ÷ | | 58 | | = | | INV | | TAN |

   ≈ 28.88658177

(3) By calculator with RPN logic:

$\boxed{32}$ $\boxed{\text{ENTER}}$ $\boxed{58}$ $\boxed{\div}$ $\boxed{\text{ARC}}$ $\boxed{\text{TAN}}$

≈ 28.88658176

To the nearest degree, $\alpha = 29°$.

$$\begin{aligned}
\beta &= 90° - \alpha \\
&= 61° \\
\gamma &= 90° \quad \text{(given)} \\
a &= 32 \quad \text{(given)} \\
b &= 58 \quad \text{(given)}
\end{aligned}$$

Notice that when solving triangles, all six parts should be stated to the correct number of significant digits—even those given in the problem.

c: (1) By table:

$$\sin \alpha = \frac{32}{c}$$

$$c = \frac{32}{\sin \alpha}$$

$$\approx \frac{32}{0.4848}$$

$$\approx 66.0066$$

(2) By calculator: $c = \sqrt{a^2 + b^2}$
Algebraic logic:

$\boxed{32}$ $\boxed{x^2}$ $\boxed{+}$ $\boxed{58}$ $\boxed{x^2}$ $\boxed{=}$ $\boxed{\sqrt{x}}$

≈ 66.24198065

RPN logic:

$\boxed{32}$ $\boxed{\text{ENTER}}$ $\boxed{\times}$ $\boxed{58}$ $\boxed{\text{ENTER}}$

$\boxed{\times}$ $\boxed{+}$ $\boxed{\sqrt{x}}$ ≈ 66.24198067

To two significant digits, $c = 66$.

As you can see, there are many ways of solving a triangle; the method you choose will depend on the accuracy you want to obtain and the type of table or calculator you have. In the rest of this chapter, it is assumed that you have access to a calculator.

ANGLE OF DEPRESSION

The solution of right triangles is necessary in a variety of situations. The first one we'll consider concerns an observer looking at an object. The *angle of depression* is the acute angle measured down from the horizontal line to the line of sight, as shown in the figure.

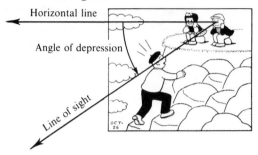

ANGLE OF ELEVATION

On the other hand, if we take the mountain climber's viewpoint and measure from the horizontal up to the line of sight, we call the angle the *angle of elevation*.

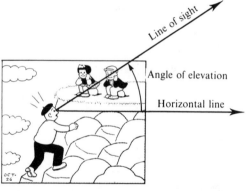

EXAMPLE 3: The angle of elevation of a tree from a point on the ground 42 m from its base is 33°. Find the height of the tree.

Solution: Let θ = angle of elevation and h = height of tree. Then

$$\tan \theta = \frac{h}{42}$$

$$h = 42 \tan 33°$$

$$\approx 42(0.6494)$$

$$\approx 27.28.$$

The tree is 27 m tall.

Notice that the answer is stated to two significant digits.

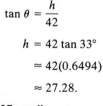

42

A second application of the solution of right triangles involves the *bearing* of a line, which is defined as an acute angle made with a north–south line. When giving the bearing of a line, we first write N or S to determine whether we measure the angle from the north or the south side of a point on the line. Then we give the measure of the angle followed by E or W, denoting which side of the north–south line we are measuring. Some examples are shown in Figure 7.5.

BEARING

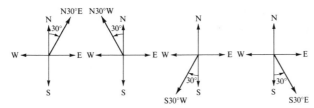

Figure 7.5 Bearing of an angle

THINK
METRIC

EXAMPLE 4: To find the width *AB* of a canyon, a surveyor measures 100 m from *A* in the direction of N42.6°W to locate point *C*. The surveyor then determines that the bearing of *CB* is N73.5°E. Find the width of the canyon if the point *B* is situated so that ∠*BAC* = 90.0°.

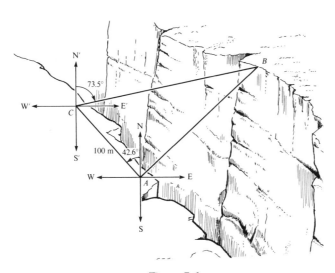

Figure 7.6

Solution: Let $\theta = \angle BCA$ in Figure 7.6. Then $\tan \theta = AB/AC$. So

$$AB = AC \tan \theta$$
$$= 100 \tan \theta.$$

$$\angle BCE' = 16.5° \quad \text{(complementary angles)}$$
$$\angle ACS' = 42.6° \quad \text{(alternate interior angles)}$$
$$\angle E'CA = 47.4° \quad \text{(complementary angles)}$$
$$\theta = \angle BCA = \angle BCE' + \angle E'CA$$
$$= 16.5° + 47.4°$$
$$= 63.9°$$

Thus,

$$AB = 100 \tan 63.9°$$
$$\approx 204.125.$$

Answer to three significant digits.

The canyon is 204 m across.

PROBLEM SET 7.5

A Problems

Solve the right triangles ($\gamma = 90°$) in Problems 1–20.

1. $a = 80; \beta = 60°$
2. $b = 37; \alpha = 69°$
3. $a = 68; b = 83$
4. $a = 29; \alpha = 76°$
5. $b = 13; \beta = 65°$
6. $a = 69; c = 73$
7. $b = 90; \beta = 13°$
8. $a = 49; \beta = 45°$
9. $a = 24; b = 29$
10. $b = 82; \alpha = 50°$
11. $c = 28.3; \alpha = 69.2°$
12. $c = 36; \alpha = 6°$
13. $\beta = 57.4°; a = 70.0$
14. $\alpha = 56.00°; b = 2350$
15. $\beta = 23°; a = 9000$
16. $b = 3100; c = 3500$
17. $\beta = 16.4°; b = 2580$
18. $\alpha = 42°; b = 350$
19. $b = 3200; c = 7700$
20. $b = 4100; c = 4300$

B Problems

21. The angle of elevation of a building from a point on the ground 30 m from its base is 38°. Find the height of the building.

22. The angle of elevation of the top of the Great Pyramid of Khufu (or Cheops) from a point on the ground 351 ft from a point directly below the top is 52.0°. Find the height of the pyramid.

23. From a cliff 150 m above the shoreline, the angle of depression of a ship is 37°. Find the distance of the ship from a point directly below the observer.

24. From a police helicopter flying at 1000 ft, a stolen car is sighted at an angle of depression of 71°. Find the distance of the car from a point directly below the helicopter.

25. To find the east–west boundary of a piece of land, a surveyor must divert his path from point C on the boundary by proceeding due south for 300 ft to a point A. Point B, which is due east of point C, is now found to be in the direction of N49°E from point A. What is the distance CB?

26. To find the distance across a river that runs east–west, a surveyor locates points P and Q on a north–south line on opposite sides of the river. She then paces out 150 ft from Q due east to a point R. Next she determines that the bearing of RP is N58°W. How far is it across the river?

27. A 16-ft ladder on level ground is leaning against a house. If the angle of elevation of the ladder is 52°, how far above the ground is the top of the ladder?

28. How far is the base of the ladder in Problem 27 from the house?

29. If the ladder in Problem 27 is moved so that the bottom is 9 ft from the house, what will be the angle of elevation?

30. Find the height of the Barrington Space Needle if the angle of elevation at 1000 ft from a point on the ground directly below the top is 58.15°.

31. The world's tallest chimney is the 5.50-million-dollar stack of the International Nickel Company. Find its height if the angle of elevation at 1000 ft from a point on the ground directly below the top of the stack is 51.36°.

32. a. The angle of elevation of a radio tower from a point on the ground 2.0 km from its base is 1.43°. How tall is the tower in meters?

    b. The angle of elevation of a radio tower from a point on the ground 2.0 miles from its base is 0.81°. How tall is the tower in feet?

    c. Do you think part a in meters or part b in feet was easier to work? Why?

33. To find the boundary of a piece of land, a surveyor must divert his path from a point A on the boundary for 500 ft in the direction S50°E. He then determines that the bearing of a point B located directly south of A is S40°W. Find the distance AB.

34. To find the distance across a river, a surveyor locates points P and Q on either side of the river. Next she measures 100 m from point Q in the direction S35°E to point R. Then she determines that point P is now in the direction of N25.0°E from point R and that $\angle PQR$ is a right angle. Find the distance across the river.

35. If the Empire State Building and the Sears Tower were situated 1000 ft apart, the angle of depression from the top of the Sears Tower to the top of the Empire State Building would be 11.53°, and the angle of depression to the foot of the Empire State Building would be 55.48°. Find the heights of the buildings.

36. On the top of the Empire State Building is a television tower. From a point 1000 ft from a point on the ground directly below the top of the tower the angle of elevation to the bottom of the tower is 51.34° and to the top of the tower is 55.81°. What is the length of the television tower?

37. Two horseback riders want to calculate the distance to a very, very exclusive restaurant (called point R). Their present location is point P. They determine that, if they travel S65.4°W for 250 ft to a new location N, the bearing of NR will be N53.7°E. How far is the restaurant from their present location if $\angle RPN$ is a right angle?

38. A wheel 5.00 ft in diameter rolls up a 15.0° incline. What is the height of the center of the wheel above the base of the incline after the wheel has completed one revolution?

39. What is the height of the center of the wheel in Problem 38 after three revolutions?

40. If the ridge in Figure 7.7 is to be 5.0 ft above point R, how far should the plate be placed from the R directly below the ridge so that the common rafter will have

"Very, very exclusive."

The *pitch* is the slope of the roof.

an angle of elevation of 14°? What will the pitch of the common rafter be if it forms this angle?

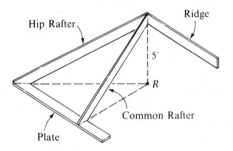

Figure 7.7 Typical roof construction

 41. Answer the questions posed in Problem 40 for an angle of elevation of 22.6°.

 42. If the distance from the earth to the sun is 92.9 million miles and the angle formed between Venus, the earth, and the sun (as shown in Figure 7.8) is 47.0°, find the distance from the sun to Venus.

Also see Problem 51.

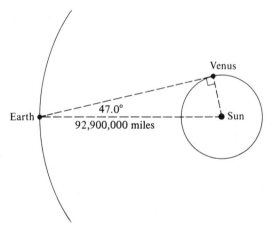

Figure 7.8

 43. Use the information in Problem 42 to find the distance from the earth to Venus.

 44. The largest ground area covered by any office building is that of the Pentagon in Arlington, Virginia. If the radius of the circumscribed circle is 783.5 ft, find the length of one side of the Pentagon.

 45. Use the information in Problem 44 to find the radius of the circle inscribed in the Pentagon.

 46. To determine the height of the building shown in Figure 7.9, we select a point *P* and find that the angle of elevation is 59.64°. We then move out a distance of 325.4 ft (on a level plane) to point *Q* and find that the angle of elevation is now 41.32°. Find the height *h* of the building.

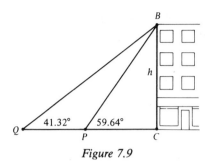

Figure 7.9

47. Using Figure 7.9, let the angle of elevation at P be α and at Q be β, and let the distance from P to Q be d. If h is the height of the building, show that

$$h = \frac{d \sin \alpha \sin \beta}{\sin (\alpha - \beta)}.$$

48. A 6.0-ft person is casting a shadow of 4.2 ft. What time is it if the sun rose at 6:15 A.M. and is directly overhead at 12:15 P.M.?
49. How long will the shadow of the person in Problem 48 be at 8:00 A.M.?
50. From the top of a tower 100 ft high, the angles of depression to two landmarks on the plane upon which the tower stands are 18.5° and 28.4°.
    a. Find the distance between the landmarks when they are on the same side of the tower.
    b. Find the distance between the landmarks when they are on opposite sides of the tower.

Mind Bogglers

51. Consult the article "Mathematical Astronomy," by Vincent J. Motto, in the February 1975 issue of the *Two Year College Mathematics Journal*. Calculate the distance from the earth to Mars, Jupiter, and Saturn.
52. Show that in every right triangle the value of c lies between $(a + b)/\sqrt{2}$ and $a + b$.
53. When curved beams are stored on top of one another, it is necessary to find the distance h when R and C are known (see Figure 7.10).
    a. Find an expression for h in terms of R and C.
    b. If $R = 1250$ ft and $C = 30.00$ ft, find h to the nearest one-hundredth of an inch.

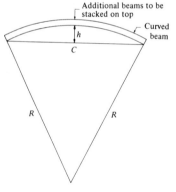

Figure 7.10 Diagram of stacked beams
This figure is not drawn to scale.

7.6 LAWS OF SINE AND COSINE

We will approach the solution of *oblique triangles* by studying the possible combinations of given information. In general, given three parts of a triangle, we will want to find the remaining three parts. But can we do so given *any* three parts? Let's list the possibilities.

Oblique triangles are triangles with no right angles.

𝕳istorical 𝕹ote

The transition from Renaissance mathematics to modern mathematics was aided in large part by the Frenchman François Viète (1540–1603). He was a lawyer who practiced mathematics as a hobby. In his book *Canon mathematicus* he solved oblique triangles by breaking them down into right triangles. He was probably the first person to develop and use the Law of Tangents.

1. SSS — We are given three sides.
2. SAS — We are given two sides and an included angle.
3. ASA or AAS — We are given two angles and a side.
4. SSA — We are given an angle and a side opposite the angle, as well as another side.
5. AAA — We are given three angles.

SSS

We'll consider these possibilities one at a time. For SSS, it is necessary for the sum of the lengths of the two smaller sides to be greater than the length of the largest side. If this is the case, a generalization of the Pythagorean Theorem, called the *Law of Cosines*, is used.

> In $\triangle ABC$ labeled as shown in Figure 7.11,
> $$c^2 = a^2 + b^2 - 2ab\cos\gamma.$$

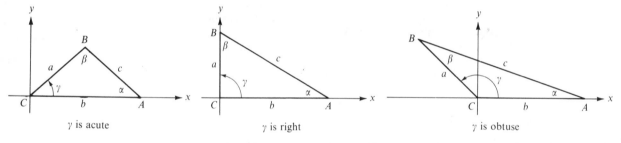

Figure 7.11 Examples of triangles

To prove the Law of Cosines, let C be an angle in standard position with A on the positive x-axis, as shown in Figure 7.11. The coordinates of the vertices are as follows.

$C(0,0)$ (since C is in standard position)

$A(b,0)$ (since A is on the x-axis a distance of b units from the origin)

$B(a\cos\gamma, a\sin\gamma)$ (Let $B(x,y)$; then by definition of the trigonometric functions, $\cos\gamma = x/a$, $\sin\gamma = y/a$, and, thus, $x = a\cos\gamma$ and $y = a\sin\gamma$)

Use the distance formula for the distance between $A(b,0)$ and $B(a\cos\gamma, a\sin\gamma)$.

$$c^2 = (a \cos \gamma - b)^2 + (a \sin \gamma - 0)^2$$
$$= a^2 \cos^2 \gamma - 2ab \cos \gamma + b^2 + a^2 \sin^2 \gamma$$
$$= a^2(\cos^2 \gamma + \sin^2 \gamma) + b^2 - 2ab \cos \gamma$$
$$= a^2 + b^2 - 2ab \cos \gamma$$

By letting A and B, respectively, be in standard position, it can also be shown that

$$a^2 = b^2 + c^2 - 2bc \cos \alpha,$$
$$b^2 = a^2 + c^2 - 2ac \cos \beta.$$

To find the angles when given three sides, solve for α, β, or γ.

LAW OF COSINES

$$a^2 = b^2 + c^2 - 2bc \cos \alpha \qquad \cos \alpha = \frac{b^2 + c^2 - a^2}{2bc}$$

$$b^2 = a^2 + c^2 - 2ac \cos \beta \qquad \cos \beta = \frac{a^2 + c^2 - b^2}{2ac}$$

$$c^2 = a^2 + b^2 - 2ab \cos \gamma \qquad \cos \gamma = \frac{a^2 + b^2 - c^2}{2ab}$$

Notice that for a right triangle, $\gamma = 90°$. This means that

$$c^2 = a^2 + b^2 - 2ab \cos 90°$$

or

$$c^2 = a^2 + b^2,$$

since $\cos 90° = 0$. But this last equation is simply the Pythagorean Theorem. Thus, the Law of Cosines is a generalization of the familar Pythagorean Theorem.

EXAMPLE 1: What is the smallest angle of a triangular patio with sides that measure 25, 18, and 21 ft?

Solution: If γ represents the smallest angle, then c (the side opposite γ) must be the smallest side, so $c = 18$. Then

$$\cos \gamma = \frac{a^2 + b^2 - c^2}{2ab}$$

$$= \frac{25^2 + 21^2 - 18^2}{2(25)(21)}.$$

$$\gamma = \cos^{-1}\left(\frac{25^2 + 21^2 - 18^2}{2(25)(21)}\right).$$

(1) By calculator with algebraic logic:

$$\boxed{25}\ \boxed{x^2}\ \boxed{+}\ \boxed{21}\ \boxed{x^2}\ \boxed{-}\ \boxed{18}\ \boxed{x^2}\ \boxed{=}\ \boxed{÷}\ \boxed{2}$$
$$\boxed{÷}\ \boxed{25}\ \boxed{÷}\ \boxed{21}\ \boxed{=}\ \boxed{INV}\ \boxed{COS} \approx 45.03565072$$

← Use this number and trigonometric tables if you have only a four-function calculator.

If you do not have a calculator and are using logarithms, you will find that the Law of Cosines will not be suitable for logarithmic calculations. However, notice that

$$\cos \gamma = \frac{a^2 + b^2 - c^2}{2ab}$$

$\cos \gamma + 1$

$$= \frac{a^2 + b^2 - c^2}{2ab} + 1$$

$$= \frac{a^2 + b^2 - c^2 + 2ab}{2ab}$$

$$= \frac{(a + b - c)(a + b + c)}{2ab}.$$

Thus,

$\cos \gamma + 1$

$$= \frac{(25 + 21 - 18)(25 + 21 + 18)}{2(25)(21)}$$

can be solved using logarithms.

If you do not have a calculator, use the Law of Tangents to solve this problem, since the Law of Cosines does not lend itself to logarithmic calculations. The Law of Tangents is discussed in Problems 30–32 of the problem set.

(2) By calculator with RPN logic:

| 25 | ENTER | × | 21 | ENTER | × | + | 18 |

| ENTER | × | − | 2 | ÷ | 25 | ÷ | 21 |

| ÷ | ARC | COS | ≈ 45.03565071

To two significant digits, the answer is 45°.

SAS

The second possibility listed for solving oblique triangles is that of being given two sides and an included angle. It is necessary that the given angle be less than 180°. Again we use the Law of Cosines for this possibility, as shown by Example 2.

EXAMPLE 2: Find c where $a = 52.0$, $b = 28.3$, and $\gamma = 28.5°$.

Solution: By the Law of Cosines:

$$c^2 = a^2 + b^2 - 2ab \cos \gamma$$
$$= (52.0)^2 + (28.3)^2 - 2(52.0)(28.3) \cos 28.5°.$$

By calculator:

$$c^2 \approx 918.355474$$
$$c \approx 30.30438044.$$

To three significant digits, $c = 30.3$.

AAA

Case 5 listed at the start of this section supposes that three angles are given. However, from what we know of similar triangles (see Figure 7.12), we

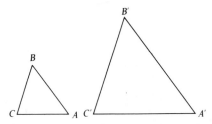

Figure 7.12 Similar triangles—similar triangles have the same shape but not necessarily the same size; that is, corresponding angles of similar triangles have equal measure

conclude that we cannot solve the triangle without knowing the length of at least one side.

ASA or AAS

Case 3 supposes that two angles and a side are given. For a triangle to be formed, the sum of the two given angles must be less than 180°, and the given side must be greater than zero. If two angles are known, you can easily find the third angle, since the sum of the three angles is 180°. The Law of Cosines is not sufficient in this case because at least two sides are needed.

Let's consider any oblique triangle, as shown in Figure 7.13.

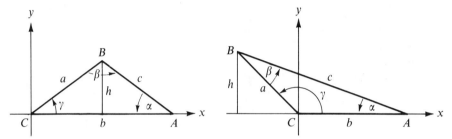

Figure 7.13   Oblique triangles

Let h = the height of the triangle with base CA. Then

$$\sin \alpha = \frac{h}{c} \quad \text{and} \quad \sin \gamma = \frac{h}{a}.$$

Solving for h, we get

$$h = c \sin \alpha \quad \text{and} \quad h = a \sin \gamma.$$

Thus,

$$c \sin \alpha = a \sin \gamma.$$

Dividing by ac, we find

$$\frac{\sin \alpha}{a} = \frac{\sin \gamma}{c}.$$

Repeat these steps for the height of the same triangle with base AB.   See Problems 27 and 28.

$$\frac{\sin \alpha}{a} = \frac{\sin \beta}{b}$$

This result is called the *Law of Sines*.

LAW OF SINES

The equation in the Law of Sines means that you can use any of the following pairs:

$$\frac{\sin \alpha}{a} = \frac{\sin \beta}{b},$$

$$\frac{\sin \alpha}{a} = \frac{\sin \gamma}{c},$$

$$\frac{\sin \beta}{b} = \frac{\sin \gamma}{c}.$$

In any $\triangle ABC$,

$$\frac{\sin \alpha}{a} = \frac{\sin \beta}{b} = \frac{\sin \gamma}{c}.$$

EXAMPLE 3: Solve the triangle in which $a = 20$, $\alpha = 38°$, and $\beta = 121°$.

Solution:
$\alpha = 38°$ (given)
$\beta = 121°$ (given)
$\gamma = 21°$ (since $\alpha + \beta + \gamma = 180°$, then $\gamma = 180° - 38° - 121° = 21°$)
$a = 20$ (given)
$b = 28$ (use the Law of Sines):

$$\frac{\sin 38°}{20} = \frac{\sin 121°}{b}; \text{ then}$$

$$b = \frac{20 \sin 121°}{\sin 38°}$$

Use tables or a calculator. $\longrightarrow$

Use logarithms or a calculator. $\longrightarrow$

$$\approx \frac{20(0.8572)}{0.6157}$$

≈ 27.85 Give answer to two significant digits.

$c = 12$ (use the Law of Sines):

$$\frac{\sin 38°}{20} = \frac{\sin 21°}{c}; \text{ then}$$

$$c \approx \frac{20 \sin 21°}{\sin 38°}$$

$$\approx 11.64$$

Notice that the answers are given to two significant digits.

SSA

The remaining case of solving oblique triangles is Case 4, in which two sides and an angle that is not an included angle are given. For convenience, we'll assume that sides a and b and angle α are given. Under what conditions will a triangle be formed? Let's consider each possibility separately.

1. Suppose that $\alpha \geq 90°$. There are two possibilities.
   i. $a \leq b$

   No triangle is formed (see Figure 7.14).

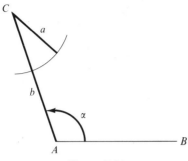

Figure 7.14

   ii. $a > b$

   One triangle is formed (see Figure 7.15). Use the Law of Sines.

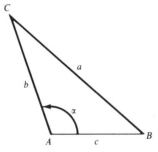

Figure 7.15

EXAMPLE 4: Let $a = 3.0$, $b = 2.0$, and $\alpha = 110°$.

Solution: $\dfrac{\sin \alpha}{a} = \dfrac{\sin \beta}{b}$

$\dfrac{\sin 110°}{3} = \dfrac{\sin \beta}{2}$

$\sin \beta = \dfrac{2}{3} \sin 110°$

$\approx \dfrac{2}{3}(0.9397)$

≈ 0.6265

$\beta \approx 38.79$

$\alpha = 110°$ (given)
$\beta = 39°$ (see work shown at left)
$\gamma = 31°$ ($\gamma = 180° - 110° - 39°$)
$a = 3.0$ (given)
$b = 2.0$ (given)
$c = 1.6$ (use the Law of Sines):

$\dfrac{\sin 110°}{3} = \dfrac{\sin \gamma}{c}$; then

$c = \dfrac{3 \sin \gamma}{\sin 110°}$

≈ 1.654

2. Suppose that $\alpha < 90°$. There are four possibilities (letting h be the altitude of $\triangle ABC$ drawn from C to AB).

   i. $a < h < b$
      No triangle is formed (see Figure 7.16).

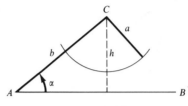

Figure 7.16

   ii. $a = h < b$
      A right triangle is formed (see Figure 7.17). Use the methods of the last section to solve the triangle.

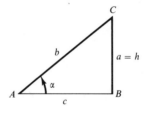

Figure 7.17

This situation is called the *ambiguous case* and is really the only special case you must watch for. All the other cases can be determined from the calculations without any special consideration.

   iii. $h < a < b$
      Two triangles are formed (see Figure 7.18).

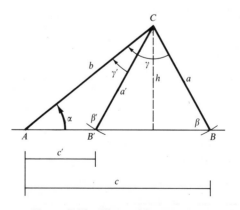

Figure 7.18   The ambiguous case

EXAMPLE 5: Let $a = 1.50$, $b = 2.00$, and $\alpha = 40.0°$.

Solution:

$$\frac{\sin \alpha}{a} = \frac{\sin \beta}{b}$$

$$\frac{\sin 40°}{1.5} = \frac{\sin \beta}{2}$$

$$\sin \beta = \frac{2 \sin 40°}{1.5}$$

$$\approx \frac{4}{3}(0.6428)$$

$$\approx 0.8570$$

$$\beta \approx 59.0°$$

But from Figure 7.18, we can see that this is only the acute-angle solution. For the obtuse-angle solution—call it β'—we find

$$\beta' = 180° - \beta$$
$$\approx 121°.$$

We finish the problem by working two calculations, which are presented side by side.

Solution 1:		Solution 2:	
$\alpha = 40.0°$	(given)	$\alpha = 40.0°$	(given)
$\beta = 59.0°$	(see above)	$\beta' = 121°$	(see above)
$\gamma = 81.0°$	($\gamma = 180° - \alpha - \beta$)	$\gamma' = 19.0°$	($\gamma' = 180° - \alpha - \beta'$)
$a = 1.50$	(given)	$a = 1.50$	(given)
$b = 2.00$	(given)	$b = 2.00$	(given)

$$c = 2.30 \quad \left(\frac{\sin \alpha}{a} = \frac{\sin \gamma}{c}; \right.$$

$$c = \frac{1.5 \sin \gamma}{\sin 40°}$$

$$\left. \approx 2.3049 \right)$$

$$c' = 0.76 \quad \left(\frac{\sin \alpha}{a} = \frac{\sin \gamma'}{c'}; \right.$$

$$c' = \frac{1.5 \sin \gamma'}{\sin 40°}$$

$$\left. \approx 0.7597 \right)$$

iv. $a \geq b$

One triangle is formed (see Figure 7.19).

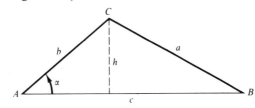

Figure 7.19

EXAMPLE 6: Let $a = 3.0$, $b = 2.0$, and $\alpha = 40°$.

Solution:

$$\frac{\sin \alpha}{a} = \frac{\sin \beta}{b}$$

$$\frac{\sin 40°}{3} = \frac{\sin \beta}{2}$$

$$\sin \beta = \frac{2}{3} \sin 40°$$

$$\approx \frac{2}{3}(0.6428)$$

$$\approx 0.4285$$

$$\beta \approx 25.374$$

$\alpha = 40°$ (given)
$\beta = 25°$ (see work shown at left)
$\gamma = 115°$ ($\gamma = 180° - \alpha - \beta$)
$a = 3.0$ (given)
$b = 2.0$ (given)

$c = 4.2$ $\left(\text{since } \sin \dfrac{40°}{3} = \sin \dfrac{\gamma}{c}\right.$

$c = \dfrac{3 \sin \gamma}{\sin 40°}$

$\left.\approx 4.2427\right)$

SUMMARY

The most important skill to be learned from this section is the ability to select the proper trigonometric law when given a particular problem. In the rest of this section, you will encounter applications of right triangles, the Law of Cosines, and the Law of Sines. A review of various types of problems may be helpful.

TO SOLVE A TRIANGLE *ABC*

Given	*Conditions on given information*	*Law to use for solution*
1. SSS	a. The sum of the lengths of the two smaller sides is less than or equal to the length of the larger side.	No solution
	b. The sum of the lengths of the two smaller sides is greater than the length of the larger side.	Law of Cosines
2. SAS	a. The angle is greater than or equal to 180°.	No solution
	b. The angle is less than 180°.	Law of Cosines
3. ASA or AAS	a. The sum of the angles is greater than or equal to 180°.	No solution.
	b. The sum of the angles is less than 180°.	Law of Sines
4. SSA	Let θ be the given angle with adjacent (adj) and opposite (opp) sides given; the height, h, is found by $$h = (\text{adj}) \sin \theta.$$	

Given	*Conditions on given information*	*Law to use for solution*
	a. $\theta \geq 90°$	
	i. opp $\leq$ adj	No solution
	ii. opp $>$ adj	Law of Sines
	b. $\theta < 90°$	
	i. opp $< h <$ adj	No solution
	ii. opp $= h <$ adj	Right-triangle solution
	iii. $h <$ opp $<$ adj	*Ambiguous case*: use the Law of Sines to find two solutions.
	iv. opp $\geq$ adj	Law of Sines
5. AAA		No solution

Remember, *when given two sides and an angle that is not an included angle, check to see whether one side is between the height of the triangle and the length of the other side. If so, there will be two solutions.*

EXAMPLE 7: On June 22, 1970, three friends were seated at a table exactly 200 ft and 285 ft, respectively, from the opposite ends of the world's longest bar, on Wharf Street in St. Louis, Missouri, as shown in Figure 7.20. They calculated the angle at their location formed by the lines drawn from either end of the bar to be 85.9°. How long is the bar?

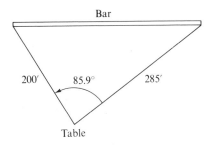

Figure 7.20

Solution: We are given SAS, so we use the Law of Cosines.

$$b^2 = a^2 + c^2 - 2ac \cos \beta$$
$$= 200^2 + 285^2 - 2(200)(285) \cos 85.9°$$
$$\approx 113{,}074 \quad \text{(by calculator)}$$
$$b \approx 336.2652$$

The length of the bar is 336 ft.

EXAMPLE 8: An airplane is 100 km N40°E of a loran station and is traveling due west at 240 kph. How long will it be (to the nearest minute) before the plane is 90 km from the loran station?

Solution: We are given SSA, so we suspect the ambiguous case.

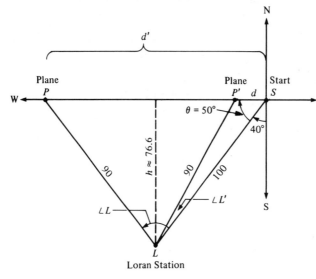

Let θ be angles $= 50°$;
100 km is the side adjacent to θ;
90 km is the side opposite to θ;

$$h = (\text{adj}) \sin \theta$$
$$= 100 \sin 50°$$
$$\approx 76.6$$
$$h < \text{opp} < \text{adj}$$
$$76.6 < 90 < 100,$$

which is the ambiguous case.

Figure 7.21

Solution I

$\angle S$: $50°$

$\angle P$: $\dfrac{\sin 50°}{90} = \dfrac{\sin P}{100}$

$$\sin P \approx 0.8512$$
$$P \approx 58.34°$$

$\angle L$: $L = 180° - S - P$
$$\approx 180° - 50° - 58.34°$$
$$\approx 71.66°$$

d: $\dfrac{\sin L}{d} = \dfrac{\sin 50°}{100}$

$$d = \dfrac{100 \sin L}{\sin 50°}$$
$$\approx 123.9$$

Solution II

$\angle S$: $50°$

$\angle P'$: $P' = 180° - P$
$$\approx 121.66°$$

$\angle L'$: $L' = 180° - S - P'$
$$\approx 180° - 50° - 121.66°$$
$$\approx 8.34°$$

d: $\dfrac{\sin L'}{d'} = \dfrac{\sin 50°}{100}$

$$d' = \dfrac{100 \sin L'}{\sin 50°}$$
$$\approx 18.93$$

To two significant figures, $d = 120$ km and 19 km; at 240 kph the plane would be 90 km from the loran station in about $\frac{1}{12}$ hr, or 5 min, and in $\frac{1}{2}$ hr, or 30 min.

PROBLEM SET 7.6

A Problems

Solve △ABC in Problems 1–12. If the triangle does not have a solution, state the reason.

1. $a = 7.0; b = 8.0; c = 2.0$
2. $a = 18, b = 25; \gamma = 30°$
3. $b = 14; c = 12; \alpha = 82°$
4. $a = 3.0; b = 4.0; \alpha = 125°$
5. $a = 5.0; b = 7.0; \alpha = 75°$
6. $a = 5.0; b = 4.0; \alpha = 125°$
7. $a = 5.0; b = 4.0; \alpha = 80°$
8. $a = 7.0; b = 9.0; \alpha = 52°$
9. $a = 10.2; b = 11.8; \alpha = 47°$
10. $a = 4.0; b = 5.0; \alpha = 56°$
11. $a = 4.5; b = 5.0; \alpha = 56°$
12. $a = 7.0; b = 5.0; \alpha = 56°$

B Problems

Solve △ABC in Problems 13–24. If the triangle does not have a solution, state the reason.

13. $a = 38; b = 41; c = 25$
14. $a = 45; b = 92; c = 41$
15. $a = 38.2; b = 14.8; \gamma = 48.2°$
16. $a = 41.0; \alpha = 45.2°; \beta = 21.5°$
17. $a = 26; b = 71; c = 88$
18. $\alpha = 48°; \beta = 105°; \gamma = 27°$
19. $a = 14.2; b = 16.3; \gamma = 35.0°$
20. $a = 14.2; b = 16.3; \gamma = 135.0°$
21. $\beta = 15.0°; \gamma = 18.0°; b = 23.5$
22. $b = 45.7; \alpha = 82.3°; \beta = 61.5°$
23. $b = 82.5; c = 52.2; \gamma = 32.1°$
24. $a = 151; b = 234; c = 416$
25. Prove that $a^2 = b^2 + c^2 - 2bc \cos \alpha$.
26. Prove that $b^2 = a^2 + c^2 - 2ac \cos \beta$.
27. Given $\triangle ABC$, show that $(\sin \alpha)/a = (\sin \beta)/b$.
28. Given $\triangle ABC$, show that $(\sin \beta)/b = (\sin \gamma)/c$.
29. Given $\triangle ABC$, show that $(\sin \alpha)/\sin \beta = a/b$.
30. Using Problem 29, show that

$$\frac{\sin \alpha - \sin \beta}{\sin \alpha + \sin \beta} = \frac{a - b}{a + b}.$$

31. Using Problem 30 and the formulas for the sum and difference of sines, show that

$$\frac{2 \cos \tfrac{1}{2}(\alpha + \beta) \sin \tfrac{1}{2}(\alpha - \beta)}{2 \sin \tfrac{1}{2}(\alpha + \beta) \cos \tfrac{1}{2}(\alpha - \beta)} = \frac{a - b}{a + b}.$$

32. On page 360, we mentioned the *Law of Tangents*, which is useful in logarithmic calculations. Using Problem 31, show that

$$\frac{\tan \tfrac{1}{2}(\alpha - \beta)}{\tan \tfrac{1}{2}(\alpha + \beta)} = \frac{a - b}{a + b}.$$

LAW OF TANGENTS

Other forms of the Law of Tangents are

$$\frac{\tan \tfrac{1}{2}(\alpha - \gamma)}{\tan \tfrac{1}{2}(\alpha + \gamma)} = \frac{a - c}{a + c}$$

and

$$\frac{\tan \tfrac{1}{2}(\beta - \gamma)}{\tan \tfrac{1}{2}(\beta + \gamma)} = \frac{b - c}{b + c}.$$

33. An artillery-gun observer must determine the distance to a target at point *T*. He knows that the target is 5.20 miles from point *I* on a nearby island. He also knows that he (at point *H*) is 4.30 miles from point *I*. If $\angle HIT$ is 68.4°, how far is he from the target?

34. A tree stands vertically on a hillside that makes an angle of 14° with the

horizontal. If the angle of elevation of the tree is 58 ft down the hill from the base is 52°, what is the height of the tree?

35. A buyer is interested in purchasing a triangular lot with vertices *LOT*, but unfortunately, the marker at point *L* has been lost. The deed indicates that *TO* is 453 ft and *LO* is 112 ft and that the angle at *L* is 82.6°. What is the distance from *L* to *T*?

36. A UFO is sighted by people in two cities 2.300 miles apart. The UFO is between and in the same vertical plane as the two cities. The angle of elevation of the UFO from the first city is 10.48° and from the second is 40.79°. At what altitude is the UFO flying? What is the actual distance of the UFO from each city?

37. At 500 ft in the direction that the Tower of Pisa is leaning, the angle of elevation is 20.24°. If the tower leans at an angle of 5.45° from the vertical, what is the length of the tower?

38. What is the angle of elevation of the leaning Tower of Pisa (described in Problem 37) if you measure from a point 500 ft in the direction exactly opposite from the way it is leaning?

39. From a blimp, the angle of depression to the top of the Eiffel Tower is 23.2° and to the bottom is 64.6°. After flying over the tower at the same height and at a distance of 1000 ft from the first location, you determine that the angle of depression to the top of the tower is now 31.4°. What is the height of the Eiffel Tower given that these measurements are in the same vertical plane?

40. The world's longest deepwater jetty is at Le Havre, France. Since access to the jetty is restricted, it was necessary for me to calculate its length by noting that it forms an angle of 85.0° with the shoreline. After pacing out 1000 ft along the line making an 85.0° angle with the jetty, I calculated the angle to the end of the jetty to be 83.6°. What is the length of the jetty?

41. The world's longest pier is at Hasa, Saudi Arabia. The length of the pier can be determined by measuring out a distance of 1.00 mile along a line on shore that makes an angle of 83.4° with the foot of the pier. At that point, the angle to the end of the pier is 88.05°. How long is the pier?

42. A level lot has dimensions as shown in Figure 7.22. If 1 acre = 43,560 sq ft, what is the total cost of treating the area for poison oak if the fee is $25 per acre?

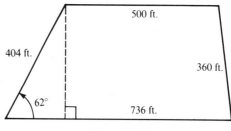

Figure 7.22

43. Show that the area *K* of $\triangle ABC$ can be found by

$$K = \tfrac{1}{2}bc \sin \alpha = \tfrac{1}{2}ac \sin \beta = \tfrac{1}{2}ab \sin \gamma$$

when two sides and an included angle are known.

44. Show that the area K of $\triangle ABC$ can be found by

$$K = \frac{a^2 \sin \beta \sin \gamma}{2 \sin \alpha}$$

when three angles and one side are known.

45. Use Problem 43 and the Law of Cosines to show that the area K of a triangle can be found by

$$K = \sqrt{s(s - a)(s - b)(s - c)}$$

when three sides are known and $s = \frac{1}{2}(a + b + c)$.

This equation is known as *Heron's*, or *Hero's*, *Formula*.

Find the area of each triangle in Problems 46–49.

46. $a = 14.2; b = 16.3; \gamma = 35.0°$
47. $B = 15.0°; C = 18.0°; b = 23.5$
48. $b = 82.5; c = 52.2; \alpha = 32.1°$
49. $a = 124; b = 325; c = 351$

Mind Bogglers

50. *Newton's Formula* involves all six parts of a triangle. It is not useful in solving a triangle, but it is helpful in checking results. Show that

$$\frac{a + b}{c} = \frac{\cos \frac{1}{2}(\alpha - \beta)}{\sin \frac{1}{2}\gamma}.$$

51. Show that the radius R of a circumscribed circle of $\triangle ABC$ satisfies the equations

$$R = \frac{a}{2 \sin \alpha} = \frac{b}{2 \sin \beta} = \frac{c}{2 \sin \gamma}.$$

52. Show that the radius r of an inscribed circle of $\triangle ABC$ satisfies the equation

$$r = \sqrt{\frac{(s - a)(s - b)(s - c)}{s}},$$

where $s = \frac{1}{2}(a + b + c)$.

7.7 SUMMARY AND REVIEW

TERMS

AAA (7.6)
AAS (7.6)
Ambiguous case (7.6)
Angle of depression (7.5)
Angle of elevation (7.5)
ASA (7.6)
Bearing of a line (7.5)
Conditional equation (7.1)
Contradiction (7.1)
Equation (7.1)

Fundamental Identities (7.1)
Identity (7.1)
Law of Cosines (7.6)
Law of Sines (7.6)
Oblique triangle (7.6)
SAS (7.6)
Solving a triangle (7.5)
SSA (7.6)
SSS (7.6)
Variable equation (7.1)

IMPORTANT IDEAS

(7.1) Eight Fundamental Identities
      Reciprocal identities

$$(1) \quad \sec \theta = \frac{1}{\cos \theta} \qquad (2) \quad \csc \theta = \frac{1}{\sin \theta} \qquad (3) \quad \cot \theta = \frac{1}{\tan \theta}$$

Ratio identities

$$(4) \quad \tan \theta = \frac{\sin \theta}{\cos \theta} \qquad (5) \quad \cot \theta = \frac{\cos \theta}{\sin \theta}$$

Pythagorean identities

$$(6) \quad \sin^2 \theta + \cos^2 \theta = 1 \qquad (7) \quad 1 + \tan^2 \theta = \sec^2 \theta$$
$$(8) \quad 1 + \cot^2 \theta = \csc^2 \theta$$

(7.2) Proving identities

(7.3) Cofunction identities

$$(9) \quad \cos (\pi/2 - \theta) = \sin \theta \qquad (10) \quad \sin (\pi/2 - \theta) = \cos \theta$$
$$(11) \quad \tan (\pi/2 - \theta) = \cot \theta$$

(7.3) Opposite angle identities

$$(12) \quad \cos (-\theta) = \cos \theta \qquad (13) \quad \sin (-\theta) = -\sin \theta \qquad (14) \quad \tan (-\theta) = -\tan \theta$$

(7.3) Sum and difference identities

$$(15) \quad \cos (\alpha + \beta) = \cos \alpha \cos \beta - \sin \alpha \sin \beta$$

$$(16) \quad \cos (\alpha - \beta) = \cos \alpha \cos \beta + \sin \alpha \sin \beta$$

$$(17) \quad \sin (\alpha + \beta) = \sin \alpha \cos \beta + \cos \alpha \sin \beta$$

$$(18) \quad \sin (\alpha - \beta) = \sin \alpha \cos \beta - \cos \alpha \sin \beta$$

$$(19) \quad \tan (\alpha + \beta) = \frac{\tan \alpha + \tan \beta}{1 - \tan \alpha \tan \beta}$$

$$(20) \quad \tan (\alpha - \beta) = \frac{\tan \alpha - \tan \beta}{1 + \tan \alpha \tan \beta}$$

(7.3) Double-angle identities

$$(21) \quad \cos 2\theta = \cos^2 \theta - \sin^2 \theta \qquad (22) \quad \sin 2\theta = 2 \sin \theta \cos \theta$$
$$= 2 \cos^2 \theta - 1$$
$$= 1 - 2 \sin^2 \theta$$

$$(23) \quad \tan 2\theta = \frac{2 \tan \theta}{1 - \tan^2 \theta}$$

(7.3) Half-angle identities

$$(24) \quad \cos \frac{1}{2} \theta = \pm \sqrt{\frac{1 + \cos \theta}{2}}$$

(25) $\sin \dfrac{1}{2}\theta = \pm \sqrt{\dfrac{1-\cos\theta}{2}}$

(26) $\tan \dfrac{1}{2}\theta = \dfrac{1-\cos\theta}{\sin\theta}$

$$= \dfrac{\sin\theta}{1+\cos\theta}$$

(7.3)   Product identities

(27) $2\cos\alpha\cos\beta = \cos(\alpha-\beta) + \cos(\alpha+\beta)$

(28) $2\sin\alpha\sin\beta = \cos(\alpha-\beta) - \cos(\alpha+\beta)$

(29) $2\sin\alpha\cos\beta = \sin(\alpha+\beta) + \sin(\alpha-\beta)$

(30) $2\cos\alpha\sin\beta = \sin(\alpha+\beta) - \sin(\alpha-\beta)$

(7.3)   Sum identities

(31) $\cos\alpha + \cos\beta = 2\cos\left(\dfrac{\alpha+\beta}{2}\right)\cos\left(\dfrac{\alpha-\beta}{2}\right)$

(32) $\cos\alpha - \cos\beta = -2\sin\left(\dfrac{\alpha+\beta}{2}\right)\sin\left(\dfrac{\alpha-\beta}{2}\right)$

(33) $\sin\alpha + \sin\beta = 2\sin\left(\dfrac{\alpha+\beta}{2}\right)\cos\left(\dfrac{\alpha-\beta}{2}\right)$

(34) $\sin\alpha - \sin\beta = 2\sin\left(\dfrac{\alpha-\beta}{2}\right)\cos\left(\dfrac{\alpha+\beta}{2}\right)$

(7.4)   Solving trigonometric equations

(7.5)   If θ is an acute angle in a right triangle, then

$$\sin\theta = \frac{\text{opposite side}}{\text{hypotenuse}}; \qquad \cos\theta = \frac{\text{adjacent side}}{\text{hypotenuse}}; \qquad \tan\theta = \frac{\text{opposite side}}{\text{adjacent side}}.$$

(7.5, 7.6)   Solving triangles

(7.6)   Law of Cosines

$$a^2 = b^2 + c^2 - 2bc\cos\alpha \qquad \cos\alpha = \frac{b^2 + c^2 - a^2}{2bc}$$

$$b^2 = a^2 + c^2 - 2ac\cos\beta \qquad \cos\beta = \frac{a^2 + c^2 - b^2}{2ac}$$

$$c^2 = a^2 + b^2 - 2ab\cos\gamma \qquad \cos\gamma = \frac{a^2 + b^2 - c^2}{2ab}$$

(7.6)   Law of Sines

$$\frac{\sin\alpha}{a} = \frac{\sin\beta}{b} = \frac{\sin\gamma}{c}$$

(7.6)   When given two sides and an angle that is not an included angle, check to see whether one side is between the height of the triangle and the length of the other side. If so, there will be two solutions.

(7.6)   Summary for solving a triangle (see pages 366–367).

REVIEW PROBLEMS

1. (7.1)   State and prove the eight fundamental identities from memory.
2. (7.1)   Find the other circular functions so that $\sin \delta = 3/5$ when $\tan \delta < 0$.
3. (7.2)   Prove that

$$\frac{1 + \tan^2 \theta}{\csc \theta} = \sec \theta \tan \theta.$$

4. (7.2)   Prove that

$$\frac{\cos \theta}{\sec \theta} - \frac{\sin \theta}{\cot \theta} = \frac{\cos \theta \cot \theta - \tan \theta}{\csc \theta}.$$

5. (7.3)   Write $\sin (x + h) - \sin x$ as a product.
6. (7.3)   Prove that

$$\frac{\sin 5\theta + \sin 3\theta}{\cos 5\theta - \cos 3\theta} = -\cot \theta.$$

7. (7.4)   Solve the equations for $0 \le \theta < 2\pi$.
   a. $2 \cos^2 \theta - \sin^2 \theta = 2$
   b. $\sin \theta \cos 2\theta - \cos \theta \sin 2\theta = 1$
   c. $4 \cos^2 2\theta + 4 \sin 2\theta = 0$

8. (7.5)   To measure the span of the Rainbow Bridge in Utah, a surveyor selected two points, P and Q, on either end of the bridge. From point Q, the surveyor measured 500 ft in the direction N38.4°E to point R. Point P was then determined to be in the direction S67.5°W. What is the span of the Rainbow Bridge if all the preceding measurements are in the same plane and $\angle PQR$ is a right angle?

9. (7.6)   A mine shaft is dug into the side of a sloping hill. The shaft is dug horizontally for 485 ft. Next, a turn is made so that the angle of elevation of the second shaft is 58.0°, forming a 58.0° angle between the shafts. The shaft is then continued for 382 ft before exiting, as shown in Figure 7.23. How far is it along a straight line from the entrance to the exit, assuming that all tunnels are in a single plane? If the slope of the hill follows the line from the entrance to the exit, what is the angle of elevation from the entrance to the exit?

10. (7.6)   Ferndale is 7 miles N50°E of Fortuna. If I leave Fortuna at noon and travel due east at 2 mph, when will I be exactly 6 miles from Ferndale?

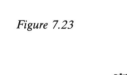

Figure 7.23

Sequences, Series, Induction, and the Binomial Theorem

8.1 ARITHMETIC AND GEOMETRIC SEQUENCES

Have you ever taken an IQ test in which you were asked to fill in the blanks to complete a given pattern? For example, try to fill in each blank space below with the result of a pattern you can see. There may be more than one answer, but you should be able to defend yours.

QUIZ

A. 1. $1, 4, 7, 10, 13,$ ____
    2. $20, 14, 8, 2, -4, -10,$ ____
    3. $a, a + d, a + 2d, a + 3d, a + 4d,$ ____
B. 4. $2, 4, 8, 16, 32,$ ____
    5. $10, 5, \frac{5}{2}, \frac{5}{4}, \frac{5}{8},$ ____
    6. $a, ar, ar^2, ar^3, ar^4,$ ____
C. 7. $1, 2, 1, 1, 2, 1, 1,$ ____
    8. $1, 1, 2, 3, 5, 8, 13,$ ____
    9. $1, 2, 4, 7, 11, 16,$ ____
    10. $1, 3, 4, 7, 11, 18,$ ____

Were some of these easier to figure out than others? The reason for the variation in difficulty of working the problems is the difference in the ways the patterns are constructed.

The easiest patterns to discover are the special types of functions called *sequences* or *progressions*.

INFINITE SEQUENCE
Remember that the set of counting numbers is

$N = \{1, 2, 3, \ldots, n, \ldots\}.$

DEFINITION: An *infinite sequence* is a function with a domain that consists of the set of counting numbers.

For convenience, we sometimes refer to an infinite sequence simply as a *sequence* or *progression*.

FINITE SEQUENCE
Sometimes we talk of a *finite sequence*, which means that we are considering as the domain the finite set

$$\{1, 2, 3, \ldots, n - 1, n\},$$

for some natural number n.

ARITHMETIC SEQUENCE
We will study two special types of sequences in this chapter. The first is an *arithmetic sequence*; this type is illustrated by the first three problems in the quiz at the beginning of this section. In an arithmetic sequence, there is a common difference between successive terms. That is, if any term is subtracted from the next term, the result is always the same, and this number is called

COMMON DIFFERENCE
the *common difference*.

Look for the common difference in Example 1.

$$4 - 1 = 3$$
$$7 - 4 = 3$$
$$10 - 7 = 3$$
$$13 - 10 = 3$$

The common difference is 3.

Thus:
$$x - 13 = 3 \quad \text{(where } x \text{ is the missing term)}$$
$$x = 16 \leftarrow \text{The next term}$$

In Example 2, the common difference is -6, so

$$x - (-10) = -6,$$

where x is the missing term. Thus,

$$x = -16$$

is the next term. In Example 3, the common difference is d and the next term is $a + 5d$.

Problems 4–6 of the opening quiz are examples of *geometric sequences.* In a geometric sequence, there is a common ratio between successive terms. If any term is divided into the next term, the result is always the same, and this number is called the *common ratio*. Look at Example 4.

$$\frac{4}{2} = 2$$

$$\frac{8}{4} = 2$$

$$\frac{16}{8} = 2$$

$$\frac{32}{16} = 2$$

The common ratio is 2.

$$\frac{x}{32} = 2 \quad \text{(where } x \text{ is the next term)}$$

Thus:
$$x = 64 \leftarrow \text{The next term}$$

In Example 5, the common ratio is $\frac{1}{2}$, so

$$\frac{x}{5/8} = \frac{1}{2},$$

EXAMPLES

1. $1, 4, 7, 10, 13,$ ____
2. $20, 14, 8, 2, -4, -10,$ ____
3. $a, a + d, a + 2d,$
   $\quad a + 3d, a + 4d,$ ____

GEOMETRIC SEQUENCE

COMMON RATIO

EXAMPLES

4. $2, 4, 8, 16, 32,$ ____
5. $10, 5, \frac{5}{2}, \frac{5}{4}, \frac{5}{8},$ ____
6. $a, ar, ar^2, ar^3, ar^4,$ ____

where x is the missing term. Thus,

$$x = \frac{5}{16}$$

is the next term. In Example 6, the common ratio is r and the next term is ar^5.

7. $1, 2, 1, 1, 2, 1, 1,$ ____
8. $1, 1, 2, 3, 5, 8, 13,$ ____
9. $1, 2, 4, 7, 11, 16,$ ____
10. $1, 3, 4, 7, 11, 18,$ ____

Problems 7–10 of the quiz illustrate patterns that are neither arithmetic nor geometric sequences. You can verify this by trying to find a common difference or a common ratio for each. Formal methods for these other types of patterns will not be developed, but rather our attention will be focused on arithmetic and geometric sequences.

A new notation is usually used when working with sequences. Remember that the domain is the set of counting numbers, so a sequence could be defined by

$$s(n) = 3n - 2, \quad \text{where } n \in \{1, 2, 3, \ldots\}.$$

Thus:

$$s(1) = 3(1) - 2 = 1$$
$$s(2) = 3(2) - 2 = 4$$
$$s(3) = 3(3) - 2 = 7$$
$$\vdots$$

However, instead of writing $s(1)$, the notation s_1 is used; in place of $s(2)$, $s_2; \ldots$; in place of $s(n)$, s_n. Thus, s_{15} means the 15th term of the sequence, and is found in the same fashion as though the notation $s(15)$ were used:

$$s_{15} = 3(15) - 2$$
$$= 43.$$

EXAMPLES: Find the first four terms of the sequence with the given nth term.

11. $s_n = 26 - 6n$

Solution: $s_1 = 26 - 6(1) = 20$
$s_2 = 26 - 6(2) = 14$
$s_3 = 26 - 6(3) = 8$
$s_4 = 26 - 6(4) = 2$

The sequence is $20, 14, 8, 2, \ldots$.

12. $s_n = (-1)^n n^2$

Solution: $s_1 = (-1)^1(1)^2 = -1$
$s_2 = (-1)^2(2)^2 = 4$
$s_3 = (-1)^3(3)^2 = -9$
$s_4 = (-1)^4(4)^2 = 16$

The sequence is $-1, 4, -9, 16, \ldots$.

13. $s_n = s_{n-1} + s_{n-2}$, $n > 3$, where $s_1 = 1$ and $s_2 = 2$

   Solution: $s_1 = 1$ (given)
   $s_2 = 2$ (given)
   $s_3 = s_2 + s_1$
       $= 2 + 1$ (by substitution)
       $= 3$
   $s_4 = s_3 + s_2$
       $= 3 + 2$
       $= 5$

   The sequence is $1, 2, 3, 5, \ldots$.

14. $s_n = 2n$

   Solution: $s_1 = 2$, $s_2 = 4$, $s_3 = 6$, $s_4 = 8$

   The sequence is $2, 4, 6, 8, \ldots$.

15. $s_n = 2n + (n - 1)(n - 2)(n - 3)(n - 4)$

   Solution: $s_1 = 2(1) + 0 = 2$
   $s_2 = 2(2) + 0 = 4$
   $s_3 = 2(3) + 0 = 6$
   $s_4 = 2(4) + 0 = 8$

   The sequence is $2, 4, 6, 8, \ldots$.

Notice from Examples 14 and 15 that if only a finite number of successive terms is known and no general term is given, then a *unique* general term cannot be given. That is, if we are given the sequence

$$2, 4, 6, 8, \underline{\quad},$$

the next term is probably 10 (if we are thinking of the general term of Example 14), but it *may* be something different. In Example 15, $s_1 = 2$, $s_2 = 4$, $s_3 = 6$, $s_4 = 8$, and

$$s_5 = 2(5) + (5 - 1)(5 - 2)(5 - 3)(5 - 4)$$
$$= 10 + (4)(3)(2)(1)$$
$$= 34.$$

In general, we are looking for the simplest general term; nevertheless, we must remember that answers are not unique *unless we are given the general term.*

PROBLEM SET 8.1

A Problems

Classify the sequences in Problems 1–15 as arithmetic, geometric, or neither, and supply the missing term.

The answers to Problems 1–15 are not unique.

1. $2, 5, 8, 11, 14, \underline{\quad}$
2. $1, 2, 1, 1, 2, 1, 1, 1, 2, 1, 1, 1, 1, \underline{\quad}$
3. $3, 6, 12, 24, 48, \underline{\quad}$

4. $5, -15, 45, -135, 405,$ ____

5. $100, 99, 97, 94, 90,$ ____

6. $1, 1, 2, 3, 5, 8, 13,$ ____

7. $p, pq, pq^2, pq^3, pq^4,$ ____

8. $97, 86, 75, 64,$ ____

9. $8, 12, 18, 26,$ ____

10. $5^5, 5^4, 5^3, 5^2,$ ____

11. $2, 5, 2, 5, 5, 2, 5, 5, 5,$ ____

12. $5, -5, -15, -25, -35,$ ____

13. $1, \ ^1/_2, \ ^1/_3, \ ^2/_3, \ ^1/_4, \ ^3/_4, \ ^1/_5, \ ^2/_5, \ ^3/_5, \ ^4/_5, \ ^1/_6,$ ____

14. $^4/_3, 2, 3, 4^1/_2,$ ____

15. $1, 8, 27, 64, 125,$ ____

Find the first three terms of the sequence with the nth term given in Problems 16–27.

16. $s_n = 4n - 3$

17. $s_n = -3 + 3n$

18. $s_n = \dfrac{10}{2^{n-1}}$

19. $s_n = a + nd$

20. $s_n = ar^{n-1}$

21. $s_n = \dfrac{n-1}{n+1}$

22. $s_n = (-1)^n$

23. $s_n = (-1)^n (n + 1)$

24. $s_n = 1 + \dfrac{1}{n}$

25. $s_n = \dfrac{n+1}{n}$

26. $s_n = 2$

27. $s_n = -5$

B Problems

28. Find the 15th term of the sequence in Problem 16.

29. Find the 102nd term of the sequence in Problem 17.

30. Find the 10th term of the sequence in Problem 18.

31. Find the 20th term of the sequence in Problem 23.

32. Find the 3rd term of the sequence $(-1)^{n+1}5^{n+1}$.

33. Find the 2nd term of the sequence $(-1)^{n-1}7^{n-1}$.

34. Find the first five terms of the sequence where $s_1 = 2$ and $s_n = 3s_{n-1}, n \geq 2$.

35. Find the first five terms of the sequence where $s_1 = 3$ and $s_n = \frac{1}{3}s_{n-1}, n \geq 2$.

The sequence expressed by Problem 36 is called the Fibonacci sequence.

36. Find the first five terms of the sequence where $s_1 = 1$, $s_2 = 1$, and $s_n = s_{n-1} + s_{n-2}, n \geq 3$.

37. Find the first five terms of the sequence where $s_1 = 1$, $s_2 = 2$, and $s_n = s_{n-1} + s_{n-2}, n \geq 3$.

Mind Bogglers

Find the next term for the sequences in Problems 38–40.

38. $1, 3, 4, 7, 11, 18, 29,$ ____

39. $225, 625, 1225, 2025,$ ____
40. $8, 5, 4, 9, 1,$ ____

8.2 ARITHMETIC SERIES

What is the total number of blocks in the diagram shown in Figure 8.1? There are ten blocks on the bottom row, nine on the next row, ..., and one on the top row. Thus, the number of blocks in each row is a term in the arithmetic sequence

$$1, 2, 3, 4, 5, 6, 7, 8, 9, 10,$$

and the total number of blocks is the sum

$$1 + 2 + 3 + 4 + 5 + 6 + 7 + 8 + 9 + 10.$$

The illustrated sum is called an *arithmetic series*.

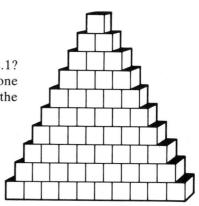

Figure 8.1

DEFINITION: The indicated sum of the terms of a finite sequence $s_1, s_2, s_3, \ldots, s_n$ is called a *series* and is denoted by

$$S_n = s_1 + s_2 + s_3 + \cdots + s_n.$$

SERIES

To find the number of blocks in Figure 8.1, consider the series

$$1 + 2 + 3 + \cdots + 9 + 10 = 55.$$

However, if the bottom row of blocks contained 100 blocks, direct addition would be very tedious, so let's investigate some other method for finding the sum of a series.

For an arithmetic sequence $s_1, s_2, s_3, s_4, \ldots, s_n$, there is a common difference d found by

$$s_n - s_{n-1} = d$$

for every $n > 1$. If we write this as

$$s_n = s_{n-1} + d$$

and let $s_1 = a$, then

$$
\begin{aligned}
s_2 &= s_1 + d \\
&= a + d \\
s_3 &= s_2 + d \\
&= (a + d) + d \\
&= a + 2d
\end{aligned}
$$

$$s_4 = s_3 + d$$
$$= (a + 2d) + d$$
$$= a + 3d$$
$$\vdots$$

The formula

$$s_n = a + (n - 1)d \longrightarrow s_n = a + (n - 1)d.$$

gives us the nth term when the first term and common difference are known.

Now we can find the sum, S_n, of the first n terms of an arithmetic sequence.

$$S_n = s_1 + s_2 + s_3 + \cdots + s_{n-2} + s_{n-1} + s_n$$
$$S_n = s_n + s_{n-1} + s_{n-2} + \cdots + s_3 + s_2 + s_1$$

We have simply reversed the order of the terms.

Add the terms of these equations to obtain

$$2S_n = (s_1 + s_n) + (s_2 + s_{n-1}) + (s_3 + s_{n-2})$$
$$+ \cdots + (s_{n-2} + s_3) + (s_{n-1} + s_2) + (s_n + s_1).$$

Now, notice that all the quantities within parentheses are equal to one another.

$$s_1 + s_n = a + [a + (n - 1)d]$$
$$= 2a + (n - 1)d$$
$$s_2 + s_{n-1} = (a + d) + [a + (n - 2)d]$$
$$= 2a + (n - 1)d \quad \text{Notice that this is the same as } s_1 + s_n.$$
$$s_3 + s_{n-2} = (a + 2d) + [a + (n - 3)d]$$
$$= 2a + (n - 1)d \quad \text{Aha! We get the same result again.}$$
$$\vdots$$

There are a total of n such sums, so

$$2S_n = n[2a + (n - 1)d]$$

$$S_n = \frac{n}{2}\left[2a + (n - 1)d\right].$$

Or, replacing $2a + (n - 1)d = a + a + [(n - 1)d]$ by $s_1 + s_n$, we have

Notice that

$$S_n = \frac{n}{2}(s_1 + s_n) \longrightarrow S_n = \frac{n}{2}(s_1 + s_n).$$

is n times the average of the first and last terms.

For the question of the number of blocks in Figure 8.1, we have

$$S_{10} = 1 + 2 + 3 + \cdots + 9 + 10$$

first last
term term
$$\downarrow \quad \downarrow$$
$$S_n = n\left(\frac{s_1 + s_n}{2}\right)$$
$$\underbrace{\qquad\qquad}$$
average of
first and
last terms

first   last
term   term
$$= 10\left(\frac{1 + 10}{2}\right)$$
↑
└number of terms

$$= 10\left(\frac{11}{2}\right)$$

$$= 55.$$

If the bottom row contained 100 blocks, then the total number of blocks, S_{100}, could be found by

$$S_{100} = 100\left(\frac{1 + 100}{2}\right)$$

$$= 50(101)$$

$$= 5050.$$

TO SUMMARIZE: If $s_1, s_2, s_3, \ldots, s_n, \ldots,$ is a finite arithmetic sequence, then

$$s_n = s_1 + (n - 1)d,$$

where d is the common difference, and

$$S_n = n\left(\frac{s_1 + s_n}{2}\right) \qquad \text{or} \qquad S_n = \frac{n}{2}[2s_1 + (n - 1)d].$$

SUM OF AN
ARITHMETIC SERIES

EXAMPLES

1. Find the sum of the first 100 even integers.
    Solution: The finite sequence is $2, 4, 6, \ldots, 200$; $s_1 = 2$, $s_{100} = 200$, and

    $$S_{100} = 100\left(\frac{2 + 200}{2}\right)$$

    $$= 100(101)$$

    $$= 10,100.$$

    Alternate Solution: You don't have to find the last term, 200; since $n = 100$, $a = 2$, and $d = 2$, you can use

    $$S_n = \frac{n}{2}[2a + (n - 1)d]$$

    $$S_{100} = \frac{100}{2}[2(2) + (99)(2)]$$

    $$= 50(4 + 198)$$

    $$= 10,100.$$

2. Find the sum of the first 50 terms of the arithmetic sequence with first term -10 and common difference 4.

Solution: $a = -10$, $n = 50$, and $d = 4$:

$$S_n = \frac{n}{2}[2a + (n - 1)d]$$

$$S_{50} = \frac{50}{2}[2(-10) + 49(4)]$$

$$= 25(-20 + 196)$$

$$= 4400.$$

PROBLEM SET 8.2

A Problems

The symbol s_n represents the general term.

Use the formula $s_n = a + (n - 1)d$ to find an expression for the general term of each arithmetic sequence.

EXAMPLE 3: $18, 14, 10, 6, \ldots$

Solution: $a = 18$, $d = 14 - 18$
$$= -4$$
Thus,

$$s_n = 18 + (n - 1)(-4)$$
$$= 18 - 4n + 4$$
$$= 22 - 4n.$$

1. $6, 11, 16, \ldots$ 2. $35, 46, 57, \ldots$
3. $-8, -1, 6, \ldots$ 4. $-1, 1, 3, \ldots$
5. $x, 2x, 3x, \ldots$ 6. $x - 5a, x - 3a, x - a$

Write out the first four terms of the arithmetic sequences in Problems 7–12 with first term a and common difference d. Also, write the general term.

7. $a = 5, d = 4$ 8. $a = 85, d = 3$
9. $a = 100, d = -5$ 10. $a = 20, d = -4$
11. $a = 5, d = x$ 12. $a = x, d = y$

B Problems

Find the indicated quantities for the given arithmetic sequences in Problems 13–20.

13. $a = 6, d = 5; s_{20}$ 14. $s_1 = 35, d = 11; s_{10}$

15. $a = 35$, $d = 11$; S_{10} 16. $a = -7$, $d = -2$; S_{100}
17. $a = -5$, $s_{30} = -63$; d 18. $s_1 = 4$, $s_6 = 24$; d
19. $s_1 = 4$, $s_6 = 24$; S_{15} 20. $a = -5$, $s_{30} = -63$; S_{10}
21. Find the sum of the first 20 terms of the arithmetic sequence with first term 100 and common difference 50.
22. Find the sum of the first 50 terms of the arithmetic sequence with first term -15 and common difference 5.
23. Find the sum of the even integers between 41 and 99.
24. Find the sum of the odd integers between 100 and 80.
25. Find the sum of the odd integers between 48 and 136.
26. Find the sum of the first n odd integers.
27. Find the sum of the first n even integers.
28. How many blocks are in a stack like Figure 8.1 where the bottom row has 28 blocks?
29. How many blocks are in a stack like Figure 8.1 where the bottom row has 87 blocks?
30. A sequence $s_1, s_2, \ldots, s_n$ is a *harmonic sequence* if its reciprocals form an arithmetic sequence. Which of the following are harmonic sequences?   HARMONIC SEQUENCE
    a. $1, {}^1/_2, {}^1/_3, {}^1/_4, {}^1/_5, \ldots$ b. ${}^1/_2, {}^1/_5, {}^1/_8, {}^1/_{11}, {}^1/_{14}, \ldots$
    c. $2, {}^2/_3, {}^2/_5, {}^2/_7, \ldots$ d. ${}^1/_5, {}^{-1}/_5, {}^{-1}/_{15}, {}^{-1}/_{25}, \ldots$
    e. ${}^3/_4, {}^1/_2, {}^1/_3, {}^2/_9, \ldots$
31. Consider the arithmetic sequence s_1, x, s_3. The number x can be found as follows:

$$x = s_1 + d$$
$$x = s_3 - d$$
$$\overline{}$$
$$2x = s_1 + s_3 \quad \text{(by adding)}$$
$$x = \frac{s_1 + s_3}{2}$$

Here, x is called the *arithmetic mean* between s_1 and s_3. Find the arithmetic   ARITHMETIC MEAN
mean between each of the given pairs of numbers.
    a. $1, 8$ b. $1, 7$ c. $-5, 3$ d. $80, 88$ e. $40, 56$
32. Find the arithmetic mean (see Problem 31) between each of the given pairs of numbers.
    a. $4, 20$ b. $4, 15$ c. $\frac{1}{2}, \frac{1}{3}$ d. $-10, -2$ e. $-\frac{2}{3}, \frac{4}{5}$
33. How many blocks are there in the solid figure shown in Figure 8.2?

Mind Boggler

34. Fill in the missing term.
    a. A, A, B, Ꭸ, C, Ɔ, D, ꓷ, ____
    b. A, C, E, G, I, K, ____
    c. A, E, F, H, I, K, L, M, N, ____
    d. CAT, THREE, ALLIGATOR, NINE, WOLF, FOUR, ELEPHANT, ____
    e. $679, 378, 168, 48, 32,$ ____

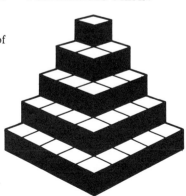

Figure 8.2

8.3   GEOMETRIC SERIES

The Peanuts cartoon illustrates a geometric progression. Charlie Brown is supposed to send a copy of the letter to six of his friends. Let's see how many people become involved in a very short time if everyone carries out the directions in the letter and the "chain" is not broken.

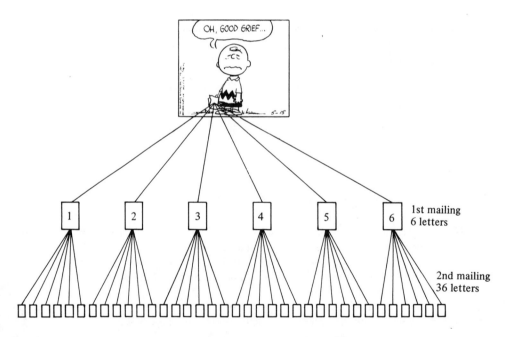

MAILING	LETTERS
2nd	36
3rd	216
4th	1,296
5th	7,776
6th	46,656
7th	279,936

8th	1,679,616
9th	10,077,696
10th	60,466,176
11th	362,797,056

In the 11th mailing, more letters will have been sent than there are people in the United States: The number of letters in only two more mailings would exceed the number of men, women, and children in the whole world.

According to the 1970 census, the population of the United States is 203,235,298.

In order to find the number of letters sent, we have to consider the series associated with the geometric sequence

$$6, 36, 216, \ldots.$$

This sequence can also be written $6^1, 6^2, 6^3, \ldots.$

We can actually add these numbers, or we can proceed as we did for arithmetic series and develop a formula that will be useful in a variety of situations.

If $s_1 = a$, then $s_2 = ar$, $s_3 = ar^2, \ldots, s_n = ar^{n-1}$, and if S_n represents the sum of the first n terms, then

$$S_n = a + ar + ar^2 + ar^3 + \cdots + ar^{n-1}$$

$$rS_n = ar + ar^2 + ar^3 + ar^4 + \cdots + ar^n.$$

We multiplied both sides by r.

Notice that except for the first and last terms, all the terms in the expressions for S_n and rS_n are the same, so that

$$rS_n - S_n = ar^n - a.$$

Now solve for S_n:

$$(r - 1)S_n = ar^n - a$$

$$S_n = \frac{ar^n - a}{r - 1} \quad (r \neq 1)$$

or

$$S_n = \frac{a(r^n - 1)}{r - 1} \quad (r \neq 1).$$

Sum of a finite geometric series

Thus, the number of letters sent in four mailings of Charlie Brown's chain letter can be calculated by letting $a = 6$, $r = 6$, and $n = 4$.

$$S_4 = \frac{6(6^4 - 1)}{6 - 1}$$

$$= \frac{6}{5}(6^4 - 1)$$

$$= 1554.$$

If you have a calculator, you can find the total number of letters sent with 13 mailings:

$$S_{13} = \frac{6(6^{13} - 1)}{6 - 1}$$

$$= \frac{6}{5}(6^{13} - 1)$$

$$= 15{,}672{,}832{,}812.$$

GEOMETRIC SERIES

General term

TO SUMMARIZE: If $s_1, s_2, \ldots, s_n$ is a finite geometric sequence, then

$$s_n = ar^{n-1},$$

where r is the common ratio, and

Sum

$$S_n = \frac{a(r^n - 1)}{r - 1}. \qquad (r \neq 1)$$

EXAMPLES

1. Find the sum of the first 6 terms of the geometric series with $a = -3$ and $r = 2$.

   Solution: $S_n = \dfrac{a(r^n - 1)}{r - 1}$

   $$S_6 = \frac{(-3)(2^6 - 1)}{2 - 1}$$

   $$= (-3)(64 - 1)$$

   $$= -189$$

2. Find the sum of the first 10 terms of the geometric series with $a = \frac{1}{2}$ and $r = \frac{1}{2}$.

   Solution: $S_n = \dfrac{a(r^n - 1)}{r - 1}$

   $$S_{10} = \frac{\frac{1}{2}[(\frac{1}{2})^{10} - 1]}{\frac{1}{2} - 1}$$

   $$= (-1)[(\tfrac{1}{2})^{10} - 1]$$

   $$= (-1)\left[\frac{1}{1024} - 1\right]$$

   $$= (-1)\left(-\frac{1023}{1024}\right)$$

   $$= \frac{1023}{1024}.$$

PROBLEM SET 8.3

A Problems

Write out the first three terms of the geometric sequences in Problems 1–8 with first term a and common ratio r.

1. $a = 5, r = 3$
2. $a = -12, r = 3$
3. $a = 1, r = -2$
4. $a = 1, r = 2$
5. $a = -15, r = {}^1/_5$
6. $a = 625, r = -{}^1/_5$
7. $a = 8, r = x$
8. $a = x, r = y$

Find a and r for the series in Problems 9–14.

9. $3, 6, 12, \ldots$
10. $7, 14, 28, \ldots$
11. $1, {}^1/_2, {}^1/_4, \ldots$
12. $100, 50, 25, \ldots$
13. $x, x^2, x^3, \ldots$
14. $xyz, xy, \ldots$

B Problems

Use the formula $s_n = ar^{n-1}$ to find an expression for the general term of each of the geometric sequences in Problems 15–20. Notice that you found a and r for each of these in Problems 9–14.

EXAMPLE 3: $50, 100, 200, \ldots$

Solution: $a = 50, r = \frac{100}{50}$

$$s_n = 50(2)^{n-1}$$
$$= 2 \cdot 5^2 \cdot 2^{n-1}$$
$$= 2^n \cdot 5^2$$

---

15. $3, 6, 12, \ldots$
16. $7, 14, 28, \ldots$
17. $1, \frac{1}{2}, \frac{1}{4}, \ldots$
18. $100, 50, 25, \ldots$
19. $x, x^2, x^3, \ldots$
20. $xyz, xy, \ldots$

Find the indicated quantities in Problems 21–27 for the given geometric sequences.

21. $a = 6, r = 3; s_5$
22. $s_1 = 100, r = \frac{1}{10}; s_{10}$
23. $s_1 = 6, r = 3; S_5$
24. $s_1 = 7, s_8 = 896; r$
25. $a = 1, r = 10; S_{10}$
26. *Calculator Problem.* $a = 3, r = \frac{1}{5}; S_8$
27. *Calculator Problem.* $a = \frac{1}{3}, r = \frac{1}{3}; S_1, S_{10}$

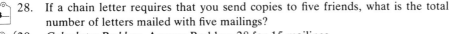

28. If a chain letter requires that you send copies to five friends, what is the total number of letters mailed with five mailings?
29. *Calculator Problem.* Answer Problem 28 for 15 mailings.
30. If a chain letter requires that you send copies to ten friends, what is the total number of letters mailed with four mailings?
31. Answer Problem 30 for ten mailings.

This means that the
population each five years is
120% of the previous total.

32. *Calculator Problem.* According to the 1970 census, the population of Hawaii is about 800,000. If the population increases 20% every five years, what will the population be in the year 2000?

33. Consider the geometric sequence s_1, x, s_3. The number x can be found by considering

$$\frac{x}{s_1} = r,$$

$$\frac{s_3}{x} = r.$$

So

$$\frac{x}{s_1} = \frac{s_3}{x}$$

$$x^2 = s_1 s_3.$$

This equation has two solutions,

$$x = \sqrt{s_1 s_3} \qquad \text{and} \qquad x = -\sqrt{s_1 s_3}.$$

GEOMETRIC MEAN

If s_1 and s_3 are both positive, then $\sqrt{s_1 s_3}$ is called the *geometric mean* of s_1 and s_3. If s_1 and s_3 are both negative, then $-\sqrt{s_1 s_3}$ is called the *geometric mean*. Find the geometric mean of each of the given pairs of numbers.

a. $1, 8$ b. $2, 8$ c. $-5, -3$ d. $-10, -2$ e. $4, 20$

Mind Boggler

34. Find three numbers with a sum equal to 9 so that these numbers form an arithmetic sequence and their squares form a geometric sequence.

8.4 INFINITE SERIES

In the previous section, we found the sum of the first n terms of a geometric series. Sometimes, it is possible to find the sum of the entire infinite geometric series. For example, we have already found the first 10 terms of the geometric series with $a = \frac{1}{2}$ and $r = \frac{1}{2}$ (Example 2 of Section 8.3, on page 388). Let's take a closer look at this series.

$$\text{Sequence: } \frac{1}{2}, \frac{1}{4}, \frac{1}{8}, \frac{1}{16}, \frac{1}{32}, \frac{1}{64}, \frac{1}{128}, \cdots$$

$$\text{Series: } \quad S_1 = \frac{1}{2}$$

$$S_2 = \frac{1}{2} + \frac{1}{4} = \frac{3}{4}$$

$$S_3 = \frac{1}{2} + \frac{1}{4} + \frac{1}{8} = \frac{7}{8}$$

$$\vdots$$

$$S_{10} = \frac{1023}{1024}$$

Does this series have a sum if we add *all* its terms? To answer this question, we consider the sum of the first n terms.

$$S_n = \frac{a(r^n - 1)}{r - 1}$$

We will rewrite this in a more convenient way.

$$S_n = \frac{a(1 - r^n)}{1 - r}$$

$$= \frac{a - ar^n}{1 - r}$$

$$= \frac{a}{1 - r} - \frac{ar^n}{1 - r}$$

For our example, $a = \frac{1}{2}$ and $r = \frac{1}{2}$, so

$$S_n = \frac{1/2}{1 - (1/2)} - \frac{(1/2)(1/2)^n}{1 - (1/2)}$$

$$= 1 - \left(\frac{1}{2}\right)^n.$$

As n becomes large, $(\frac{1}{2})^n$ becomes close to zero and S_n becomes close to one. Using the notation of Chapter 4, this can be written

$$\lim_{n \to \infty} S_n = 1.$$

Of course, not all infinite geometric series have limits. Charlie Brown's chain letter problem gave rise to a geometric series with $a = 6$ and $r = 6$. Thus,

$$S_n = \frac{a}{1 - r} - \frac{ar^n}{1 - r}$$

$$= \frac{6}{1 - 6} - \frac{6(6^n)}{1 - 6}$$

$$= -\frac{6}{5} + \frac{6}{5}(6^n).$$

We see that S_n becomes large without bound as n becomes large.

In general, it can be shown that if $|r| < 1$, then $r^n \to 0$ as $n \to \infty$. This means that

$$S_n \to \frac{a}{1 - r}.$$

The notation $|r| < 1$ is another way of saying $-1 < r < 1$.

SUM OF INFINITE GEOMETRIC SERIES

This means that if n is large enough, then S_n can be made as close to $a/(1 - r)$ as we please. If $|r| \geq 1$, then the infinite geometric series has no sum.

If $s_1, s_2, \ldots, s_n, \ldots$, is an infinite geometric sequence with $s_1 = a$ and common ratio r so that $|r| < 1$, then

$$\lim_{n \to \infty} S_n = \frac{a}{1 - r},$$

and this limit is the sum of the infinite geometric series

$$s_1 + s_2 + s_3 + \cdots + s_n + \cdots.$$

It might be interesting to do a calculator exercise for Example 1.

n	S_n
1	100
2	150(100 + 50)
3	175(100 + 50 + 25)
4	187.5(100 + 50 + 25 + 12.5)
5	193.75 (use formula on
6	196.875 page 388)
7	198.4375
8	199.21875
9	199.609375
10	199.8046875
⋮	⋮
20	199.8046875
⋮	⋮
30	199.9999998
31	199.9999999
32	200.0000000

This means that for all values greater than 31, the calculator can no longer distinguish between the *actual* value and the *limiting* value of 200.

EXAMPLES: Find the sum of the infinite geometric series, if possible.

1. $100 + 50 + 25 + \cdots$

    Solution: $a = 100$, $r = \frac{1}{2}$

    $$\lim_{n \to \infty} S_n = \frac{100}{1 - (1/2)}$$
    $$= 200$$

2. $-5 + 10 - 20 + \cdots$

    Solution: $a = -5$, $r = -2$; since $|r| \geq 1$, this infinite series does not have a sum.

3. The repeating decimal $0.727272\ldots$ is a rational number and can therefore be written as the quotient of two integers. To find this representation, we write the decimal as a geometric series.

    $$0.727272\ldots = 0.72 + 0.0072 + 0.000072 + \cdots$$
    $$= 0.72 + 0.72(0.01) + 0.72(0.0001) + \cdots$$
    $$= 0.72 + 0.72(0.01) + 0.72(0.01)^2 + \cdots$$

    We see that $a = 0.72$, $r = 0.01$, and

    $$\lim_{n \to \infty} S_n = \frac{0.72}{1 - 0.01} \quad \text{(since } |r| < 1\text{)}$$

    Remember, we want to write this out as a fraction, not as a decimal.

    $$= \frac{0.72}{0.99}$$
    $$= \frac{72/100}{99/100}$$
    $$= \frac{72}{99}$$
    $$= \frac{8}{11}$$

4. Many results can be proved by using infinite series. For example, consider the unit square shown in Figure 8.3. The area of the square is 1 sq unit.

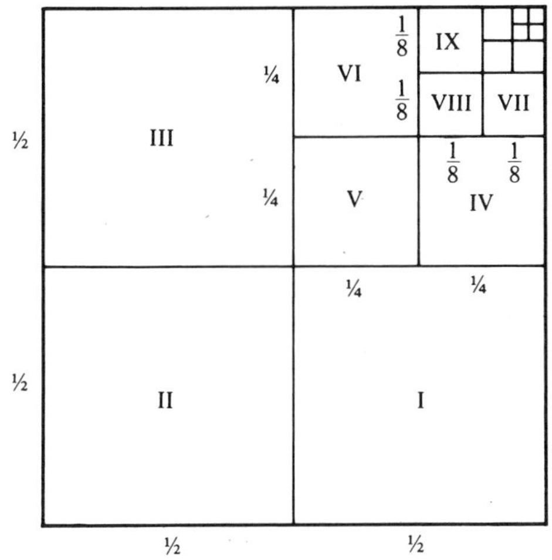

Figure 8.3

Suppose we find the area as follows:

a. Divide the square into quarters.
b. Divide one quarter into quarters.
c. Divide one of these smaller quarters into quarters.
d. Continue this process.

What is the total area of squares I, II, III, IV, . . . ?

Solution:

$$\text{total area} = [\text{I} + \text{II} + \text{III}] + [\text{IV} + \text{V} + \text{VI}] + [\text{VII} + \cdots]$$

By I we mean the area of square I, by II the area of square II, and so on.

$$= \left[\left(\frac{1}{2}\cdot\frac{1}{2}\right) + \left(\frac{1}{2}\cdot\frac{1}{2}\right) + \left(\frac{1}{2}\cdot\frac{1}{2}\right)\right]$$

$$+ \left[\left(\frac{1}{4}\cdot\frac{1}{4}\right) + \left(\frac{1}{4}\cdot\frac{1}{4}\right) + \left(\frac{1}{4}\cdot\frac{1}{4}\right)\right]$$

$$+ \left[\left(\frac{1}{8}\cdot\frac{1}{8}\right) + \left(\frac{1}{8}\cdot\frac{1}{8}\right) + \left(\frac{1}{8}\cdot\frac{1}{8}\right)\right] + \cdots$$

$$= \frac{3}{2^2} + \frac{3}{4^2} + \frac{3}{8^2} + \frac{3}{16^2} + \cdots$$

$$= \frac{3}{2^2} + \frac{3}{2^4} + \frac{3}{2^6} + \frac{3}{2^8} + \cdots$$

$$= \frac{3}{2^2}\left(1 + \frac{1}{2^2} + \frac{1}{2^4} + \frac{1}{2^6} + \cdots\right)$$

$$= \frac{3}{2^2}\left[1 + \left(\frac{1}{2^2}\right)^1 + \left(\frac{1}{2^2}\right)^2 + \left(\frac{1}{2^2}\right)^3 + \cdots\right]$$

But

$$1 + \frac{1}{2^2} + \left(\frac{1}{2^2}\right)^2 + \left(\frac{1}{2^2}\right)^3 + \cdots$$

is an infinite geometric series with $a = 1$, $r = 1/2^2$, so

$$\lim_{n \to \infty} S_n = \frac{1}{1 - (1/2^2)} = \frac{1}{3/4} = \frac{4}{3}.$$

The total area is

$$\frac{3}{2^2}\left(\frac{4}{3}\right) = 1,$$

which agrees with the known result.

5. We began this section by showing

$$\frac{1}{2} + \frac{1}{4} + \frac{1}{8} + \frac{1}{16} + \cdots = 1$$

since $a = \frac{1}{2}$, $r = \frac{1}{2}$, and

$$\lim_{n \to \infty} S_n = \frac{a}{1 - r}$$

$$= \frac{1/2}{1 - (1/2)}$$

$$= 1.$$

We can form a geometric model for this series. Construct squares of sides $^1/_2, {}^1/_4, {}^1/_8, {}^1/_{16}, \ldots$ on a Cartesian coordinate system, as shown in Figure 8.4. Next, draw the line L connecting the upper right vertices of the squares.

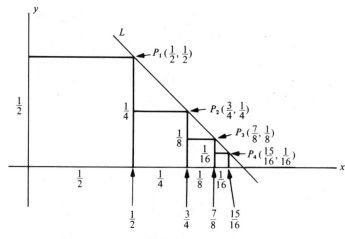

Figure 8.4

The coordinates of these points are $P_1(^1/_2, \, ^1/_2), P_2(^3/_4, \, ^1/_4), P_3(^7/_8, \, ^1/_8), \ldots$. Where does this line cross the x-axis?

Solution: The slope of L is

$$m = \frac{(1/4) - (1/2)}{(3/4) - (1/2)} = \frac{-1/4}{1/4} = -1$$

The equation of L is found by the point-slope formula:

$$y - \frac{1}{2} = -1\left(x - \frac{1}{2}\right)$$

$$x + y - 1 = 0,$$

which has an x-intercept of 1. This shows geometrically that the series

$$\frac{1}{2} + \frac{1}{4} + \frac{1}{8} + \cdots = 1,$$

as seen in Figure 8.4.

PROBLEM SET 8.4

A Problems

Find the sum, if possible, of the infinite geometric series in Problems 1–6.

1. $1 + \frac{1}{2} + \frac{1}{4} + \cdots$
2. $1000 + 500 + 250 + \cdots$
3. $100 + 50 + 25 + \cdots$
4. $-20 + 10 - 5 + \cdots$
5. $-45 - 15 - 5 + \cdots$
6. $-216 - 36 - 6 + \cdots$

Represent each repeating decimal in Problems 7–14 as the quotient of two integers by considering an infinite geometric series.

7. $0.\overline{4}$
8. $0.\overline{5}$
9. $0.\overline{27}$
10. $0.\overline{18}$
11. $2.\overline{45}$
12. $0.4\overline{18}$
13. $0.4\overline{182}$
14. 0.2185

Recall that $0.\overline{4} = 0.444\ldots$. The bar indicates the digits in the decimal that repeat.

B Problems

Find the sum, if possible, of the infinite geometric series in Problems 15–24.

15. $\frac{1}{3} + \frac{1}{9} + \frac{1}{27} + \cdots$
16. $\frac{1}{4} + \frac{1}{16} + \frac{1}{64} + \cdots$
17. $1 + (1.08)^{-1} + (1.08)^{-2} + \cdots$
18. $1 + (1.10)^{-1} + (1.10)^{-2} + \cdots$
19. $2 + \sqrt{2} + 1 + \cdots$
20. $3 + \sqrt{3} + 1 + \cdots$
21. $(1 + \sqrt{2}) + 1 + (-1 + \sqrt{2}) + \cdots$

Hint: find the distance traveled from the time the ball hits the ground the first time and then add 10 ft.

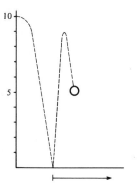

20% of $100 = $20
20% of $20 = $4

Recall that a unit square has sides of length one.

22. $(\sqrt{2} - 1) + 1 + (\sqrt{2} + 1) + \cdots$

23. $5.03\overline{1}$

24. $2.25\overline{34}$

25. A new type of SUPERBALL advertises that it will rebound $^9/_{10}$ of its original height. If it is dropped from a height of 10 ft, how far will the ball travel before coming to rest?

26. Repeat Problem 25 for a ball that rebounds $\frac{2}{3}$ of its original height.

27. Suppose that a piece of machinery costing $10,000 depreciates 20% of its present value each year. That is, the first year $10,000(0.20) = $2000 is depreciated. The second year's depreciation is

$$\$8000(0.20) = \$1600$$

since the value the second year is $10,000 - $2000 = $8000. The third year's depreciation is

$$\$6400(0.20) = \$1280.$$

If the depreciation is calculated this way indefinitely, what is the total depreciation?

28. Winnie Winner wins $100 in a pie-baking contest run by the Hi-Do Pie Co. The company gives Winnie the $100. However, the tax collector wants 20% of the $100. Winnie pays the tax. But then she realizes that she didn't really win a $100 prize and tells her story to the Hi-Do Co. The friendly Hi-Do Co. gives Winnie the $20 she paid in taxes. Unfortunately, the tax collector now wants 20% of the $20. She pays the tax again and then goes back to the Hi-Do Co. with her story. Assume that this can go on indefinitely. How much money does the Hi-Do Co. have to give Winnie so that she will really win $100? How much does she pay in taxes?

29. Repeat Example 4 where the side of the square is a.

30. A unit square *ABCD* is cut out of construction paper. Another square *EFGH* with side $\frac{1}{2}$ unit is cut out of paper and placed on top of *ABCD*, as shown in Figure 8.5. Next, squares with sides $\frac{1}{4}, \frac{1}{8}, \frac{1}{16}, \ldots$ are cut out and placed as shown in the figure. What is the total area of all the squares?

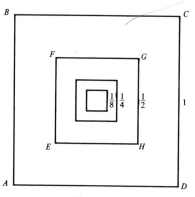

Figure 8.5

31. Repeat Problem 30 for square *ABCD* with side of length *a*, *EFGH* with side of length *a*/2, and so on.

Mind Bogglers

32. Consider a sequence of equilateral triangles, as shown in Figure 8.6, where the sides of △*ABC* are 1, the sides of △*ADE* and △*EFB* are $\frac{1}{2}$, and so on. If this process is repeated forever, what is the perimeter of the triangles? What is the total area of the triangles?

Hint for Problem 32: the height of an equilateral triangle is $\frac{1}{2}a\sqrt{3}$, where *a* is the length of the side.

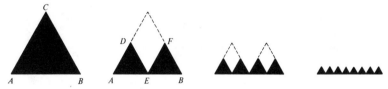

Figure 8.6

33. Square *ABCD* has sides of length *a*. Square *EFGH* is formed by connecting the midpoints of the sides of the first square, as shown in Figure 8.7. Assume that the pattern of shaded regions in the square is continued indefinitely. What is the area of the shaded regions?

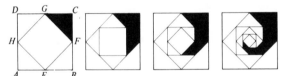

Figure 8.7

34. An equilateral triangle of side *a* is cut out of paper, as shown in Figure 8.8a. Next, three equilateral triangles, each of side *a*/3, are cut out and placed in the middle of each side of the first triangle, as shown in Figure 8.8b. Twelve equilateral triangles, each of side *a*/9, are now placed halfway along each of the sides of this figure, as shown in Figure 8.8c. Figure 8.8d shows the result of adding 48 equilateral triangles, each of side *a*/27, to the previous figure. Assume that this procedure is repeated indefinitely. Find:
a. The perimeter of the figure we obtain
b. The area of the figure we obtain

Figure 8.8

35. We are given a square carpet of side 1 m, as shown in Figure 8.9a. The carpet is divided into nine equal squares and the center square is cut out, as shown in Figure 8.9b. Each of the other eight squares is divided into nine equal squares and the center square is cut out (Figure 8.9c). Assume this pattern is repeated indefinitely. Find the total area that is removed from the original.

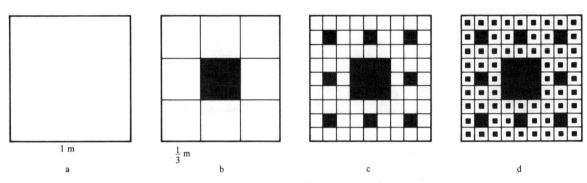

Figure 8.9

36. a. Find 0.333 ... as the quotient of two integers by considering an infinite geometric series.
     b. Find 0.999 ... as the quotient of two integers by considering an infinite geometric series.
     c. Are your results to parts a and b consistent? That is, is your answer to part b three times your answer to part a?

8.5 SUMMATION, FACTORIAL, AND PASCAL

In this section, we introduce some notation that is useful not only when working with series, but also in other mathematics, particularly calculus. While introducing this notation, we will also be able to review many of the ideas of arithmetic and geometric series.

SUMMATION NOTATION

Recall that a finite sequence is a function with a domain that consists of the set of counting numbers less than or equal to some given natural number. If we consider the function $s(k) = 2k$ and $N = \{1, 2, 3, 4\}$, we can write the following table:

k	$s(k) = 2k$
1	2
2	4
3	6
4	8

The sum of the terms of this finite arithmetic sequence is the sum of the series

$$2 + 4 + 6 + 8.$$

We can denote this sum by using the symbol $\sum$ (called sigma) as follows:

This is the last natural number in the domain.

$$\sum_{k=1}^{4} 2k = 2 + 4 + 6 + 8$$

This is the function being evaluated.

This is the first natural number in the domain.

$\sum$ refers to a series, not the sum of the series in general. We use

$$\sum_{j=1}^{n} a_i$$

in two senses here.

This symbol means we evaluate the function for each number in the domain and *add* the resulting terms.

Thus,

$$\sum_{k=1}^{4} 2k = 2 + 4 + 6 + 8 = 20.$$

20 is actually the numerical value of the indicated sum $2 + 4 + 6 + 8$.

EXAMPLE 1: Let $s(k) = (2k + 1)$ and $N = \{3, 4, 5, 6\}$.

	k	$s(k) = 2k + 1$
First natural number $\longrightarrow$ in domain; $k = 3$	3	7
	4	9
	5	11
Last natural number $\longrightarrow$ in domain; $k = 6$	6	13

$\sum$ means we add these values

Thus,

$$\sum_{k=3}^{6} (2k + 1) = 7 + 9 + 11 + 13 = 40.$$

$$k = 4 \quad k = 5 \quad k = 6$$

This is obtained by letting $k = 3$ and evaluating $(2k + 1)$.

The sum of this series can be found by

$$S_n = \frac{a(r^n - 1)}{r - 1}$$

where $a = \frac{1}{2}$, $r = \frac{1}{2}$.

$$S_n = \frac{\frac{1}{2}[(\frac{1}{2})^n - 1]}{\frac{1}{2} - 1}$$

$$= -\left[\frac{1}{2^n} - 1\right]$$

$$= -\left[\frac{1 - 2^n}{2^n}\right]$$

$$= \frac{2^n - 1}{2^n}$$

Therefore,

$$\sum_{k=1}^{n} \frac{1}{2^k} = \frac{2^n - 1}{2^n}.$$

EXAMPLE 2: Let $s(k) = 1/2^k$ and $N = \{1, 2, 3, 4, \ldots, n\}$.

$$\sum_{k=1}^{n} \frac{1}{2^k} = \frac{1}{2} + \frac{1}{4} + \frac{1}{8} + \cdots + \frac{1}{2^n}$$

$$\qquad\quad \uparrow \qquad \uparrow \qquad \uparrow \qquad\qquad\quad \uparrow$$
$$\quad k = 1 \quad k = 2 \quad k = 3 \qquad\qquad k = n$$

EXAMPLE 3: Write the sum of the arithmetic series

$$S_n = s_1 + s_2 + \cdots + s_n$$

using $\sum$ notation.

Solution: $\quad S_n = \sum_{k=1}^{n} s_k$

By formula (from page 383),

$$S_n = \sum_{k=1}^{n} s_k = \frac{n}{2}(s_1 + s_n).$$

EXAMPLE 4: Write the sum of the finite geometric series

$$S_n = a + ar + ar^2 + \cdots + ar^{n-1}$$

using $\sum$ notation.

Solution: $\quad S_n = \sum_{k=1}^{n} ar^{k-1}$

By formula (from page 388),

$$S_n = \sum_{k=1}^{n} ar^{k-1} = \frac{a(r^n - 1)}{r - 1} \quad (r \neq 1).$$

EXAMPLE 5: Write the sum of the infinite geometric series

$$S_n = a + ar + ar^2 + \cdots,$$

where $|r| < 1$.

Solution: $\lim_{n \to \infty} S_n = \lim_{n \to \infty} \sum_{k=1}^{n} ar^{k-1}$

By formula,

$$\lim_{n \to \infty} S_n = \lim_{n \to \infty} \sum_{k=1}^{n} ar^{k-1} = \frac{a}{1 - r}.$$

FACTORIAL NOTATION

There are occasions when we want to consider the *product* of the first n natural numbers. We denote this by $n!$ and call it *n-factorial*.

DEFINITION: The symbol $n! = 1 \cdot 2 \cdot 3 \cdot \cdots \cdot (n-1) \cdot n$ and is called *n-factorial* (*n* a natural number). Also, we define

$$0! = 1 \quad \text{and} \quad 1! = 1.$$

FACTORIAL

EXAMPLES

$$0! = 1$$
$$1! = 1$$
$$2! = 1 \cdot 2 = 2$$
$$3! = 1 \cdot 2 \cdot 3 = 6$$
$$4! = 1 \cdot 2 \cdot 3 \cdot 4 = 24$$
$$5! = 1 \cdot 2 \cdot 3 \cdot 4 \cdot 5 = 120$$
$$6! = 1 \cdot 2 \cdot 3 \cdot 4 \cdot 5 \cdot 6 = 720$$
$$7! = 1 \cdot 2 \cdot 3 \cdot 4 \cdot 5 \cdot 6 \cdot 7 = 5040$$
$$8! = 1 \cdot 2 \cdot 3 \cdot \cdots \cdot 7 \cdot 8 = 40{,}320$$
$$9! = 1 \cdot 2 \cdot 3 \cdot \cdots \cdot 8 \cdot 9 = 362{,}880$$
$$10! = 1 \cdot 2 \cdot 3 \cdot \cdots \cdot 9 \cdot 10 = 3{,}628{,}800$$

Historical Note

Factorial notation was first used by Christian Kramp in 1808.

EXAMPLES: Evaluate the given expressions.

6. $5! - 4! = 120 - 24$
    $= 96$

7. $(5 - 4)! = 1!$
    $= 1$

8. $\dfrac{8!}{4!} = \dfrac{8 \cdot 7 \cdot 6 \cdot 5 \cdot \cancel{4} \cdot \cancel{3} \cdot \cancel{2} \cdot \cancel{1}}{\cancel{4} \cdot \cancel{3} \cdot \cancel{2} \cdot \cancel{1}}$

    $= 8 \cdot 7 \cdot 6 \cdot 5$

    $= 1680$

9. $\left(\dfrac{8}{4}\right)! = 2!$

    $= 2$

10. $\dfrac{10!}{8!} = \dfrac{10 \cdot 9 \cdot \cancel{8!}}{\cancel{8!}}$

    $= 90$

Notice that
$$10! = 10 \cdot 9!$$
$$= 10 \cdot 9 \cdot 8!$$
$$= 10 \cdot 9 \cdot 8 \cdot 7!$$
and so on.

PASCAL'S TRIANGLE NOTATION

Consider the following pattern, called *Pascal's triangle*:

```
row 0:                             1
row 1:                         1       1
row 2:                     1       2       1
row 3:                 1       3       3       1
row 4:             1       4       6       4       1
row 5:         1       5      10      10       5       1
row 6:     1       6      15      20      15       6       1
                                   ⋮
```

There are many interesting relationships associated with this pattern, but we are concerned with an expression representing the entries in the pattern. Do you see how to generate additional rows of the triangle?

1. Each row begins and ends with a 1.
2. Notice that we began counting the rows with row 0. This is because after row 0, the second entry in the row is the same as the row number. Thus, row 7 begins 1 7 ....
3. The triangle is symmetric about the middle. This means that the entries of each row are the same at the beginning and the end. Thus, row 7 ends with ... 7 1.
4. To find new entries, we can simply add the two entries just above in the preceding row. Thus, row 7 is found by looking at row 6.

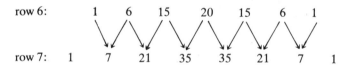

```
row 6:        1       6      15      20      15       6       1

row 7:    1       7      21      35      35      21       7       1
```

If we write Pascal's triangle in the form of rows and columns, as shown in Figure 8.10, we see that we can locate any entry in the triangle by using an ordered pair.

Column / Row	0	1	2	3	4	⋯
0	1					
1	1	1				
2	1	2	1			
3	1	3	3	1		
4	1	4	6	4	1	
⋮						

Figure 8.10   Pascal's triangle

We write $\binom{n}{r}$ to represent the element in the nth row and rth column of the triangle. Therefore:

$$\binom{0}{0} = 1$$

$$\binom{1}{0} = 1 \qquad \binom{1}{1} = 1$$

$$\binom{2}{0} = 1 \qquad \binom{2}{1} = 2 \qquad \binom{2}{2} = 1$$

$$\binom{3}{0} = 1 \qquad \binom{3}{1} = 3 \qquad \binom{3}{2} = 3 \qquad \binom{3}{3} = 1$$

$$\vdots$$

In general,

$$\binom{n}{r} = \frac{n!}{r!(n-r)!}.$$

You should verify this formula for the values shown in the triangle.

This means that if we want to find the entry in row 8 and column 3, we find

For example,

$$\binom{8}{3} = \frac{8!}{3!(8-3)!}$$

$$= \frac{8!}{3!5!}$$

$$= \frac{8 \cdot 7 \cdot 6 \cdot 5!}{3 \cdot 2 \cdot 1 \cdot 5!}$$

$$= 56.$$

$$\binom{3}{2} = \frac{3!}{2!(3-2)!}$$

$$= \frac{3!}{2!1!}$$

$$= 3$$

EXAMPLES: Evaluate the given expressions.

11. $\binom{52}{2}$

Solution: $\binom{52}{2} = \dfrac{52!}{2!(52-2)!}$

$$= \frac{52!}{2!50!}$$

$$= \frac{52 \cdot 51 \cdot 50!}{2 \cdot 1 \cdot 50!}$$

$$= 1326$$

12. $\binom{n}{n}$

   Solution: $\binom{n}{n} = \dfrac{n!}{n!(n-n)!}$

   $= \dfrac{n!}{n!\,0!}$

   $= 1$

13. $\binom{n}{n-1}$

   Solution: $\binom{n}{n-1} = \dfrac{n!}{(n-1)!\,[n-(n-1)!]}$

   $= \dfrac{n!}{(n-1)!\,1!}$

   $= \dfrac{n(n-1)!}{(n-1)!}$

   $= n$

PROBLEM SET 8.5

A Problems

Evaluate the expressions in Problems 1–24.

1. $A = 4! - 2!$

2. $B = 5! - 3!$

3. $C = (4-2)!$

4. $D = (6-3)!$

5. $E = \dfrac{9!}{7!}$

6. $F = \dfrac{10!}{6!}$

7. $G = \dfrac{12!}{10!}$

8. $H = \dfrac{10!}{4!\,6!}$

9. $I = \dfrac{12!}{3!\,(12-3)!}$

10. $J = \dfrac{52!}{3!\,(52-3)!}$

11. $L = \displaystyle\sum_{k=2}^{6} k$

12. $M = \displaystyle\sum_{m=1}^{4} m^2$

13. $N = \displaystyle\sum_{n=0}^{6} (2n+1)$

14. $P = \displaystyle\sum_{p=1}^{6} 2p$

15. $Q = \binom{8}{1}$

16. $R = \binom{5}{4}$

17. $S = \binom{8}{2}$

18. $K = \binom{52}{3}$

19. $T = \binom{8}{3}$

20. $U = \binom{7}{4}$

21. $V = \binom{8}{4}$

22. $W = \binom{5}{5}$

23. $X = \binom{52}{2}$

24. $Y = \binom{10}{1}$

25. $Z = \binom{1000}{1 \cdot}$

26. Replace each set of parentheses below with the letter from Problems 1–25 corresponding to the value you obtained in those problems. The letter "oh" has already been filled in for you.

(56)(210)(72)   (220)(49)(5040)(220)(49)(220)(56)(72)!

(49)(O)   (O)(56)(210)(72)(5)

(8)(35)(72)(28)(56)(220)(O)(49)   (210)(22)(28)

(72)(70)(72)(5)   (30)(O)(70)(72)(6)

(28)(O)   (42)(5)(O)(5040)(O)(35)(49)(6)(20)(10)

(56)(210)(72)   (28)(42)(220)(5)(220)(56)   (O)(5040)   (30)(22)(49).

B Problems

27. Write
$$\frac{1}{2} + \frac{1}{4} + \frac{1}{8} + \cdots + \frac{1}{128}$$
using summation notation.

28. Write $2 + 4 + 6 + \cdots + 100$ using summation notation.

29. Write $1 + 6 + 36 + 216 + 1296$ using summation notation.

30. Write $5 + 15 + 45 + 135 + 405$ using summation notation.

31. Write out
$$\sum_{j=1}^{r} a_j b_j$$
without summation notation.

32. In Problem 31, let $b_j = k$ and show that
$$\sum_{j=1}^{r} k a_j = k \sum_{j=1}^{r} a_j.$$

Historical Quote

This is a quotation of the German mathematician David Hilbert (1862–1943). At the Paris Congress in 1900 he proposed 23 problems that he believed should be those occupying the attention of 20th-century mathematicians. He stated: "If we obtain an idea of the probable development of mathematical knowledge in the immediate future, we must let the unsettled questions pass before our minds and look over the problems which the science of today sets and whose solution we expect from the future." Mathematicians are still working on these problems today, and roughly half of them are still unsolved. As Hilbert said in proposing his problems, "As long as a branch of science offers an abundance of problems, so long is it alive." (From *A History of Mathematics*, by Carl Boyer. New York: Wiley, 1968.)

33. In Problem 32, let $a_j = 1$ and show that

$$\sum_{j=1}^{r} k = kr.$$

34. Show that

$$\sum_{k=1}^{n} (a_k + b_k) = \sum_{k=1}^{n} a_k + \sum_{k=1}^{n} b_k.$$

35. Examine Pascal's triangle carefully. Look for patterns and answer the following questions:
    a. What is the second number in row 100?
    b. What is the next-to-last number in row 200?

36. Which rows of Pascal's triangle contain only odd numbers?

37. Consider $11^0, 11^1, 11^2, 11^3, \ldots$, and explain how the powers of 11 are related to Pascal's triangle. Does this pattern come to an end?

Mind Bogglers

38. How are the numbers below related to Pascal's triangle?
    a. Square numbers

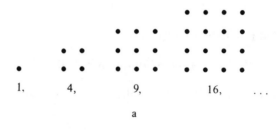

1,   4,   9,   16,   . . .

a

    b. Triangular numbers

1,   3,   6,   10,   . . .

b

39. Show that

$$\binom{n-1}{r-1} + \binom{n-1}{r} = \binom{n}{r}.$$

40. Show that

$$\binom{n}{r} = \binom{n}{n-r}.$$

8.6 BINOMIAL THEOREM

In mathematics, it is frequently necessary to expand $(a + b)^n$. If n is very large, direct calculation is rather tedious so we try to find an easy pattern that will not only help us find $(a + b)^n$, but also will allow us to find any given term in that expansion.

Let's consider the powers of $(a + b)$, which we find by direct multiplication:

$$(a + b)^0 = 1$$
$$(a + b)^1 = 1 \cdot a + 1 \cdot b$$
$$(a + b)^2 = 1 \cdot a^2 + 2 \cdot ab + 1 \cdot b^2$$
$$(a + b)^3 = 1 \cdot a^3 + 3 \cdot a^2b + 3 \cdot ab^2 + 1 \cdot b^3$$
$$(a + b)^4 = 1 \cdot a^4 + 4 \cdot a^3b + 6 \cdot a^2b^2 + 4 \cdot ab^3 + 1 \cdot b^4$$
$$(a + b)^5 = 1 \cdot a^5 + 5 \cdot a^4b + 10 \cdot a^3b^2 + 10 \cdot a^2b^3 + 5 \cdot ab^4 + 1 \cdot b^5$$
$$\vdots$$

If we ignore the coefficients and focus our attention only on the variables, we see a pattern:

$(a + b)^1$:	a	b			
$(a + b)^2$:	a^2	ab	b^2		
$(a + b)^3$:	a^3	a^2b	ab^2	b^3	
$(a + b)^4$:	a^4	a^3b	a^2b^2	ab^3	b^4

$$\vdots$$

Do you see it? As we go from left to right, the powers of a decrease and the powers of b increase. Notice that the sum of the exponents for each term is the same as the original power

$(a + b)^n$:

$$a^n \qquad a^{n-1}b \qquad a^{n-2}b^2 \quad \cdots \quad a^{n-r}b^r \quad \cdots \quad a^2b^{n-2} \qquad ab^{n-1} \qquad b^n$$

Next, we consider the coefficients.

$(a + b)^0$:				1				
$(a + b)^1$:			1		1			
$(a + b)^2$:			1	2		1		
$(a + b)^3$:		1	3		3		1	
$(a + b)^4$:	1	4	6		4		1	
$(a + b)^5$:	1	5	10	10	5		1	

$$\vdots$$

Historical Note

The binomial theorem was discovered in 1665 by Sir Isaac Newton (1642–1727), perhaps the greatest mathematician of all time. He was a genius of the highest order but was often absent-minded. As a boy, Newton was supposedly sent to the barn to cut a hole at the bottom of the barn door for cats to go in and out. According to the story, he cut two holes, a large one for the cat and a small one for the kittens.

His statement of the binomial theorem is awkward by today's standards, but remember that he found it by a laborious trial-and-error procedure:

$$``\overline{P + PQ}\Big|\frac{m}{n}$$

$$= P\frac{m}{n} + \frac{m}{n}AQ$$

$$+ \frac{m - n}{2n}BQ$$

$$+ \frac{m - 2n}{3n}CQ$$

$$+ \frac{m - 3n}{4n}DQ + \text{etc.},$$

where $P + PQ$ stands for a Quantity whose Root or Power or whose Root of a Power is to be found, P being the first term of that quantity, Q being the remaining terms divided by the first term and m/n the numerical Index of the powers of $P + PQ \ldots$. Finally, in place of the terms that occur in the course of the work in the Quotient, I shall use A, B, C, D, etc. Thus, A stands for the first term $P(m/n)$; B for the second term $(m/n)AQ$; and so on." (From D. E. Smith, *Source Book in Mathematics* (1929), pp. 224–228.)

BINOMIAL THEOREM

Remember,

$$\binom{n}{k} = \frac{n!}{k!(n-k)!}.$$

Do you see the pattern? The coefficients are the numbers in Pascal's triangle. Using the notation we introduced in the last section for Pascal's triangle, we have

$$(a + b)^n = \binom{n}{0}a^n + \binom{n}{1}a^{n-1}b + \binom{n}{2}a^{n-2}b^2 + \cdots + \binom{n}{r}a^{n-r}b^r$$

$$+ \cdots + \binom{n}{n-2}a^2b^{n-2} + \binom{n}{n-1}ab^{n-1} + \binom{n}{n}b^n.$$

Using summation notation, we obtain a more compact form, called the *Binomial Theorem*.

BINOMIAL THEOREM: For any positive integer n,

$$(a + b)^n = \sum_{k=0}^{n} \binom{n}{k}a^{n-k}b^k.$$

The proof of the Binomial Theorem is by mathematical induction, which is discussed in the next section.

EXAMPLES

1. Find $(x + y)^8$.
    Solution: For smaller powers, use Pascal's triangle to obtain the coefficients in the expansion. Thus,

    $$(x + y)^8 = x^8 + 8x^7y + 28x^6y^2 + 56x^5y^3 + 70x^4y^4$$
    $$+ 56x^3y^5 + 28x^2y^6 + 8xy^7 + y^8.$$

2. Find $(x - 2y)^4$.
    Solution: In this example, $a = x$ and $b = -2y$ and the coefficients are in Pascal's triangle.

    $$(x - 2y)^4 = 1 \cdot x^4 + 4 \cdot x^3(-2y) + 6 \cdot x^2(-2y)^2 + 4 \cdot x(-2y)^3 + 1 \cdot (-2y)^4$$
    $$= x^4 - 8x^3y + 24x^2y^2 - 32xy^3 + 16y^4$$

3. Find $(a + b)^{15}$.
    Solution: The power is rather large so we use the Binomial Theorem to find the coefficients.

    $$(a + b)^{15} = \binom{15}{0}a^{15} + \binom{15}{1}a^{14}b + \binom{15}{2}a^{13}b^2 + \cdots + \binom{15}{14}ab^{14} + \binom{15}{15}b^{15}$$

    $$= \frac{15!}{0!\,15!}a^{15} + \frac{15!}{1!\,14!}a^{14}b + \frac{15!}{2!\,13!}a^{13}b^2 + \cdots + \frac{15!}{14!\,1!}ab^{14} + \frac{15!}{15!\,0!}b^{15}$$

    $$= a^{15} + 15a^{14}b + 105a^{13}b^2 + \cdots + 15ab^{14} + b^{15}$$

4. Find the coefficient of the term $x^2 y^{10}$ in the expansion of $(x + 2y)^{12}$.
    Solution: $n = 12$, $k = 10$, $a = x$, and $b = 2y$; thus, the term we seek is

$$\binom{12}{10} x^2 (2y)^{10} = \frac{12!}{10! \, 2!} (2)^{10} x^2 y^{10}.$$

The coefficient is $66(1024) = 67{,}584$.

PROBLEM SET 8.6

A Problems

In Problems 1–6, expand using the Binomial Theorem.

1. $(x + y)^5$ 2. $(x - y)^6$
3. $(x + 2)^5$ 4. $(x - 2)^6$
5. $(a + b)^4$ 6. $(2x + 3y)^4$
7. Find the coefficient of $a^5 b^6$ in $(a - b)^{11}$.
8. Find the coefficient of $a^{14} b$ in $(a^2 - 2b)^8$.
9. Find the coefficient of $a^{10} b^4$ in $(a + b)^{14}$.

B Problems

Find the first four terms in the expansions given in Problems 10–19.

10. $(x - y)^{15}$ 11. $(x + 2y)^{16}$
12. $(x + \sqrt{2})^8$ 13. $(x - 2y)^{12}$
14. $(1 - 0.03)^{13}$ 15. $(1 - 0.02)^{12}$
16. $(1.04)^{10}$ 17. $(1.07)^8$

Hint for Problem 16: write
$1.04 = 1 + 0.04$.

18. $(0.95)^8$ 19. $(0.98)^7$

Hint for Problem 18: write
$0.95 = 1 - 0.05$.

$ 20. If \$1 is invested at 8% interest compounded annually for 10 years, what is the amount present? (Give answer to nearest cent.)

$ 21. If \$1 is invested at 9% interest compounded annually for 20 years, what is the amount present to the nearest cent?

Recall that

$$A = P(1 + r)^n,$$

where A is the amount present after n years when P dollars are invested at r-percent interest compounded annually.

Hint for Problem 22: expand $(a + b)^n$, where $a = 1$ and $b = 1$.

22. Show that

$$\binom{n}{0} + \binom{n}{1} + \binom{n}{2} + \cdots + \binom{n}{n-1} + \binom{n}{n} = 2^n.$$

This says that the sum of the entries of the nth row of Pascal's triangle is 2^n.

Mind Bogglers

Hint for Problem 23: find $(1 + 1)^1$, $(1 + 1/2)^2$, $(1 + 1/3)^3$, $(1 + 1/4)^4$, $(1 + 1/5)^5$, and look for a pattern.

23. Make a conjecture about

$$\lim_{n \to \infty} \left(1 + \frac{1}{n}\right)^n. \quad e^{\wedge}$$

24. Show that for every integer n,

$$\sum_{j=1}^{n} (-1)^j \binom{n}{j} = 0.$$

25. Show that for every positive integer n,

$$\sum_{j=0}^{n} 2^j \binom{n}{j} = 3^n.$$

8.7 MATHEMATICAL INDUCTION

Mathematical induction is an important method of proof in mathematics. It bridges the gap between conjecture and proof. For example, in the last section, the Binomial Theorem was stated without proof. Mathematical induction offers a method for proving this result for all positive integers. Let's begin with a simple example. Suppose you want to know the sum of the first n odd integers. You could begin by looking for a pattern.

Since this is a simple example, it can be worked by finding the sum of an arithmetic series. (But not all the problems we will work by mathematical induction will be arithmetic or geometric series.) Consider

$$1 + 3 + 5 + \cdots + (2n - 1),$$

which is an arithmetic series with sum

$$S_n = n\left[\frac{1 + (2n - 1)}{2}\right]$$

$$= n\left[\frac{2n}{2}\right]$$

$$= n^2.$$

$$
\begin{array}{ll}
1 & = 1 \\
1 + 3 & = 4 \\
1 + 3 + 5 & = 9 \\
1 + 3 + 5 + 7 & = 16 \\
1 + 3 + 5 + 7 + 9 & = 25
\end{array}
$$

Do you see a pattern? It appears that the sum of the first n odd numbers is n^2 since $1 = 1^2$, $4 = 2^2$, $9 = 3^2$, $16 = 4^2$, and so on. We now want to *prove deductively* that

$$1 + 3 + 5 + \cdots + \underbrace{(2n - 1)}_{\uparrow} = n^2$$

$$\underbrace{}_{n\text{th odd number}}$$

is true for all positive integers n. How can we proceed? We use a method called *mathematical induction*, which is used to prove certain propositions about the positive integers. The proposition is denoted by $P(n)$. For example, in this case we would let $P(n)$ be the proposition

$$P(n): \qquad 1 + 3 + 5 + \cdots + (2n - 1) = n^2.$$

We then see that:

$$P(1): \quad 1 = 1^2$$
$$P(2): \quad 1 + 3 = 2^2$$
$$P(3): \quad 1 + 3 + 5 = 3^2$$
$$P(4): \quad 1 + 3 + 5 + 7 = 4^2$$

$$\vdots$$

$$P(100): \quad 1 + 3 + 5 + \cdots + 199 = 100^2$$
$$P(x - 1): \quad 1 + 3 + 5 + \cdots + (2x - 3) = (x - 1)^2$$
$$P(x): \quad 1 + 3 + 5 + \cdots + (2x - 1) = x^2$$
$$P(x + 1): \quad 1 + 3 + 5 + \cdots + (2x + 1) = (x + 1)^2$$

These statements, of course, are what we want to show; that is, we want to show that $P(n)$ is true for *all n* when $n = 1, n = 2, \ldots$ (*n* a positive integer). Let S denote the set of positive integers for which $P(n)$ is true. If we can show that 1 is in S and that if k is in S, then $k + 1$ is in S, then we know that all positive integers are in S. This leads to a statement of the *principle of mathematical induction* (PMI).

PRINCIPLE OF MATHEMATICAL INDUCTION: If a given proposition $P(n)$ is true for $P(1)$ and if the truth of $P(k)$ implies the truth of $P(k + 1)$, then $P(n)$ is true for all positive integers.

MATHEMATICAL INDUCTION (PMI)

We thus have the following procedure for proof by mathematical induction:

1. Prove $P(1)$ is true.
2. Assume $P(k)$ is true.
3. Prove $P(k + 1)$ is true.
4. Conclude that $P(n)$ is true for all positive integers n.

Students often have a certain uneasiness when they first use the principle of mathematical induction as a method of proof. Suppose this principle is used with a stack of large dominoes set up as shown in the cartoon. How can the man in the cartoon be certain of knocking over all the dominoes?

1. He would have to be able to knock over the first one.
2. He would have to have the dominoes arranged so that *if* the *k*th domino falls, then the next one, the $(k + 1)$st, will also fall. That is, each domino is set up so that, if it falls, it causes the next one to fall. We have set up a kind of "chain reaction" here. The first domino falls; this knocks over the next one (the second domino); the second one knocks over the next one (the third domino); the third one knocks over the next one; and so on. This continues until all the dominoes are knocked over.

Returning to our example, we want to prove that

$$1 + 3 + 5 + \cdots + (2n - 1) = n^2$$

is true for all positive integers *n*.

Step 1. Prove $P(1)$ true: $1 = 1^2$ is true.
Step 2. Assume $P(k)$ true: $1 + 3 + 5 + \cdots + (2k - 1) = k^2$
Step 3. Prove $P(k + 1)$ true.
 To prove:

$$1 + 3 + 5 + \cdots + [2(k + 1) - 1] = (k + 1)^2$$

or

$$1 + 3 + 5 + \cdots + (2k + 1) = (k + 1)^2$$

Notice that

$$2(k + 1) - 1$$
$$= 2k + 2 - 1$$
$$= 2k + 1$$

This is the next term in the sequence.

The left side is now exactly like the left side of the statement we want to prove.

Statements	Reasons
1. $1 + 3 + 5 + \cdots + (2k - 1) = k^2$	1. By hypothesis (Step 2)
2. $1 + 3 + 5 + \cdots + (2k - 1) + (2k + 1)$ $= k^2 + (2k + 1)$	2. Add $(2k + 1)$ to both sides
3. $1 + 3 + 5 + \cdots + (2k - 1) + (2k + 1)$ $= k^2 + 2k + 1$	3. Associative
4. $1 + 3 + 5 + \cdots + (2k - 1) + (2k + 1)$ $= (k + 1)^2$	4. Factoring (distributive)

Step 4. The proposition $P(n)$ is true for all positive integers *n* by PMI.

EXAMPLES: Prove or disprove each.

1. $2 + 4 + 6 + \cdots + 2n = n(n + 1)$

   Step 1.   Prove $P(1)$ true: $2 \overset{?}{=} 1(1 + 1)$
   $$2 = 2; \quad \text{it is true}$$
   Step 2.   Assume $P(k)$.
   Hypothesis: $2 + 4 + 6 + \cdots + 2k = k(k + 1)$
   Step 3.   Prove $P(k + 1)$.
   To prove: $2 + 4 + 6 + \cdots + 2(k + 1) = (k + 1)(k + 2)$

Statements	*Reasons*
1. $2 + 4 + 6 + \cdots + 2k = k(k + 1)$	1. Hypothesis (Step 2)
2. $2 + 4 + 6 + \cdots + 2k + 2(k + 1)$	2. Add $2(k + 1)$ to both sides
$\quad = k(k + 1) + 2(k + 1)$	
3. $\quad = (k + 1)(k + 2)$	3. Factoring

   Step 4.   The proposition is true for all positive integers n by PMI.

2. $n^3 + 2n$ is divisible by 3.   A number is divisible by 3 if it has a factor of 3.
   Step 1.   Prove $P(1)$: $1^3 + 2 \cdot 1 = 3$, which is divisible by 3; and, since $1^3 + 2 \cdot 1 = 3$, $P(1)$ is true.
   Step 2.   Assume $P(k)$.
   Hypothesis: $k^3 + 2k$ is divisible by 3
   Step 3.   Prove $P(k + 1)$.
   To prove: $(k + 1)^3 + 2(k + 1)$ is divisible by 3

Statements	*Reasons*
1. $(k + 1)^3 + 2(k + 1)$ $= k^3 + 3k^2 + 3k + 1 + 2k + 2$	1. Distributive, associative, and commutative axioms
2. $= (3k^2 + 3k + 3) + (k^3 + 2k)$	2. Commutative and associative
3. $= 3(k^2 + k + 1) + (k^3 + 2k)$	3. Distributive
4. $3(k^2 + k + 1)$ is divisible by 3	4. Definition of divisibility by 3
5. $k^3 + 2k$ is divisible by 3	5. Hypothesis
6. $(k + 1)^3 + 2(k + 1)$ is divisible by 3	6. Both terms are divisible by 3; therefore, the sum is divisible by 3

   Step 4.   The proposition $P(n)$ is true for all positive integers n by PMI.

3. $n + 1$ is prime
   Step 1.   Prove $P(1)$: $1 + 1 = 2$ is a prime
   Step 2.   Assume $P(k)$.
   Hypothesis: $k + 1$ is a prime

Example 3 shows that even though we made an assumption in Step 2, it isn't going to change anything. If the proposition we are trying to prove is not true, making the assumption in Step 2 that it is true will NOT enable us to prove it true.

Step 3. Prove $P(k + 1)$.

To prove: $(k + 1) + 1$ is a prime; this is not possible since $(k + 1) + 1 = k + 2$, which is not prime whenever k is an even positive integer.

Step 4. Any conclusion? You can't conclude that it is false—only that induction doesn't work. But it is, in fact, false, and a counter-example is found by letting $n = 3$; then

$$n + 1 = 4 \quad \text{is not prime.}$$

4. $1 \cdot 2 \cdot 3 \cdot 4 \cdots \cdot n < 0$

Students often slip into the habit of skipping either the first or second step in a proof by mathematical induction. This is dangerous and it is important to check every step. Suppose a careless person didn't verify the first step. Then the results could be as follows:

Step 2. Assume $P(k)$.

Hypothesis: $1 \cdot 2 \cdot 3 \cdots \cdot k < 0$

Step 3. Prove $P(k + 1)$.

To prove: $1 \cdot 2 \cdot 3 \cdots \cdot k \cdot (k + 1) < 0$

Proof: $1 \cdot 2 \cdot 3 \cdots \cdot k < 0$ by hypothesis; $k + 1$ is positive since k is a positive integer

$$\underbrace{1 \cdot 2 \cdot 3 \cdots \cdot k}_{\text{negative}} \cdot \underbrace{(k + 1)}_{\text{positive}} < 0$$

Step 3 is proved.

Step 4. The proposition is not true for all positive integers, since the first step, $1 < 0$, does not hold.

PROBLEM SET 8.7

A Problems

1. Prove that
$$1 + 2 + 3 + \cdots + n = \frac{n(n + 1)}{2}$$
for all positive integers n.

2. Prove that
$$1^2 + 2^2 + 3^2 + \cdots + n^2 = \frac{n(n + 1)(2n + 1)}{6}$$
for all positive integers n.

3. Prove that
$$\sum_{i=1}^{n} (2i - 1)^2 = \frac{n(2n - 1)(2n + 1)}{3}$$
for all positive integers n.

4. Prove that

$$\sum_{j=1}^{n} j^3 = \frac{n^2(n+1)^2}{4}$$

for all positive integers n.

5. Prove that

$$2^2 + 4^2 + 6^2 + \cdots + (2n)^2 = \frac{2n(n+1)(2n+1)}{3}$$

for all positive integers n.

6. Prove that

$$1 \cdot 2 + 2 \cdot 3 + 3 \cdot 4 + \cdots + n(n+1) = \frac{n(n+1)(n+2)}{3}$$

for all positive integers n.

7. Prove that

$$1 \cdot 3 + 2 \cdot 4 + 3 \cdot 5 + \cdots + n(n+2) = \frac{n(n+1)(2n+7)}{6}$$

for all positive integers n.

8. Prove that

$$1 + r + r^2 + \cdots + r^n = \frac{r^{n+1} - 1}{r - 1}$$

for all positive integers n.

9. Prove that $n^5 - n$ is divisible by 5 for all positive integers n.
10. Prove that $n(n+1)(n+2)$ is divisible by 6 for all positive integers n.
11. Prove that $(1 + n)^2 \geq 1 + n^2$ for all positive integers n.

B Problems

12. Notice the following.

$$1^3 = 1^2$$
$$1^3 + 2^3 = 3^2$$
$$1^3 + 2^3 + 3^3 = 6^2$$
$$1^3 + 2^3 + 3^3 + 4^3 = 10^2$$

Make a conjecture based on the above pattern and then prove or disprove your conjecture.

13. Notice the following.

$$1 = 1$$
$$1 + 4 = 5$$
$$1 + 4 + 7 = 12$$
$$1 + 4 + 7 + 10 = 22$$

Make a conjecture based on the above pattern and then prove or disprove your conjecture.

Define $b^{n+1} = b^n \cdot b$ and $b^0 = 1$. Use this definition to prove the properties of exponents in Problems 14–17 for all positive integers n.

14. $b^m \cdot b^n = b^{m+n}$

15. $(b^m)^n = b^{mn}$

16. $(ab)^n = a^n b^n$

17. $\left(\dfrac{a}{b}\right)^n = \dfrac{a^n}{b^n}$

18. Prove that

$$\frac{1}{2} + \frac{1}{3} + \frac{1}{4} + \frac{1}{5} + \cdots + \frac{1}{2^n} < n$$

for all positive integers.

19. Prove that $2^n > n$ for all positive integers.
20. Prove the generalized distributive property:

$$a(b_1 + b_2 + b_3 + \cdots + b_n) = ab_1 + ab_2 + ab_3 + \cdots + ab_n.$$

21. Prove the generalized triangle inequality:

$$|a_1 + a_2 + \cdots + a_n| \le |a_1| + |a_2| + \cdots + |a_n|.$$

22. Prove that

$$\binom{k}{r} + \binom{k}{r-1} = \binom{k+1}{r}$$

using the formula

$$\binom{n}{k} = \frac{n!}{k!(n-k)!}.$$

23. The Binomial Theorem can be proved for any positive integer n by using mathematical induction. Fill in the missing steps and reasons. *To prove*:

$$(a + b)^n = \sum_{j=0}^{n} \binom{n}{j} a^{n-j} b^j$$

$$= \binom{n}{0} a^n + \binom{n}{1} a^{n-1} b + \binom{n}{2} a^{n-2} b^2 + \cdots + \binom{n}{r} a^{n-r} b^r$$

$$+ \cdots + \binom{n}{n-2} a^2 b^{n-2} + \binom{n}{n-1} ab^{n-1} + \binom{n}{n} b^n$$

Step 1. Prove it true for $n = 1$.
   a. Fill in these details.
Step 2. Assume it true for $n = k$.
   b. Fill in the statement of the hypothesis.
Step 3. Prove it true for $n = k + 1$.

$$\text{To prove: } (a + b)^{k+1} = \sum_{j=0}^{k+1} \binom{k+1}{j} a^{k+1-j} b^j$$

Proof:

c. Fill in the statement of the hypothesis.

d. Fill in these details; the final simplified form is

$$(a + b)^{k+1} = a^{k+1} + \cdots + \left[\binom{k}{r} + \binom{k}{r-1}\right]a^{k-r+1}b^r + \cdots + b^{k+1}.$$

e. Use Problem 22 to complete the proof.

Mind Bogglers

24. What is wrong with the following "proof," which results in $1 = 2$?

$$(a + b)^n = a^n + na^{n-1}b + \frac{n(n-1)}{2!}a^{n-2}b^2 + \cdots + nab^{n-1} + b^n$$

(from the Binomial Theorem)

Let $n = 0$. Then

$$(a + b)^0 = a^0 + 0 + 0 + \cdots + 0 + b^0$$

(substitution and zero multiplication)

$1 = 1 + 0 + 0 + \cdots + 0 + 1$ (definition of zero exponent)

$1 = 2$

25. What is wrong with the following "proof" that all horses are the same color? Let $P(n)$ be the proposition that every set of n horses is the same color.

Step 1. Prove $P(1)$. Clearly, every set of one horse has the property that all horses in the set are the same color.

Step 2. Assume $P(k)$. Assume that for each set of k horses, all the horses in the set have the same color.

Step 3. Prove $P(k + 1)$. Prove that for each set of $k + 1$ horses, all horses in the set have the same color. Number the horses $h_1, h_2, h_3, \ldots, h_{k+1}$. The subset $\{h_1, h_2, \ldots, h_k\}$ all have the same color by hypothesis. Replace h_k by h_{k+1}. The resulting set is a set of k horses of the same color by hypothesis. Now, h_1 and h_k are the same color, and h_1 and h_{k+1} are the same color, so all $k + 1$ horses are the same color.

Step 4. All horses are the same color by PMI.

Historical Note

The first appearance of Problem 25 was in an unpublished manuscript of Thomas Aquinas proving the existence of God, with "horses" replaced by "angels."

8.8 SUMMARY AND REVIEW

TERMS

Arithmetic sequence (8.1)
Arithmetic series (8.2)
Binomial Theorem (8.6)
Common difference (8.1)
Common ratio (8.1)
Factorial (8.5)
Finite sequence (8.1)
Geometric sequence (8.1)
Geometric series (8.3)

Infinite geometric series (8.4)
Infinite sequence (8.1)
Mathematical induction (8.7)
Pascal's triangle (8.5)
Progression (8.1)
Sequence (8.1)
Series (8.2)
Summation (8.5)

IMPORTANT IDEAS

(8.1, 8.2)   An arithmetic sequence is a function with a domain that consists of the set of counting numbers so that if any term is subtracted from the next term, the result is always a constant, called the *common difference, d*.

nth term: $s_n = s_1 + (n - 1)d$

sum: $S_n = n\left(\dfrac{s_1 + s_n}{2}\right)$ or $\dfrac{n}{2}[2s_1 + (n - 1)d]$

(8.1, 8.3)   A geometric sequence is a function with a domain that consists of the set of counting numbers so that if any term is divided into the next term, the result is always a constant, called the *common ratio, r*. If a is the first term, then:

nth term: $s_n = ar^{n-1}$

sum: $S_n = \dfrac{a(r^n - 1)}{r - 1}$ $(r \neq 1)$

(8.4)   If $s_1, s_2, \ldots, s_n, \ldots$, is an infinite geometric sequence with $s_1 = a$ and common ratio r so that $|r| < 1$, then

$$\lim_{n \to \infty} S_n = \frac{a}{1 - r}.$$

(8.5) n-factorial, written $n!$, is defined by

$$n! = 1 \cdot 2 \cdot 3 \cdots (n - 1) \cdot n; \qquad 0! = 1; \qquad 1! = 1.$$

PASCAL'S TRIANGLE

```
            1
          1   1
        1   2   1
      1   3   3   1
    1   4   6   4   1
  1   5  10  10   5   1
1   6  15  20  15   6   1
```

(8.5) $\displaystyle \binom{n}{r} = \frac{n!}{r!(n - r)!}$

where $\displaystyle \binom{n}{r}$ represents the number in the nth row and rth column of Pascal's triangle.

(8.6)   Binomial Theorem: for any positive integer n,

$$(a + b)^n = \sum_{k=0}^{n} \binom{n}{k} a^{n-k} b^k.$$

(8.7)   Principle of mathematical induction: if a given proposition $P(n)$ is true for $P(1)$ and if the truth of $P(k)$ implies the truth of $P(k + 1)$, then $P(n)$ is true for all positive integers n.

REVIEW PROBLEMS

1. (8.1, 8.2, 8.3)   Classify each sequence as arithmetic, geometric, or neither. Also, find an expression for the general term if it is an arithmetic or geometric sequence. If it is neither, find the pattern and give the next two terms.
    a. $1, 11, 21, 31, \ldots$ b. $1, 11, 111, 1111, \ldots$

c. $1, 11, 121, 1331, \ldots$ d. $54, 18, 6, 2, \ldots$

e. $1, 4, 9, 16, \ldots$

2. (8.2, 8.3)   Find the sum of the first ten terms of the arithmetic and geometric sequences in Problem 1.

3. (8.2, 8.3)   Find the missing quantities.

   a. $a = 5, r = 2; s_{10}, S_5$

   b. $a = 50, d = -5; s_{10}, S_5$

   c. $s_1 = 2, s_{10} = 20; d, S_{10}$

   d. $s_1 = 512, s_{10} = 1; r, S_{10}$

4. a. (8.4) Find $-2.\overline{18}$ as the quotient of two integers by considering an infinite geometric series.

   b. (8.4) Find the sum of the infinite series $100 + 50 + 25 + \cdots$.

5. (8.2, 8.3, 8.5)   Evaluate:

   a. $\left(\dfrac{8}{4}\right)!$ b. $\dbinom{8}{4}$ c. $\dbinom{p}{q}$ d. $\displaystyle\sum_{k=1}^{100} [5 + (k - 1)4]$ e. $\displaystyle\sum_{k=1}^{10} 2(3)^{k-1}$

6. (8.6)   Find:

   a. $(x - y)^5$ b. $(2x + y)^5$

   c. The coefficient of $x^8 y^4$ in the expansion of $(x + 2y)^{12}$

7. (8.2, 8.3)   Suppose someone tells you she has traced her family tree back ten generations. How many people does she have on her family tree if there were no intermarriages?

Count her mother and father as the first generation back.

8. (8.3)   A certain type of bacterium divides into two bacteria every 20 min. If there are 1024 bacteria in the culture now, how many will there be in 24 hr, assuming that no bacteria die? Leave your answer in exponential form.

9. (8.2, 8.3)   Three numbers that sum to 21 form a geometric sequence. If 1 is added to the first number and 4 is subtracted from the third number, the resulting numbers form an arithmetic sequence. What are the numbers of the original geometric sequence?

10. (8.7)   Prove that $4 + 8 + 12 + \cdots + 4n = 2n(n + 1)$ for all positive integers n by mathematical induction.

Analytic Geometry

9

9.1   INTRODUCTION TO CONICS

In Chapter 2, we looked at quadratic functions of the form

$$y = ax^2 + bx + c \quad (a \neq 0).$$

Not all second-degree equations represent functions. For example,

$$y = x^2$$

is a function, but

$$x = y^2$$

is not, even though both are second-degree equations.

We shall now consider the general second-degree equation

$$Ax^2 + Bxy + Cy^2 + Dx + Ey + F = 0$$

for any constants A, B, C, D, E, and F. If $A = B = C = 0$, then the equation is not quadratic but linear (first-degree); and if $A \neq 0$ or $B \neq 0$ or $C \neq 0$ (or all three not zero), the equation is quadratic.

Historically, second-degree equations in two variables were first considered in a geometrical context and were called *conic sections* because the curves they represent can be described as the intersections of a right circular cone and a plane. There are three general ways a plane can intersect a cone, as shown in Figure 9.1.

CONIC SECTIONS

The cone is thought of as extending infinitely in both directions. The part of the cone on each side of the vertex is called a *nappe*.

Parabola
The plane is parallel to one of the generators of the curve.

Ellipse
The plane intersects only one nappe. A circle is a special ellipse in which the plane is perpendicular to the axis of the cone.

Hyperbola
The plane intersects both nappes of the cone.

Figure 9.1   Conic sections

Let's consider the parabola again.

GEOMETRICAL DEFINITION OF A PARABOLA

FOCUS, DIRECTRIX

DEFINITION: A parabola is a set of all points in the plane equidistant from a given point (called the *focus*) and a given line (called the *directrix*).

AXIS OF A PARABOLA

To obtain a parabola from this definition, we'll use the special type of graph paper shown in Figure 9.2, where F is the focus and L is the directrix. We plot points in the plane equidistant from the focus and the directrix. Then, we draw a line through the focus and perpendicular to the directrix. This line is called the *axis* of the parabola. Let V be the point on this line halfway between the focus and the directrix. This is the point of the parabola nearest

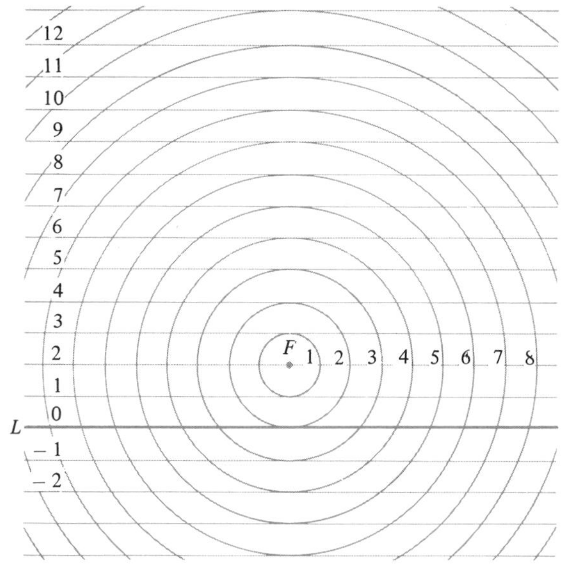

Figure 9.2   Parabola graph paper

to either the focus or the directrix. It is called the *vertex*, or *center*, of the parabola. We plot other points equidistant from F and L, as shown in Figure 9.3.

The vertex and the center of a parabola refer to the same point.

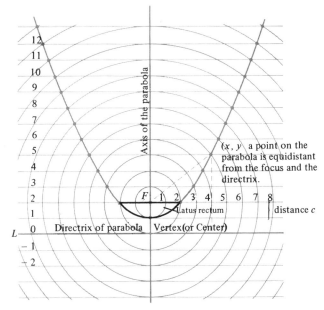

Figure 9.3   Parabola graphed from definition

In Figure 9.3, let c be the distance from the center to the focus. Notice that the distance from the center to the directrix is also c. Consider the segment that passes through the focus perpendicular to the axis and with endpoints on the parabola. This segment has length $4c$ and is called the *latus rectum*.

LATUS RECTUM

To obtain the equation of a parabola, first consider a special case—a parabola with focus $F(0, c)$ and directrix $y = -c$, where c is any positive number. This parabola must have its center at the origin (remember that the center is halfway between the focus and the directrix) and must open up, as shown in Figure 9.4.

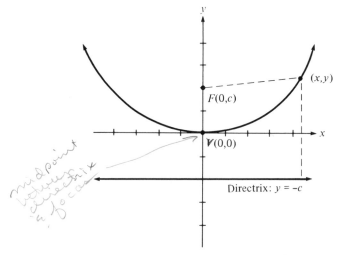

Figure 9.4   Graph of the parabola $x^2 = 4cy$

Let (x, y) be any point on the parabola. Then, from the definition of a parabola,

distance from (x, y) to $(0, c)$ = distance from (x, y) to directrix

or

$$\sqrt{(x - 0)^2 + (y - c)^2} = y + c.$$

Square both sides:

Simplifying, we get

$$x^2 + (y - c)^2 = (y + c)^2$$
$$x^2 + y^2 - 2cy + c^2 = y^2 + 2cy + c^2$$
$$x^2 = 4cy$$

$$x^2 = 4cy,$$

which is the equation of this parabola.

You can repeat the argument above for parabolas that have center at the origin and open down, left, and right to obtain the results summarized in the table. A positive number c, the distance from the focus to the center, is assumed given

Parabola	Focus	Directrix	Center	Equation
Opens up	$(0, c)$	$y = -c$	$(0, 0)$	$x^2 = 4cy$
Opens down	$(0, -c)$	$y = c$	$(0, 0)$	$x^2 = -4cy$
Opens right	$(c, 0)$	$x = -c$	$(0, 0)$	$y^2 = 4cx$
Opens left	$(-c, 0)$	$x = c$	$(0, 0)$	$y^2 = -4cx$

STANDARD FORM
EQUATIONS FOR
PARABOLAS

EXAMPLES

1. Graph $x^2 = 8y$.

    Solution: Recognize this equation as a parabola that opens up. The center is $(0, 0)$, and notice by inspection that

    $$4c = 8$$
    $$c = 2.$$

    Thus, the focus is $(0, 2)$. After plotting the center, $V(0, 0)$, and the focus, $F(0, 2)$, the only question is the "fatness" of the parabola. Remember that the *length of the latus rectum is* $4c$, so in this case it is 8. Since a parabola is symmetric with respect to its axis, draw a segment of length 8 with the midpoint at F. Using these three points (the center and the endpoints of the latus rectum), sketch the parabola, as shown in Figure 9.5.

    In Chapter 2, we plotted a couple of points or found the y-intercept. The method in these examples is more useful and efficient for determining the graph of a parabola.

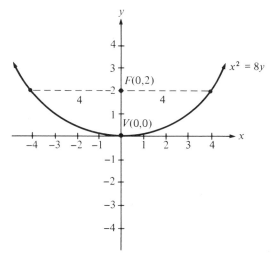

Figure 9.5   Graph of the parabola $x^2 = 8y$

2. Graph $y^2 = -12x$.

   Solution: Recognize this equation as a parabola that opens left. The center is $(0, 0)$ and

   $$4c = 12$$
   $$c = 3$$

   (recall that c is positive), so the focus is $(-3, 0)$. The length of the latus rectum is 12 and we draw the parabola as shown in Figure 9.6.

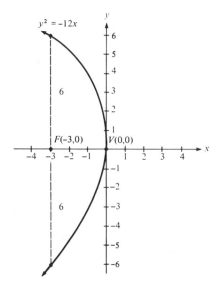

Figure 9.6   Graph of the parabola $y^2 = -12x$

3. Graph $2y^2 - 5x = 0$.

   Solution: You must first put the equation into standard form by solving for the second-degree term.

   $$y^2 = \frac{5}{2}x$$

   The center is $(0, 0)$, and

   $$4c = \frac{5}{2}$$

   $$c = \frac{5}{8}.$$

   Thus, the parabola opens to the right, the focus is $(5/8, 0)$, and the length of the latus rectum is $5/2$, as shown in Figure 9.7.

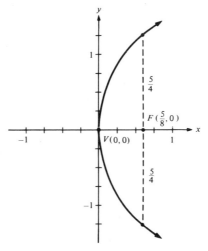

Figure 9.7   Graph of the parabola $2y^2 - 5x = 0$

4. Find the equation of the parabola with directrix $y = 4$ and focus at $F(0, -4)$.

    Solution: This curve is a parabola that opens down with center at the origin, as shown in Figure 9.8. Notice that $c = 4$ and $4c = 16$. Since the equation is of the form

    $$x^2 = -4cy,$$

    we see that the desired equation is

    $$x^2 = -16y.$$

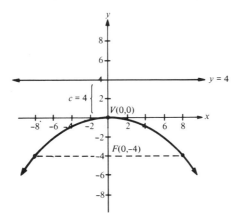

Figure 9.8   Graph of the parabola with focus at $(0, -4)$ and directrix $y = 4$

5. Find the equation of the parabola with directrix $x = 3$ and focus at $F(-3, 0)$.

Solution: This curve is a parabola that opens to the left and is of the form

$$y^2 = -4cx.$$

Since $c = 3$, the desired equation is

$$y^2 = -12x,$$

and the graph is drawn as shown in Figure 9.9.

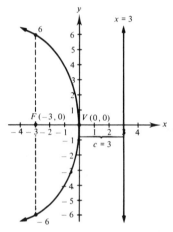

Figure 9.9 Graph of the parabola with directrix $x = 3$ and focus $(-3, 0)$

We are very limited in the parabolas we've been considering, since we assumed that the center is at the origin and the directrix is parallel to one of the coordinate axes. Suppose, however, that you're given a parabola with center at (h, k) and a directrix parallel to one of the coordinate axes. In Section 1.5, we showed that we could translate the axes to (h, k) by a substitution,

$$x' = x - h$$
$$y' = y - k.$$

So we can write the equations of translated parabolas as follows:

These are the general equations for parabolas and should be memorized.

Parabola	Focus	Directrix	Center	Equation
Opens up	$(h, k + c)$	$y = k - c$	(h, k)	$(x - h)^2 = 4c(y - k)$
Opens down	$(h, k - c)$	$y = k + c$	(h, k)	$(x - h)^2 = -4c(y - k)$
Opens right	$(h + c, k)$	$x = h - c$	(h, k)	$(y - k)^2 = 4c(x - h)$
Opens left	$(h - c, k)$	$x = h + c$	(h, k)	$(y - k)^2 = -4c(x - h)$

EXAMPLES

6. Sketch $x^2 + 4y + 8x + 4 = 0$.
    Solution:
    Step 1. Associate together the terms involving the variable that is squared.

    $$x^2 + 8x = -4y - 4$$

    Step 2. Complete the square for the variable that is squared.

    coefficient is 1
    $\downarrow$

    $$x^2 + 8x + \left(\frac{1}{2} \cdot \mathbf{8}\right)^2 = -4y - 4 + \left(\frac{1}{2} \cdot \mathbf{8}\right)^2$$
    $\uparrow$

    $\frac{1}{2}$ of this coefficient, squared, is added to both sides

    $$x^2 + 8x + \mathbf{16} = -4y - 4 + \mathbf{16}$$
    $$(x + 4)^2 = -4y + 12$$

    If necessary, review Section 2.4, page 85, for the procedure of completing the square.

    Step 3. Factor out the coefficient of the first-degree term.

    $$(x + 4)^2 = -4(y - 3)$$

    Step 4. Determine the center by inspection. Plot (h, k); in this example, the center is $(-4, 3)$. (See Figure 9.10.)
    Step 5. Determine the focus. By inspection, $4c = 4$, $c = 1$, and the parabola opens down from the center, as shown in Figure 9.10.
    Step 6. Plot the endpoints of the latus rectum; $4c = 4$. Draw the parabola as shown in Figure 9.10.

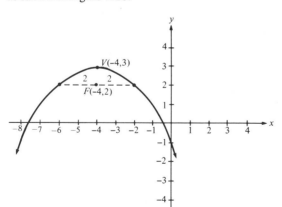

Figure 9.10   Graph of $x^2 + 4y + 8x + 4 = 0$

7. Sketch $2y^2 + 6y + 5x + 10 = 0$.
    Solution: $2y^2 + 6y = \mathbf{-5x - 10}$

    $$y^2 + 3y = -\frac{5}{2}x - 5$$

    Divide both sides by 2 so the leading coefficient is 1; then complete the square.

$$y^2 + 3y + \frac{9}{4} = -\frac{5}{2}x - 5 + \frac{9}{4}$$

$$\left(y + \frac{3}{2}\right)^2 = -\frac{5}{2}x - \frac{11}{4}$$

$$\left(y + \frac{3}{2}\right)^2 = -\frac{5}{2}\left(x + \frac{11}{10}\right)$$

The center is $(-\frac{11}{10}, -\frac{3}{2})$ and $4c = \frac{5}{2}$, $c = \frac{5}{8}$. We sketch the curve as shown in Figure 9.11.

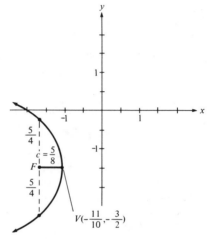

Figure 9.11   Graph of $2y^2 + 6y + 5x + 10 = 0$

8. Find the equation of the parabola with focus at $(4, -3)$ and directrix the line $x + 2 = 0$.

    Solution: Sketch the given information, as shown in Figure 9.12. The center

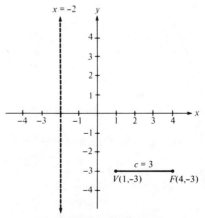

Figure 9.12

is $(1, -3)$ since it must be equidistant from F and the directrix. Note that $c = 3$. Thus, substitute into the equation

$$(y - k)^2 = 4c(x - h)$$

since the parabola opens to the right. The desired equation is

$$(y + 3)^2 = 12(x - 1).$$

9. Find the equation of the parabola with center at $(2, -1)$, axis parallel to the y-axis, and passing through $(3, 2)$.

    Solution: Sketch the given information, as shown in Figure 9.13.

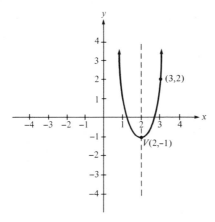

Figure 9.13

The parabola opens up and thus has the form

$$(x - h)^2 = 4c(y - k).$$

Since the center is $(2, -1)$,

$$(x - 2)^2 = 4c(y + 1).$$

Also, since it passes through $(3, 2)$, this point must satisfy the equation:

$$(3 - 2)^2 = 4c(2 + 1).$$

Solving for c, we find

$$c = \frac{1}{12}.$$

Therefore, the desired equation is

$$(x - 2)^2 = \frac{1}{3}(y + 1).$$

PROBLEM SET 9.1

A Problems

Sketch the curves in Problems 1–12. Label the focus F and the vertex V.

1. $y^2 = 8x$
2. $y^2 = -20x$
3. $4x^2 = 10y$
4. $2x^2 = -4y$
5. $2x^2 + 5y = 0$
6. $3y^2 - 15x = 0$
7. $4y^2 + 3x = 12$
8. $y^2 - 4x + 10y + 13 = 0$
9. $y^2 + 4x - 3y + 1 = 0$
10. $x^2 + 9y - 6x + 18 = 0$
11. $2y^2 + 8y - 20x + 148 = 0$
12. $9x^2 + 6x + 18y - 23 = 0$

Find the equation of each curve in Problems 13–20. Sketch the curve.

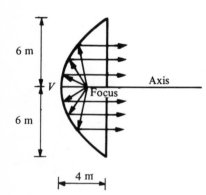

6 m

6 m

V

Focus

Axis

4 m

13. Directrix $x = 0$; focus at $(5, 0)$
14. Directrix $y = 0$; focus at $(0, -3)$
15. Directrix $x - 3 = 0$; vertex at $(-1, 2)$
16. Directrix $y + 4 = 0$; vertex at $(4, -1)$
17. Vertex at $(-2, -3)$; focus at $(-2, 3)$
18. Vertex at $(-3, 4)$; focus at $(1, 4)$
19. The set of all points with distances from $(4, 3)$ that equal their distances from $(0, 3)$
20. The set of all points with distances from $(4, 3)$ that equal their distances from $(-2, 1)$
21. Find the equation of the parabola with center at $(-3, 2)$, axis parallel to the y-axis, and which passes through $(-2, -1)$.
22. Find the equation of the parabola with center at $(4, 2)$, axis parallel to the x-axis, and which passes through $(-3, -4)$.

A *parabolic reflector* (or *parabolic mirror*) has the property that, if a source of light is placed at the *focus* of the mirror, the light rays will reflect from the mirror as rays parallel to the axis. Such a parabolic reflector is used in automobile headlamps to deliver an intense concentrated beam of light. For its source of light, the headlamp has a filament to which electric wiring connections from the battery furnish a filament voltage. For safer night driving, we do not want all light rays to be

B Problems

 23. If the path of a baseball is parabolic and is 200 ft wide at the base and 50 ft high in the center, write the equation that gives the path of the baseball if we let the origin be the point of departure for the ball.

 24. A parabolic archway has the dimensions shown in Figure 9.14. Find the equation of the parabolic portion.

 25. A radar antenna is constructed so that a cross section through its axis is a parabola with the receiver at the focus. Find the focus if the antenna is 12 m across and its depth is 4 m.

 26. If the diameter of a parabolic reflector is 16 cm and the depth is 8 cm, find the focus.

 27. Derive the equation of a parabola with $F(0, -c)$, where c is a positive number and the directrix is the line $y = c$.

 28. Derive the equation of a parabola with $F(c, 0)$, where c is a positive number and the directrix is the line $x = -c$.

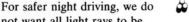 29. Derive the equation of a parabola with $F(-c, 0)$, where c is a positive number and the directrix is the line $x = c$.

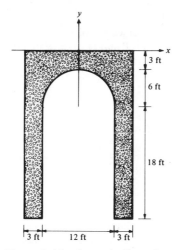

Figure 9.14   A parabolic archway

 30. Show that the length of the latus rectum for the parabola $y^2 = 4cx$ is $4c$ $(c > 0)$.

Find the points of intersection (if any) for each line and parabola in Problems 31–34. Show the result both algebraically and geometrically.

31. $\begin{cases} y = 2x + 10 \\ y = x^2 + 4x + 7 \end{cases}$ 32. $\begin{cases} y = -2x + 4 \\ y = x^2 - 12x + 25 \end{cases}$

33. $\begin{cases} 2x + y - 7 = 0 \\ y^2 - 6y - 4x + 17 = 0 \end{cases}$ 34. $\begin{cases} 3x + 2y - 5 = 0 \\ y^2 + 4y + 3x - 4 = 0 \end{cases}$

 35. Douglas Dodge, a physical education expert, has made a study and determined that a woman who runs the 100-m dash reaches her peak at age 20. Her time T, for a 100-m dash at a particular age, A, is

$$T = c_1(A - 20)^2 + c_2$$

for constants c_1 and c_2. If Wyomia Tyus ran the race when she was 16 years old in 11.4 sec and when she was 20 in 11.0 sec, predict her time for running the race when she is 40 years old. Graph this function.

36. According to another physical education expert, Greg O'Neil, the peak age for a woman runner is 24. Using this information, answer the questions posed in Problem 35.

Mind Bogglers

 37. Let $L: Ax + By + C = 0$ be any nonvertical line and let $P(x_0, y_0)$ be any point not on the line.
   a. Find the slope of L.
   b. Let L' be a line through P perpendicular to L. Find the slope of L'.

parallel to the axis. Some light must be aimed far down the road; some light must be spread to the side of the road; and some light must be distributed upward to illuminate high objects such as bridges or overhead signs. How do we spread and aim light beams to these different directions? If we deliberately *offset* the filament from the focus, we change the beam entirely. In our current four-lamp system, we have one-filament sealed-beam units in two headlamps and two-filament sealed-beam units in the other two. The position of the filaments in relation to the reflector accomplishes most of the desired illumination patterns. The rest of the light spreading is taken care of by the design of the lenses. These contain flutes and prisms designed to bend light rays.

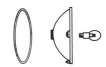

c. Find the equation of L'.

d. Let Q be the point of intersection of L and L'. Find Q.

e. Find the distance between P and Q to show that the distance from a point to a line is

$$d = \frac{|Ax_0 + By_0 + C|}{\sqrt{A^2 + B^2}}.$$

Hint for Problems 38 and 39: use the definition of a parabola and Problem 37.

38. Find the equation of the parabola with focus at $(4, -3)$ and directrix $x - y + 3 = 0$.

39. Find the equation of the parabola with focus at $(3, -5)$ and directrix $12x - 5y + 4 = 0$.

9.2 ELLIPSES

The second conic we'll consider in this chapter is called an *ellipse*.

ELLIPSE

DEFINITION: An *ellipse* is the set of all points in the plane such that, for each point on the curve, the sum of its distances from two fixed points is a constant.

The fixed points are called the *foci* (plural for *focus*). To see what an ellipse looks like, we will use the special type of graph paper shown in Figure 9.15, where F_1 and F_2 are the foci.

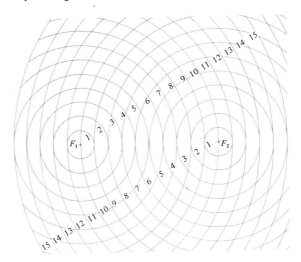

Figure 9.15

Let the given constant be 12. Plot all the points in the plane so that the sum of their distances from the foci is 12. For example, if a point is 8 units from F_1, then it is 4 units from F_2, and you can plot the points P_1 and P_2 as shown in Figure 9.16.

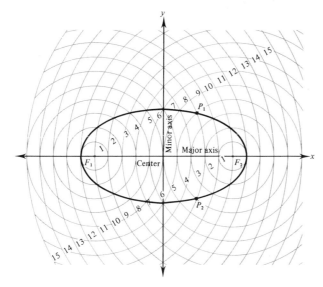

Figure 9.16 Ellipse

The line passing through F_1 and F_2 is called the *major axis*. The *center* is the midpoint of the segment $\overleftrightarrow{F_1F_2}$. The line passing through the center perpendicular to the major axis is called the *minor axis*. The ellipse is symmetric with respect to both the major and minor axes.

To find the equation of an ellipse, first consider a special case where the center is at the origin. Let the distance from the center to a focus be the positive number c; that is, let $F_1(-c, 0)$ and $F_2(c, 0)$ be the foci and let the constant distance be $2a$, as shown in Figure 9.17. Notice that the center of this

MAJOR AXIS, CENTER

MINOR AXIS

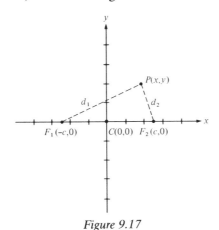

Figure 9.17

ellipse is $(0, 0)$. Let $P(x, y)$ be any point on the ellipse. By using the distance formula and the definition of an ellipse, we obtain

$$d_1 + d_2 = 2a,$$

or $$\sqrt{(x + c)^2 + (y - 0)^2} + \sqrt{(x - c)^2 + (y - 0)^2} = 2a.$$

Simplifying, we get

Isolate one radical.

$$\sqrt{(x + c)^2 + y^2} = 2a - \sqrt{(x - c)^2 + y^2}$$

Square both sides.

$$(x + c)^2 + y^2 = 4a^2 - 4a\sqrt{(x - c)^2 + y^2} + (x - c)^2 + y^2$$

$$x^2 + 2cx + c^2 + y^2 = 4a^2 - 4a\sqrt{(x - c)^2 + y^2} + x^2 - 2cx + c^2 + y^2$$

Since $a \neq 0$, divide by $4a$.

$$4a\sqrt{(x - c)^2 + y^2} = 4a^2 - 4cx$$

$$\sqrt{(x - c)^2 + y^2} = a - \frac{c}{a}x$$

Square both sides again.

$$(x - c)^2 + y^2 = \left(a - \frac{c}{a}x\right)^2$$

$$x^2 - 2cx + c^2 + y^2 = a^2 - 2cx + \frac{c^2}{a^2}x^2$$

$$x^2 + y^2 = a^2 - c^2 + \frac{c^2}{a^2}x^2$$

$$x^2 - \frac{c^2}{a^2}x^2 + y^2 = a^2 - c^2$$

$$\left(1 - \frac{c^2}{a^2}\right)x^2 + y^2 = a^2 - c^2$$

$$\frac{(a^2 - c^2)}{a^2}x^2 + y^2 = a^2 - c^2$$

Divide both sides by $a^2 - c^2$.

$$\frac{x^2}{a^2} + \frac{y^2}{a^2 - c^2} = 1.$$

Letting $b^2 = a^2 - c^2$, we find

$$\frac{x^2}{a^2} + \frac{y^2}{b^2} = 1.$$

If $x = 0$, the y-intercepts are obtained:

$$\frac{y^2}{b^2} = 1.$$

$$y = \pm b.$$

If $y = 0$, the x-intercepts are obtained: $x = \pm a$. The intercepts on the major axis are called the *vertices* of the ellipse.

VERTICES

Ellipse	Foci	Constant distance	Center	Equation
Horizontal	$(-c, 0), (c, 0)$	$2a$	$(0, 0)$	$\dfrac{x^2}{a^2} + \dfrac{y^2}{b^2} = 1$
Vertical	$(0, c), (0, -c)$	$2a$	$(0, 0)$	$\dfrac{y^2}{a^2} + \dfrac{x^2}{b^2} = 1$

where $b^2 = a^2 - c^2$ or $c^2 = a^2 - b^2$

STANDARD FORM EQUATIONS FOR ELLIPSES

EXAMPLE 1: Sketch

$$\frac{x^2}{9} + \frac{y^2}{4} = 1.$$

Solution: The center of the ellipse is $(0, 0)$. The x-intercepts are ± 3 (these are the vertices) and the y-intercepts are ± 2. Sketch the ellipse, as shown in Figure 9.18.

The foci can also be found, since

$$c^2 = a^2 - b^2$$
$$c^2 = 9 - 4$$
$$c = \pm\sqrt{5}.$$

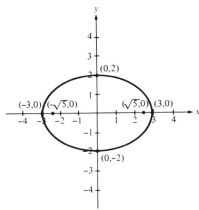

Figure 9.18   Graph of $\dfrac{x^2}{9} + \dfrac{y^2}{4} = 1$

The equation of the ellipse with major axis vertical, $F_1(0, c)$, $F_2(0, -c)$, and constant distance $2a$ is found in a similar fashion. Simplifying the equation as before, we find

$$\frac{y^2}{a^2} + \frac{x^2}{b^2} = 1,$$

where $b^2 = a^2 - c^2$.

Notice that in both cases, a^2 must be larger than both c^2 and b^2. If it were not, we would have a square number equal to a negative, which does not exist in the set of real numbers.

EXAMPLE 2: Sketch

$$\frac{x^2}{4} + \frac{y^2}{9} = 1.$$

The foci are found by

$$c^2 = a^2 - b^2$$
$$c^2 = 9 - 4$$
$$c = \pm\sqrt{5}.$$

Solution: $a^2 = 9$ and $b^2 = 4$, which is an ellipse with major axis vertical. The x-intercepts are ±2 and the y-intercepts are ±3 (these are the vertices). The sketch is shown in Figure 9.19.

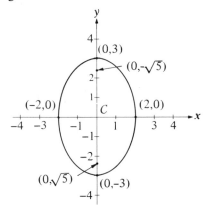

Remember, the foci are always on the major axis, so we plot $(0, \sqrt{5})$ and $(0, -\sqrt{5})$ for the foci.

Figure 9.19   Graph of $\dfrac{x^2}{4} + \dfrac{y^2}{9} = 1$

The equations of the ellipses with center at (h, k) should be memorized. The segment from the center to a vertex on the major axis is called a *semi-major axis* and has length a; the segment from the center to an intercept on the minor axis is called a *semi-minor axis* and has length b.

If we let the center of the ellipse be at (h, k), we can translate the coordinate axes to obtain the results given in the table.

| Ellipse | Center | Foci | Intercepts | | Equation |
			Major axis	Minor axis	
Horizontal	(h, k)	$(h + c, k)$ $(h - c, k)$	$(h - a, k)$ $(h + a, k)$	$(h, k + b)$ $(h, k - b)$	$\dfrac{(x - h)^2}{a^2} + \dfrac{(y - k)^2}{b^2} = 1$
Vertical	(h, k)	$(h, k + c)$ $(h, k - c)$	$(h, k + a)$ $(h, k - a)$	$(h - b, k)$ $(h + b, k)$	$\dfrac{(y - k)^2}{a^2} + \dfrac{(x - h)^2}{b^2} = 1$

EXAMPLES

3. Graph

$$\frac{(x-3)^2}{25} + \frac{(y-1)^2}{16} = 1.$$

Solution:

Step 1.   Plot the center (h, k). By inspection, the center of this ellipse is (3, 1). This becomes the center of a new translated coordinate system. The vertices and foci are now measured with reference to the new origin at (3, 1).

Step 2.   Plot the x- and y-intercepts. These are ± 5 and ± 4, respectively. Remember to measure these distances from (3, 1), as shown in Figure 9.20.

The foci are found by

$$c^2 = a^2 - b^2$$
$$= 25 - 16$$
$$= 9.$$

The distance from the center to either focus is 3, so the coordinates of the foci are (6, 1) and (0, 1).

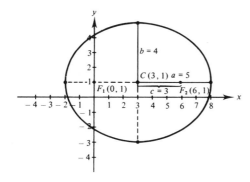

Figure 9.20   Graph of $\dfrac{(x-3)^2}{25} + \dfrac{(y-1)^2}{16} = 1$

4. Graph $3x^2 + 4y^2 + 24x - 16y + 52 = 0$.

Solution:

Step 1.   Associate together the x and the y terms:

$$(3x^2 + 24x) + (4y^2 - 16y) = -52.$$

Step 2.   We have to complete the square in both x and y. This requires that the coefficients of the squared terms be 1. We can accomplish this by factoring:

$$3(x^2 + 8x \quad) + 4(y^2 - 4y \quad) = -52.$$

Next, complete the square for both x and y, being sure to add the same number to both sides:

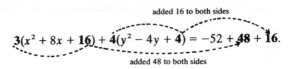

$$3(x^2 + 8x + 16) + 4(y^2 - 4y + 4) = -52 + 48 + 16.$$

added 16 to both sides

added 48 to both sides

Step 3. Factor:

$$3(x + 4)^2 + 4(y - 2)^2 = 12.$$

Step 4. Divide both sides by 12:

$$\frac{(x + 4)^2}{4} + \frac{(y - 2)^2}{3} = 1.$$

Step 5. Plot the center (h, k). By inspection, you can see the center is $(-4, 2)$. The vertices are at ± 2, and the length of the semi-minor axis is $\sqrt{3}$, as shown in Figure 9.21.

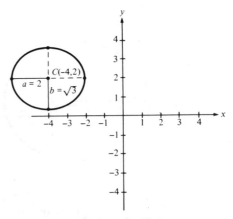

Figure 9.21 Graph of $3x^2 + 4y^2 + 24x - 16y + 52 = 0$

We have seen that some ellipses are more circular and some more flat than others. A measure of the amount of flatness of an ellipse is called its *eccentricity*, which is defined as

ECCENTRICITY, ε

$$\varepsilon = \frac{c}{a}.$$

Notice that

$$\varepsilon = \frac{c}{a} = \frac{\sqrt{a^2 - b^2}}{a} = \sqrt{\frac{a^2 - b^2}{a^2}} = \sqrt{1 - \left(\frac{b}{a}\right)^2}.$$

Since $c < a$, we see that ε is between 0 and 1. If $a = b$, then $\varepsilon = 0$ and the conic is a *circle*. If the ratio b/a is small, then the ellipse is very flat. Thus, for an ellipse,

CIRCLE

$$0 \le \varepsilon < 1$$

and ε measures the amount of roundness of the ellipse. The closer that ε is to 1, the more flat the ellipse becomes.

Let's take a closer look at the case where $a = b$. This common distance is called the *radius* of the circle and is denoted by r. Thus, when $a = b = r$, we have

$$\frac{(x - h)^2}{r^2} + \frac{(y - k)^2}{r^2} = 1,$$

RADIUS

which is the equation we found in Section 2.3.

EXAMPLES

5. Graph $x^2 + y^2 + 6x - 14y + 22 = 0$.
   Solution: Complete the square in x and y.

$$(x^2 + 6x \quad) + (y^2 - 14y \quad) = -22$$
$$(x^2 + 6x + 9) + (y^2 - 14y + 49) = -22 + 9 + 49$$
$$(x + 3)^2 + (y - 7)^2 = 36$$

This is a circle with center at $(-3, 7)$ and radius 6, as shown in Figure 9.22.

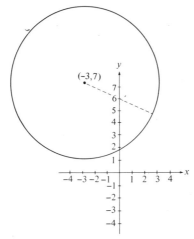

Figure 9.22   Graph of $x^2 + y^2 + 6x - 14y + 22 = 0$

6. Find the equation of the circle passing through the points $(-3, 4)$, $(4, 5)$, and $(1, -4)$. Sketch the graph.
   Solution: Now, $(x - h)^2 + (y - k)^2 = r^2$ can be written as

$$x^2 + y^2 + C_1 x + C_2 y + C_3 = 0$$

for some constants C_1, C_2, and C_3. Since the points lie on the circle, they satisfy the equation. Thus,

$$(-3, 4): \quad -3C_1 + 4C_2 + C_3 = -25,$$
$$(4, 5): \quad 4C_1 + 5C_2 + C_3 = -41,$$
$$(1, -4): \quad C_1 - 4C_2 + C_3 = -17.$$

This is a system of three equations and three unknowns that can be solved simultaneously to find $C_1 = -2$, $C_2 = -2$, and $C_3 = -23$. Thus,

$$x^2 + y^2 - 2x - 2y - 23 = 0$$
$$(x^2 - 2x + 1) + (y^2 - 2y + 1) = 23 + 1 + 1$$
$$(x - 1)^2 + (y - 1)^2 = 25.$$

The graph is shown in Figure 9.23.

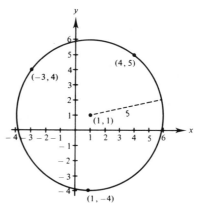

Figure 9.23

7. Find the equation of the ellipse with vertices at $(\pm 5, 0)$ and foci at $(\pm 3, 0)$.
    Solution: By inspection, the ellipse is horizontal and is centered at the origin, where $a = 5$ and $c = 3$. Thus

$$c^2 = a^2 - b^2$$
$$9 = 25 - b^2$$
$$b^2 = 16.$$

The equation is

$$\frac{x^2}{25} + \frac{y^2}{16} = 1.$$

8. Find the equation of the ellipse with vertices at $(3, 2)$ and $(3, -4)$ and foci at $(3, \sqrt{5} - 1)$ and $(3, -\sqrt{5} - 1)$.
    Solution: By inspection, the ellipse is vertical and is centered at $(3, -1)$, where $a = 3$, $c = \sqrt{5}$ (see Figure 9.24). Thus,

$$c^2 = a^2 - b^2$$
$$5 = 9 - b^2$$
$$b^2 = 4.$$

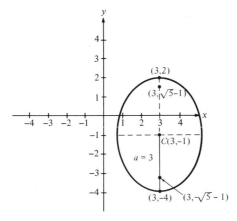

Figure 9.24 Sketch for Example 8

The equation is

$$\frac{(y + 1)^2}{9} + \frac{(x - 3)^2}{4} = 1.$$

9. Find the equation of the set of all points with the sum of distances from $(-3, 2)$ and $(5, 2)$ equal to 16.
   Solution: By inspection, the ellipse is horizontal and is centered at $(1, 2)$. We are given

$$2a = 16$$
$$a = 8,$$

and we see that $c = 4$. Thus,

$$c^2 = a^2 - b^2$$
$$16 = 64 - b^2$$
$$b^2 = 48.$$

The equation is

$$\frac{(x - 1)^2}{64} + \frac{(y - 2)^2}{48} = 1.$$

10. Find the equation of the ellipse with foci at $(-3, 6)$ and $(-3, 2)$ with $\varepsilon = \frac{1}{5}$.
    Solution: By inspection, the ellipse is vertical and is centered at $(-3, 4)$ with $c = 2$. Since

$$\varepsilon = \frac{c}{a} = \frac{1}{5}$$

and $c = 2$, then

$$\frac{2}{a} = \frac{1}{5},$$

Just because

$$\frac{c}{a} = \frac{1}{5},$$

you cannot assume that $c = 1$ and $a = 5$; all you know is that the *ratio* of c to a is $1/5$.

which implies $a = 10$. Also,

$$c^2 = a^2 - b^2$$
$$4 = 100 - b^2$$
$$b^2 = 96.$$

Thus, the equation is

$$\frac{(y - 4)^2}{100} + \frac{(x + 3)^2}{96} = 1.$$

PROBLEM SET 9.2

A Problems

Sketch the curves in Problems 1–16.

1. $x^2 + y^2 = 1$

2. $\dfrac{x^2}{4} + \dfrac{y^2}{9} = 1$

3. $\dfrac{x^2}{25} + \dfrac{y^2}{36} = 1$

4. $x^2 + \dfrac{y^2}{9} = 1$

5. $4x^2 + 9y^2 = 36$

6. $25x^2 + 16y^2 = 400$

7. $36x^2 + 25y^2 = 900$

8. $3x^2 + 2y^2 = 6$

9. $4x^2 + 3y^2 = 12$

10. $5x^2 + 10y^2 = 7$

11. $(x - 2)^2 + (y + 3)^2 = 25$

12. $\dfrac{(x + 3)^2}{81} + \dfrac{(y - 1)^2}{49} = 1$

13. $\dfrac{(x - 3)^2}{16} + \dfrac{(y - 2)^2}{9} = 1$

14. $\dfrac{(x + 2)^2}{25} + \dfrac{(y + 4)^2}{9} = 1$

15. $3(x + 1)^2 + 4(y - 1)^2 = 12$

16. $10(x - 5)^2 + 6(y + 2)^2 = 60$

Find the equations of the curves in Problems 17–23. Sketch the curves.

17. The set of points 6 units from the point $(4, 5)$
18. The set of points such that the sum of the distances from $(-6, 0)$ and $(6, 0)$ is 20
19. The ellipse with vertices at $(0, 7)$ and $(0, -7)$ and foci at $(0, 5)$ and $(0, -5)$
20. The ellipse with vertices at $(4, 3)$ and $(4, -5)$ and foci at $(4, 2)$ and $(4, -4)$
21. The ellipse with foci at $(-4, -3)$ and $(2, -3)$ with eccentricity $4/5$
22. The circle passing through $(2, 2)$, $(-2, -6)$, and $(5, 1)$
23. The ellipse passing through $(5, 2)$ and $(3, \sqrt{5})$ with axes along the coordinate axes

B Problems

Sketch the curves in Problems 24–29.

24. $x^2 + 4x + y^2 + 6y - 12 = 0$
25. $9x^2 + 4y^2 - 18x + 16y - 11 = 0$
26. $16x^2 + 9y^2 + 96x - 36y + 36 = 0$
27. $x^2 + 25y^2 + 14x + 150y + 273 = 0$
28. $3x^2 + 4y^2 + 2x - 8y + 4 = 0$
29. $144x^2 + 72y^2 - 72x + 48y - 7 = 0$
30. Derive the equation of the ellipse with foci at $(0, c)$ and $(0, -c)$ and constant distance $2a$. Let $b^2 = a^2 - c^2$. Show all your work.
31. Derive the equation of a circle with center (h, k) and radius r by using the distance formula.
32. *Calculator Problem.* The orbit of the earth around the sun is elliptic with the sun at one of the foci. If the semimajor axis of the ellipse is $9.3 \cdot 10^7$ miles and the eccentricity is about 0.017, determine the greatest and least distances of the earth from the sun, correct to two significant digits.
33. *Calculator Problem.* Repeat Problem 32 for Mars assuming the semimajor axis is $1.4 \cdot 10^8$ miles and the eccentricity is 0.093.

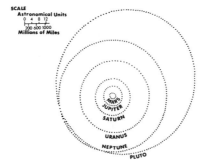

34. *Calculator Problem.* A stone tunnel is to be constructed such that the opening is a semielliptic arch, as shown in Figure 9.25. It is necessary to know the height at 4-ft intervals from the center. That is, how high is the tunnel at 4, 8, 12, 14, 16, and 20 ft from the center?

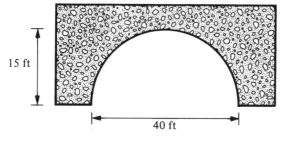

Figure 9.25   Semielliptic arch

The orbits of the planets are elliptical in shape. If the sun is placed at one of the foci of a giant ellipse, the orbit of the earth is an ellipse. The *perihelion* is the point where the planet comes closest to the sun and the *aphelion* is the farthest distance the planet is from the sun. The eccentricity for the planets measures the amount of roundness of the orbit. The eccentricity of a circle is 0 and that of a parabola is 1. The eccentricity for each of the planets in our solar system is given below.

Planet	Eccentricity
Mercury	0.194
Venus	0.007
Earth	0.017
Mars	0.093
Jupiter	0.048
Saturn	0.056
Uranus	0.047
Neptune	0.009
Pluto	0.249

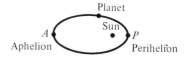

The orbits of a satellite can be calculated from Kepler's Third Law:

$$\frac{\text{mass of (planet + satellite)}}{\text{mass of (sun + planet + satellite)}}$$

$$= \frac{(\text{semimajor axis of satellite orbit})}{(\text{semimajor axis of planet orbit})^3}$$

$$\times \frac{(\text{period of planet})^2}{(\text{period of satellite})^2}$$

35. A whispering gallery is a building with an elliptically shaped dome roof. Such a building has the interesting acoustical property that a whisper at one focus can be heard at the other. If a whispering gallery has the dimensions shown in Figure 9.26, how far from the walls are the foci?

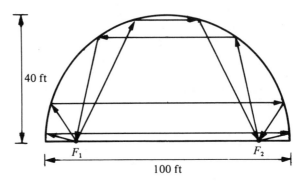

Figure 9.26 Whispering gallery

36. Find the points of intersection (if any) of the ellipse $16(x - 2)^2 + 9(y + 1)^2 = 144$ and the line $4x - 3y - 23 = 0$. Graph both equations.

37. Find the points of intersection (if any) of the circle $x^2 + y^2 + 4x + 6y + 4 = 0$ and the parabola $x^2 + 4x + 8y + 4 = 0$. Graph both equations.

DIRECTRICES

 38. If we are given an ellipse with foci at $(-c, 0)$ and $(c, 0)$ and vertices at $(-a, 0)$ and $(a, 0)$, we define the *directrices* of the ellipse as the lines $x = a/\varepsilon$ and $x = -a/\varepsilon$. Show that an ellipse is the set of all points with distances from $F(c, 0)$ equal to ε times their distances from the line $x = a/\varepsilon$ $(a > 0, c > 0)$.

LATUS RECTUM

 39. A line segment through a focus parallel to a directrix (see Problem 38) and cut off by the ellipse is called the *latus rectum*. Show that the length of the latus rectum of the following ellipse is $2b^2/a$:

$$\frac{x^2}{a^2} + \frac{y^2}{b^2} = 1.$$

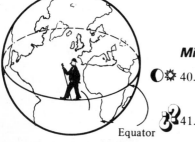

Figure 9.27 A steel band around the earth at the equator

Mind Bogglers

40. Do some research to find the masses of the planets and their satellites (an encyclopedia will have this information). Verify Kepler's Third Law as given on page 445.

41. Figure 9.27 brings to mind a question posed in *The Education of T. C. Mits*, by Lillian R. Lieber (New York: Norton, 1944). Suppose there is a steel band fitting tightly around the equator of the earth. Now suppose that you remove it, cut it at one place, and then splice in an additional piece 10 ft long, so that the new band is 10 ft longer than the original one. If you now replace it on the equator, it will

fit more loosely, won't it? The question is: How large a space will there now be between the band and the earth? Will it be large enough for

a. a piece of tissue paper to just slip through?
b. a man to crawl through on hands and knees?
c. a man 6 ft tall to walk through?

The diameter of the earth is about 8000 miles.

9.3   HYPERBOLAS

The last of the conic sections to be considered has a definition similar to that of the ellipse.

DEFINITION: A *hyperbola* is the set of all points in the plane such that for each point on the curve, the difference of its distances from two fixed points is a constant.

HYPERBOLA

The fixed points are called the *foci*. A hyperbola with foci at F_1 and F_2, where the given constant is 8, is shown in Figure 9.28. The line passing

FOCI

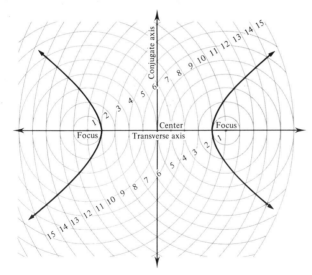

Figure 9.28   Graph of a hyperbola from the definition

through the foci is called the *transverse axis*. The *center* is the midpoint of the segment connecting the foci. The line passing through the center perpendicular to the transverse axis is called the *conjugate axis*. The hyperbola is symmetric with respect to both the transverse and the conjugate axes.

TRANSVERSE AXIS, CENTER

CONJUGATE AXIS

If we use the definition, we can derive the equation for a hyperbola with foci at $(-c, 0)$ and $(c, 0)$ with constant distance $2a$. If (x, y) is any point on the curve, then

$$\left| \sqrt{(x + c)^2 + (y - 0)^2} - \sqrt{(x - c)^2 + (y - 0)^2} \right| = 2a.$$

The procedure here is the same as that shown for the ellipse so the details are left as a problem.

Notice that

$$c^2 = a^2 - b^2$$

for the ellipse and

$$c^2 = a^2 + b^2$$

for the hyperbola. For the ellipse, it is necessary that

$$a^2 > b^2$$

but for the hyperbola, there is no restriction on the relative sizes for a and b.

STANDARD FORM EQUATIONS FOR THE HYPERBOLA

Simplifying, we obtain

$$\frac{x^2}{a^2} - \frac{y^2}{c^2 - a^2} = 1.$$

If $b^2 = c^2 - a^2$, then

$$\frac{x^2}{a^2} - \frac{y^2}{b^2} = 1,$$

which is the standard form equation. Repeat the argument for a hyperbola with foci $(0, c)$ and $(0, -c)$, and you will obtain the other standard form equation for a hyperbola with a vertical transverse axis.

Hyperbola	Foci	Constant distance	Center	Equation
Horizontal	$(-c, 0), (c, 0)$	$2a$	$(0, 0)$	$\dfrac{x^2}{a^2} - \dfrac{y^2}{b^2} = 1$
Vertical	$(0, c), (0, -c)$	$2a$	$(0, 0)$	$\dfrac{y^2}{a^2} - \dfrac{x^2}{b^2} = 1$

where $b^2 = c^2 - a^2$ or $c^2 = a^2 + b^2$

As with the other conics, we will sketch a hyperbola by determining some information about the curve directly from the equation by inspection. The points of intersection of the hyperbola with the transverse axis are called the *vertices*. For

$$\frac{x^2}{a^2} - \frac{y^2}{b^2} = 1 \qquad \text{and} \qquad \frac{y^2}{a^2} - \frac{x^2}{b^2} = 1$$

we see that the vertices occur at $(a, 0)$, $(-a, 0)$ and $(0, a)$, $(0, -a)$, respectively. The number $2a$ is the *length of the transverse axis*. The hyperbola does not intersect the conjugate axis, but if we plot the points $(0, b)$, $(0, -b)$ and $(-b, 0)$, $(b, 0)$, respectively, we determine a segment on the conjugate axis called the *length of the conjugate axis*.

EXAMPLE 1: Sketch

$$\frac{x^2}{4} - \frac{y^2}{9} = 1.$$

Solution: The center of the hyperbola is $(0, 0)$, $a = 2$, and $b = 3$. We plot the vertices at ± 2, as shown in Figure 9.29. The transverse axis is along the x-axis and the conjugate axis is along the y-axis. Plot the length of the conjugate axis at ± 3. We call these points the *pseudovertices*, since the curve does not actually pass through these points. Next, we draw lines through the vertices and pseudovertices parallel to the axes of the hyperbola. These lines form what we will call the *central rectangle*. The diagonal lines passing through the corners of the central rectangle are slant asymptotes for the hyperbola, as shown in Figure 9.30; they aid in sketching the hyperbola.

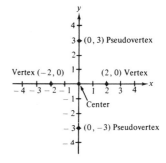

Figure 9.29

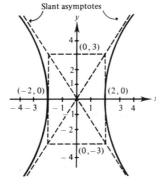

Figure 9.30   Graph of $\dfrac{x^2}{4} - \dfrac{y^2}{9} = 1$

The equations of the slant asymptotes for the hyperbola given by the equation

$$\frac{x^2}{a^2} - \frac{y^2}{b^2} = 1$$

are

$$y = \frac{b}{a}x \qquad \text{and} \qquad y = -\frac{b}{a}x.$$

To justify this result, solve the equation of the hyperbola for y:

$$\frac{y^2}{b^2} = \frac{x^2}{a^2} - 1$$

$$y^2 = b^2 \left(\frac{x^2 - a^2}{a^2} \right)$$

$$y = \pm \frac{b}{a} \sqrt{x^2 - a^2}.$$

Now write

$$y = \pm \frac{b}{a} \sqrt{x^2 \left(1 - \frac{a^2}{x^2}\right)}$$

$$y = \pm \frac{bx}{a} \sqrt{1 - \frac{a^2}{x^2}}$$

in order to see that

$$\lim_{|x| \to \infty} \sqrt{1 - \frac{a^2}{x^2}} = 1.$$

So, as $|x|$ becomes large,

$$y = \pm \frac{b}{a} x.$$

This argument is even more convincing if you find an expression for the vertical distance between the line

$$y = \frac{b}{a} x$$

and the hyperbola

$$y = \frac{b}{a} \sqrt{x^2 - a^2}$$

in Quadrant I. You are asked to do this in Problem 27.

If the center of the hyperbola is (h, k), the following equations for a hyperbola are obtained:

Hyperbola	Center	Foci	Vertices	Pseudo-vertices	Equations
Horizontal	(h, k)	$(h + c, k)$ $(h - c, k)$	$(h - a, k)$ $(h + a, k)$	$(h, k + b)$ $(h, k - b)$	$\dfrac{(x - h)^2}{a^2} - \dfrac{(y - k)^2}{b^2} = 1$
Vertical	(h, k)	$(h, k + c)$ $(h, k - c)$	$(h, k + a)$ $(h, k - a)$	$(h - b, k)$ $(h + b, k)$	$\dfrac{(y - k)^2}{a^2} - \dfrac{(x - h)^2}{b^2} = 1$

EXAMPLES

2. Sketch $16x^2 - 9y^2 - 128x - 18y + 103 = 0$.

    Solution: Complete the square in both x and y.

    $$(16x^2 - 128x) + (-9y^2 - 18y) = -103$$

    $$16(x^2 - 8x \quad\ ) - 9(y^2 + 2y \quad\ ) = -103$$

    $$16(x^2 - 8x + 16) - 9(y^2 + 2y + 1) = -103 + 256 - 9$$

    $$16(x - 4)^2 - 9(y + 1)^2 = 144$$

    $$\frac{(x - 4)^2}{9} - \frac{(y + 1)^2}{16} = 1$$

You can also find the foci, since

$$c^2 = a^2 + b^2.$$

Thus,

$$c^2 = 9 + 16$$
$$c = \pm 5.$$

The graph is shown in Figure 9.31.

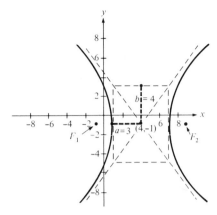

Figure 9.31 Sketch of $16x^2 - 9y^2 - 128x - 18y + 103 = 0$

3. Find the equation of the hyperbola with vertices at $(2, 4)$ and $(2, -2)$ and foci at $(2, 6)$ and $(2, -4)$.

Solution: We plot the given points, as shown in Figure 9.32. Notice that the

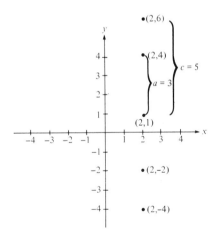

Figure 9.32

center of the hyperbola is $(2, 1)$ since it is the midpoint of the segment connecting the foci. Also, $c = 5$ and $a = 3$. Thus, since

$$c^2 = a^2 + b^2,$$

we have

$$25 = 9 + b^2$$
$$b^2 = 16$$

and the equation is

$$\frac{(y-1)^2}{9} - \frac{(x-2)^2}{16} = 1.$$

The eccentricity of the hyperbola and parabola are defined by the same equation that was used for the ellipse, namely

$$\varepsilon = \frac{c}{a}.$$

Remember that for the ellipse, $0 \le \varepsilon < 1$; however, for the hyperbola, $c > a$ so $\varepsilon > 1$, and for the parabola, $c = a$ so $\varepsilon = 1$.

4. Find the equation of the hyperbola with foci at $(-3, 2)$ and $(5, 2)$ and with eccentricity $3/2$.
    Solution: The center of the hyperbola is $(1, 2)$ and $c = 4$. Also, since

    $$\varepsilon = \frac{c}{a} = \frac{3}{2},$$

    we have

    $$\frac{4}{a} = \frac{3}{2},$$

    which implies

    $$a = \frac{8}{3}.$$

    And since $c^2 = a^2 + b^2$,

    $$16 = \frac{64}{9} + b^2$$

    $$b^2 = \frac{80}{9}.$$

    Thus, the equation is

    $$\frac{(x-1)^2}{64/9} - \frac{(y-2)^2}{80/9} = 1.$$

5. Find the set of points such that the difference of their distances from $(6, 2)$ and $(6, -5)$ is always 3.
    Solution: The center of the hyperbola is $(6, -\frac{3}{2})$ and $c = \frac{7}{2}$. Also, $2a = 3$, so $a = \frac{3}{2}$. Thus, since

    $$c^2 = a^2 + b^2,$$

    $$\frac{49}{4} = \frac{9}{4} + b^2$$

    $$b^2 = 10.$$

    The equation is

    $$\frac{\left(y + \frac{3}{2}\right)^2}{\frac{9}{4}} - \frac{(x-6)^2}{10} = 1.$$

PROBLEM SET 9.3

A Problems

Sketch the curves in Problems 1–10.

1. $x^2 - y^2 = 1$

2. $\dfrac{x^2}{4} - \dfrac{y^2}{9} = 1$

3. $\dfrac{x^2}{9} - \dfrac{y^2}{4} = 1$

4. $\dfrac{x^2}{16} - \dfrac{y^2}{25} = 1$

5. $\dfrac{x^2}{36} - \dfrac{y^2}{9} = 1$

6. $36y^2 - 25x^2 = 900$

7. $3x^2 - 4y^2 = 12$

8. $3x^2 - 4y^2 = 5$

9. $\dfrac{(x-2)^2}{4} - \dfrac{(y+3)^2}{16} = 1$

10. $\dfrac{(x+3)^2}{8} - \dfrac{(y-1)^2}{5} = 1$

Find the equations of the curves in Problems 11–16.

11. The hyperbola with vertices at $(0, 5)$ and $(0, -5)$ and foci at $(0, 7)$ and $(0, -7)$
12. The set of points such that the difference of their distances from $(-6, 0)$ and $(6, 0)$ is 10
13. The hyperbola with foci at $(5, 0)$ and $(-5, 0)$ and eccentricity 5
14. The hyperbola with vertices at $(4, 4)$ and $(4, 8)$ and foci at $(4, 3)$ and $(4, 9)$
15. The set of points such that the difference of their distances from $(4, -3)$ and $(-4, -3)$ is 6
16. The hyperbola with vertices at $(-2, 0)$ and $(6, 0)$ passing through $(10, 3)$

B Problems

Sketch the curves in Problems 17–24.

EXAMPLES

6. $\dfrac{x^2}{4} + \dfrac{y^2}{9} = 0$

   Solution: There is only one point that satisfies this equation—namely $(0, 0)$. This is called a *degenerate ellipse.*

7. $\dfrac{x^2}{4} - \dfrac{y^2}{9} = 0$

   Solution: We can factor the equation into

   $$\left(\frac{x}{2} - \frac{y}{3}\right)\left(\frac{x}{2} + \frac{y}{3}\right) = 0.$$

Notice that you cannot put the equations in Examples 6 and 7 into standard form so we must approach the problem differently. These are examples of *degenerate conics.*

Review Problems 69–72 on page 120 in which we discussed graphs of factored equations.

Notice that the graph of

$$\frac{x^2}{4} - \frac{y^2}{9} = 0$$

is the same as the asymptotes for the hyperbola

$$\frac{x^2}{4} - \frac{y^2}{9} = 1.$$

For this reason it is sometimes called a *degenerate hyperbola*.

From algebra, recall that a product is zero if and only if one (or both) of the factors is zero. This means that

$$\frac{x}{2} - \frac{y}{3} = 0 \qquad \text{or} \qquad \frac{x}{2} + \frac{y}{3} = 0,$$

as shown in Figure 9.33.

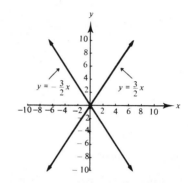

Figure 9.33   Graph of $\dfrac{x^2}{4} - \dfrac{y^2}{9} = 0$

---

17. $5(x - 2)^2 - 2(y + 3)^2 = 10$
18. $4(x + 4)^2 - 3(y + 3)^2 = -12$
19. $3x^2 - 5y^2 + 18x + 10y - 8 = 0$
20. $9x^2 - 18x - 11 = 4y^2 + 16y$
21. $4y^2 - 8y + 4 = 3x^2 - 2x$
22. $x^2 - 4x + y^2 + 6y - 12 = 0$
23. $x^2 - y^2 = 2x + 4y - 3$
24. $3x^2 - 4y^2 = 12x + 80y + 88$
25. Derive the equation of the hyperbola with foci at $(-c, 0)$ and $(c, 0)$ and constant distance $2a$. Let $b^2 = c^2 - a^2$. Show all your work.
26. Derive the equation of the hyperbola with foci at $(0, c)$ and $(0, -c)$ and constant distance $2a$. Let $b^2 = c^2 - a^2$. Show all your work.
27. Let d represent the vertical distance between the hyperbola in the first quadrant,

$$y = \frac{b}{a}\sqrt{x^2 - a^2},$$

and the line

$$y = \frac{b}{a}x$$

in the first quadrant. Show that

$$\lim_{|x| \to \infty} d = 0.$$

28. Repeat Problem 27 for Quadrant IV.
29. Show that the length of the diagonal of the central rectangle of a hyperbola is $2c$.
30. Given the hyperbola

$$\frac{x^2}{a^2} - \frac{y^2}{b^2} = 1,$$

we define the *directrices* of the hyperbola as the lines   DIRECTRICES

$$x = \frac{a}{\varepsilon} \quad \text{and} \quad x = -\frac{a}{\varepsilon}.$$

Show that the hyperbola is the set of all points with distances from $F(c, 0)$ that are equal to ε times their distances from the line $x = a/\varepsilon$.

31. A line through a focus parallel to a directrix and cut off by the hyperbola is called the *latus rectum*. Show that the length of the latus rectum of the following hyperbola is $2b^2/a$:   LATUS RECTUM

$$\frac{x^2}{a^2} - \frac{y^2}{b^2} = 1.$$

Mind Bogglers

32. The equation of a hyperbola may be used to find the location of an explosion (or, in military terms, the location of an enemy weapon) with a process called *range finding*. If an explosion occurs at some point P and the precise times the sound of the explosion reaches two listening points, F_1 and F_2, can be recorded, then the location of the explosion can be determined. If the difference of times is multiplied by the velocity of sound, the result is the difference of the distances of P from F_1 and F_2. To locate the point of the explosion, a third listening station, F_3, is used and the procedure is repeated for F_2 and F_3. The intersection of the two resulting hyperbolas gives the exact position of the explosion. Do some research into this problem of range finding and write a report or give a presentation to the class.

The points F_1 and F_2 represent the foci of a hyperbola, and the difference of the distances is the number $2a$ for the hyperbola.

33. One application of hyperbolas is to navigation. *Loran* is a system by which a ship or aircraft can determine its position by radar and radio signals sent from known

Problem 33:

Hyperbolas are used in some techniques of navigation, and of satellite tracking, where the difference of distances from two fixed points is observed (a ship's distances from two fixed stations on the shore, a satellite's distances from two antennas). If A and B are land stations and the difference in distance from them is electronically observed on the ship, its location must be on a certain hyperbola with A and B as foci. Also, if the difference in distance from B and a third station C is observed, the ship must be on a certain hyperbola with B and C as foci; thus, its position is on the intersection of the two hyperbolic curves. The *two sets of hyperbolas serve as a coordinate system*, comparable to the Cartesian system, where two sets of straight lines intersect at right angles.

stations, as described in the marginal note. Do some research into this problem and write a report or give a presentation to the class.

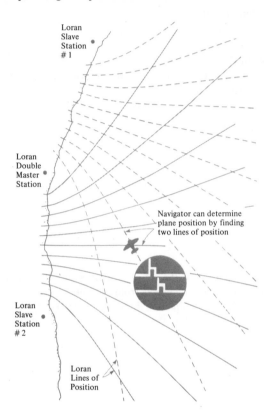

Loran
Slave
Station
1

Loran
Double
Master
Station

Navigator can determine
plane position by finding
two lines of position

Loran
Slave
Station
2

Loran
Lines of
Position

9.4 ROTATIONS

All the curves considered in this chapter can be characterized by the general second-degree equation

$$Ax^2 + Bxy + Cy^2 + Dx + Ey + F = 0.$$

Notice that the xy term has not appeared before. The presence of this term indicates that the conic has been rotated.

It is important to be able to recognize the curve by inspection of the equation before we begin. The first thing to notice is whether or not there is an xy term.

If $B = 0$ (that is, no xy term), then:

TYPE OF CURVE	DEGREE OF EQUATION	DEGREE IN x	DEGREE IN y	RELATIONSHIP TO GENERAL EQUATION
1. Line	First	First	First	$A = C = 0$
2. Parabola	Second	First	Second	$A = 0$ and $C \neq 0$
		Second	First	$A \neq 0$ and $C = 0$
3. Ellipse	Second	Second	Second	A and C have same sign
4. Circle	Second	Second	Second	$A = C$
5. Hyperbola	Second	Second	Second	A and C have opposite signs

If $B \neq 0$ (that is, there is an xy term), then:

DISCRIMINANT	TYPE OF CURVE
$B^2 - 4AC < 0$	Ellipse (or its degenerate case)
$B^2 - 4AC = 0$	Parabola (or its degenerate case)
$B^2 - 4AC > 0$	Hyperbola (or its degenerate case)

The proof of this depends upon the fact that the expression $B^2 - 4AC$ is unchanged when the axes are rotated through any angle. You are asked to show this in the problem set at the end of this section.

EXAMPLES: Identify each curve.

1. $x^2 + 4xy + 4y^2 = 9$
    Solution: $B^2 - 4AC = 16 - 4(1)(4) = 0$; parabola

2. $2x^2 + 3xy + y^2 = 25$
    Solution: $B^2 - 4AC = 9 - 4(2)(1) > 0$; hyperbola

3. $x^2 + xy + y^2 - 8x + 8y = 0$
    Solution: $B^2 - 4AC = 1 - 4(1)(1) < 0$; ellipse

4. $xy = 5$
    Solution: $B^2 - 4AC = 1 - 4(0)(0) > 0$; hyperbola

To graph a conic that has been rotated, we have to introduce a new coordinate axis for which the curve is in standard position. How can we do this?

This question can be answered by considering how you were able to read the preceding paragraph. What did you do? Probably, you turned the book until the paragraph was right-side up. We do the same for a rotated conic.

If we rotate the axis through an angle θ ($0 < \theta < 90°$), the relationship between the old coordinates (x, y) and the new coordinates (x', y') can be found by considering Figure 9.34. Let O be the origin and P be a point with

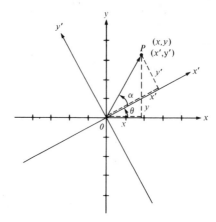

Figure 9.34   Rotation of axes

coordinates (x, y) relative to the old coordinate system and (x', y') relative to the new rotated coordinate system. Let θ be the amount of rotation and let α be the angle between the x'-axis and $|OP|$. Then, using the definition of sine and cosine, we have

$$x = |OP| \cos (\theta + \alpha), \qquad x' = |OP| \cos \alpha,$$
$$y = |OP| \sin (\theta + \alpha), \qquad y' = |OP| \sin \alpha.$$

Now,

$$
\begin{aligned}
x = |OP| \cos (\theta + \alpha) &= |OP|[\cos \theta \cos \alpha - \sin \theta \sin \alpha] \\
&= |OP| \cos \theta \cos \alpha - |OP| \sin \theta \sin \alpha \\
&= (|OP| \cos \alpha) \cos \theta - (|OP| \sin \alpha) \sin \theta \\
&= x' \cos \theta - y' \sin \theta.
\end{aligned}
$$

Also,

$$
\begin{aligned}
y = |OP| \sin (\theta + \alpha) &= |OP| \sin \theta \cos \alpha + |OP| \cos \theta \sin \alpha \\
&= x' \sin \theta + y' \cos \theta.
\end{aligned}
$$

Therefore:

$$x = x' \cos \theta - y' \sin \theta \qquad y = x' \sin \theta + y' \cos \theta$$

It is important that we rotate the axes the correct amount for each given case. That is, the new axes should be rotated the same amount as the given conic so that it will be in standard position after the rotation. To find out how much to rotate the axes, we substitute

$$x = x' \cos \theta - y' \sin \theta, \qquad y = x' \sin \theta + y' \cos \theta$$

into

$$Ax^2 + Bxy + Cy^2 + Dx + Ey + F = 0 \quad (B \neq 0).$$

We obtain (after a lot of simplifying)

$$
\begin{aligned}
(A \cos^2 \theta &+ B \cos \theta \sin \theta + C \sin^2 \theta)x'^2 \\
&+ [B(\cos^2 \theta - \sin^2 \theta) + 2(C - A)\sin \theta \cos \theta]x'y' \\
&+ (A \sin^2 \theta - B \sin \theta \cos \theta + \cos^2 \theta)y'^2 \\
&+ (D \cos \theta + E \sin \theta)x' + (-D \sin \theta + E \cos \theta)y' \\
&+ F = 0.
\end{aligned}
$$

This looks terrible, but we want to choose θ so that the $x'y' = 0$. That is,

$$B(\cos^2 \theta - \sin^2 \theta) + 2(C - A)\sin \theta \cos \theta = 0$$

Using double-angle identities, $B \neq 0, \theta \neq 0$

or

$$B \cos 2\theta + (C - A)\sin 2\theta = 0$$

$$B \cos 2\theta = (A - C)\sin 2\theta$$

$$\frac{\cos 2\theta}{\sin 2\theta} = \frac{A - C}{B}.$$

Simplifying, we obtain

$$\cot 2\theta = \frac{A - C}{B}.$$

This equation is more like it since it is not too difficult to work with.

EXAMPLES: Find the appropriate rotation so that the given curve will be in standard position relative to the rotated axes. Also, find the x and y values in the new coordinate system.

5. $xy = 6$

Solution:

$$\cot 2\theta = \frac{A - C}{B}$$

$$= \frac{0 - 0}{1}$$

$$= 0$$

Thus, $2\theta = 90°$ and $\theta = 45°$.

$$x = x' \cos \theta - y' \sin \theta \qquad y = x' \sin \theta + y' \cos \theta$$

Since $\cos 45° = \sin 45°$
$= 1/\sqrt{2}$

$$= x'\left(\frac{1}{\sqrt{2}}\right) - y'\left(\frac{1}{\sqrt{2}}\right) \qquad = x'\left(\frac{1}{\sqrt{2}}\right) + y'\left(\frac{1}{\sqrt{2}}\right)$$

$$= \frac{1}{\sqrt{2}}(x' - y') \qquad\qquad = \frac{1}{\sqrt{2}}(x' + y')$$

6. $7x^2 - 6\sqrt{3}xy + 13y^2 - 16 = 0$
    Solution:

$$\cot 2\theta = \frac{A - C}{B}$$

$$= \frac{7 - 13}{-6\sqrt{3}}$$

$$= \frac{1}{\sqrt{3}}$$

Thus, $2\theta = 60°$ and $\theta = 30°$.

$$x = x' \cos \theta - y' \sin \theta \qquad y = x' \sin \theta + y' \cos \theta$$

Since $\cos 30° = \sqrt{3}/2$ and
$\sin 30° = \frac{1}{2}$

$$= x'\left(\frac{\sqrt{3}}{2}\right) - y'\left(\frac{1}{2}\right) \qquad = \frac{1}{2}(x' + \sqrt{3}y')$$

$$= \frac{1}{2}(\sqrt{3}x' - y')$$

7. $x^2 - 4xy + 4y^2 + 5\sqrt{5}y - 10 = 0$

Solution: $\cot 2\theta = \dfrac{1 - 4}{-4} = \dfrac{3}{4}$

This is not an exact value so
we have to use the following
identities from Chapter 7:

$$\cot 2\theta = \frac{x}{y},$$

$$\cos 2\theta = \frac{x}{\sqrt{x^2 + y^2}},$$

$$\cos \theta = \pm\sqrt{\frac{1 + \cos 2\theta}{2}},$$

$$\sin \theta = \pm\sqrt{\frac{1 - \cos 2\theta}{2}},$$

$$\tan \theta = \frac{\sin \theta}{\cos \theta}.$$

If $\cot 2\theta = \frac{3}{4}$, then $\cos 2\theta = \frac{3}{5}$ and

$$\cos \theta = \sqrt{\frac{1 + (3/5)}{2}} \qquad \sin \theta = \sqrt{\frac{1 - (3/5)}{2}}$$

$$= \frac{2}{\sqrt{5}} \qquad\qquad = \frac{1}{\sqrt{5}}$$

Hence:

$$x = x' \cos \theta - y' \sin \theta \qquad y = x' \sin \theta + y' \cos \theta$$

$$= x'\left(\frac{2}{\sqrt{5}}\right) - y'\left(\frac{1}{\sqrt{5}}\right) \qquad = x'\left(\frac{1}{\sqrt{5}}\right) + y'\left(\frac{2}{\sqrt{5}}\right)$$

$$= \frac{1}{\sqrt{5}}(2x' - y') \qquad\qquad = \frac{1}{\sqrt{5}}(x' + 2y')$$

To find the rotation, find

$$\tan \theta = \frac{\sin \theta}{\cos \theta}$$

$$= \frac{1/\sqrt{5}}{2/\sqrt{5}}$$

$$= \frac{1}{2}.$$

Recall that the slope of the x'-axis is the tangent of the angle of inclination. Since θ is the angle of inclination, we draw the x'-axis so that it has a "rise" of one unit and a "run" of two units. Notice that we found the rotated axes without ever consulting a table of trigonometric values.

TO SKETCH A ROTATED CONIC:

1. Determine the nature of the conic by calculating $B^2 - 4AC$.
2. Find the angle of rotation.
3. Find x and y in the new coordinate system.
4. Substitute the values found in Step 3 into the given equation and simplify.
5. Sketch the resulting equation relative to the new x'- and y'-axes. You may have to complete the square if it is not centered at the origin.

EXAMPLES

8. Sketch $xy = 6$.

   Solution: This curve is a hyperbola since $B^2 - 4AC = 1 - 0 > 0$. From Example 5, $\theta = 45°$ and

   $$x = \frac{1}{\sqrt{2}}(x' - y'), \qquad y = \frac{1}{\sqrt{2}}(x' + y').$$

   Substituting into the original equation, we get

   $$\left[\frac{1}{\sqrt{2}}(x' - y')\right]\left[\frac{1}{\sqrt{2}}(x' + y')\right] = 6,$$

   and simplifying gives

   $$x'^2 - y'^2 = 12$$

   $$\frac{x'^2}{12} - \frac{y'^2}{12} = 1.$$

THE STRING METHOD OF CONSTRUCTING CONICS:

Circle

1. Given center C and fixed radius r.
2. Fix a string at C. How long should the string be?
3. If the pencil keeps the string taut while tracing out a curve, the curve is a circle.

Parabola

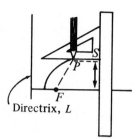

Directrix, *L*

1. Given focus *F* and directrix *L*.
2. Fix a ruler parallel to *L* at an arbitrary distance.
3. Fix a string at *F* and at a point *S* on a triangle that moves along the ruler. How long does the string have to be?
4. If the pencil keeps the string taut, the curve is a parabola.

Ellipse

F_1 F_2

1. Given foci F_1 and F_2 and fixed length $2a$.
2. Fix a string at F_1 and F_2. How long should the string be?
3. If the pencil keeps the string taut while tracing out a curve, the curve is an ellipse.

The sketch is shown in Figure 9.35.

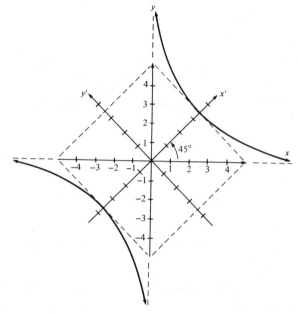

Figure 9.35   Graph of $xy = 6$

9. Sketch $7x^2 - 6\sqrt{3}xy + 13y^2 - 16 = 0$.
    Solution: This curve is an ellipse since $B^2 - 4AC = 36(3) - 4(7)(13) < 0$. From Example 6, $\theta = 30°$ and

$$x = \frac{1}{2}(\sqrt{3}x' - y'), \qquad y = \frac{1}{2}(x' + \sqrt{3}y').$$

Substituting into the original equation, we get

$$7\left(\frac{1}{2}\right)^2 (\sqrt{3}x' - y')^2 - 6\sqrt{3}\left(\frac{1}{2}\right)(\sqrt{3}x' - y')\left(\frac{1}{2}\right)(x' + \sqrt{3}y')$$

$$+ 13\left(\frac{1}{2}\right)^2 (x' + \sqrt{3}y')^2 - 16 = 0,$$

and simplifying gives

$$\frac{x'^2}{4} + \frac{y'^2}{1} = 1.$$

The sketch is shown in Figure 9.36.

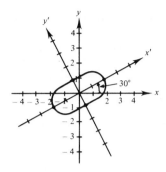

Figure 9.36 Graph of $7x^2 - 6\sqrt{3}xy + 13y^2 - 16 = 0$

10. Sketch $x^2 - 4xy + 4y^2 + 5\sqrt{5}y - 10 = 0$.

Solution: This curve is a parabola since $B^2 - 4AC = 16 - 4(1)(4) = 0$. From Example 7, the rotation is given by $\tan\theta = \frac{1}{2}$ and

$$x = \frac{1}{\sqrt{5}}(2x' - y'), \qquad y = \frac{1}{\sqrt{5}}(x' + 2y').$$

Substituting, we get

$$\frac{1}{5}(2x' - y')^2 - 4\left(\frac{1}{5}\right)(2x' - y')(x' + 2y') + 4\left(\frac{1}{5}\right)(x' + 2y')^2$$
$$+ 5\sqrt{5}\left(\frac{1}{\sqrt{5}}\right)(x' + 2y') - 10 = 0,$$

and simplifying gives

$$y'^2 + 2y' = -x' + 2.$$

Completing the square,

$$(y' + 1)^2 = -(x' - 3).$$

The sketch is shown in Figure 9.37.

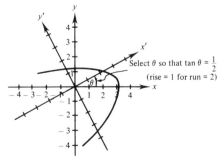

Select θ so that $\tan\theta = \frac{1}{2}$
(rise = 1 for run = 2)

Figure 9.37 Graph of $x^2 - 4xy + 4y^2 + 5\sqrt{5}y - 10 = 0$

Hyperbola

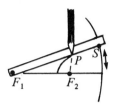

1. Given foci F_1 and F_2 and fixed length $2a$.
2. A beam compass is set to describe an arbitrary circle with center at F_1.
3. A string is fastened to F_2 and a point S on the beam. How long does the string have to be?
4. If the pencil keeps the string taut while S traces the circle, the curve is a hyperbola.

As you've seen, all the conics can be reduced to a standard form equation by means of a translation or rotation. It is important to remember these equations and some basic information about these curves. The important ideas are summarized in Table 9.1.

TABLE 9.1   Conic Summary

	Parabola	*Ellipse*	*Hyperbola*
Definition	All points equidistant from a given point and a given line	All points with the sum of distances from two fixed points constant	All points with the difference of distances from two fixed points constant
Equations	Up $\quad x^2 = 4cy$ Down $\quad x^2 = -4cy$ Right $\quad y^2 = 4cx$ Left $\quad y^2 = -4cx$	$c^2 = a^2 - b^2$ Horizontal axis: $$\frac{x^2}{a^2} + \frac{y^2}{b^2} = 1$$ Vertical axis: $$\frac{y^2}{a^2} + \frac{x^2}{b^2} = 1$$	$c^2 = a^2 + b^2$ Horizontal axis: $$\frac{x^2}{a^2} - \frac{y^2}{b^2} = 1$$ Vertical axis: $$\frac{y^2}{a^2} - \frac{x^2}{b^2} = 1$$
Recognition	Second-degree equation; linear in one variable, quadratic in the other variable	Second-degree equation; coefficients of x^2 and y^2 have same sign	Second-degree equation; coefficients of x^2 and y^2 have different sign
Distance from center to foci	c	c	c
Distance from center to vertex	0	a	a
Eccentricity	$\varepsilon = 1$	$0 \le \varepsilon < 1$	$\varepsilon > 1$
Directrix	Perpendicular to axis c units from the center (one directrix)	Perpendicular to major axis $\pm a/\varepsilon$ units from center (two directrices)	Perpendicular to transverse axis $\pm a/\varepsilon$ units from center (two directrices)

SUMMARY

The equations for the translation of axes to the point (h, k) are

$$x' = x - h \quad \text{and} \quad y' = y - k.$$

The equations for the rotation of axes through an angle θ are

$$x = x' \cos \theta - y' \sin \theta \quad \text{and} \quad y = x' \sin \theta + y' \cos \theta.$$

The amount of rotation, θ, for the curve $Ax^2 + Bxy + Cy^2 + Dx + Ey + F = 0$ is found by

$$\cot 2\theta = \frac{A - C}{B}.$$

To graph the general second-degree equation

$$Ax^2 + Bxy + Cy^2 + Dx + Ey + F = 0,$$

follow the steps shown in Figure 9.38.

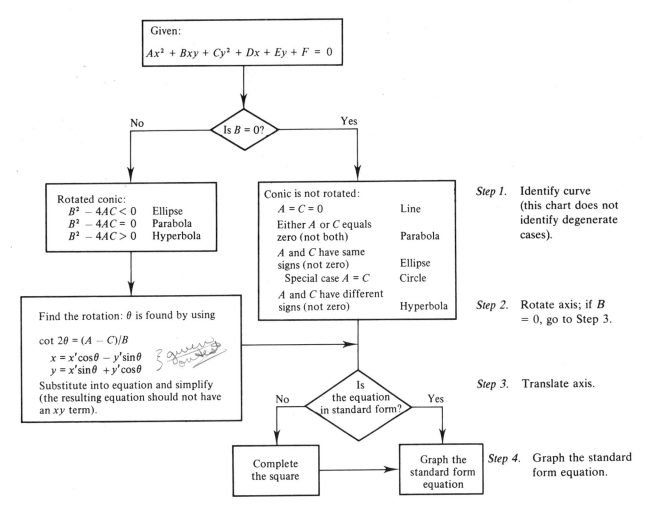

Figure 9.38   Procedure for graphing conics

PROBLEM SET 9.4

A Problems

In Problems 1–12, identify and sketch the curve.

1. $y^2 - 4x + 2y + 21 = 0$

2. $x^2 + 4x + 12y + 64 = 0$

3. $\dfrac{(x-3)^2}{4} - \dfrac{(y+2)}{6} = 1$

4. $x^2 + y^2 + 2x - 4y - 20 = 0$

5. $\dfrac{x}{9} + \dfrac{y}{25} = 1$

6. $x^2 - 4y^2 - 6x - 8y - 11 = 0$

7. $y^2 - 6y - 4x + 5 = 0$

8. $(x+3)^2 + (y-2)^2 = 0$

9. $9x^2 + 25y^2 - 54x - 200y + 256 = 0$

10. $xy = 4$

11. $x^2 + 2xy + y^2 + 12\sqrt{2}x - 6 = 0$

12. $8x^2 - 4xy + 5y^2 = 36$

13. Identify each of the curves given by the equations in Problems 15–21 below.

14. Identify each of the curves given by the equations in Problems 22–27 below.

B Problems

Problems 15–27 can be worked in two parts.

a. Find the appropriate rotation so that the given curve will be in standard position relative to the rotated axes. Also, find the x and y values in the new coordinate system.

b. Sketch the curve.

15. $xy = 8$

16. $xy = -1$

17. $13x^2 - 10xy + 13y^2 - 72 = 0$

18. $5x^2 - 26xy + 5y^2 + 72 = 0$

19. $x^2 + 4xy + 4y^2 + 10\sqrt{5}x = 9$

20. $5x^2 - 4xy + 8y^2 = 36$

21. $23x^2 + 26\sqrt{3}xy - 3y^2 - 144 = 0$

22. $3x^2 + 2\sqrt{3}xy + y^2 + 16x - 16\sqrt{3}y = 0$

23. $24x^2 + 16\sqrt{3}xy + 8y^2 - x + \sqrt{3}y - 8 = 0$

24. $3x^2 - 2\sqrt{3}xy + y^2 + 24x + 24\sqrt{3}y = 0$

25. $13x^2 - 6\sqrt{3}xy + 7y^2 + (16\sqrt{3} - 8)x + (-16 - 8\sqrt{3})y + 16 = 0$

26. $3x^2 - 10xy + 3y^2 - 32 = 0$

27. $5x^2 - 3xy + y^2 + 65x - 25y + 203 = 0$

Sketch the graphs of Problems 28–38 on a single set of coordinate axes. The result is a picture of a familiar object. Note the limitations on the domain or range.

28. $4x^2 + 16y = 0$: $y \ge -4$

29. $2x - y - 10 = 0$: $-12 \ge y \ge -14$

30. $2x - y - 8 = 0$: $-12 \ge y \ge -13.6$

31. $x^2 + 256(y + 12)^2 \le 64$: $2x - 8 \le y \le 2x - 10$

32. $4x^2 - 3y^2 - 24y - 112 = 0$: $-12 \le y \le -4$

33. $16x^2 + 96x + 16y^2 + 480y + 3708 = 0$

34. $x^2 + y^2 - 3y = 0$: $y \le 2$

35. $100x^2 - 7y^2 + 98y - 368 = 0$: $2 \le y \le 12$

36. $x^2 + 8(y - 12)^2 = 2$: $y \ge 12$

37. $x^2 + 64(y + 4)^2 = 16$: $y \ge -4$

38. $9x^2 + 2y^2 - 48y + 270 = 0$: $y \ge 12$

Mind Bogglers

39. In Problem 29 on page 49 and Problems 69–72 on pages 120–121, we wrote equations for the letters A, E, F, H, I, J, K, L, M, N, T, U, V, W, X, Y, and Z. The remaining letters cannot be written as functions but can be written using the conic sections. For example, the letter B can be written as the product of

   a. a parabola through $(0, 0)$ and $(0, 2)$ with vertex $(2, 1)$:

   $$(y - 1)^2 = -\frac{1}{2}(x - 2);$$

   b. a parabola through $(0, 2)$ and $(0, 4)$ with vertex $(2, 3)$:

   $$(y - 3)^2 = -\frac{1}{2}(x - 2);$$

   c. a line: $x = 0$.
   Forming the product, we have

   $$x(2y^2 + x - 4y)(2y^2 + x - 12y + 16) = 0.$$

   Write equations for the letters C, D, G, O, P, Q, R, and S.

40. Write an equation for the following message:

   MATH

41. Write an equation for your name. For example,

   KARL

   has the equation

   $$[x(x + y - 2)(2x - 3y + 6)][(y - 2)(2x - y)(2x + y - 8)]$$
   $$\times [x(2y^2 + x - 4y)(x + y)][xy] = 0.$$

42. Graph the message given by the following equation:

   $$[x(2y^2 + x - 12y + 16)(x + y - 2)][y(x - 2)(y - 4)]$$

   $$\times \left[xy(y - 4)\left(\frac{x\sqrt{2 - y} - 3\sqrt{2 - y}}{\sqrt{2 - y}} \right)\left(\frac{y\sqrt{x - 3} - 2\sqrt{x - 3}}{\sqrt{x - 3}} \right) \right]$$

   $$\times [x(y - 2)(x - 4)][(x - 2)(y - 4)] = 0.$$

43. Let $Ax^2 + Bxy + Cy^2 + Dx + Ey + F = 0$. Show that $B^2 - 4AC = B'^2 - 4A'C'$ for any angle θ through which the axes may be rotated and that A', B', and C' are the values given on page 459. Use this fact to prove that (if the graph exists)

   $$B^2 - 4AC = 0, \quad \text{the graph is a parabola;}$$
   $$B^2 - 4AC < 0, \quad \text{the graph is an ellipse;}$$
   $$B^2 - 4AC > 0, \quad \text{the graph is a hyperbola.}$$

Recall the domain and range for this type of problem:
$0 \le x \le 4, 0 \le y \le 4$.

Answers to this problem are not unique.

Remember our agreements concerning messages:

1. Each letter should be enclosed in brackets.
2. Each new set of brackets implies a shift to the right five units.

Note the function of the radicals in this letter:

$$\frac{x\sqrt{2 - y} - 3\sqrt{2 - y}}{\sqrt{2 - y}}$$

$$= (x - 3)\left(\frac{\sqrt{2 - y}}{\sqrt{2 - y}} \right)$$

The second factor is 1 if $2 - y > 0$, so what is needed here is the line segment $x = 3$ for $y < 2$.

9.5 PARAMETRIC EQUATIONS

"THERE WAS NOTHING WHERE I CAME FROM, AND NOW YOU TELL ME THERE'S NOTHING UP AHEAD!"

Suppose the farmer in the cartoon keeps track of the motorist as he drives off and plots his position using the corner of his fence as the origin of a coordinate system. He obtains the following information by calculating the x and y distances in terms of the time elapsed (t, in seconds) after the motorist leaves:

t	0	1	2	3	4
x	1	4	7	10	13
y	0	2	4	6	8

From the table, we see that at time $t = 4$ sec, the car is located at the point $(13, 8)$. We can graph these points as shown in Figure 9.39. Notice that the t

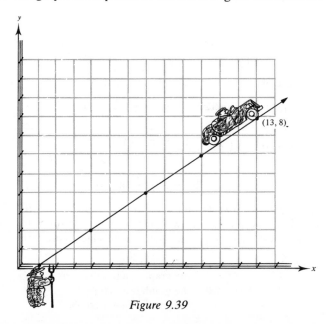

Figure 9.39

scale does not appear in the graph. The quantity t is called a *parameter*. The points in Figure 9.39 seem to lie along a line. Suppose we can find functions g and h so that

$$x = g(t)$$
$$y = h(t).$$

PARAMETER

These equations are called the *parametric equations* of a line. For this example, since the graph appears to be linear, we can find the linear relationship for t and x using the points $(0, 1)$ and $(1, 4)$ to obtain

$$x = 1 + 3t.$$

Also, for t and y you could use $(0, 0)$ and $(1, 2)$ to obtain

$$y = 2t.$$

Thus,

$$x = 1 + 3t$$
$$y = 2t$$

are the parametric equations for the line shown in Figure 9.39.

PARAMETRIC EQUATIONS

You would use the two-point form of the equation of a line to find $x = 1 + 3t$. We chose the points $(0, 1)$ and $(1, 4)$, but any two points from the table could be chosen.

> **DEFINITION:** A *parameter* is an arbitrary constant in a mathematical expression. Equations in which coordinates are each expressed in terms of a parameter are called *parametric equations*.

PARAMETER

PARAMETRIC EQUATIONS

EXAMPLE 1: Plot the curve represented by the parametric equations

$$x = \cos \theta$$
$$y = \sin \theta.$$

Solution: The parameter is θ and you can generate a table of values by using Table 4, exact values, or a calculator.

θ	0°	15°	30°	45°	60°	75°	90°	120°	...
x	1.00	0.97	0.87	0.71	0.50	0.26	0.00	−0.50	...
y	0.00	0.26	0.50	0.71	0.87	0.97	1.00	0.87	...

These points are plotted in Figure 9.40. If the plotted points are connected, you can see that the curve is a circle.

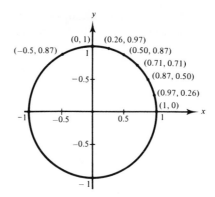

Figure 9.40   Graph of $x = \cos \theta$, $y = \sin \theta$

It is possible to recognize the parametric equations in Example 1 as a unit circle if you square both sides of the equation and add:

$$x^2 = \cos^2 \theta$$

$$y^2 = \sin^2 \theta$$

$$x^2 + y^2 = \cos^2 \theta + \sin^2 \theta.$$

So

$$x^2 + y^2 = 1.$$

EXAMPLES: Eliminate the parameter for the parametric equations given.

2. $x = t + 2$, $y = t^2 + 2t - 1$

Solution: Solve the first equation for t: $t = x - 2$. Substitute into the second equation:

$$\begin{aligned} y &= (x - 2)^2 + 2(x - 2) - 1 \\ &= x^2 - 4x + 4 + 2x - 4 - 1 \\ &= x^2 - 2x - 1. \end{aligned}$$

You can recognize this as a parabola. It can now be graphed by completing the square or by using the parametric equations and plotting points.

3. $x = t^2 - 3t + 1$, $y = -t^2 + 2t + 3$

Solution: It is not as easy to solve one of these equations for t as it was in Example 2. You can, however, add one equation to the other:

$$x + y = -t + 4 \qquad \text{or} \qquad t = 4 - x - y.$$

This can be substituted into either equation to give:

$$x = (4 - x - y)^2 - 3(4 - x - y) + 1$$
$$= 16 - 4x - 4y - 4x + x^2 + xy$$
$$- 4y + xy + y^2 - 12 + 3x + 3y + 1$$
$$= x^2 + y^2 + 2xy - 5x - 5y + 5$$
$$= x^2 + 2xy + y^2 - 5x - 5y + 5.$$

This is a rotated parabola since $B^2 - 4AC = 4 - 4(1)(1) = 0$.

4. $x = 2^t$, $y = 2^{t+1}$

   Solution:

$$\frac{y}{x} = \frac{2^{t+1}}{2^t}$$
$$= 2^{t+1-t}$$
$$= 2$$

Thus,
$$y = 2x.$$

Let's consider the parametric equations given in Example 4 a little more closely. We can plot a curve by using the parametric equations or by eliminating the parameter. Let's plot the equations in Example 4 both ways.

The values on this table are found by substitution. For example, if $t = 0$ then

$$x = 2^0$$
$$= 1$$

and

$$y = 2^{0+1}$$
$$= 2.$$

1. Parametric form:

t	0	1	2	3
x	1	2	4	8
y	2	4	8	16

Can $x = 3$? Solve

$$3 = 2^t$$
$$\log 3 = t \log 2$$
$$t = \frac{\log 3}{\log 2}$$
$$\approx 1.5850,$$

and then $y = 2^{1.5850+1} \approx 6$.
Can $x = 0$? No, since $0 = 2^t$ has no solution.
Can $x = -1$? No, since $2^t > 0$.

The graph is shown in Figure 9.41.

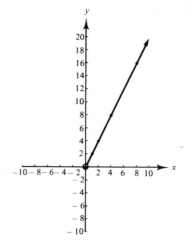

Figure 9.41   Graph of $x = 2^t$, $y = 2^{t+1}$

2. Eliminate the parameter as shown in Example 4:

$$y = 2x.$$

Can $x = 3$? Solve $y = 2(3) = 6$.
Can $x = 0$? Solve $y = 2(0) = 0$.
Can $x = -1$? Solve $y = 2(-1) = -2$.
The graph is shown in Figure 9.42.

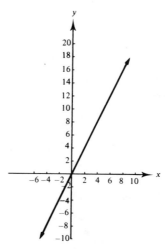

Figure 9.42   Graph of $y = 2x$

Of course, what you notice by comparing Figures 9.41 and 9.42 is that you must be very careful about the domain for x when eliminating the parameter. It should also be noted that sometimes it is impossible to eliminate the parameter in any simple way, as in the equations

$$x = 3t^5 - 4t^2 + 7t + 11$$
$$y = 4t^{13} + 12t^5 + 6t^4 + 16.$$

We therefore have to be able to graph parametric equations using both the method of plotting points and the method of eliminating the parameter.

PROBLEM SET 9.5

A Problems

Plot the curves in Problems 1–10 by plotting points.

1. $x = 4t,\ y = -2t$
2. $x = t + 1,\ y = 2t$
3. $x = 2t,\ y = t^2 + t + 1$
4. $x = 3t,\ y = t^2 - t + 6$
5. $x = 3\cos\theta,\ y = 3\sin\theta$
6. $x = 5\cos\theta,\ y = 2\sin\theta$
7. $x = t^2 + 2t + 3,\ y = t^2 + t - 4$
8. $x = t^2 + 3t - 4,\ y = 2t^2 + 4t - 1$
9. $x = 3^t,\ y = 3^{t+1}$
10. $x = 2^t,\ y = 2^{1-t}$

Eliminate the parameter in Problems 11–20 and plot the resulting equations.

11. $x = 4t,\ y = -2t$
12. $x = t + 1,\ y = 2t$
13. $x = 2t,\ y = t^2 + t + 1$
14. $x = 3t,\ y = t^2 - t + 6$
15. $x = 3\cos\theta,\ y = 3\sin\theta$
16. $x = 5\cos\theta,\ y = 2\sin\theta$
17. $x = t^2 + 2t + 3,\ y = t^2 + t - 4$
18. $x = t^2 + 3t - 4,\ y = 2t^2 + 4t - 1$
19. $x = 3^t,\ y = 3^{t+1}$
20. $x = 2^t,\ y = 2^{1-t}$

B Problems

Plot the curves in Problems 21–30 by any convenient method.

21. $x = 60t,\ y = 80t - 16t^2$
22. $x = 10\cos t,\ y = 10\sin t$
23. $x = \theta - \sin\theta,\ y = 1 - \cos\theta$
24. $x = \theta + \sin\theta,\ y = 1 - \cos\theta$
25. $x = 4\tan 2t,\ y = 3\sec 2t$
26. $x = 2\tan 2t,\ y = 4\sec 2t$
27. $x = 1 + \cos t,\ y = 3 - \sin t$
28. $x = 2 - \sin t,\ y = -3t\cos t$
29. $x = 3\cos\theta + \cos 3\theta,\ y = 3\sin\theta - \sin 3\theta$
30. $x = \cos t + t\sin t,\ y = \sin t - t\cos t$

Hint for Problem 31: consider the illustration below.

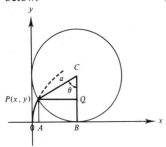

Find the coordinates of $P(x, y)$. Notice that

$x = |OA|,$

$y = |PA|.$

Find x and y in terms of θ, the amount of rotation in radians.

Hint for Problem 32: consider the illustration below.

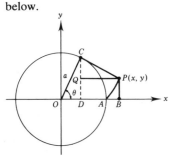

Find the coordinates of $P(x, y)$. Notice that

$x = |OB|,$

$y = |PB|.$

Find x and y in terms of θ, the amount of rotation in radians.

POLAR COORDINATE SYSTEM
POLE

Mind Bogglers

31. Suppose a light is attached to the edge of a bike wheel. The path of the light is shown in Figure 9.43. If the radius of the wheel is a, find the equation for the path of the light. Such a curve is called a *cycloid*.

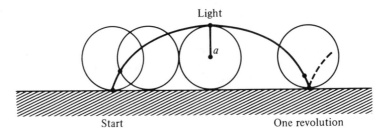

Figure 9.43   Graph of a cycloid

32.   Suppose a string is wound around a circle of radius a. The string is then unwound in the plane of the circle while it is held tight, as shown in Figure 9.44. Find the equation for this curve, called the *involute of a circle*.

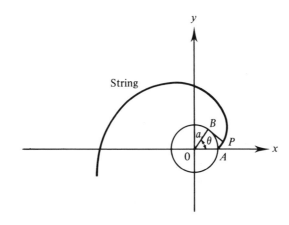

Figure 9.44   Graph of the involute of a circle

9.6 POLAR COORDINATES

Up to this point in the book, we've used a rectangular coordinate system. Now we'll consider a different system, called the *polar coordinate system*. In this system, we fix a point O, called the *origin* or *pole*, and measure a point in the plane by an ordered pair $P(r, \theta)$, where θ measures the angle from the positive x-axis and r represents the directed distance from the pole to the point P. Both r and θ can be any real number. When plotting points, first measure an angle θ and then measure out a length r along the ray from the pole through P $(\overrightarrow{OP})$. If θ is positive, the angle is measured in a counter-clockwise direction, and if θ is negative, it is measured in a clockwise direction. If r is positive, it is measured along the ray $\overrightarrow{OP}$, and if it is negative, the distance is measured along a ray that has just the opposite direction as $\overrightarrow{OP}$.

Historical Note

Polar coordinates were introduced by Jacob Bernoulli in 1691.

EXAMPLE 1: Figure 9.45 shows several plotted points; make sure you understand how each one can be plotted if you're given the ordered pair.

If you want to understand this section, you *must* study this example to see how each point is plotted.

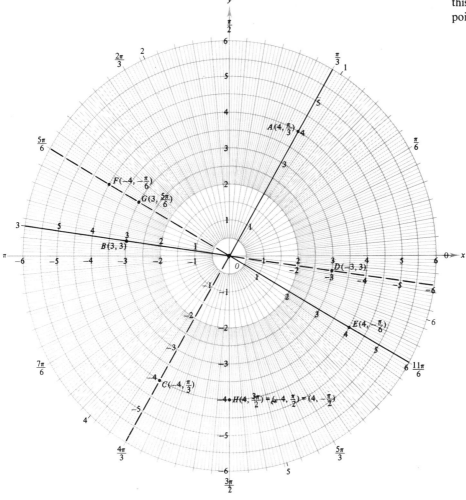

Figure 9.45   Several polar form points

In rectangular coordinates, every ordered pair (x, y) is associated in a one-to-one fashion with the points in the plane.

Every point in polar form has two primary representations:

(r, θ),

where

$$0 \leq \theta < 2\pi,$$

and

$(-r, \pi + \theta)$,

where

$$0 \leq \pi + \theta < 2\pi.$$

$$\frac{\pi}{4} + \pi = \frac{5\pi}{4}$$

$$\frac{5\pi}{4} + \pi = \frac{9\pi}{4},$$

but $(5, 9\pi/4)$ is not a primary representation of the point since $9\pi/4 > 2\pi$.

One thing you may notice immediately from Figure 9.45 is that ordered pairs in polar form are not associated with points in the plane in a one-to-one fashion. Indeed, given any point in the plane, there are infinitely many ordered pairs associated with the point in polar form. If you are given a point (r, θ) in polar coordinates, then $(-r, \theta + \pi)$ also represents that point. There are also infinitely many others, all of which have the same first component and second components that are multiples of 2π added to the angle. We call (r, θ) and $(-r, \theta + \pi)$ the *primary representations of the point* if the angles θ and $\theta + \pi$ are between 0 and 2π. Thus,

$$
\begin{array}{ll}
(r, \theta) & (-r, \theta + \pi) \\
(r, \theta + 2\pi) & (-r, \theta + 3\pi) \\
(r, \theta - 2\pi) & (-r, \theta - \pi) \\
\quad\vdots & \quad\vdots \\
(r, \theta + 2k\pi) & (-r, \theta + (2k + 1)\pi)
\end{array}
$$

all name the same point for any integer k.

EXAMPLES: Give both primary representations for the given points.

2. $(3, \pi/4)$ has primary representations $(3, \pi/4)$ and $(-3, 5\pi/4)$

3. $(5, 5\pi/4)$ has primary representations $(5, 5\pi/4)$ and $(-5, \pi/4)$

4. $(-6, -2\pi/3)$ has primary representations $(6, \pi/3)$ and $(-6, 4\pi/3)$

5. $(9, 5)$ has primary representations $(9, 5)$ and $(9, 5 + \pi)$; a point like $(9, 5 + \pi)$ is usually approximated by writing $(9, 8.14)$

6. $(9, 7)$ has primary representations $(9, 7 - 2\pi)$ or $(9, 0.72)$ and $(-9, 7 - \pi)$ or $(-9, 3.86)$

The relationship between the two coordinate systems can easily be found by using the definition of the trigonometric functions (see Figure 9.46).

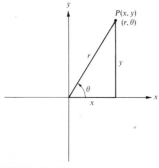

Figure 9.46 Relationship between rectangular and polar coordinates

1. To change from polar to rectangular:
$$x = r \cos \theta;$$
$$y = r \sin \theta.$$

2. To change from rectangular to polar:
$$r = \sqrt{x^2 + y^2};$$

$$\tan \theta = \frac{y}{x}.$$

If r and θ are related by an equation, we can speak of the *graph of the equation*. For example, $r = 5$ is the equation of a circle with center at the origin and radius 5. Also, $\theta = \pi/3$ is the equation of a line passing through the origin (see Figure 9.47).

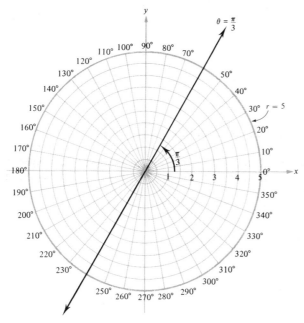

Figure 9.47   Graphs of $r = 5$ and $\theta = \pi/3$

As with other equations we've graphed in this book, we'll begin graphing polar form curves by plotting some points. However, first you must be able to recognize whether or not a point in polar form satisfies a given equation.

EXAMPLES: Show that the given points lie on the graph of

$$r = \frac{2}{1 - \cos \theta}.$$

7. $(2, \pi/2)$
   Solution: Substituting, we have

$$2 \overset{?}{=} \frac{2}{1 - \cos (\pi/2)}$$

$$\overset{?}{=} \frac{2}{1 - 0}$$

$$2 = 2.$$

Thus, the point is on the curve, since it satisfies the equation

8. $(-2, 3\pi/2)$
    Solution: Substituting, we have

    $$-2 \overset{?}{=} \frac{2}{1 - \cos{(3\pi/2)}}$$

    $$\overset{?}{=} \frac{2}{1 - 0}$$

    $$-2 \neq 2.$$

    The equation is not satisfied, but we cannot say that the point is not on the curve. Indeed, we see from Example 7 that it is on the curve, since $(-2, 3\pi/2)$ and $(2, \pi/2)$ name the same point! So, even if one primary representation of a point does not satisfy the equation, we must still check the other one.

9. $(-1, 2\pi)$
    Solution: Substituting, we have

    $$-1 \overset{?}{=} \frac{2}{1 - \cos 2\pi}$$

    $$\overset{?}{=} \frac{2}{1 - 1},$$

    which is undefined. Checking the other representation of the same point—namely $(1, 3\pi)$—we get

    $$1 \overset{?}{=} \frac{2}{1 - \cos 3\pi}$$

    $$\overset{?}{=} \frac{2}{1 - (-1)}$$

    $$1 = 1.$$

    Thus, the point is on the curve.

The above examples lead us to the following definition:

DEFINITION: A point other than the pole is on a polar form curve if and only if at least one of its primary representations satisfies the given equation.

We now turn our attention to some polar form graphing.

EXAMPLE 10: Graph $r = 2(1 - \cos \theta)$.

Solution: First construct a table of values by choosing values for θ and approximating the corresponding values for r.

θ	0	$\pi/6$	$\pi/3$	$\pi/2$	$2\pi/3$	$5\pi/6$	π	$7\pi/6$	$4\pi/3$	$3\pi/2$	$5\pi/3$	$11\pi/6$
r (approx.)	0	0.27	1	2	3	3.7	4	3.7	3	2	1	0.27

These points are plotted and then connected, as shown in Figure 9.48.

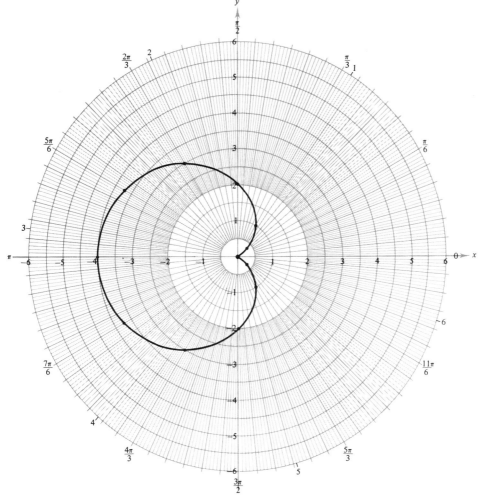

Figure 9.48   Graph of $r = 2(1 - \cos \theta)$

CARDIOID

The curve given in Example 10 is called a *cardioid* because of its heart shape. The graphs of $r = a(1 + \cos\theta)$, $r = a(1 - \sin\theta)$, and $r = a(1 + \sin\theta)$ are also cardioids except that the curves are turned in different directions.

EXAMPLE 11: Graph $r = 4\sin 2\theta$.

Solution:

θ	0	$\pi/12$	$\pi/6$	$\pi/4$	$\pi/3$	$5\pi/12$	$\pi/2$	$7\pi/12$	$2\pi/3$	$3\pi/4$	$5\pi/6$	$11\pi/12$	π
r (approx.)	0	2	3.5	4	3.5	2	0	-2	-3.5	-4	-3.5	2	0

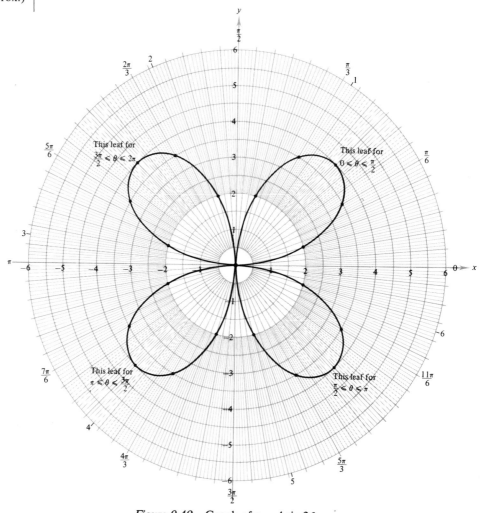

Figure 9.49   Graph of $r = 4\sin 2\theta$

For values of θ between π and 2π, the table repeats. The points (r, θ)—*not* $(r, 2\theta)$—are plotted and used to sketch the curve shown in Figure 9.49.

The curve in Example 11 is called a *rose curve.* The graphs of $r = a \cos n\theta$ and $r = a \sin n\theta$ (n a positive integer) are also rose curves. If $n = 1$, the rose is a curve with one petal and is circular; if n is odd, the rose is n-leaved; if n is even, the rose is $2n$-leaved. Some other rose curves are shown in Figure 9.50.

ROSE CURVE

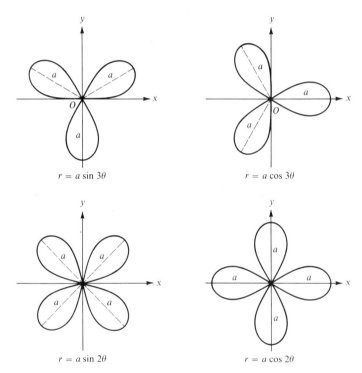

$r = a \sin 3\theta$

$r = a \cos 3\theta$

$r = a \sin 2\theta$

$r = a \cos 2\theta$

Figure 9.50   Rose curves

EXAMPLE 12: Graph $r^2 = 9 \sin 2\theta$.

Solution: When calculating a table of values if r is squared, be sure to obtain two values. For example, if $\theta = \pi/4$, then $\sin 2\theta = 1$, $r^2 = 9$, and $r = 3$ or -3.

θ	0	$\pi/12$	$\pi/6$	$\pi/4$	$\pi/3$	$5\pi/12$	$\pi/2$
r (approx.)	0	± 2.1	± 2.8	± 3	± 2.8	± 2.1	0

For $\pi/2 < \theta < \pi$, there are no values for r since $\sin 2\theta$ is negative. For $\pi \leq \theta \leq 2\pi$, the values r repeat the sequence given above so these points are plotted and then connected, as shown in Figure 9.51.

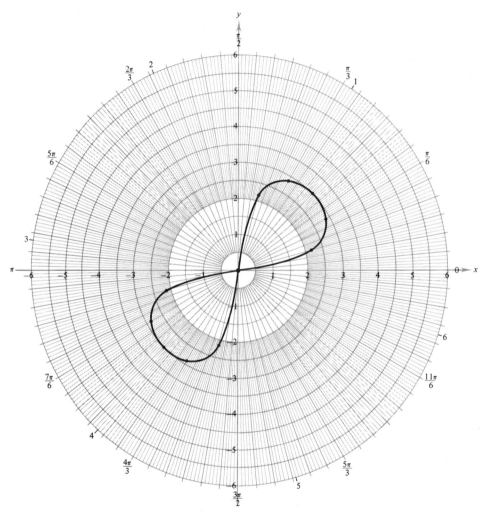

Figure 9.51   Graph of $r^2 = 9 \sin 2\theta$

LEMNISCATE

The curve in Example 12 is called a *lemniscate*. The graphs of $r^2 = a^2 \cos 2\theta$ and $r^2 = a^2 \sin 2\theta$ are lemniscates. For the one with the cosine function, the figure eight is centered on the *x*-axis.

PROBLEM SET 9.6

A Problems

In Problems 1–10, plot the given polar form points. Give both primary representations and the rectangular coordinates.

1. $(4, \pi/4)$ 2. $(6, \pi/3)$ 3. $(5, 2\pi/3)$ 4. $(3, -\pi/6)$
5. $(3/2, -5\pi/6)$ 6. $(5, -\pi/2)$ 7. $(-4, 4)$ 8. $(4, 10)$
9. $(-4, 5\pi)$ 10. $(-3, -7\pi/3)$

In Problems 11–20, plot the given rectangular form points. Give both primary representations in polar form.

11. $(5, 5)$ 12. $(-1, \sqrt{3})$ 13. $(2, -2\sqrt{3})$ 14. $(-2, -2)$
15. $(3, -3)$ 16. $(-6, 6)$ 17. $(-\sqrt{3}, 1)$ 18. $(4, 3)$
19. $(-12, 5)$ 20. $(3, 7)$
21. Derive the equations for changing from polar coordinates to rectangular coordinates.
22. Derive the equations for changing from rectangular coordinates to polar coordinates.

In Problems 23–27, tell whether the given points lie on the curve $r = 5/(1 - \sin \theta)$.

23. $(10, \pi/6)$ 24. $(5, \pi/2)$ 25. $(-10, 5\pi/6)$ 26. $(-10/3, 5\pi/6)$
27. $(20 + 10\sqrt{3}, \pi/3)$

In Problems 28–32, tell whether the given points lie on the curve $r = 2(1 - \cos \theta)$.

28. $(1, \pi/3)$ 29. $(1, -\pi/3)$ 30. $(-1, \pi/3)$ 31. $(-2, \pi/2)$
32. $(-2 - \sqrt{2}, \pi/4)$

Find three ordered pairs satisfying each of the equations in Problems 33–44. Give both primary representations for each point.

33. $r^2 = 9 \cos \theta$ 34. $r^2 = 9 \cos 2\theta$ 35. $r = 3\theta$
36. $r = 5\theta$ 37. $r\theta = 4$ 38. $r = 2 - 3 \sin \theta$
39. $r = 2(1 + \cos \theta)$ 40. $r = 2 \cos \theta$ 41. $r = \tan \theta$

42. $r = 6 \cos 3\theta$ 43. $r = \dfrac{8}{1 - 2 \cos \theta}$ 44. $\dfrac{r}{1 - \sin \theta} = 2$

B Problems

Sketch the curves given in Problems 45–59.

45. $r = 2(1 + \cos \theta)$ 46. $r = 3(1 - \sin \theta)$ 47. $r = 4(1 + \sin \theta)$
48. $r = 4 \cos 2\theta$ 49. $r = 5 \sin 3\theta$ 50. $r = 3 \cos 3\theta$
51. $r = 2 \cos \theta$ 52. $r^2 = 9 \cos 2\theta$ 53. $r^2 = 16 \cos 2\theta$
54. $r^2 = 16 \sin 2\theta$ 55. $r = 5 \sin (\pi/6)$ 56. $r = 9 \cos (\pi/3)$
57. $r = \theta$ 58. $r = 3\theta$ 59. $r = \tan \theta$

Mind Boggler

60. Consider the graph of $r = \sin(a\theta/b)$, where a/b is reduced (a and b are integers, $b \neq 0$). If $b = 1$, then $r = \sin a\theta$, which was discussed in Section 9.6. The rose curve is described by the following tables.

a	b	Petal Length	Number of Primary Petals (Not a repeat of one already graphed) $B = \dfrac{P}{F}$	Flower Length (maximum interval in which the polar coordinates do not repeat themselves) F
odd	odd	$\dfrac{b}{a}\pi$	a	$b\pi$
a or b even		$\dfrac{b}{a}\pi$	$2a$	$2b\pi$

a. Find P, N, and F for the following curves and draw the graphs:

i. $r = \sin \dfrac{1}{5}\theta$ ii. $r = \sin \dfrac{2}{5}\theta$

iii. $r = \sin \dfrac{3}{5}\theta$ iv. $r = \sin \dfrac{4}{5}\theta$

v. $r = \sin \theta$ vi. $r = \sin \dfrac{6}{5}\theta$

vii. $r = \sin \dfrac{7}{5}\theta$ viii. $r = \sin \dfrac{8}{5}\theta$

ix. $r = \sin \dfrac{9}{5}\theta$ x. $r = \sin 2\theta$

b. Can you draw a rose with 6 petals?*

9.7 SUMMARY AND REVIEW

TERMS

Asymptote (9.3)
Cardioid (9.6)
Conic section (9.1)
Conjugate axis (9.3)
Directrix (9.1)
Discriminant (9.4)

Eccentricity (9.2)
Ellipse (9.2)
ε (9.2)
Foci (9.2, 9.3)
Focus (9.1)
Hyperbola (9.3)

*My thanks to Clyde Russell of Santa Rosa Junior College for this problem.

Latus rectum (9.1)

Lemniscate (9.6)

Major axis (9.2)

Minor axis (9.2)

Parabola (9.1)

Parameter (9.5)

Parametric equations (9.5)

Polar coordinate system (9.6)

Pole (9.6)

Primary representations of a point (9.6)

Rose curve (9.6)

Rotation (9.4)

Transverse axis (9.3)

Vertex (9.1)

IMPORTANT IDEAS

(9.1–9.4)   Identify a conic section as a line, parabola, ellipse, circle, or hyperbola by inspection.

(9.1–9.4)   Given a quadratic equation, sketch the graph.

(9.1–9.4)   Given a conic section (or information about the graph), write the equation.

(9.1–9.4)   Conic summary (see pages 464–465).

(9.5)   Graph parametric equations by plotting points and by eliminating the parameter (if possible).

(9.6)   To change from rectangular coordinates (x, y) to polar coordinates, use

$$r = \sqrt{x^2 + y^2}; \qquad \tan \theta = \frac{y}{x}.$$

(9.6)   To change from polar coordinates (r, θ) to rectangular coordinates, use

$$x = r \cos \theta; \qquad y = r \sin \theta.$$

(9.6)   The primary representations of a point in polar coordinates are (r, θ) where $0 \le \theta < 2\pi$, and $(-r, \theta + \pi)$, where $0 \le \theta + \pi < 2\pi$. The other representations add multiples of 2π to the angles.

REVIEW PROBLEMS

1. (9.1–9.4) Identify the given curve as a line, parabola, ellipse, circle, or hyperbola.

    a. $xy + x^2 - 3x = 5$

    b. $x^2 + y^2 + xy + 3x - y = 3$

    c. $3x - 2y^2 - 4y + 7 = 0$

    d. $\dfrac{x}{16} + \dfrac{y}{4} = 1$

    e. $x^2 + 2xy + y^2 = 10$

    f. $25x^2 + 16y^2 = 400$

    g. $(x - 1)(y + 1) = 7$

    h. $25(x - 2)^2 + 25(y + 1)^2 = 400$

    i. $x = 2 + 5t, y = -1 - 3t$

    j. $2x^2 - y^2 + 4xy - 2x + 3y = 6$

2. (9.6) Write both primary representations for the polar form points; also, graph each point.

    a. $(5, \sqrt{75})$ b. $(3, -2\pi/3)$ c. $(-2, 2)$ d. $(-5, 9.4247)$

    e. *Calculator Problem.* Change the polar forms in parts a–d into rectangular form.

3. Write the equations of the described curves.
    a. (9.2) The ellipse with the center at $(4, 1)$, a focus at $(5, 1)$, and a semimajor axis 2
    b. (9.3) The set of points with the difference of distances from $(-3, 4)$ and $(-7, 4)$ equal to 2
4. Write the equations of the described curves.
    a. (9.1) The parabola with vertex at $(6, 3)$ and directrix $x = 1$
    b. (9.3) The hyperbola with center at $(-5, 4)$, a focus at $(0, 4)$, and eccentricity 2

Graph the curves in Problems 5–9.

5. a. (9.2) $25x^2 + 16y^2 = 400$ b. (9.3) $x^2 - y^2 + x - y = 3$
6. a. (9.1) $x^2 = y$ b. (9.2) $x^2 + y^2 = 4x + 2y - 3$
7. a. (9.5) $x = 3 \cos \theta, y = 5 \sin \theta$ b. (9.5) $x = t^2 + 3t - 1, y = t^2 + 2t + 5$
8. a. (9.6) $r = 4 \cos 5\theta$ b. (9.6) $r = 2 + 2 \cos \theta$
9. (9.4) $x(x - y) = y(y - x) - 1$
10. a. (9.4) What is the angle of rotation for

$$5x^2 + 4xy + 5y^2 + 3x - 2y + 5 = 0?$$

    b. (9.4) What is the tangent of the angle of rotation for

$$4x^2 + 4xy + y^2 + 3x - 2y + 7 = 0?$$

Appendix A

Sets of Numbers

This book assumes a certain familiarity with some basic sets of numbers in mathematics. This appendix briefly reviews operations and algebra using these sets.

COUNTING NUMBERS

$$N = \{1, 2, 3, 4, \ldots\}$$

INTEGERS

$$Z = \{\ldots, -3, -2, -1, 0, 1, 2, 3, \ldots\}$$

The integers account for the majority of our computations. The essential notion used in defining the operations on integers is absolute value.

$$|n| = \begin{cases} n & \text{if } n \geq 0 \\ -n & \text{if } n < 0 \end{cases}$$

ABSOLUTE VALUE

Addition of Integers

Procedure for adding integers

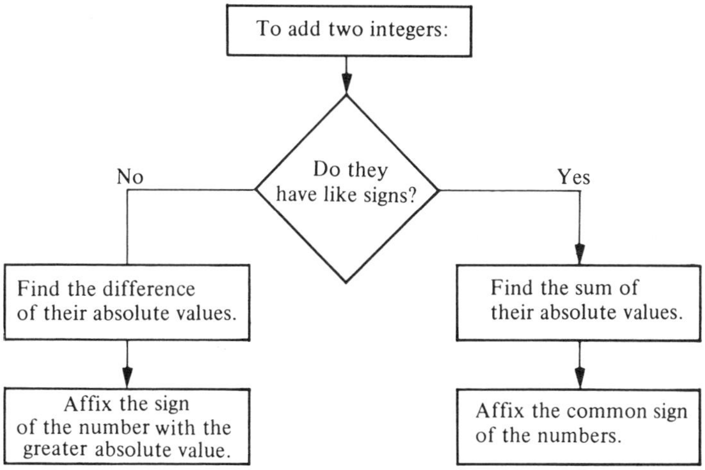

Figure A.1   Procedure for adding integers

487

Subtraction of Integers

Subtraction is defined as an addition:

Procedure for subtracting integers

$$a - b = a + (-b)$$

Subtracting a number is equivalent to adding its opposite.

EXAMPLES

1. $(-5) + (-7) = -(5 + 7)$ (like signs; common sign)
    $\qquad\qquad\quad = -12$ (add)

2. $(-5) + (+7) = +(7 - 5)$ (unlike signs; sign of the 7)
    $\qquad\qquad\quad = +2$ (subtract)

3. $(+5) + (-7) = -(7 - 5)$ (unlike signs; sign of the 7)
    $\qquad\qquad\quad = -2$ (subtract)

4. $(-5) + (-7) = -(5 + 7)$ (like signs; common sign)
    $\qquad\qquad\quad = -12$ (add)

5. $(-6) - (-9) = (-6) + (+9)$ (add opposite)
    $\qquad\qquad\quad = +(9 - 6)$ (unlike signs; sign of 9)
    $\qquad\qquad\quad = +3$ (subtract)

6. $(-18) - (+11) = (-18) + (-11)$ (add opposite)
    $\qquad\qquad\qquad = -(18 + 11)$ (like signs; common sign)
    $\qquad\qquad\qquad = -29$ (add)

7. $(-21) + (+5) - (-8) - (+12)$
    $= (-21) + (+5) + (+8) + (-12)$ (add opposites)
    $= [(-21) + (-12)] + [(+5) + (+8)]$ (group terms with same sign)
    $= (-33) + (+13)$ (add like signs)
    $= -(33 - 13)$ (unlike signs; sign of 33)
    $= -20$ (subtract)

Multiplication of Integers

Procedure for multiplying integers

Figure A.2 shows a flowchart to use when multiplying two integers.

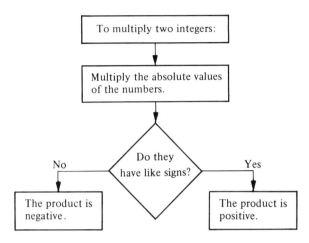

Figure A.2 Procedure for multiplying integers

Division of Integers

Similar to the case of subtraction, division requires no special rules since it is defined as multiplication:

$$a \div b = \frac{a}{b} = a \cdot \frac{1}{b}$$

Dividing by a number is equivalent to multiplying by its reciprocal.

Procedure for dividing integers

EXAMPLES

8. $(-3)(2) = -6$ (unlike signs; product negative)

The 2 is assumed to be positive if no sign appears.

9. $(-7)(-5) = 35$ (like signs; product positive)

10. $(-2)(-3)(-5) = 6(-5)$ (group the first two numbers; like signs; product positive)

             $= -30$ (unlike signs; product negative)

11. $2 - 3(6 - 1) = 2 + (-3)[6 + (-1)]$

                $= 2 + (-3)[5]$

                $= 2 + (-15)$

                $= -13$

With several operations there is an order of operations agreement:

RATIONAL NUMBERS

1. First, carry out operations in parentheses.
2. Next, carry out multiplications and divisions in the order of left to right.
3. Finally, carry out additions and subtractions in the order of left to right.

$$Q = \left\{ \frac{a}{b} \,\middle|\, a \text{ and } b \text{ are integers, } b \neq 0 \right\}$$

The rational numbers also account for many of our computations. The essential notion when working with rational numbers is the idea of reducing a fraction.

Fundamental Property

$$\frac{ak}{bk} = \frac{a}{b} \quad (b \neq 0, k \neq 0)$$

A fraction is reduced if the numerator and denominator have no common factors except 1.

Signs of a Fraction

Let p and q be positive integers. Since

$$\frac{p}{q} = \frac{-p}{-q} \quad \text{and} \quad \frac{-p}{q} = \frac{p}{-q},$$

$$\frac{p}{q} = -\frac{-p}{q} = -\frac{p}{-q},$$

$$\frac{-p}{q} = -\frac{p}{q} = -\frac{-p}{-q},$$

all possible forms can be written as p/q or $-p/q$. These forms are called the *standard forms* of a fraction.

Equality of Rational Numbers

$$\frac{a}{b} = \frac{c}{d} \quad \text{if and only if} \quad ad = bc$$

Multiplication of Rational Numbers

$$\frac{a}{b} \cdot \frac{c}{d} = \frac{a \cdot c}{b \cdot d} \quad (b, d \neq 0)$$

Division of Rational Numbers

$$\frac{a}{b} \div \frac{c}{d} = \frac{a \cdot d}{b \cdot c} \quad (b, c, d \neq 0)$$

EXAMPLES

12. $\dfrac{2}{3} \cdot \dfrac{5}{7} = \dfrac{2 \cdot 5}{3 \cdot 7} = \dfrac{10}{21}$

13. $\dfrac{2}{3} \cdot \dfrac{9}{7} = \dfrac{18}{21} = \dfrac{6 \cdot 3}{7 \cdot 3} = \dfrac{6}{7}$

14. $\dfrac{3}{5} \cdot \dfrac{10}{27} = \dfrac{3 \cdot 2 \cdot 5}{5 \cdot 3 \cdot 3 \cdot 3} = \dfrac{2}{9}$

15. $\dfrac{2}{3} \div \dfrac{4}{5} = \dfrac{2 \cdot 5}{3 \cdot 4} = \dfrac{5}{6}$

16. $\dfrac{5}{8} \div 6 = \dfrac{5}{8} \cdot \dfrac{1}{6} = \dfrac{15}{4}$ $\left(\dfrac{15}{4} \text{ is reduced since there are no common factors of } \right.$
$\left. 15 \text{ and } 4 \text{ except } 1 \right)$

Notice in Example 13 that factoring was done after multiplying so that the result could be reduced. It is often easier to factor before multiplying:

$$\dfrac{2}{3} \cdot \dfrac{9}{7} = \dfrac{2 \cdot 3 \cdot 3}{3 \cdot 7} = \dfrac{6}{7}.$$

The cancellation marks do not mean that we are crossing out the factors, but rather rewriting them as factors of 1.

Addition and Subtraction of Rational Numbers

$$\dfrac{a}{b} \pm \dfrac{c}{b} = \dfrac{a \pm c}{b} \quad (b \neq 0)$$

$$\dfrac{a}{b} \pm \dfrac{c}{d} = \dfrac{ad \pm bc}{bd} \quad (b, d \neq 0)$$

The addition and subtraction of rational numbers is accomplished by finding a common denominator, as shown in Examples 17–19.

EXAMPLES

17. $\dfrac{3}{14} + \dfrac{8}{21} = \dfrac{3}{14} \cdot \dfrac{3}{3} + \dfrac{8}{21} \cdot \dfrac{2}{2}$

$\quad = \dfrac{9}{42} + \dfrac{16}{42}$

$\quad = \dfrac{25}{42}$

18. $\dfrac{5}{8} - \dfrac{7}{12} = \dfrac{5}{8} \cdot \dfrac{3}{3} + \dfrac{-7}{12} \cdot \dfrac{2}{2}$

$\quad = \dfrac{15}{24} + \dfrac{-14}{24}$

$\quad = \dfrac{1}{24}$

The common denominator is found by factoring:

prime factors

$14 = 2 \cdot \quad 7$
$21 = \quad 3 \cdot 7$
$\overline{\qquad \qquad 2 \cdot 3 \cdot 7 = 42}$

one representative of *each* different factor | common denominator

$8 = 2^3$
$12 = 2^2 \cdot 3$

$2^3 \cdot 3 = 24$

If representatives of factors have different exponents, take the one with the largest exponent as the representative.

$2 = 2$
$12 = 2^2 \cdot 3$
$3 = 3$
$10 = 2 \cdot 5$

$2^2 \cdot 3 \cdot 5 = 60$

common denominator

$i = \sqrt{-1}$

19. $\dfrac{3}{2} + \dfrac{5}{12} - \left(\dfrac{2}{3} - \dfrac{1}{10}\right) = \dfrac{3}{2} \cdot \dfrac{30}{30} + \dfrac{5}{12} \cdot \dfrac{5}{5} + \dfrac{-2}{3} \cdot \dfrac{20}{20} + \dfrac{1}{10} \cdot \dfrac{6}{6}$

$= \dfrac{90}{60} + \dfrac{25}{60} + \dfrac{-40}{60} + \dfrac{6}{60}$

$= \dfrac{81}{60}$

$= \dfrac{3 \cdot 3 \cdot 3 \cdot \cancel{3}}{2 \cdot 2 \cdot \cancel{3} \cdot 5}$

$= \dfrac{27}{20}$

IRRATIONAL NUMBERS

$Q' = \{x \mid x \text{ is a nonrepeating, nonterminating decimal}\}$

REAL NUMBERS

$R = \{x \mid x \text{ is a rational or irrational number}\}$

COMPLEX NUMBERS

$C = \{z \mid z = a + bi, \text{ where } a \text{ and } b \text{ are real numbers and } i^2 = -1\}$

If $b = 0$, then we have the subset of the complex numbers $a + 0i = a$, which is the set of real numbers.

If $a = 0$ with $b \neq 0$, then we have a *pure imaginary number* $0 + bi = bi$.

If $a = 0$ with $b = 1$, then we have the *imaginary unit* $0 + 1i = i$.

To work with complex numbers, we must define equality, along with the usual arithmetic operations. Let $a + bi$ and $c + di$ be complex numbers. (Remember that a, b, c, and d are real numbers.)

PROPERTIES OF
COMPLEX NUMBERS

Equality: $a + bi = c + di$ if and only if $a = c$ and $b = d$

Addition: $(a + bi) + (c + di) = (a + c) + (b + d)i$

Subtraction: $(a + bi) - (c + di) = (a - c) + (b - d)i$

Multiplication: $(a + bi)(c + di) = (ac - bd) + (ad + bc)i$

Division: $\dfrac{a + bi}{c + di} = \dfrac{(ac + bd) + (bc - ad)i}{c^2 + d^2} = \dfrac{ac + bd}{c^2 + d^2} + \dfrac{bc - ad}{c^2 + d^2}i$

It is not necessary to memorize these definitions; we can deal with two complex numbers as we would any binomials as long as we remember that $i^2 = -1$.

EXAMPLES: Simplify each of the given complex numbers. (*Simplify* means write in the form $a + bi$.)

20. $(4 + 5i) + (3 + 4i) = 7 + 9i$

21. $(2 - i) - (3 - 5i) = -1 + 4i$

22. $(5 - 2i) + (3 + 2i) = 8$ or $8 + 0i$

23. $(4 + 3i) - (4 + 2i) = i$ or $0 + i$

24. $(2 + 3i)(4 + 2i) = 8 + 16i + 6i^2$
$$= 8 + 16i - 6$$
$$= 2 + 16i$$

25. $(a + bi)(c + di) = ac + adi + bci + bdi^2$
$$= ac + (ad + bc)i - bd$$
$$= (ac - bd) + (ad + bc)i$$

26. $(4 - 3i)(4 + 3i) = 16 - 9i^2$
$$= 16 + 9$$
$$= 25$$

Notice that the result of Example 25 corresponds to the definition of multiplication given above, but in this example the calculations were carried out in the usual way while remembering that $i^2 = -1$.

Example 26 gives a clue for dividing complex numbers. The definition would be difficult to remember, so instead we use the idea of *conjugates*. The complex numbers $a + bi$ and $a - bi$ are called *complex conjugates*, and each is the conjugate of the other:

$$(a + bi)(a - bi) = a^2 - b^2i^2$$
$$= a^2 + b^2,$$

COMPLEX CONJUGATES

which is a real number. Thus, simplify a quotient by using the conjugate of the denominator, as illustrated by Examples 27–29.

EXAMPLES: Simplify.

multiply by 1

27. $\dfrac{15 - 5i}{2 - i} = \dfrac{15 - 5i}{2 - i} \cdot \dfrac{2 + i}{2 + i} = \dfrac{30 + 5i - 5i^2}{4 - i^2}$

conjugates

$$= \dfrac{35 + 5i}{5}$$
$$= 7 + i$$

Recall that if $^{10}/_5 = 2$, we can check by multiplying $5 \cdot 2 = 10$. For Example 27, we can check by multiplying $(2 - i)(7 + i)$:

$$(2 - i)(7 + i) = 14 - 5i - i^2$$
$$= 15 - 5i.$$

28. $\dfrac{6 + 5i}{2 + 3i} = \dfrac{6 + 5i}{2 + 3i} \cdot \dfrac{2 - 3i}{2 - 3i}$

$= \dfrac{12 - 8i - 15i^2}{4 - 9i^2}$

$= \dfrac{27}{13} - \dfrac{8}{13}i$

29. $\dfrac{a + bi}{c + di} = \dfrac{a + bi}{c + di} \cdot \dfrac{c - di}{c - di}$

$= \dfrac{ac - adi + bci - bdi^2}{c^2 - d^2i^2}$

$= \dfrac{(ac + bd) + (bc - ad)i}{c^2 + d^2}$

Notice that this result corresponds to the definition of division given on page 492.

PROBLEM SET A.1

A Problems

Simplify the expressions in Problems 1–40.

1. a. $8 + (-3)$ b. $(-9) + 5$ c. $(-10) + (-4)$ d. $(162) + (-27)$
2. a. $-15 + 8$ b. $(-14) + 27$ c. $42 + (-121)$ d. $62 + (-62)$
3. a. $10 - 7$ b. $7 - 10$ c. $6 - (-4)$ d. $(-5) - (-10)$
4. a. $5 - (-5)$ b. $7 - (-3)$ c. $0 - (-15)$ d. $(-46) - (-46)$
5. a. $(-7) - (-18)$ b. $(-4) - 8$ c. $62 - (-112)$ d. $5 - (-416)$

6. a. $14(-5)$ b. $(-14)(-5)$ c. $\dfrac{-12}{4}$ d. $\dfrac{-63}{-9}$

7. a. $\dfrac{12}{-4}$ b. $-6 \cdot \dfrac{14}{-2}$ c. $\dfrac{-528}{-4}$ d. $(-1)^3$

8. a. $(-6)(14)$ b. $(-2)(-3)$ c. $\dfrac{34}{-2}$ d. $10 \cdot \dfrac{-8}{-2}$

9. a. $(-12) + [-4 + (-3)]$ b. $(-5) + (-6 + 10)$
10. a. $[5 + (-7)] + [5 + (-4)]$ b. $[6 + (-8)] + (6 + 8)$
11. a. $-5(8 - 12)$ b. $(-5 \cdot 8) - 12$
12. a. $14 + (-10) - [8 - 11]$ b. $6 - [(-8) - 5]$
13. a. $[48 \div (-6)] \div (-2)$ b. $48 \div [(-6) \div (-2)]$

14. a. $\dfrac{2}{3} + \dfrac{7}{9}$ b. $\dfrac{-5}{7} + \dfrac{4}{3}$

15. a. $\dfrac{-12}{35} - \dfrac{8}{15}$ b. $\dfrac{1}{2} + \dfrac{1}{3} + \dfrac{1}{5}$

16. a. $\left(\dfrac{4}{5} + 2\right) + \dfrac{-7}{25}$ b. $2 + \dfrac{1}{2}$

17. a. $\dfrac{7}{9} - \dfrac{2}{3}$ b. $\dfrac{4}{7} - \dfrac{-5}{9}$

18. a. $\dfrac{-3}{5} - \dfrac{-6}{9}$ b. $\left(\dfrac{3}{10} - \dfrac{7}{100}\right) - \dfrac{9}{1000}$

19. a. $\dfrac{3}{10} - \left(\dfrac{7}{100} - \dfrac{9}{1000}\right)$ b. $5 - \dfrac{1}{5}$

20. a. $\dfrac{2}{3} \cdot \dfrac{5}{7}$ b. $\dfrac{-1}{8} \cdot \dfrac{2}{-5}$

21. a. $5 \cdot \dfrac{6}{7}$ b. $\dfrac{47}{-5} \cdot \dfrac{-15}{26}$

22. a. $\left(\dfrac{-5}{7} \cdot \dfrac{-14}{25}\right) \cdot \dfrac{15}{-21}$ b. $7 \cdot \dfrac{3}{4} - \dfrac{1}{4} \cdot 3$

23. a. $\dfrac{4}{9} \div \dfrac{2}{3}$ b. $\dfrac{-2}{9} \div \dfrac{6}{7}$

24. a. $\dfrac{105}{-11} \div \dfrac{-15}{33}$ b. $\left(\dfrac{5}{7} \div \dfrac{-14}{25}\right) \div \dfrac{15}{21}$

25. a. $\dfrac{5}{7} \div \left(\dfrac{-14}{25} \div \dfrac{15}{21}\right)$ b. $6 \div \dfrac{1}{6} + \dfrac{3}{5} - 5$

26. a. $(3 + 3i) + (5 + 4i)$ b. $(6 - 2i) + (5 + 3i)$
27. a. $(4 - 2i) - (3 + 4i)$ b. $5 - (2 - 3i)$
28. a. $5i - (5 + 5i)$ b. $(5 - 3i) - (5 + 2i)$
29. a. $(3 + 4i) - (7 + 4i)$ b. $-2(-4 + 5i)$
30. a. $4(2 - i) - 3(-1 - i)$ b. $6(3 + 2i) + 4(-2 - 3i)$
31. a. $i(2 + 3i)$ b. $i(5 - 2i)$
32. a. $(3 - i)(2 + i)$ b. $(4 - i)(2 + i)$
33. a. $(5 - 2i)(5 + 2i)$ b. $(8 - 5i)(8 + 5i)$
34. a. $(3 - 5i)(3 + 5i)$ b. $(7 - 9i)(7 + 9i)$
35. a. $-i^2$ b. $-i^3$
36. a. i^3 b. i^4
37. a. $-i^4$ b. $-i^5$
38. a. $-i^6$ b. i^6
39. a. i^{11} b. i^{236}
40. a. $-i^{1980}$ b. i^{1976}

𝕳istorical 𝕹ote

The idea of the square root of a negative number is one that bothered early mathematicians. The Hindus Mahavira (850) and Bhaskara (1150) were aware of the problem, but the first to consider it seriously was Girolamo Cardano (see the historical note on page 52), who referred to it as sophistic (which means clever or plausible but unsound and tending to mislead). Descartes (see the historical note on page 10) classified numbers as *real* and *imaginary* in 1637, and Euler (see the historical note on page 9) was the first to use the letter *i* and the name *complex number*.

B Problems

Simplify the expressions in Problems 41–54.

41. $(6 - 2i)^2$

42. $(3 + 3i)^2$

43. $(4 + 5i)^2$

44. $(3 - 5i)^3$

45. $\dfrac{-3}{1 + i}$

46. $\dfrac{5}{4 - i}$

47. $\dfrac{2}{i}$

48. $\dfrac{5}{i}$

49. $\dfrac{3i}{5 - 2i}$

50. $\dfrac{-2i}{3 + i}$

51. $\dfrac{5 + 3i}{4 - i}$

52. $\dfrac{4 - 2i}{3 + i}$

53. $\dfrac{1 - 6i}{1 + 6i}$

54. $\dfrac{2 + 7i}{2 - 7i}$

Appendix B

Equations, Inequalities, and Intervals

LINEAR EQUATIONS

A *linear equation in one variable* is an equation equivalent to

$$ax + b = 0,$$

LINEAR EQUATION IN ONE VARIABLE

where x is a variable and a and b are any real numbers. An *open equation* is an equation containing a variable that may be either true or false, depending upon the replacement for the variable. A *root* or a *solution* is a replacement for the variable that makes the equation true. The *solution set* of an open equation is the set of all solutions of the equation. To *solve an equation* means to find its solution set.

OPEN EQUATION

ROOT OF AN EQUATION
SOLUTION SET

Linear equations are solved by using the following principles:

ADDITION PRINCIPLE: If $a = b$, then $a + c = b + c$.

MULTIPLICATION PRINCIPLE: If $a = b$, then $ac = bc$.

---

a, b, and c are any real numbers

EXAMPLES: Solve the equations.

1. $4x + 2 = 5x - 7$
    $4x + 2 + (-5x) = 5x - 7 + (-5x)$
    $-x + 2 = -7$
    $-x + 2 + (-2) = -7 + (-2)$
    $-x = -9$
    $x = 9$ Solution set: $\{9\}$

2. $4(x - 3) + 5x = 5(8 + x)$
    $4x - 12 + 5x = 40 + 5x$
    $9x - 12 = 40 + 5x$
    $4x - 12 = 40$
    $4x = 52$
    $x = 13$ Solution set: $\{13\}$

Use the following steps in solving a linear equation:

Step 1. Use the distributive property to clear the equation of parentheses if necessary.

Step 2. Combine similar terms.

Step 3. Use the addition principle to write an equivalent equation so that the terms involving the variable are on one side and the constant terms are on the other side.

Step 4. Use the multiplication principle to write an equivalent equation so that the variable is isolated on one side.

3. $5(x + 1) + 3x = 8x + 7$
   $\mathbf{5x + 5} + 3x = 8x + 7$
   $\mathbf{8x} + 5 = 8x + 7$
   $5 = 7$ Solution set: $\varnothing$

QUADRATIC EQUATIONS

A quadratic equation in one variable is an equation equivalent to

$$ax^2 + bx + c = 0 \quad (a \neq 0),$$

where x is a variable and a, b, and c are real numbers. Quadratic equations are solved by using the following principles:

a, b, and c are any real numbers

> PRINCIPLE OF ZERO PRODUCTS:
>
> $\quad ab = 0$ if and only if $a = 0$ or $b = 0$
>
> QUADRATIC FORMULA: If $ax^2 + bx + c = 0$, $a \neq 0$, then
>
> $$x = \frac{-b \pm \sqrt{b^2 - 4ac}}{2a}.$$

Use the following steps in solving a quadratic equation:

Step 1. Put the quadratic into the standard form

$$ax^2 + bx + c = 0.$$

Step 2. Factor, if possible, and use the principle of zero products to solve the resulting linear equations.

Step 3. If it is not factorable, solve by using the quadratic formula.

EXAMPLES

4. $\quad\quad\quad x^2 = 4$
   $\quad\quad x^2 - 4 = 0$
   $(x - 2)(x + 2) = 0$
   $\quad\quad\quad x = 2, \quad x = -2$ Solution set: $\{2, -2\}$

5. $\quad\quad\quad x^2 - 4 = 3x$
   $\quad\quad x^2 - 3x - 4 = 0$
   $(x - 4)(x + 1) = 0$
   $\quad\quad\quad x = 4 \quad$ or $\quad -1$ Solution set: $\{4, -1\}$

6. $x^2 + 2x - 4 = 0$

$$x = \frac{-2 \pm \sqrt{4 - 4(1)(-4)}}{2}$$

$$= \frac{-2 \pm 2\sqrt{5}}{2}$$

$$= -1 \pm \sqrt{5}$$ Solution set: $\{-1 + \sqrt{5}, -1 - \sqrt{5}\}$

The quadratic formula is derived and discussed in Section 2.5. See that section for further examples.

INTERVAL NOTATION

For any two real numbers a and b, exactly one of the following is true:

1. $a = b$ 2. $a < b$ 3. $a > b$

Inequalities can be used to represent intervals on the real number line. For example, the inequality $x < 3$ denotes the interval of the real number line shown in Figure B.1.

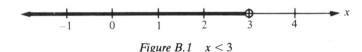

Figure B.1 $x < 3$

If the number x is between two values a and b $(a < b)$, then we write

$$a < x < b,$$

which can be represented as shown in Figure B.2.

Figure B.2 $a < x < b$

There are four possibilities for representing an interval of values between a and b, depending on the inclusion of the endpoints. For this reason, the following terminology and notation is introduced:

$a < x < b$ open interval (a, b)

$a \leq x \leq b$ closed interval $[a, b]$

$a \leq x < b$ half-open interval $[a, b)$

$a < x \leq b$ half-open interval $(a, b]$

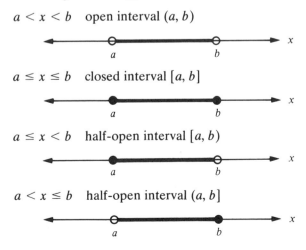

The symbol (a, b) for open interval notation should not be confused with ordered-pair notation. Its usage should be clear in the context in which it is used—on a number line.

EXAMPLES

7. Write interval notation for the number lines.

    a.   [1, 3]

    b.   (2, 5]

    c.   [0, 4)

    d.   (−1, 4)

8. Write interval notation for the given inequalities

    a. $-6 \leq x < 5$ $[-6, 5)$
    b. $7 < x < 8$ $(7, 8)$
    c. $-9 \leq x \leq -2$ $[-9, -2]$
    d. $-5 < x \leq 0$ $(-5, 0]$

Sometimes intervals are of unlimited extent in one or both directions. In such cases, we use the symbol ∞, as shown in Examples 9–13.

EXAMPLES

9. $x > 2$ $(2, \infty)$

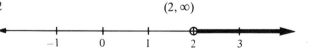

10. $x \geq 2$ $[2, \infty)$

11. $x < 2$ $\qquad\qquad$ $(-\infty, 2)$

12. $x \leq 2$ $\qquad\qquad$ $(-\infty, 2]$

13. x is any real number $\qquad$ $(-\infty, \infty)$

LINEAR INEQUALITIES

A linear inequality in one variable is an inequality equivalent to

$$ax + b < 0, \qquad ax + b \leq 0, \qquad ax + b > 0, \qquad \text{or} \qquad ax + b \geq 0,$$

where x is a variable and a and b are any real numbers. Linear inequalities are solved by using the following principles:

ADDITION PRINCIPLE: If $a < b$, then $a + c < b + c$ for any real numbers a, b, and c.

POSITIVE MULTIPLICATION PRINCIPLE: If $a < b$, then $ac < bc$ if c is a positive real number.

NEGATIVE MULTIPLICATION PRINCIPLE: If $a < b$, then $ac > bc$ if c is a negative real number.

These principles also hold for $\leq$, $>$, and $\geq$.

By comparing these principles with those of equality, you can see that the procedure is the same as with a linear equation except that, if you multiply or divide by a negative number, the order of the inequality is reversed.

EXAMPLES: Solve the given inequalities.

14. $\quad 5x - 3 \geq 7$

$\quad\quad 5x - 3 + \mathbf{3} \geq 7 + \mathbf{3}$

$\quad\quad\quad\quad 5x \geq 10$

$\quad\quad\quad\quad\quad x \geq 2 \quad$ Solution set: $[2, \infty)$

15. $2(4 - 3t) < 5t - 14$
     $8 - 6t < 5t - 14$
     $-6t < 5t - 22$
     $-11t < -22$
     $t > 2$ (order of the inequality is reversed because
                 we divided by a negative number)
     Solution set: $(2, \infty)$

Sometimes, two inequality statements are combined, as with

$$-3 \leq 2x + 1 \leq 5.$$

This means that $-3 \leq 2x + 1$ *and* $2x + 1 \leq 5$. It also means that $2x + 1$ is between -3 and 5. This second interpretation allows us to work the inequality as shown in Examples 16 and 17.

EXAMPLES: Solve the inequalities.

16. $-3 \leq 2x + 1 \leq 5$

   $-3 - \mathbf{1} \leq 2x + 1 - \mathbf{1} \leq 5 - \mathbf{1}$

       $-4 \leq 2x \leq 4$

       $\dfrac{-4}{\mathbf{2}} \leq \dfrac{2x}{2} \leq \dfrac{4}{\mathbf{2}}$

       $-2 \leq x \leq 2$ Solution set: $[-2, 2]$

17. $-5 \leq 1 - 3x \leq 10$

     $-6 \leq -3x \leq 9$

     $\dfrac{-6}{-\mathbf{3}} \geq \dfrac{-3x}{-3} \geq \dfrac{9}{-\mathbf{3}}$ (inequality reversed)

      $2 \geq x \geq -3$

However, $2 \geq x \geq -3$ is not proper because the left endpoint should be the smaller value, so the answer is rewritten as

     $-3 \leq x \leq 2$ Solution set: $[-3, 2]$

QUADRATIC INEQUALITIES

A quadratic inequality in one variable is an inequality equivalent to

$$ax^2 + bx + c < 0 \quad (a \neq 0),$$

$<$ can be replaced by $\leq$, $>$, and $\geq$

where x is a variable and a, b, and c are any real numbers.

The procedure for solving a quadratic inequality is similar to that for solving a quadratic equality. First, use the properties of inequality to obtain a zero on one side of the inequality. The next step is to factor, if possible, the quadratic expression on the left. For example,

$$x^2 - 4 \geq 0$$

$$(x - 2)(x + 2) \geq 0.$$

We want to find the values of x that make the inequality valid. The values for which each of the factors is zero are called the *critical values* of x. The critical values for this example are 2 and -2. For *every other value of x* the inequality is either positive or negative. Next, plot the critical values on a number line, as shown in Figure B.3. These critical values divide the number line into inter-

CRITICAL VALUES

Figure B.3 Number line with critical values

vals separated by the zeros of this expression. Next, choose a sample value from each of the intervals. Evaluate each factor to determine the sign only—it is not necessary to do the arithmetic to find its value.

Sample Value	Factor	Sign of Factor
This is *your* choice. ↓	This is done mentally. ↓	
$x = -100$	$x - 2 = -100 - 2$	$-$
	$x + 2 = -100 + 2$	$-$
$x = 0$	$x - 2 = 0 - 2$	$-$
	$x + 2 = 0 + 2$	$+$
$x = 100$	$x - 2 = 100 - 2$	$+$
	$x + 2 = 100 + 2$	$+$

The above procedure can be summarized as shown in Figure B.4. By using the properties of multiplication, we see that

$(x - 2)(x + 2)$	is positive in the interval	$x < -2;$
$(x - 2)(x + 2)$	is negative in the interval	$-2 < x < 2;$
$(x - 2)(x + 2)$	is positive in the interval	$x > 2.$

Since we want

$$(x - 2)(x + 2) \geq 0,$$

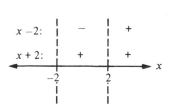

Figure B.4 Procedure for solving $x^2 - 4 \geq 0$

we see that the solution set is in the interval

$$x < -2 \quad \text{or} \quad x > 2$$

and includes endpoints of those intervals. In general, if the given inequality is of the form $\leq$ or $\geq$, the endpoints are included, and if it has the form $<$ or $>$, the endpoints are excluded. This whole procedure can be summarized as shown in Figure B.5. Thus, the solution set is $x \leq -2$ or $x \geq 2$.

Using interval notation, this solution set can be written

$$(-\infty, -2] \cup [2, \infty).$$

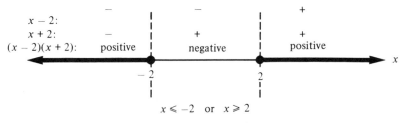

Figure B.5 Solution of $x^2 - 4 \geq 0$

EXAMPLE 18: Solve the inequality $x^2 \geq 4 - 3x$.

Solution:

$$x^2 + 3x - 4 \geq 0$$

factor:

$$(x + 4)(x - 1) \geq 0$$

Plot the critical values on a number line:

Check the signs of the factors for each interval:

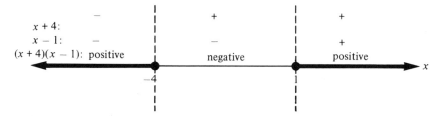

Use the properties of multiplication of real numbers to obtain the solution shown above. In this case, there are two intervals $x \leq -4$ or $x \geq 1$ that satisfy the inequality. Thus, the solution set is $(-\infty, -4] \cup [1, \infty)$.

In the case of the inequalities $>$ or $<$, the solution set does not include the critical values.

EXAMPLE 19: Solve the inequality $2x^2 < 5 - 9x$.

Solution: $\qquad 2x^2 + 9x - 5 < 0$

$\qquad\qquad (2x - 1)(x + 5) < 0$

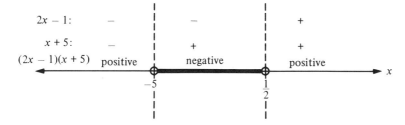

Solution set: $-5 < x < \frac{1}{2}$ or $(-5, \frac{1}{2})$

EXAMPLE 20: $x^2 + 2x - 4 < 0$

Solution: The term on the left is in simplified form and cannot be factored. Therefore, we proceed by considering $x^2 + 2x - 4$ as a single factor. To find the critical values, we find the values for which the factor is zero.

$$x^2 + 2x - 4 = 0$$

$$x = \frac{-2 \pm \sqrt{4 - 4(1)(-4)}}{2}$$

$$= \frac{-2 \pm 2\sqrt{5}}{2}$$

$$= -1 \pm \sqrt{5}$$

Plot the critical values and check the sign of the factor in each interval:

We see that the solution set is $-1 - \sqrt{5} < x < -1 + \sqrt{5}$ or $(-1 - \sqrt{5}, -1 + \sqrt{5})$.

PROBLEM SET B.1

A Problems

Solve the equations or inequalities given in Problems 1–28.

1. $3x - 7 = -1$
2. $18 = 3(x + 4)$
3. $x(8 + 3) = x + 20$
4. $2x + (5 + 3x) = 15$
5. $3x + 5 - x = 2x - 2$
6. $2 - (x - 3) = 6 + 4(x - 4)$
7. $(x - 1)(x + 2) = x^2 + 3$
8. $4(x - 7) + 3(x + 2) = 5x - 7$
9. $3(2x - 1) - (x + 3) = -1$
10. $5(2x - 1) + 7(2x - 1) = 4x - 2$
11. $x + 7 \geq -3$
12. $10 < 5 + y$
13. $2 > -s$
14. $-t \leq -3$
15. $3a + 4 \leq a + 2$
16. $4b - 3 > 2b - 13$
17. $3(4 - m) \geq 6m$
18. $5(m - 3) \leq 13m + 5$
19. $4a - 7 > 3(a + 1)$
20. $6(2y - 1) \leq 3(y + 4)$
21. $2(4 - 3x) > 4(3 - x)$
22. $3(2 - 5y) \leq 5(y + 2) + 4$
23. $5 \leq x + 4 \leq 7$
24. $1 \leq x + 5 \leq 8$
25. $2 \leq 2x \leq 8$
26. $5 \leq 5x \leq 15$
27. $5 \leq 1 - 4x \leq 13$
28. $13 \leq 4 - 3x \leq 19$

B Problems

For Problems 29–44, find the solution set.

29. $x^2 + 2x - 3 > 0$
30. $x^2 - x - 6 > 0$
31. $6x^2 - x - 12 < 0$
32. $8x^2 + 2x - 3 < 0$
33. $x^2 \geq 4$
34. $x^2 \geq 9$
35. $5x - 6 \geq x^2$
36. $4 \geq x^2 + 3x$
37. $6x^2 + 32 + 4x < 9x^2$
38. $2x^2 + 3x - 3 < x^2 + 6x$
39. $x^2 + 2x - 1 < 0$
40. $x^2 - 2x - 2 < 0$
41. $x^2 + 3x - 7 \geq 0$
42. $2x^2 + 4x + 5 \geq 0$
43. $x^2 - 2x - 6 \leq 0$
44. $x^2 - 8x + 13 > 0$

45. In a certain chemistry experiment, a reaction will occur only if the temperature is between 20° and 30° C. Find the range of temperature in Fahrenheit degrees, where

$$C = \frac{5}{9}(F - 32).$$

46. The IQ for the majority of the population is in the range

$$90 \leq IQ \leq 110,$$

where IQ is found according to the formula

$$IQ = \frac{MA \cdot 100}{CA},$$

MA is mental age, and CA is chronological age. If a certain group of 14-year-olds is tested, what is the range of mental ages for this group?

$ 47. The cost for manufacturing a certain type of candle is given by $C = 200 + 23x$, where x is the number of candles produced and C is the cost in cents. The revenue is the amount of income resulting from the sales. If $R = 48x$ represents the revenue, and the company want to make a profit, then R must be greater than C. Find the number of candles that must be produced to realize a profit.

Note: 200 represents the *fixed expenses* (they remain the same and do not change regardless of the number of candles produced), whereas $23x$ represents the *variable expenses* (they vary depending on the number of candles produced).

48. An architect is designing a facing for a building that uses both equilateral triangles and squares. If the side of the triangle is 2 ft longer than the side of the square, and the perimeter of the square is greater than the perimeter of the triangle, what are the possible lengths for the side of the square?

49. Current postal regulations state that no package may be sent if its combined length, width, and height exceed 72 in. What are the dimensions of a box to be mailed with equal height and width if the length is four times the height?

50. If an object is shot up from the ground with an initial velocity of 256 fps, its distance in feet above the ground at the end of t seconds is given by $d = 128t - 16t^2$ (neglecting air resistance). Find the length of time for which $d \geq 240$.

51. Repeat Problem 50 where $d \geq 192$.
52. For Problem 50, determine the length of time the projectile would be in the air.

$ 53. *Calculator Problem.* An investor wants to subdivide a piece of property and knows that the area of each lot must be at least 10,000 sq ft. If each lot is 100 ft longer than it is wide, what is the minimum width for each lot? (Hint: don't forget that the width must be greater than zero.)

54. The huge ten-story-high Echo satellite was designed to reflect radio waves back to earth. To be a good reflector, the spherical satellite required a diameter that was at least as large as the wavelength λ of the wave reflected—that is,

$$\frac{D}{\lambda} \geq 1.$$

If it is a spherical satellite, then the surface area is given by

$$S = 4\pi r^2$$

and the volume is

$$V = \frac{4}{3}\pi r^3.$$

Determine the minimum diameter D (in meters) needed for the satellite to be a good reflector of waves with frequency of 10^7 Hz. The length of a wave is the distance traveled by a series of waves during a given time divided by the frequency or number of waves propagated during that time. That is,

$$\lambda = \frac{c}{f},$$

where c is the speed of light, $3 \cdot 10^8$ mps, and f is the frequency. Also, find the surface area and volume of the Echo satellite.

The *hertz* (Hz) is a unit indicating the frequency in cycles per second (cps); 1 Hz = 1 cps.

Appendix C

Significant Digits

Whenever we work with measurements, the quantities are necessarily approximations. The digits known to be correct in a number obtained by a measurement are called *significant digits*. The digits 1, 2, 3, 4, 5, 6, 7, 8, and 9 are always significant, whereas the digit 0 may or may not be significant.

1. Zeros that come between two other digits are significant, as in 203 or 10.04.
2. If the only function of the zero is to place the decimal point, it is not significant, as in

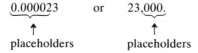

0.000023   or   23,000.

↑   ↑

placeholders   placeholders

If it does more than fix the decimal point, it is significant, as in

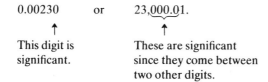

0.00230   or   23,000.01.

↑   ↑

This digit is   These are significant
significant.   since they come between
                       two other digits.

This second rule can, of course, result in certain ambiguities, such as in 23,000 (measured to the *exact* unit). To avoid such confusion, we use scientific notation in this case.

$$2.3 \times 10^4 \qquad \text{two significant digits}$$
$$2.3000 \times 10^4 \quad \text{five significant digits}$$

EXAMPLES

1. Two significant digits: 46, 0.00083, 4.0×10^1, 0.050

2. Three significant digits: 523, 403, 4.00×10^2, 0.000800

3. Four significant digits: 600.1, 4.000×10^1, 0.0002345

When we are doing calculations with approximate numbers (particularly when using a calculator), it is often necessary to round off results.

TO ROUND OFF NUMBERS:

1. Increase the last retained digit by 1 if the residue is larger than 5.
2. Retain the last digit unchanged if the residue is smaller than 5.
3. Retain the last digit unchanged if even, or increase it by 1 if odd, if the residue is exactly 5.

Elaborate rules for computation of approximate data can be developed (when it is necessary for some applications, such as in chemistry), but there are two simple rules that will work satisfactorily for the material in this book.

ADDITION–SUBTRACTION: Add or subtract in the usual fashion, and then round off the result so that the last digit retained is in the column farthest to the right in which both given numbers have significant digits.

MULTIPLICATION–DIVISION: Multiply or divide in the usual fashion, and then round off the result to the smaller number of significant digits found in either of the given numbers.

These agreements are particularly important when we are using a calculator, since the results obtained will look much more accurate on the calculator than they actually are. Suppose we want to calculate

$$b = \frac{50}{\tan 35°}.$$

1. By division: from Table 4, $\tan 35° = 0.7002$. Thus,

$$b = \frac{50}{0.7002} = 71.408169090\ldots.$$

2. By multiplication:

$$b = \frac{50}{\tan 35°} = 50 \cot 35°$$
$$= 50(1.428) \quad \text{(from Table 4)}$$
$$= 71.400.$$

3. By calculator with algebraic logic (set to degrees):

| 50 | ÷ | 35 | tan | = |

By calculator with RPN logic:

| 50 | ENTER | 35 | tan | ÷ |

The answer is now displayed:

$$b = \frac{50}{\tan 35°} = 71.40740034.$$

Notice that all the answers above differ. We now use the multiplication–division rule.

50: This number has one or two significant digits; there is no ambiguity if we write 5×10^1 or 5.0×10^1. *In this book*, if the given data include a number with a doubtful degree of accuracy, *we will assume the maximum degree of accuracy*. Thus, 50 has two significant digits.

tan 35°: From Table 4, we find this number to 4 significant digits; on a calculator you may have 8, 10, or 12 significant digits (depending on the calculator).

The result of this division is correct to two significant digits, namely,

$$b = \frac{50}{\tan 35°} = 71,$$

which agrees with all the above methods of solution.

  In solving triangles in this text, we will assume a certain relationship in the accuracy of the measurement between the sides and the angles.

ACCURACY IN SIDES	ACCURACY IN ANGLES
Two significant digits	Nearest degree
Three significant digits	Nearest tenth of a degree
Four significant digits	Nearest hundredth of a degree

  This chart means that if the data include one side given with two significant digits and another with three significant digits, the angle would be computed to the nearest degree. If one side is given to four significant digits and an angle to the nearest tenth of a degree, then the other sides would be given to three significant digits and the angles computed to the nearest tenth of a degree. In general, results should not be more accurate than the least accurate item of the given data.

  If you have access only to a four-function calculator, you can use Table 4 in conjunction with your calculator. For example, to find b, you first find tan 35° = 0.7002 and then calculate

$$b = \frac{50}{0.7002} \quad \boxed{50} \quad \boxed{\div} \quad \boxed{0.7002} \quad \boxed{=}$$

or

$$= 71.41 \quad \boxed{50} \quad \boxed{\text{ENTER}} \quad \boxed{0.7002} \quad \boxed{\div}$$

or, to two significant digits, $b = 71$. Of course, one of the great advantages of calculators is that they enable us to work with much greater accuracy without having to use interpolation or tables.

See Problems 19–31 of Problem Set 5.4 (on page 237) for some practice problems with significant digits.

Appendix D

Tables

TABLE 1   Squares and Square Roots (0–199)

No.	Sq.	Sq. Rt.	Factors	No.	Sq.	Sq. Rt.	Factors
1	1	1.000		51	2,601	7.141	$3 \cdot 17$
2	4	1.414	2	52	2,704	7.211	$2^2 \cdot 13$
3	9	1.732	3	53	2,809	7.280	53
4	16	2.000	2^2	54	2,916	7.348	$2 \cdot 3^3$
5	25	2.236	5	55	3,025	7.416	$5 \cdot 11$
6	36	2.449	$2 \cdot 3$	56	3,136	7.483	$2^3 \cdot 7$
7	49	2.646	7	57	3,249	7.550	$3 \cdot 19$
8	64	2.828	2^3	58	3,364	7.616	$2 \cdot 29$
9	81	3.000	3^2	59	3,481	7.681	59
10	100	3.162	$2 \cdot 5$	60	3,600	7.746	$2^2 \cdot 3 \cdot 5$
11	121	3.317	11	61	3,721	7.810	61
12	144	3.464	$2^2 \cdot 3$	62	3,844	7.874	$2 \cdot 31$
13	169	3.606	13	63	3,969	7.937	$3^2 \cdot 7$
14	196	3.742	$2 \cdot 7$	64	4,096	8.000	2^6
15	225	3.873	$3 \cdot 5$	65	4,225	8.062	$5 \cdot 13$
16	256	4.000	2^4	66	4,356	8.124	$2 \cdot 3 \cdot 11$
17	289	4.123	17	67	4,489	8.185	67
18	324	4.243	$2 \cdot 3^2$	68	4,624	8.246	$2^2 \cdot 17$
19	361	4.359	19	69	4,761	8.307	$3 \cdot 23$
20	400	4.472	$2^2 \cdot 5$	70	4,900	8.367	$2 \cdot 5 \cdot 7$
21	441	4.583	$3 \cdot 7$	71	5,041	8.426	71
22	484	4.690	$2 \cdot 11$	72	5,184	8.485	$2^3 \cdot 3^2$
23	529	4.796	23	73	5,329	8.544	73
24	576	4.899	$2^3 \cdot 3$	74	5,476	8.602	$2 \cdot 37$
25	625	5.000	5^2	75	5,625	8.660	$3 \cdot 5^2$
26	676	5.099	$2 \cdot 13$	76	5,776	8.718	$2^2 \cdot 19$
27	729	5.196	3^3	77	5,929	8.775	$7 \cdot 11$
28	784	5.292	$2^2 \cdot 7$	78	6,084	8.832	$2 \cdot 3 \cdot 13$
29	841	5.385	29	79	6,241	8.888	79
30	900	5.477	$2 \cdot 3 \cdot 5$	80	6,400	8.944	$2^4 \cdot 5$
31	961	5.568	31	81	6,561	9.000	3^4
32	1,024	5.657	2^5	82	6,724	9.055	$2 \cdot 41$
33	1,089	5.745	$3 \cdot 11$	83	6,889	9.110	83
34	1,156	5.831	$2 \cdot 17$	84	7,056	9.165	$2^2 \cdot 3 \cdot 7$
35	1,225	5.916	$5 \cdot 7$	85	7,225	9.220	$5 \cdot 17$
36	1,296	6.000	$2^2 \cdot 3^2$	86	7,396	9.274	$2 \cdot 43$
37	1,369	6.083	37	87	7,569	9.327	$3 \cdot 29$
38	1,444	6.164	$2 \cdot 19$	88	7,744	9.381	$2^3 \cdot 11$
39	1,521	6.245	$3 \cdot 13$	89	7,921	9.434	89
40	1,600	6.325	$2^3 \cdot 5$	90	8,100	9.487	$2 \cdot 3^2 \cdot 5$
41	1,681	6.403	41	91	8,281	9.539	$7 \cdot 13$
42	1,764	6.481	$2 \cdot 3 \cdot 7$	92	8,464	9.592	$2^2 \cdot 23$
43	1,849	6.557	43	93	8,649	9.644	$3 \cdot 31$
44	1,936	6.633	$2^2 \cdot 11$	94	8,836	9.695	$2 \cdot 47$
45	2,025	6.708	$3^2 \cdot 5$	95	9,025	9.747	$5 \cdot 19$
46	2,116	6.782	$2 \cdot 23$	96	9,216	9.798	$2^5 \cdot 3$
47	2,209	6.856	47	97	9,409	9.849	97
48	2,304	6.928	$2^4 \cdot 3$	98	9,604	9.899	$2 \cdot 7^2$
49	2,401	7.000	7^2	99	9,801	9.950	$3^2 \cdot 11$
50	2,500	7.071	$2 \cdot 5^2$	100	10,000	10.000	$2^2 \cdot 5^2$

TABLE 2 Powers of *e*

x	e^x	e^{-x}	x	e^x	e^{-x}	x	e^x	e^{-x}
0.00	1.000	1.000	0.50	1.649	0.607	1.00	2.718	0.368
0.01	1.010	0.990	0.51	1.665	0.600	1.01	2.746	0.364
0.02	1.020	0.980	0.52	1.682	0.595	1.02	2.773	0.361
0.03	1.031	0.970	0.53	1.699	0.589	1.03	2.801	0.357
0.04	1.041	0.961	0.54	1.716	0.583	1.04	2.829	0.353
0.05	1.051	0.951	0.55	1.733	0.577	1.05	2.858	0.350
0.06	1.062	0.942	0.56	1.751	0.571	1.06	2.886	0.346
0.07	1.073	0.932	0.57	1.768	0.566	1.07	2.915	0.343
0.08	1.083	0.923	0.58	1.786	0.560	1.08	2.945	0.340
0.09	1.094	0.914	0.59	1.804	0.554	1.09	2.974	0.336
0.10	1.105	0.905	0.60	1.822	0.549	1.10	3.004	0.333
0.11	1.116	0.896	0.61	1.840	0.543	1.11	3.034	0.330
0.12	1.127	0.887	0.62	1.859	0.538	1.12	3.065	0.326
0.13	1.139	0.878	0.63	1.878	0.533	1.13	3.096	0.323
0.14	1.150	0.869	0.64	1.896	0.527	1.14	3.127	0.320
0.15	1.162	0.861	0.65	1.916	0.522	1.15	3.158	0.317
0.16	1.174	0.852	0.66	1.935	0.517	1.16	3.190	0.313
0.17	1.185	0.844	0.67	1.954	0.512	1.17	3.222	0.310
0.18	1.197	0.835	0.68	1.974	0.507	1.18	3.254	0.307
0.19	1.209	0.827	0.69	1.994	0.502	1.19	3.287	0.304
0.20	1.221	0.819	0.70	2.014	0.497	1.20	3.320	0.301
0.21	1.234	0.811	0.71	2.034	0.492	1.21	3.353	0.298
0.22	1.246	0.803	0.72	2.054	0.487	1.22	3.387	0.295
0.23	1.259	0.795	0.73	2.075	0.482	1.23	3.421	0.292
0.24	1.271	0.787	0.74	2.096	0.477	1.24	3.456	0.289
0.25	1.284	0.779	0.75	2.117	0.472	1.25	3.490	0.287
0.26	1.297	0.771	0.76	2.138	0.468	1.26	3.525	0.284
0.27	1.310	0.763	0.77	2.160	0.463	1.27	3.561	0.281
0.28	1.323	0.756	0.78	2.182	0.458	1.28	3.597	0.278
0.29	1.336	0.748	0.79	2.203	0.454	1.29	3.633	0.275
0.30	1.350	0.741	0.80	2.226	0.449	1.30	3.669	0.273
0.31	1.363	0.733	0.81	2.248	0.445	1.31	3.706	0.270
0.32	1.377	0.726	0.82	2.270	0.440	1.32	3.743	0.267
0.33	1.391	0.719	0.83	2.293	0.436	1.33	3.781	0.264
0.34	1.405	0.712	0.84	2.316	0.432	1.34	3.819	0.262
0.35	1.419	0.705	0.85	2.340	0.427	1.35	3.857	0.259
0.36	1.433	0.698	0.86	2.363	0.423	1.36	3.896	0.257
0.37	1.448	0.691	0.87	2.387	0.419	1.37	3.935	0.254
0.38	1.462	0.684	0.88	2.411	0.415	1.38	3.975	0.252
0.39	1.477	0.677	0.89	2.435	0.411	1.39	4.015	0.249
0.40	1.492	0.670	0.90	2.460	0.407	1.40	4.055	0.247
0.41	1.507	0.664	0.91	2.484	0.403	1.41	4.096	0.244
0.42	1.522	0.657	0.92	2.509	0.399	1.42	4.137	0.242
0.43	1.537	0.651	0.93	2.535	0.395	1.43	4.179	0.239
0.44	1.553	0.644	0.94	2.560	0.391	1.44	4.221	0.237
0.45	1.568	0.638	0.95	2.586	0.387	1.45	4.263	0.235
0.46	1.584	0.631	0.96	2.612	0.383	1.46	4.306	0.232
0.47	1.600	0.625	0.97	2.638	0.379	1.47	4.349	0.230
0.48	1.616	0.619	0.98	2.664	0.375	1.48	4.393	0.228
0.49	1.632	0.613	0.99	2.691	0.372	1.49	4.437	0.225

TABLE 2 Powers of e (continued)

x	e^x	e^{-x}	x	e^x	e^{-x}	x	e^x	e^{-x}
1.50	4.482	0.223	2.00	7.389	0.135	2.50	12.182	0.082
1.51	4.527	0.221	2.01	7.463	0.134	2.51	12.305	0.081
1.52	4.572	0.219	2.02	7.538	0.133	2.52	12.429	0.080
1.53	4.618	0.217	2.03	7.614	0.131	2.53	12.554	0.080
1.54	4.665	0.214	2.04	7.691	0.130	2.54	12.680	0.079
1.55	4.712	0.212	2.05	7.768	0.129	2.55	12.807	0.078
1.56	4.759	0.210	2.06	7.846	0.127	2.56	12.936	0.077
1.57	4.807	0.208	2.07	7.925	0.126	2.57	13.066	0.077
1.58	4.855	0.206	2.08	8.004	0.125	2.58	13.197	0.076
1.59	4.904	0.204	2.09	8.085	0.124	2.59	13.330	0.075
1.60	4.953	0.202	2.10	8.166	0.122	2.60	13.464	0.074
1.61	5.003	0.200	2.11	8.248	0.121	2.61	13.599	0.074
1.62	5.053	0.198	2.12	8.331	0.120	2.62	13.736	0.073
1.63	5.104	0.196	2.13	8.415	0.119	2.63	13.874	0.072
1.64	5.155	0.194	2.14	8.499	0.118	2.64	14.013	0.071
1.65	5.207	0.192	2.15	8.585	0.116	2.65	14.154	0.071
1.66	5.259	0.190	2.16	8.671	0.115	2.66	14.296	0.070
1.67	5.312	0.188	2.17	8.758	0.114	2.67	14.440	0.069
1.68	5.366	0.186	2.18	8.846	0.113	2.68	14.585	0.069
1.69	5.420	0.185	2.19	8.935	0.112	2.69	14.732	0.068
1.70	5.474	0.183	2.20	9.025	0.111	2.70	14.880	0.067
1.71	5.529	0.181	2.21	9.116	0.110	2.71	15.029	0.067
1.72	5.585	0.179	2.22	9.207	0.109	2.72	15.180	0.066
1.73	5.641	0.177	2.23	9.300	0.108	2.73	15.333	0.065
1.74	5.697	0.176	2.24	9.393	0.106	2.74	15.487	0.065
1.75	5.755	0.174	2.25	9.488	0.105	2.75	15.643	0.064
1.76	5.812	0.172	2.26	9.583	0.104	2.76	15.800	0.063
1.77	5.871	0.170	2.27	9.679	0.103	2.77	15.959	0.063
1.78	5.930	0.169	2.28	9.777	0.102	2.78	16.119	0.062
1.79	5.989	0.167	2.29	9.875	0.101	2.79	16.281	0.061
1.80	6.050	0.165	2.30	9.974	0.100	2.80	16.445	0.061
1.81	6.110	0.164	2.31	10.074	0.099	2.81	16.610	0.060
1.82	6.172	0.162	2.32	10.176	0.098	2.82	16.777	0.060
1.83	6.234	0.160	2.33	10.278	0.097	2.83	16.945	0.059
1.84	6.297	0.159	2.34	10.381	0.096	2.84	17.116	0.058
1.85	6.360	0.157	2.35	10.486	0.095	2.85	17.288	0.058
1.86	6.424	0.156	2.36	10.591	0.094	2.86	17.462	0.057
1.87	6.488	0.154	2.37	10.697	0.093	2.87	17.637	0.057
1.88	6.553	0.153	2.38	10.805	0.093	2.88	17.814	0.056
1.89	6.619	0.151	2.39	10.913	0.092	2.89	17.993	0.056
1.90	6.686	0.150	2.40	11.023	0.091	2.90	18.174	0.055
1.91	6.753	0.148	2.41	11.134	0.090	2.91	18.357	0.054
1.92	6.821	0.147	2.42	11.246	0.089	2.92	18.541	0.054
1.93	6.890	0.145	2.43	11.359	0.088	2.93	18.728	0.053
1.94	6.959	0.144	2.44	11.473	0.087	2.94	18.916	0.053
1.95	7.029	0.142	2.45	11.588	0.086	2.95	19.106	0.052
1.96	7.099	0.141	2.46	11.705	0.085	2.96	19.298	0.052
1.97	7.171	0.139	2.47	11.822	0.085	2.97	19.492	0.051
1.98	7.243	0.138	2.48	11.941	0.084	2.98	19.688	0.051
1.99	7.316	0.137	2.49	12.061	0.083	2.99	19.886	0.050
						3.00	20.086	0.050

TABLE 3 Logarithms of Numbers

N	0	1	2	3	4	5	6	7	8	9
10	0000	0043	0086	0128	0170	0212	0253	0294	0334	0374
11	0414	0453	0492	0531	0569	0607	0645	0682	0719	0755
12	0792	0828	0864	0899	0934	0969	1004	1038	1072	1106
13	1139	1173	1206	1239	1271	1303	1335	1367	1399	1430
14	1461	1492	1523	1553	1584	1614	1644	1673	1703	1732
15	1761	1790	1818	1847	1875	1903	1931	1959	1987	2014
16	2041	2068	2095	2122	2148	2175	2201	2227	2253	2279
17	2304	2330	2355	2380	2405	2430	2455	2480	2504	2529
18	2553	2577	2601	2625	2648	2672	2695	2718	2742	2765
19	2788	2810	2833	2856	2878	2900	2923	2945	2967	2989
20	3010	3032	3054	3075	3096	3118	3139	3160	3181	3201
21	3222	3243	3263	3284	3304	3324	3345	3365	3385	3404
22	3424	3444	3464	3483	3502	3522	3541	3560	3579	3598
23	3617	3636	3655	3674	3692	3711	3729	3747	3766	3784
24	3802	3820	3838	3856	3874	3892	3909	3927	3945	3962
25	3979	3997	4014	4031	4048	4065	4082	4099	4116	4133
26	4150	4166	4183	4200	4216	4232	4249	4265	4281	4298
27	4314	4330	4346	4362	4378	4393	4409	4425	4440	4456
28	4472	4487	4502	4518	4533	4548	4564	4579	4594	4609
29	4624	4639	4654	4669	4683	4698	4713	4728	4742	4757
30	4771	4786	4800	4814	4829	4843	4857	4871	4886	4900
31	4914	4928	4942	4955	4969	4983	4997	5011	5024	5038
32	5051	5065	5079	5092	5105	5119	5132	5145	5159	5172
33	5185	5198	5211	5224	5237	5250	5263	5276	5289	5302
34	5315	5328	5340	5353	5366	5378	5391	5403	5416	5428
35	5441	5453	5465	5478	5490	5502	5514	5527	5539	5551
36	5563	5575	5587	5599	5611	5623	5635	5647	5658	5670
37	5682	5694	5705	5717	5729	5740	5752	5763	5775	5786
38	5798	5809	5821	5832	5843	5855	5866	5877	5888	5899
39	5911	5922	5933	5944	5955	5966	5977	5988	5999	6010
40	6021	6031	6042	6053	6064	6075	6085	6096	6107	6117
41	6128	6138	6149	6160	6170	6180	6191	6201	6212	6222
42	6232	6243	6253	6263	6274	6284	6294	6304	6314	6325
43	6335	6345	6355	6365	6375	6385	6395	6405	6415	6425
44	6435	6444	6454	6464	6474	6484	6493	6503	6513	6522
45	6532	6542	6551	6561	6571	6580	6590	6599	6609	6618
46	6628	6637	6646	6656	6665	6675	6684	6693	6702	6712
47	6721	6730	6739	6749	6758	6767	6776	6785	6794	6803
48	6812	6821	6830	6839	6848	6857	6866	6875	6884	6893
49	6902	6911	6920	6928	6937	6946	6955	6964	6972	6981
50	6990	6998	7007	7016	7024	7033	7042	7050	7059	7067
51	7076	7084	7093	7101	7110	7118	7126	7135	7143	7152
52	7160	7168	7177	7185	7193	7202	7210	7218	7226	7235
53	7243	7251	7259	7267	7275	7284	7292	7300	7308	7316
54	7324	7332	7340	7348	7356	7364	7372	7380	7388	7396
N	0	1	2	3	4	5	6	7	8	9

TABLE 3   Logarithms of Numbers (*continued*)

N	0	1	2	3	4	5	6	7	8	9
55	7404	7412	7419	7427	7435	7443	7451	7459	7466	7474
56	7482	7490	7497	7505	7513	7520	7528	7536	7543	7551
57	7559	7566	7574	7582	7589	7597	7604	7612	7619	7627
58	7634	7642	7649	7657	7664	7672	7679	7686	7694	7701
59	7709	7716	7723	7731	7738	7745	7752	7760	7767	7774
60	7782	7789	7796	7803	7810	7818	7825	7832	7839	7846
61	7853	7860	7868	7875	7882	7889	7896	7903	7910	7917
62	7924	7931	7938	7945	7952	7959	7966	7973	7980	7987
63	7993	8000	8007	8014	8021	8028	8035	8041	8048	8055
64	8062	8069	8075	8082	8089	8096	8102	8109	8116	8122
65	8129	8136	8142	8149	8156	8162	8169	8176	8182	8189
66	8195	8202	8209	8215	8222	8228	8235	8241	8248	8254
67	8261	8267	8274	8280	8287	8293	8299	8306	8312	8319
68	8325	8331	8338	8344	8351	8357	8363	8370	8376	8382
69	8388	8395	8401	8407	8414	8420	8426	8432	8439	8445
70	8451	8457	8463	8470	8476	8482	8488	8494	8500	8506
71	8513	8519	8525	8531	8537	8543	8549	8555	8561	8567
72	8573	8579	8585	8591	8597	8603	8609	8615	8621	8627
73	8633	8639	8645	8651	8657	8663	8669	8675	8681	8686
74	8692	8698	8704	8710	8716	8722	8727	8733	8739	8745
75	8751	8756	8762	8768	8774	8779	8785	8791	8797	8802
76	8808	8814	8820	8825	8831	8837	8842	8848	8854	8859
77	8865	8871	8876	8882	8887	8893	8899	8904	8910	8915
78	8921	8927	8932	8938	8943	8949	8954	8960	8965	8971
79	8976	8982	8987	8993	8998	9004	9009	9015	9020	9025
80	9031	9036	9042	9047	9053	9058	9063	9069	9074	9079
81	9085	9090	9096	9101	9106	9112	9117	9122	9128	9133
82	9138	9143	9149	9154	9159	9165	9170	9175	9180	9186
83	9191	9196	9201	9206	9212	9217	9222	9227	9232	9238
84	9243	9248	9253	9258	9263	9269	9274	9279	9284	9289
85	9294	9299	9304	9309	9315	9320	9325	9330	9335	9340
86	9345	9350	9355	9360	9365	9370	9375	9380	9385	9390
87	9395	9400	9405	9410	9415	9420	9425	9430	9435	9440
88	9445	9450	9455	9460	9465	9469	9474	9479	9484	9489
89	9494	9499	9504	9509	9513	9518	9523	9528	9533	9538
90	9542	9547	9552	9557	9562	9566	9571	9576	9581	9586
91	9590	9595	9600	9605	9609	9614	9619	9624	9628	9633
92	9638	9643	9647	9652	9657	9661	9666	9671	9675	9680
93	9685	9689	9694	9699	9703	9708	9713	9717	9722	9727
94	9731	9736	9741	9745	9750	9754	9759	9763	9768	9773
95	9777	9782	9786	9791	9795	9800	9805	9809	9814	9818
96	9823	9827	9832	9836	9841	9845	9850	9854	9859	9863
97	9868	9872	9877	9881	9886	9890	9894	9899	9903	9908
98	9912	9917	9921	9926	9930	9934	9939	9943	9948	9952
99	9956	9961	9965	9969	9974	9978	9983	9987	9991	9996
N	0	1	2	3	4	5	6	7	8	9

TABLE 4   Trigonometric Functions (Degrees)

Deg.	Sin	Tan	*Cot	Cos		Deg.	Sin	Tan	Cot	Cos	
0.0	0.00000	0.00000	∞	1.0000	**90.0**	**6.0**	0.10453	0.10510	9.514	0.9945	**84.0**
.1	.00175	.00175	573.0	1.0000	89.9	.1	.10626	.10687	9.357	.9943	83.9
.2	.00349	.00349	286.5	1.0000	.8	.2	.10800	.10863	9.205	.9942	.8
.3	.00524	.00524	191.0	1.0000	.7	.3	.10973	.11040	9.058	.9940	.7
.4	.00698	.00698	143.24	1.0000	.6	.4	.11147	.11217	8.915	.9938	.6
.5	.00873	.00873	114.59	1.0000	.5	.5	.11320	.11394	8.777	.9936	.5
.6	.01047	.01047	95.49	0.9999	.4	.6	.11494	.11570	8.643	.9934	.4
.7	.01222	.01222	81.85	.9999	.3	.7	.11667	.11747	8.513	.9932	.3
.8	.01396	.01396	71.62	.9999	.2	.8	.11840	.11924	8.386	.9930	.2
.9	.01571	.01571	63.66	.9999	89.1	.9	.12014	.12101	8.264	.9928	83.1
1.0	0.01745	0.01746	57.29	0.9998	**89.0**	**7.0**	0.12187	0.12278	8.144	0.9925	**83.0**
.1	.01920	.01920	52.08	.9998	88.9	.1	.12360	.12456	8.028	.9923	82.9
.2	.02094	.02095	47.74	.9998	.8	.2	.12533	.12633	7.916	.9921	.8
.3	.02269	.02269	44.07	.9997	.7	.3	.12706	.12810	7.806	.9919	.7
.4	.02443	.02444	40.92	.9997	.6	.4	.12880	.12988	7.700	.9917	.6
.5	.02618	.02619	38.19	.9997	.5	.5	.13053	.13165	7.596	.9914	.5
.6	.02792	.02793	35.80	.9996	.4	.6	.13226	.13343	7.495	.9912	.4
.7	.02967	.02968	33.69	.9996	.3	.7	.13399	.13521	7.396	.9910	.3
.8	.03141	.03143	31.82	.9995	.2	.8	.13572	.13698	7.300	.9907	.2
.9	.03316	.03317	30.14	.9995	88.1	.9	.13744	.13876	7.207	.9905	82.1
2.0	0.03490	0.03492	28.64	0.9994	**88.0**	**8.0**	0.13917	0.14054	7.115	0.9903	**82.0**
.1	.03664	.03667	27.27	.9993	87.9	.1	.14090	.14232	7.026	.9900	81.9
.2	.03839	.03842	26.03	.9993	.8	.2	.14263	.14410	6.940	.9898	.8
.3	.04013	.04016	24.90	.9992	.7	.3	.14436	.14588	6.855	.9895	.7
.4	.04188	.04191	23.86	.9991	.6	.4	.14608	.14767	6.772	.9893	.6
.5	.04362	.04366	22.90	.9990	.5	.5	.14781	.14945	6.691	.9890	.5
.6	.04536	.04541	22.02	.9990	.4	.6	.14954	.15124	6.612	.9888	.4
.7	.04711	.04716	21.20	.9989	.3	.7	.15126	.15302	6.535	.9885	.3
.8	.04885	.04891	20.45	.9988	.2	.8	.15299	.15481	6.460	.9882	.2
.9	.05059	.05066	19.74	.9987	87.1	.9	.15471	.15660	6.386	.9880	81.1
3.0	0.05234	0.05241	19.081	0.9986	**87.0**	**9.0**	0.15643	0.15838	6.314	0.9877	**81.0**
.1	.05408	.05416	18.464	.9985	86.9	.1	.15816	.16017	6.243	.9874	80.9
.2	.05582	.05591	17.886	.9984	.8	.2	.15988	.16196	6.174	.9871	.8
.3	.05756	.05766	17.343	.9983	.7	.3	.16160	.16376	6.107	.9869	.7
.4	.05931	.05941	16.832	.9982	.6	.4	.16333	.16555	6.041	.9866	.6
.5	.06105	.06116	16.350	.9981	.5	.5	.16505	.16734	5.976	.9863	.5
.6	.06279	.06291	15.895	.9980	.4	.6	.16677	.16914	5.912	.9860	.4
.7	.06453	.06467	15.464	.9979	.3	.7	.16849	.17093	5.850	.9857	.3
.8	.06627	.06642	15.056	.9978	.2	.8	.17021	.17273	5.789	.9854	.2
.9	.06802	.06817	14.669	.9977	86.1	.9	.17193	.17453	5.730	.9851	80.1
4.0	0.06976	0.06993	14.301	0.9976	**86.0**	**10.0**	0.1736	0.1763	5.671	0.9848	**80.0**
.1	.07150	.07168	13.951	.9974	85.9	.1	.1754	.1781	5.614	.9845	79.9
.2	.07324	.07344	13.617	.9973	.8	.2	.1771	.1799	5.558	.9842	.8
.3	.07498	.07519	13.300	.9972	.7	.3	.1788	.1817	5.503	.9839	.7
.4	.07672	.07695	12.996	.9971	.6	.4	.1805	.1835	5.449	.9836	.6
.5	.07846	.07870	12.706	.9969	.5	.5	.1822	.1853	5.396	.9833	.5
.6	.08020	.08046	12.429	.9968	.4	.6	.1840	.1871	5.343	.9829	.4
.7	.08194	.08221	12.163	.9966	.3	.7	.1857	.1890	5.292	.9826	.3
.8	.08368	.08397	11.909	.9965	.2	.8	.1874	.1908	5.242	.9823	.2
.9	.08542	.08573	11.664	.9963	85.1	.9	.1891	.1926	5.193	.9820	79.1
5.0	0.08716	0.08749	11.430	0.9962	**85.0**	**11.0**	0.1908	0.1944	5.145	0.9816	**79.0**
.1	.08889	.08925	11.205	.9960	84.9	.1	.1925	.1962	5.097	.9813	78.9
.2	.09063	.09101	10.988	.9959	.8	.2	.1942	.1980	5.050	.9810	.8
.3	.09237	.09277	10.780	.9957	.7	.3	.1959	.1998	5.005	.9806	.7
.4	.09411	.09453	10.579	.9956	.6	.4	.1977	.2016	4.959	.9803	.6
.5	.09585	.09629	10.385	.9954	.5	.5	.1994	.2035	4.915	.9799	.5
.6	.09758	.09805	10.199	.9952	.4	.6	.2011	.2053	4.872	.9796	.4
.7	.09932	.09981	10.019	.9951	.3	.7	.2028	.2071	4.829	.9792	.3
.8	.10106	.10158	9.845	.9949	.2	.8	.2045	.2089	4.787	.9789	.2
.9	.10279	.10334	9.677	.9947	84.1	.9	.2062	.2107	4.745	.9785	78.1
6.0	0.10453	0.10510	9.514	0.9945	**84.0**	**12.0**	0.2079	0.2126	4.705	0.9781	**78.0**
	Cos	Cot	*Tan	Sin	Deg.		Cos	Cot	Tan	Sin	Deg.

*Interpolation in this section of the table is inaccurate.

TABLE 4   Trigonometric Functions (Degrees) (*continued*)

Deg.	Sin	Tan	Cot	Cos	
12.0	0.2079	0.2126	4.705	0.9781	**78.0**
.1	.2096	.2144	4.665	.9778	77.9
.2	.2113	.2162	4.625	.9774	.8
.3	.2130	.2180	4.586	.9770	.7
.4	.2147	.2199	4.548	.9767	.6
.5	.2164	.2217	4.511	.9763	.5
.6	.2181	.2235	4.474	.9759	.4
.7	.2198	.2254	4.437	.9755	.3
.8	.2215	.2272	4.402	.9751	.2
.9	.2233	.2290	4.366	.9748	77.1
13.0	0.2250	0.2309	4.331	0.9744	**77.0**
.1	.2267	.2327	4.297	.9740	76.9
.2	.2284	.2345	4.264	.9736	.8
.3	.2300	.2364	4.230	.9732	.7
.4	.2317	.2382	4.198	.9728	.6
.5	.2334	.2401	4.165	.9724	.5
.6	.2351	.2419	4.134	.9720	.4
.7	.2368	.2438	4.102	.9715	.3
.8	.2385	.2456	4.071	.9711	.2
.9	.2402	.2475	4.041	.9707	76.1
14.0	0.2419	0.2493	4.011	0.9703	**76.0**
.1	.2436	.2512	3.981	.9699	75.9
.2	.2453	.2530	3.952	.9694	.8
.3	.2470	.2549	3.923	.9690	.7
.4	.2487	.2568	3.895	.9686	.6
.5	.2504	.2586	3.867	.9681	.5
.6	.2521	.2605	3.839	.9677	.4
.7	.2538	.2623	3.812	.9673	.3
.8	.2554	.2642	3.785	.9668	.2
.9	.2571	.2661	3.758	.9664	75.1
15.0	0.2588	0.2679	3.732	0.9659	**75.0**
.1	.2605	.2698	3.706	.9655	74.9
.2	.2622	.2717	3.681	.9650	.8
.3	.2639	.2736	3.655	.9646	.7
.4	.2656	.2754	3.630	.9641	.6
.5	.2672	.2773	3.606	.9636	.5
.6	.2689	.2792	3.582	.9632	.4
.7	.2706	.2811	3.558	.9627	.3
.8	.2723	.2830	3.534	.9622	.2
.9	.2740	.2849	3.511	.9617	74.1
16.0	0.2756	0.2867	3.487	0.9613	**74.0**
.1	.2773	.2886	3.465	.9608	73.9
.2	.2790	.2905	3.442	.9603	.8
.3	.2807	.2924	3.420	.9598	.7
.4	.2823	.2943	3.398	.9593	.6
.5	.2840	.2962	3.376	.9588	.5
.6	.2857	.2981	3.354	.9583	.4
.7	.2874	.3000	3.333	.9578	.3
.8	.2890	.3019	3.312	.9573	.2
.9	.2907	.3038	3.291	.9568	73.1
17.0	0.2924	0.3057	3.271	0.9563	**73.0**
.1	.2940	.3076	3.251	.9558	72.9
.2	.2957	.3096	3.230	.9553	.8
.3	.2974	.3115	3.211	.9548	.7
.4	.2990	.3134	3.191	.9542	.6
.5	.3007	.3153	3.172	.9537	.5
.6	.3024	.3172	3.152	.9532	.4
.7	.3040	.3191	3.133	.9527	.3
.8	.3057	.3211	3.115	.9521	.2
.9	.3074	.3230	3.096	.9516	72.1
18.0	0.3090	0.3249	3.078	0.9511	**72.0**
	Cos	Cot	Tan	Sin	Deg.

Deg.	Sin	Tan	Cot	Cos	
18.0	0.3090	0.3249	3.078	0.9511	**72.0**
.1	.3107	.3269	3.060	.9505	71.9
.2	.3123	.3288	3.042	.9500	.8
.3	.3140	.3307	3.024	.9494	.7
.4	.3156	.3327	3.006	.9489	.6
.5	.3173	.3346	2.989	.9483	.5
.6	.3190	.3365	2.971	.9478	.4
.7	.3206	.3385	2.954	.9472	.3
.8	.3223	.3404	2.937	.9466	.2
.9	.3239	.3424	2.921	.9461	71.1
19.0	0.3256	0.3443	2.904	0.9455	**71.0**
.1	.3272	.3463	2.888	.9449	70.9
.2	.3289	.3482	2.872	.9444	.8
.3	.3305	.3502	2.856	.9438	.7
.4	.3322	.3522	2.840	.9432	.6
.5	.3338	.3541	2.824	.9426	.5
.6	.3355	.3561	2.808	.9421	.4
.7	.3371	.3581	2.793	.9415	.3
.8	.3387	.3600	2.778	.9409	.2
.9	.3404	.3620	2.762	.9403	70.1
20.0	0.3420	0.3640	2.747	0.9397	**70.0**
.1	.3437	.3659	2.733	.9391	69.9
.2	.3453	.3679	2.718	.9385	.8
.3	.3469	.3699	2.703	.9379	.7
.4	.3486	.3719	2.689	.9373	.6
.5	.3502	.3739	2.675	.9367	.5
.6	.3518	.3759	2.660	.9361	.4
.7	.3535	.3779	2.646	.9354	.3
.8	.3551	.3799	2.633	.9348	.2
.9	.3567	.3819	2.619	.9342	69.1
21.0	0.3584	0.3839	2.605	0.9336	**69.0**
.1	.3600	.3859	2.592	.9330	68.9
.2	.3616	.3879	2.578	.9323	.8
.3	.3633	.3899	2.565	.9317	.7
.4	.3649	.3919	2.552	.9311	.6
.5	.3665	.3939	2.539	.9304	.5
.6	.3681	.3959	2.526	.9298	.4
.7	.3697	.3979	2.513	.9291	.3
.8	.3714	.4000	2.500	.9285	.2
.9	.3730	.4020	2.488	.9278	68.1
22.0	0.3746	0.4040	2.475	0.9272	**68.0**
.1	.3762	.4061	2.463	.9265	67.9
.2	.3778	.4081	2.450	.9259	.8
.3	.3795	.4101	2.438	.9252	.7
.4	.3811	.4122	2.426	.9245	.6
.5	.3827	.4142	2.414	.9239	.5
.6	.3843	.4163	2.402	.9232	.4
.7	.3859	.4183	2.391	.9225	.3
.8	.3875	.4204	2.379	.9219	.2
.9	.3891	.4224	2.367	.9212	67.1
23.0	0.3907	0.4245	2.356	0.9205	**67.0**
.1	.3923	.4265	2.344	.9198	66.9
.2	.3939	.4286	2.333	.9191	.8
.3	.3955	.4307	2.322	.9184	.7
.4	.3971	.4327	2.311	.9178	.6
.5	.3987	.4348	2.300	.9171	.5
.6	.4003	.4369	2.289	.9164	.4
.7	.4019	.4390	2.278	.9157	.3
.8	.4035	.4411	2.267	.9150	.2
.9	.4051	.4431	2.257	.9143	66.1
24.0	0.4067	0.4452	2.246	0.9135	**66.0**
	Cos	Cot	Tan	Sin	Deg.

TABLE 4 Trigonometric Functions (Degrees) (*continued*)

Deg.	Sin	Tan	Cot	Cos	
24.0	0.4067	0.4452	2.246	0.9135	**66.0**
.1	.4083	.4473	2.236	.9128	65.9
.2	.4099	.4494	2.225	.9121	.8
.3	.4115	.4515	2.215	.9114	.7
.4	.4131	.4536	2.204	.9107	.6
.5	.4147	.4557	2.194	.9100	.5
.6	.4163	.4578	2.184	.9092	.4
.7	.4179	.4599	2.174	.9085	.3
.8	.4195	.4621	2.164	.9078	.2
.9	.4210	.4642	2.154	.9070	65.1
25.0	0.4226	0.4663	2.145	0.9063	**65.0**
.1	.4242	.4684	2.135	.9056	64.9
.2	.4258	.4706	2.125	.9048	.8
.3	.4274	.4727	2.116	.9041	.7
.4	.4289	.4748	2.106	.9033	.6
.5	.4305	.4770	2.097	.9026	.5
.6	.4321	.4791	2.087	.9018	.4
.7	.4337	.4813	2.078	.9011	.3
.8	.4352	.4834	2.069	.9003	.2
.9	.4368	.4856	2.059	.8996	64.1
26.0	0.4384	0.4877	2.050	0.8988	**64.0**
.1	.4399	.4899	2.041	.8980	63.9
.2	.4415	.4921	2.032	.8973	.8
.3	.4431	.4942	2.023	.8965	.7
.4	.4446	.4964	2.014	.8957	.6
.5	.4462	.4986	2.006	.8949	.5
.6	.4478	.5008	1.997	.8942	.4
.7	.4493	5029	1.988	.8934	.3
.8	.4509	.5051	1.980	.8926	.2
.9	.4524	.5073	1.971	.8918	63.1
27.0	0.4540	0.5095	1.963	0.8910	**63.0**
.1	.4555	.5117	1.954	.8902	62.9
.2	.4571	.5139	1.946	.8894	.8
.3	.4586	.5161	1.937	.8886	.7
.4	.4602	.5184	1.929	.8878	.6
.5	.4617	.5206	1 021	.8870	.5
.6	.4633	.5228	1.913	.8862	.4
.7	.4648	.5250	1.905	.8854	.3
.8	.4664	.5272	1.897	.8846	.2
.9	.4679	.5295	1.889	.8838	62.1
28.0	0.4695	0.5317	1.881	0.8829	**62.0**
.1	.4710	.5340	1.873	.8821	61.9
.2	.4726	.5362	1.865	.8813	.8
.3	.4741	.5384	1.857	.8805	.7
.4	.4756	.5407	1.849	.8796	.6
.5	.4772	.5430	1.842	.8788	.5
.6	.4787	.5452	1.834	.8780	.4
.7	.4802	.5475	1.827	.8771	.3
.8	.4818	.5498	1.819	.8763	.2
.9	.4833	.5520	1.811	.8755	61.1
29.0	0.4848	0.5543	1.804	0.8746	**61.0**
.1	.4863	.5566	1.797	.8738	60.9
.2	.4879	.5589	1.789	.8729	.8
.3	.4894	.5612	1.782	.8721	.7
.4	.4909	.5635	1.775	.8712	.6
.5	.4924	.5658	1.767	.8704	.5
.6	.4939	.5681	1.760	.8695	.4
.7	.4955	.5704	1.753	.8686	.3
.8	.4970	.5727	1.746	.8678	.2
.9	.4985	.5750	1.739	.8669	60.1
30.0	0.5000	0.5774	1.732	0.8660	**60.0**
	Cos	Cot	Tan	Sin	Deg.

Deg.	Sin	Tan	Cot	Cos	
30.0	0.5000	0.5774	1.7321	0.8660	**60.0**
.1	.5015	.5797	1.7251	.8652	59.9
.2	.5030	.5820	1.7182	.8643	.8
.3	.5045	.5844	1.7113	.8634	.7
.4	.5060	.5867	1.7045	.8625	.6
.5	.5075	.5890	1.6977	.8616	.5
.6	.5090	.5914	1.6909	.8607	.4
.7	.5105	.5938	1.6842	.8599	.3
.8	.5120	.5961	1.6775	.8590	.2
.9	.5135	.5985	1.6709	.8581	59.1
31.0	0.5150	0.6009	1.6643	0.8572	**59.0**
.1	.5165	.6032	1.6577	.8563	58.9
.2	.5180	.6056	1.6512	.8554	.8
.3	.5195	.6080	1.6447	.8545	.7
.4	.5210	.6104	1.6383	.8536	.6
.5	.5225	.6128	1.6319	.8526	.5
.6	.5240	.6152	1.6255	.8517	.4
.7	.5255	.6176	1.6191	.8508	.3
.8	.5270	.6200	1.6128	.8499	.2
.9	.5284	.6224	1.6066	.8490	58.1
32.0	0.5299	0.6249	1.6003	0.8480	**58.0**
.1	.5314	.6273	1.5941	.8471	57.9
.2	.5329	.6297	1.5880	.8462	.8
.3	.5344	.6322	1.5818	.8453	.7
.4	.5358	.6346	1.5757	.8443	.6
.5	.5373	.6371	1.5697	.8434	.5
.6	.5388	.6395	1.5637	.8425	.4
.7	.5402	.6420	1.5577	.8415	.3
.8	.5417	.6445	1.5517	.8406	.2
.9	.5432	.6469	1.5458	.8396	57.1
33.0	0.5446	0 6494	1.5399	0.8387	**57.0**
.1	.5461	.6519	1.5340	.8377	56.9
.2	.5476	.6544	1.5282	.8368	.8
.3	.5490	.6569	1.5224	.8358	.7
.4	.5505	.6594	1.5166	.8348	.6
.5	.5519	.6619	1.5108	.8339	.5
.6	.5534	.6644	1.5051	.8329	.4
.7	.5548	6669	1.4994	.8320	.3
.8	.5563	.6694	1.4938	.8310	.2
.9	.5577	.6720	1.4882	.8300	56.1
34.0	0.5592	0.6745	1.4826	0.8290	**56.0**
.1	.5606	.6771	1.4770	.8281	55.9
.2	.5621	.6796	1.4715	.8271	.8
.3	.5635	.6822	1.4659	.8261	.7
.4	.5650	.6847	1.4605	.8251	.6
.5	.5664	.6873	1.4550	.8241	.5
.6	.5678	.6899	1.4496	.8231	.4
.7	.5693	.6924	1.4442	.8221	.3
.8	.5707	.6950	1.4388	.8211	.2
.9	.5721	.6976	1.4335	.8202	55.1
35.0	0.5736	0.7002	1.4281	0.8192	**55.0**
.1	.5750	.7028	1.4229	.8181	54.9
.2	.5764	.7054	1.4176	.8171	.8
.3	.5779	.7080	1.4124	.8161	.7
.4	.5793	.7107	1.4071	.8151	.6
.5	.5807	.7133	1.4019	.8141	.5
.6	.5821	.7159	1.3968	.8131	.4
.7	.5835	.7186	1.3916	.8121	.3
.8	.5850	.7212	1.3865	.8111	.2
.9	.5864	.7239	1.3814	.8100	54.1
36.0	0.5878	0.7265	1.3764	0.8090	**54.0**
	Cos	Cot	Tan	Sin	Deg.

TABLE 4 Trigonometric Functions (Degrees) (*continued*)

Deg.	Sin	Tan	Cot	Cos		Deg.	Sin	Tan	Cot	Cos	
36.0	0.5878	0.7265	1.3764	0.8090	**54.0**	**40.5**	0.6494	0.8541	1.1708	0.7604	**49.5**
.1	.5892	.7292	1.3713	.8080	53.9	.6	.6508	.8571	1.1667	.7593	.4
.2	.5906	.7319	1.3663	.8070	.8	.7	.6521	.8601	1.1626	.7581	.3
.3	.5920	.7346	1.3613	.8059	.7	.8	.6534	.8632	1.1585	.7570	.2
.4	.5934	.7373	1.3564	.8049	.6	.9	.6547	.8662	1.1544	.7559	49.1
.5	.5948	.7400	1.3514	.8039	.5	**41.0**	0.6561	0.8693	1.1504	0.7547	**49.0**
.6	.5962	.7427	1.3465	.8028	.4	.1	.6574	.8724	1.1463	.7536	48.9
.7	.5976	.7454	1.3416	.8018	.3	.2	.6587	.8754	1.1423	.7524	.8
.8	.5990	.7481	1.3367	.8007	.2	.3	.6600	.8785	1.1383	.7513	.7
.9	.6004	.7508	1.3319	.7997	53.1	.4	.6613	.8816	1.1343	.7501	.6
37.0	0.6018	0.7536	1.3270	0.7986	**53.0**	.5	.6626	.8847	1.1303	.7490	.5
.1	.6032	.7563	1.3222	.7976	52.9	.6	.6639	.8878	1.1263	.7478	.4
.2	.6046	.7590	1.3175	.7965	.8	.7	.6652	.8910	1.1224	.7466	.3
.3	.6060	.7618	1.3127	.7955	.7	.8	.6665	.8941	1.1184	.7455	.2
.4	.6074	.7646	1.3079	.7944	.6	.9	.6678	.8972	1.1145	.7443	48.1
.5	.6088	.7673	1.3032	.7934	.5	**42.0**	0.6691	0.9004	1.1106	0.7431	**48.0**
.6	.6101	.7701	1.2985	.7923	.4	.1	.6704	.9036	1.1067	.7420	47.9
.7	.6115	.7729	1.2938	.7912	.3	.2	.6717	.9067	1.1028	.7408	.8
.8	.6129	.7757	1.2892	.7902	.2	.3	.6730	.9099	1.0990	.7396	.7
.9	.6143	.7785	1.2846	.7891	52.1	.4	.6743	.9131	1.0951	.7385	.6
38.0	0.6157	0.7813	1.2799	0.7880	**52.0**	.5	.6756	.9163	1.0913	.7373	.5
.1	.6170	.7841	1.2753	.7869	51.9	.6	.6769	.9195	1.0875	.7361	.4
.2	.6184	.7869	1.2708	.7859	.8	.7	.6782	.9228	1.0837	.7349	.3
.3	.6198	.7898	1.2662	.7848	.7	.8	.6794	.9260	1.0799	.7337	.2
.4	.6211	.7926	1.2617	.7837	.6	.9	.6807	.9293	1.0761	.7325	47.1
.5	.6225	.7954	1.2572	.7826	.5	**43.0**	0.6820	0.9325	1.0724	0.7314	**47.0**
.6	.6239	.7983	1.2527	.7815	.4	.1	.6833	.9358	1.0686	.7302	46.9
.7	.6252	.8012	1.2482	.7804	.3	.2	.6845	.9391	1.0649	.7290	.8
.8	.6266	.8040	1.2437	.7793	.2	.3	.6858	.9424	1.0612	.7278	.7
.9	.6280	.8069	1.2393	.7782	51.1	.4	.6871	.9457	1.0575	.7266	.6
39.0	0.6293	0.8098	1.2349	0.7771	**51.0**	.5	.6884	.9490	1.0538	.7254	.5
.1	.6307	.8127	1.2305	.7760	50.9	.6	.6896	.9523	1.0501	.7242	.4
.2	.6320	.8156	1.2261	.7749	.8	.7	.6909	.9556	1.0464	.7230	.3
.3	.6334	.8185	1.2218	.7738	.7	.8	.6921	.9590	1.0428	.7218	.2
.4	.6347	.8214	1.2174	.7727	.6	.9	.6934	.9623	1.0392	.7206	46.1
.5	.6361	.8243	1.2131	.7716	.5	**44.0**	0.6947	0.9657	1.0355	0.7193	**46.0**
.6	.6374	.8273	1.2088	.7705	.4	.1	.6959	.9691	1.0319	.7181	45.9
.7	.6388	.8302	1.2045	.7694	.3	.2	.6972	.9725	1.0283	.7169	.8
.8	.6401	.8332	1.2002	.7683	.2	.3	.6984	.9759	1.0247	.7157	.7
.9	.6414	.8361	1.1960	.7672	50.1	.4	.6997	.9793	1.0212	.7145	.6
40.0	0.6428	0.8391	1.1918	0.7660	**50.0**	.5	.7009	.9827	1.0176	.7133	.5
.1	.6441	.8421	1.1875	.7649	49.9	.6	.7022	.9861	1.0141	.7120	.4
.2	.6455	.8451	1.1833	.7638	.8	.7	.7034	.9896	1.0105	.7108	.3
.3	.6468	.8481	1.1792	.7627	.7	.8	.7046	.9930	1.0070	.7096	.2
.4	.6481	.8511	1.1750	.7615	.6	.9	.7059	.9965	1.0035	.7083	45.1
40.5	0.6494	0.8541	1.1708	0.7604	**49.5**	**45.0**	0.7071	1.0000	1.0000	0.7071	**45.0**
	Cos	Cot	Tan	Sin	Deg.		Cos	Cot	Tan	Sin	Deg.

TABLE 5 Trigonometric Functions (Radians)

Rad.	Sin	Tan	Cot	Cos	Rad	Sin	Tan	Cot	Cos
.00	.00000	.00000	∞	1.00000	**.50**	.47943	.54630	1.8305	.87758
.01	.01000	.01000	99.997	0.99995	.51	.48818	.55936	1.7878	.87274
.02	.02000	.02000	49.993	.99980	.52	.49688	.57256	1.7465	.86782
.03	.03000	.03001	33.323	.99955	.53	.50553	.58592	1.7067	.86281
.04	.03999	.04002	24.987	.99920	.54	.51414	.59943	1.6683	.85771
.05	.04998	.05004	19.983	.99875	.55	.52269	.61311	1.6310	.85252
.06	.05996	.06007	16.647	.99820	.56	.53119	.62695	1.5950	.84726
.07	.06994	.07011	14.262	.99755	.57	.53963	.64097	1.5601	.84190
.08	.07991	.08017	12.473	.99680	.58	.54802	.65517	1.5263	.83646
.09	.08988	.09024	11.081	.99595	.59	.55636	.66956	1.4935	.83094
.10	.09983	.10033	9.9666	.99500	**.60**	.56464	.68414	1.4617	.82534
.11	.10978	.11045	9.0542	.99396	.61	.57287	.69892	1.4308	.81965
.12	.11971	.12058	8.2933	.99281	.62	.58104	.71391	1.4007	.81388
.13	.12963	.13074	7.6489	.99156	.63	.58914	.72911	1.3715	.80803
.14	.13954	.14092	7.0961	.99022	.64	.59720	.74454	1.3431	.80210
.15	.14944	.15114	6.6166	.98877	.65	.60519	.76020	1.3154	.79608
.16	.15932	.16138	6.1966	.98723	.66	.61312	.77610	1.2885	.78999
.17	.16918	.17166	5.8256	.98558	.67	.62099	.79225	1.2622	.78382
.18	.17903	.18197	5.4954	.98384	.68	.62879	.80866	1.2366	.77757
.19	.18886	.19232	5.1997	.98200	.69	.63654	.82534	1.2116	.77125
.20	.19867	.20271	4.9332	**.98007**	**.70**	.64422	.84229	1.1872	.76484
.21	.20846	.21314	4.6917	.97803	.71	.65183	.85953	1.1634	.75836
.22	.21823	.22362	4.4719	.97590	.72	.65938	.87707	1.1402	.75181
.23	.22798	.23414	4.2709	.97367	.73	.66687	.89492	1.1174	.74517
.24	.23770	.24472	4.0864	.97134	.74	.67429	.91309	1.0952	.73847
.25	.24740	.25534	3.9163	.96891	.75	.68164	.93160	1.0734	.73169
.26	.25708	.26602	3.7591	.96639	.76	.68892	.95045	1.0521	.72484
.27	.26673	.27676	3.6133	.96377	.77	.69614	.96967	1.0313	.71791
.28	.27636	.28755	3.4776	.96106	.78	.70328	.98926	1.0109	.71091
.29	.28595	.29841	3.3511	.95824	.79	.71035	1.0092	.99084	.70385
.30	.29552	.30934	3.2327	.95534	**.80**	.71736	1.0296	.97121	.69671
.31	.30506	.32033	3.1218	.95233	.81	.72429	1.0505	.95197	.68950
.32	.31457	.33139	3.0176	.94924	.82	.73115	1.0717	.93309	.68222
.33	.32404	.34252	2.9195	.94604	.83	.73793	1.0934	.91455	.67488
.34	.33349	.35374	2.8270	.94275	.84	.74464	1.1156	.89635	.66746
.35	.34290	.36503	2.7395	.93937	.85	.75128	1.1383	.87848	.65998
.36	.35227	.37640	2.6567	.93590	.86	.75784	1.1616	.86091	.65244
.37	.36162	.38786	2.5782	.93233	.87	.76433	1.1853	.84365	.64483
.38	.37092	.39941	2.5037	.92866	.88	.77074	1.2097	.82668	.63715
.39	.38019	.41105	2.4328	.92491	.89	.77707	1.2346	.80998	.62941
.40	.38942	.42279	2.3652	.92106	**.90**	.78333	1.2602	.79355	.62161
.41	.39861	.43463	2.3008	.91712	.91	.78950	1.2864	.77738	.61375
.42	.40776	.44657	2.2393	.91309	.92	.79560	1.3133	.76146	.60582
.43	.41687	.45862	2.1804	.90897	.93	.80162	1.3409	.74578	.59783
.44	.42594	.47078	2.1241	.90475	.94	.80756	1.3692	.73034	.58979
.45	.43497	.48306	2.0702	.90045	.95	.81342	1.3984	.71511	.58168
.46	.44395	.49545	2.0184	.89605	.96	.81919	1.4284	.70010	.57352
.47	.45289	.50797	1.9686	.89157	.97	.82489	1.4592	.68531	.56530
.48	.46178	.52061	1.9208	.88699	.98	.83050	1.4910	.67071	.55702
.49	.47063	.53339	1.8748	.88233	.99	.83603	1.5237	.65631	.54869
.50	.47943	.54630	1.8305	.87758	**1.00**	.84147	1.5574	.64209	.54030
Rad.	Sin	Tan	Cot	Cos	Rad.	Sin	Tan	Cot	Cos

TABLE 5   Trigonometric Functions (Radians) (*continued*)

Rad.	Sin	Tan	Cot	Cos	Rad.	Sin	Tan	Cot	Cos
1.00	.84147	1.5574	.64209	.54030	**1.50**	.99749	14.101	.07091	.07074
1.01	.84683	1.5922	.62806	.53186	1.51	.99815	16.428	.06087	.06076
1.02	.85211	1.6281	.61420	.52337	1.52	.99871	19.670	.05084	.05077
1.03	.85730	1.6652	.60051	.51482	1.53	.99917	24.498	.04082	.04079
1.04	.86240	1.7036	.58699	.50622	1.54	.99953	32.461	.03081	.03079
1.05	.86742	1.7433	.57362	.49757	1.55	.99978	48.078	.02080	.02079
1.06	.87236	1.7844	.56040	.48887	1.56	.99994	92.621	.01080	.01080
1.07	.87720	1.8270	.54734	.48012	1.57	1.00000	1255.8	.00080	.00080
1.08	.88196	1.8712	.53441	.47133	1.58	.99996	−108.65	−.00920	−.00920
1.09	.88663	1.9171	.52162	.46249	1.59	.99982	−52.067	−.01921	−.01920
1.10	.89121	1.9648	.50897	.45360	**1.60**	.99957	−34.233	−.02921	−.02920
1.11	.89570	2.0143	.49644	.44466	1.61	.99923	−25.495	−.03922	−.03919
1.12	.90010	2.0660	.48404	.43568	1.62	.99879	−20.307	−.04924	−.04918
1.13	.90441	2.1198	.47175	.42666	1.63	.99825	−16.871	−.05927	−.05917
1.14	.90863	2.1759	.45959	.41759	1.64	.99761	−14.427	−.06931	−.06915
1.15	.91276	2.2345	.44753	.40849	1.65	.99687	−12.599	−.07937	−.07912
1.16	.91680	2.2958	.43558	.39934	1.66	.99602	−11.181	−.08944	−.08909
1.17	.92075	2.3600	.42373	.39015	1.67	.99508	−10.047	−.09953	−.09904
1.18	.92461	2.4273	.41199	.38092	1.68	.99404	− 9.1208	−.10964	−.10899
1.19	.92837	2.4979	.40034	.37166	1.69	.99290	− 8.3492	−.11977	−.11892
1.20	.93204	2.5722	.38878	.36236	**1.70**	.99166	− 7.6966	−.12993	−.12884
1.21	.93562	2.6503	.37731	.35302	1.71	.99033	− 7.1373	−.14011	−.13875
1.22	.93910	2.7328	.36593	.34365	1.72	.98889	− 6.6524	−.15032	−.14865
1.23	.94249	2.8198	.35463	.33424	1.73	.98735	− 6.2281	−.16056	−.15853
1.24	.94578	2.9119	.34341	.32480	1.74	.98572	− 5.8535	−.17084	−.16840
1.25	.94898	3.0096	.33227	.31532	1.75	.98399	− 5.5204	−.18115	−.17825
1.26	.95209	3.1133	.32121	.30582	1.76	.98215	− 5.2221	−.19149	−.18808
1.27	.95510	3.2236	.31021	.29628	1.77	.98022	− 4.9534	−.20188	−.19789
1.28	.95802	3.3413	.29928	.28672	1.78	.97820	− 4.7101	−.21231	−.20768·
1.29	.96084	3.4672	.28842	.27712	1.79	.97607	− 4.4887	−.22278	−.21745
1.30	.96356	3.6021	.27762	.26750	**1.80**	.97385	− 4.2863	−.23330	−.22720
1.31	.96618	3.7471	.26687	.25785	1.81	.97153	− 4.1005	−.24387	−.23693
1.32	.96872	3.9033	.25619	.24818	1.82	.96911	− 3.9294	−.25449	−.24663
1.33	.97115	4.0723	.24556	.23848	1.83	.96659	− 3.7712	−.26517	−.25631
1.34	.97348	4.2556	.23498	.22875	1.84	.96398	− 3.6245	−.27590	−.26596
1.35	.97572	4.4552	.22446	.21901	1.85	.96128	− 3.4881	−.28669	−.27559·
1.36	.97786	4.6734	.21399	.20924	1.86	.95847	− 3.3608	−.29755	−·28519
1.37	.97991	4.9131	.20354	.19945	1.87	.95557	− 2.2419	−.30846	−.29476
1.38	.98185	5.1774	.19315	.18964	1.88	.95258	− 3.1304	−.31945	−.30430
1.39	.98370	5.4707	.18279	.17981	1.89	.94949	− 3.0257	−.33051	−.31381
1.40	.98545	5.7979	.17248	.16997	**1.90**	.94630	− 2.9271	−.34164	−.32329
1.41	.98710	6.1654	.16220	.16010	1.91	.94302	− 2.8341	−.35284	−.33274
1.42	.98865	6.5811	.15195	.15023	1.92	.93965	− 2.7463	−.36413	−.34215
1.43	.99010	7.0555	.14173	.14033	1.93	.93618	− 2.6632	−.37549	−.35153
1.44	.99146	7.6018	.13155	.13042	1.94	.93262	− 2.5843	−.38695	−.36087
1.45	.99271	8.2381	.12139	.12050	1.95	.92896	− 2.5095	−.39849	−.37018
1.46	.99387	8.9886	.11125	.11057	1.96	.92521	− 2.4383	−.41012	−.37945
1.47	.99492	9.8874	.10114	.10063	1.97	.92137	− 2.3705	−.42185	−.38868
1.48	.99588	10.983	.09105	.09067	1.98	.91744	− 2.3058	−.43368	−.39788
1.49	.99674	12.350	.08097	.08071	1.99	.91341	− 2.2441	−.44562	−.40703
1.50	.99749	14.101	.07091	.07074	**2.00**	.90930	− 2.1850	−.45766	−.41615
Rad.	Sin	Tan	Cot	Cos	Rad.	Sin	Tan	Cot	Cos

Appendix E

Answers to Selected Problems

PROBLEM SET 1.1, PAGE 9

1. Both **3.** Both **5.** Neither **7.** Relation **9.** Both **11.** Relation **13.** Both
15. a. $0.92 b. $0.45 c. $0.21 d. $1.31 **17.** a. $0.20 b. $e(1974) - e(1944)$
19. a. 3 b. 0 c. 3 d. 12 **21.** a. -5 b. -7 c. -11 d. -15

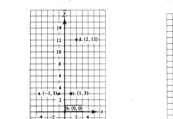

23. a. 5 b. 25 c. 40 d. 50 **25.** a. $0.018 b. $0.0125 c. $0.058

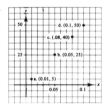

d. $\dfrac{s(1944 + h) - s(1944)}{h}$ **27.** At $t = 1.4$ sec,

$d = 147.86$ m; at $t = 2.6$ sec, $d = 259.40$ m; a restriction
on t would impose a limitation on the domain values;
$0 < t < 23.09$

29. $P(5, f(5))$, $Q(x_0, f(x_0))$ **31.** It has exactly 5 elements.

PROBLEM SET 1.2, PAGE 18

1. a. 1 b. 5 c. -5 d. $2\sqrt{5} + 1$ e. $2\pi + 1$ **3.** a. $2w + 1$ b. $2w^2 - 1$
c. $2t^2 - 1$ d. $2v^2 - 1$ e. $2m + 1$ **5.** a. $3 + 2\sqrt{2}$ b. $5 + 4\sqrt{2}$ c. $2t^2 + 12t + 17$
d. $2t^2 + 4t + 3$ e. $2m^2 - 4m + 1$ **7.** a. $-8x^2 - 26x - 13$ b. $-8t - 14$ **9.** a. 2
b. $4t + 2h$ **11.** a. $w^2 - 1$ b. $h^2 - 1$ c. $w^2 + 2wh + h^2 - 1$ d. $w^2 + h^2 - 2$
13. a. $x^4 - 1$ b. $x - 1$ c. $x^2 + 2xh + h^2 - 1$ d. $x^2 - 1$ **15.** a. $2xh + h^2$

b. $2x + h$ **17.** $10x + 5h$ **19.** $\dfrac{-1}{x(x + h)}$ **21.** $\dfrac{-2x - h}{(x^2 + 2xh + h^2 + 1)(x^2 + 1)}$

525

23.

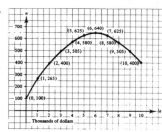

25. a. 45   b. 60   c. 75
d. 82.5   e. $90 - 15h$ f. 88.5
27. a. 0.9   b. 0.8   c. 0.4
d. 0   e. Answers vary. The rates are
declining to the point where the rate is 0 for
more than 30 min. This seems to indicate
that the maximum list size is 25 words,
regardless of the length of time spent
($10 \leq t \leq 60$).

29. a. 512   b. 192   c. 80   d. $64h + 16h^2$ e. $32xh + 16h^2$ **31.** a. 16,250
b. The average number of marriages per year between the years 1970 and $1970 + h$.

33. a. $\dfrac{P(1930) - P(1830)}{100} = 10,000,000;$ $\dfrac{P(1960) - P(1930)}{30} = 33,333,333;$ $\dfrac{P(1975) - P(1960)}{15}$

$= 66,666,667$ b. 1%; 3.3%; 6.7%

PROBLEM SET 1.3, PAGE 29

1. a. 33   b. -7 c. 5   d. $\dfrac{5}{17}$ e. 7   **3.** a. 0   b. Not defined   c. 1500   d. $\dfrac{201}{10,000}$

or 0.0201   e. $-\dfrac{7}{4}$ **5.** a. 7   b. 5   c. 24   d. 4   e. 3   **7.** a. $x^2 + 2x - 2$

b. $-x^2 + 2x - 4$ c. $2x^3 - 3x^2 + 2x - 3$ d. $\dfrac{2x - 3}{x^2 + 1}$ e. Reals   **9.** a. $\dfrac{x^3 - x^2 - x + 1}{x - 2}$

b. $\dfrac{-x^3 + 5x^2 - x - 7}{x - 2}$ c. $(2x - 3)(x + 1)^2$ d. $\dfrac{2x - 3}{(x - 2)^2}$ e. Reals, $x \neq 2$; also $x \neq -1$ for part d

11. a. $2x^2 - 1$ b. $4x^2 - 12x + 10$ **13.** a. $x^4 - 2x^3 - 3x^2 + 4x + 4$ b. $x^4 - x^2 - 2$

15.

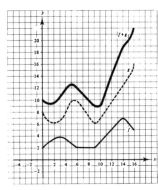

17.

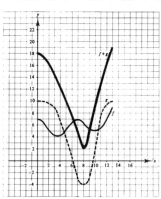

19.

21. a. $4x^2 - 4x + 1$ b. $6x + 3$ c. $36x^2 + 36x + 9$ d. $36x^2 + 36x + 9$ **23.** a. x

b. $\dfrac{1}{2}x^2 - \dfrac{3}{2}$ c. $x^2 + 1$ d. $x^2 + 1$ **25.** a. $6x - 13$ b. $2x - 3$ c. $6x - 7$ d. $6x - 7$

27. a. x^2 b. x^6 c. x^6 d. x^6 **29.** a. 144π b. $36\pi t^2$ c. $\left\{ t \middle| 0 < t < \dfrac{8}{3} \right\}$

31. Answers vary   **33.** Answers vary   **35.** Answers vary

PROBLEM SET 1.4, PAGE 39

1. 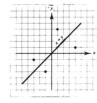   **3.** 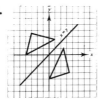   **5.**   **7.**

9. $f^{-1}(x) = x - 3$ **11.** $g^{-1}(x) = \dfrac{x}{5}$ **13.** Inverse is $y = \pm\sqrt{x + 5}$ (not a function)

15. $f^{-1}(x) = x$ **17.** $f^{-1}(x) = \dfrac{2x + 1}{x}$ **19.** Inverses   **21.** Not inverses

23. Not inverses   **25.** Inverses   **27.** Not inverses

In Problems 29–38, there may be more than one way of defining F.

29. a. $f^{-1} = \{(5, 4), (3, 6), (1, 7), (4, 2)\}$ b.
c. A function   d. $F = f$

31. a. $f^{-1} = \left\{ (x, y) \middle| y = \dfrac{1}{2}x - \dfrac{3}{2} \right\}$ b.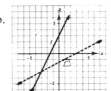
c. A function
d. $F = f$

33. a. $f^{-1}(x) = x - 5$ b.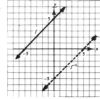
c. A function
d. $F = f$

35. a. $y = \pm\sqrt{-2x}$
c. Not a function
d. $F = f$ where $x \geq 0$;
$F^{-1}(x) = \sqrt{-2x}$, where
$x \leq 0$

b. 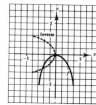   e.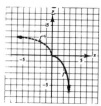

37. a. $f^{-1}(x) = x^2$, $x > 0$
c. A function
d. $F = f$

b.

39. a. 1   b. 3   c. 6   d. −4   e. −3   **41.** a. 0   b. 2   c. 6   d. 7   e. 4
43. a. 0   b. 4   c. 0   d. 4   **45.** a. 0   b. 0   c. undefined   d. 3.5

PROBLEM SET 1.5, PAGE 48

1. a. $(0, -3)$ b.

3. a. $(4, 0)$ b.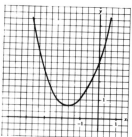

5. a. $(5, 1)$ b.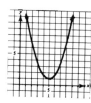

7. a. $\left(-\dfrac{5}{3}, \dfrac{2}{3}\right)$ b.

9. a. $\left(-\dfrac{9}{7}, \dfrac{4}{7}\right)$ b.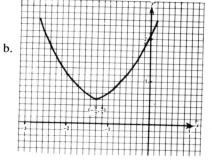

11.

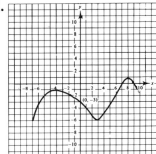

13.

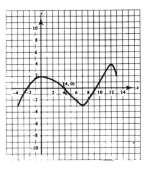

15.

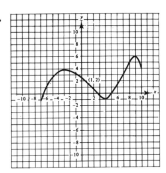

17.

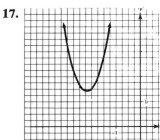

19.

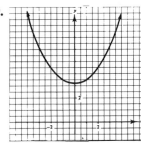

21.

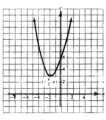

23.

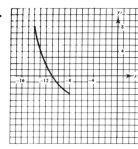

25.

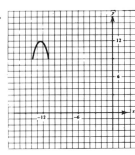

27.

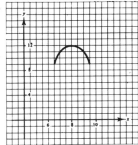

REVIEW PROBLEMS, PAGE 51

1. Answers vary. A function can be described as a set, a rule, or a mapping. **2.** a. Neither
b. Relation   c. Relation   d. Both   e. Both   **3.** a. -7 b. $3w + 2$ c. $3m^2 + 3n^2 + 2$
d. $3x^2 + 6x - 1$ e. 3   **4.** a. $10x + 1 + 5h$ b. $125x^4 + 50x^3 + 60x^2 + 11x + 7$ **5.** a. $x^2 + 5x - 2$

b. $x^2 - 5x + 2$ c. $5x^3 - 2x^2$ d. $\dfrac{x^2}{5x - 2}$ e. $25x^2 - 20x + 4$ **6.** a. $5x^2 - 2$ b. x^4

c. $20x^2 + 20x + 3$ d. $20x^2 + 20x + 3$ e. $8x + 7$

7.

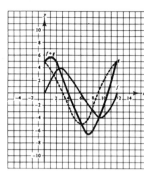

8. a. Let $F = f$, where the domain of F is $\{0, 1, 2\}$. Then

$F^{-1} = \{(1, 0), (4, 1), (9, 2)\}$ b. $f^{-1}(x) = -\dfrac{2}{3}x$

c. $f^{-1}(x) = -\dfrac{2}{3}(x - 3)$ d. Let $F = f$, where the

domain of F is $x \geq 0$. Then $F^{-1}(x) = \sqrt{-\dfrac{2}{3}x}$, $x \leq 0$

e. Let $F = f$, where the domain of F is $x \geq 0$.

Then $F^{-1}(x) = \sqrt{2 - \dfrac{2}{3}x}$, $x \leq 3$.

9. a.

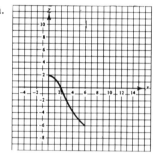

b. and c.

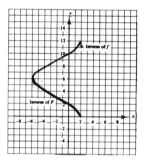

d. 0   e. 3

10.

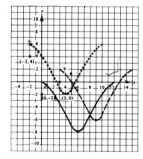

PROBLEM SET 2.1, PAGE 62

1–8. The graphs form a picture of a sailboat.

9.

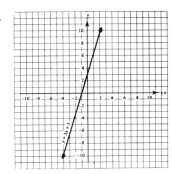

11.

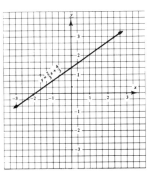

13.

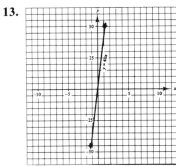

15.

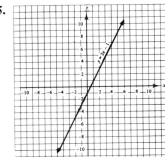

17.

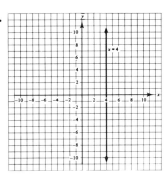

19.

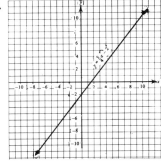

21.

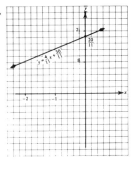

23–30. The graphs form a picture of an envelope.   **31.** $5x - y + 6 = 0$ **33.** $y = 0$

35. $3x - y - 3 = 0$ **37.** $x - 2y + 3 = 0$

39. $x - 2y + 2 = 0$ **41.** $x - 4 = 0$

43. $y - 6 = 0$

45.
$$\begin{cases} y_1 = mx_1 + b \\ y_2 = mx_2 + b \end{cases}$$

$$y_1 - y_2 = m(x_1 - x_2)$$

$$m = \frac{y_1 - y_2}{x_1 - x_2} = \frac{y_2 - y_1}{x_2 - x_1}$$

$$y_1 = \left(\frac{y_2 - y_1}{x_2 - x_1}\right) x_1 + b$$

$$b = y_1 - \left(\frac{y_2 - y_1}{x_2 - x_1}\right) x_1$$

Thus,

$$y = \left(\frac{y_2 - y_1}{x_2 - x_1}\right) x + y_1 - \left(\frac{y_2 - y_1}{x_2 - x_1}\right) x_1$$

$$y - y_1 = \left(\frac{y_2 - y_1}{x_2 - x_1}\right) (x - x_1)$$

47. a. $A(x_0, x_0{}^2)$; $B(x_0 + \Delta x, x_0{}^2 + 2x_0\Delta x + (\Delta x)^2)$ b. $2x_0 + \Delta x$ **49.** a. $A(x_0, x_0{}^2 + 2)$;
$B(x_0 + \Delta x, x_0{}^2 + 2x_0\Delta x + (\Delta x)^2 + 2)$ b. $2x_0 + \Delta x$ **51.** Estimated population in New York State in
1980 is 19.6 million and in 2000 it is 22.4 million.   **53.** $60p - n - 10 = 0$
55. $1000p - 3n + 1100 = 0$

PROBLEM SET 2.2, PAGE 74

1. 43   **3.** 76   **5.** a. 8   b. 7   c. 15   d. 25   e. 0   **7.** $u^2 + 7$ **9.** $\pi^2 - 9$

11. $2\pi - 5$ **13.** m if $m \geq 0$; $-m$ if $m < 0$ **15.** $\{-5, 2\}$ **17.** $\{-1, 9\}$ **19.** $-1 \leq x \leq \frac{7}{3}$

21. $\left\{\frac{2}{3}, 6\right\}$ **23.** $x < -\frac{6}{5}$ or $x > 2$ **25.** $\{2\}$

27.

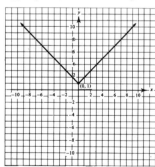

29.

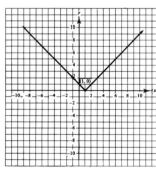

31.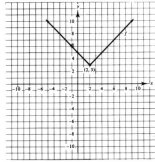

33.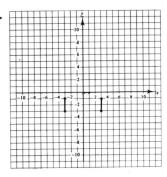

35.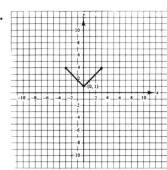

37.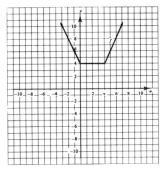

39. $|ab| = \sqrt{(ab)^2}$ definition of square root

   $= \sqrt{a^2 b^2}$ property of exponents

   $= \sqrt{a^2} \sqrt{b^2}$ property of square roots

   $= |a| |b|$ definition of square root

41. $|x - 3| < 1$

   $-1 < x - 3 < 1$

   $-2 < 2x - 6 < 2$

   $-2 + 7 < 2x - 6 + 7 < 2 + 7$

   $5 < 2x + 1 < 9$

   $5 < f(x) < 9$

43. $|x - 3| < \dfrac{1}{10}$

   $-\dfrac{1}{10} < x - 3 < \dfrac{1}{10}$

   $-\dfrac{2}{10} < 2x - 6 < \dfrac{2}{10}$

   $-\dfrac{2}{10} + 7 < 2x + 1 < \dfrac{2}{10} + 7$

   $6.8 < 2x + 1 < 7.2$

   $6.8 < f(x) < 7.2$

45. $|f(x) - 7| = |2x + 1 - 7| = |2x - 6| = 2|x - 3| < 2d$. We want $2d < \dfrac{1}{100}$. Choose $d < \dfrac{1}{200}$.

PROBLEM SET 2.3, PAGE 80

1.

3.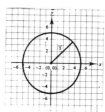

5.

7. ∅

9.

11. 5   **13.** $2\sqrt{5}$ **15.** $\sqrt{37}$ **17.**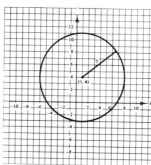

19.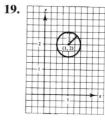

21.

23.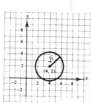

25. ∅

27.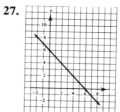

29. $\sqrt{(c-a)^2 + (d-b)^2}$ **31.** $\sqrt{x^2 + [f(x)]^2}$ **33.** $\dfrac{5}{3}\sqrt{11}$ **35.** 4

37. $x^2 - 4x + y^2 - 6y - 36 = 0$ **39.** $16x^2 + 25y^2 = 400$ **41.** $140x^2 - 4y^2 = 35$

PROBLEM SET 2.4, PAGE 89

1.
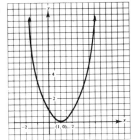

3.
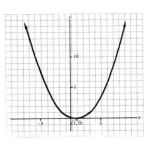

5.
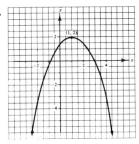

7.
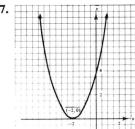

9.
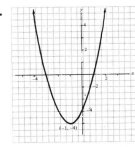

11.
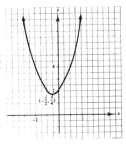

13.
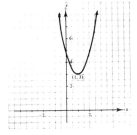

15.
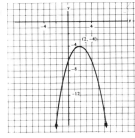

17.
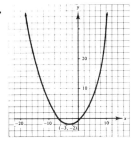

19. 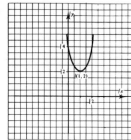 **21.** 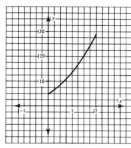 **23.**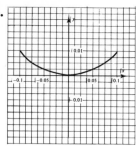

25.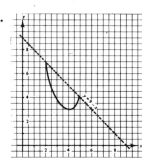

27. Vertex $(-1, 3)$; opens down; maximum value, $y = 3$, occurs when $x = -1$ **29.** Vertex $(8, -15)$; opens down; maximum value, $y = -15$, occurs when $x = 8$.   **31.** Vertex $\left(-\frac{1}{3}, \frac{2}{3}\right)$; opens down; maximum value $y = \frac{2}{3}$, occurs when $x = -\frac{1}{3}$.

33. Maximum height is 1117 ft; reached after 2 sec   **35.** Maximum daily profit is \$480; reached when 6 items per day are produced   **37.** a. $2x - y = 0$ c. Answers vary; they are the same at about 30 mph.

PROBLEM SET 2.5, PAGE 100

1. $d = 49$; 2 real roots   **3.** $d = 0$; 1 real root   **5.** $d = 49$; 2 real roots   **7.** $d = 9$; 2 real roots   **9.** $d = 1 + 8\sqrt{2}$; 2 real roots   **11.** $\left\{-3, \frac{5}{2}\right\}$ **13.** $\left\{\frac{\sqrt{5}}{2}, -\frac{\sqrt{5}}{2}\right\}$ **15.** $\left\{\frac{3 + i}{2}, \frac{3 - i}{2}\right\}$

17. $\left\{\frac{3}{2}\right\}$ **19.** $\left\{\frac{3 \pm \sqrt{17}}{4}\right\}$ **21.** $\left\{\frac{-1 \pm \sqrt{1 + 8w}}{4}\right\}$ **23.** $\frac{-1 \pm \sqrt{-3y - 5}}{3}$ **25.** $\left\{\frac{1 \pm |t|i}{2}\right\}$

27. $\frac{-3 \pm \sqrt{8y - 23}}{4}$ **29.** $3 \pm \sqrt{4y - y^2}$

31. $f(x) = ax^2 + bx + c, \ a \neq 0$

$$= a\left(x^2 + \frac{b}{a}x + \frac{c}{a}\right)$$

$$= a\left[x^2 - \left(-\frac{b}{a}\right)x + \frac{c}{a}\right]$$

$$= a[x^2 - (r_1 + r_2)x + r_1 r_2]$$

$$= a(x - r_1)(x - r_2)$$

33. Either 2 or 10 radios

35. a. 1066 ft b. About 11 sec. c.

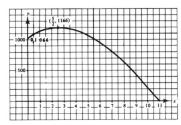

37. $x = \frac{1}{2} y^2 + 200y$; brick, 14.7 in.; steel, 20.8 in.; aluminum, 29.4 in.; concrete, 19.0 in.

REVIEW PROBLEMS, PAGE 105

1. a.

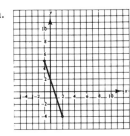

b.

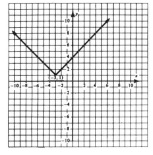

2. a.

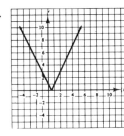

b.

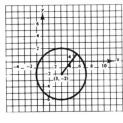

3. a–c.

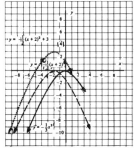

4. a. $5x + 2y - 1 = 0$ b. $2x - y + 7 = 0$ c. $6x + 7y - 8 = 0$ d. $x - 4 = 0$

5. a. $\left\{2, \dfrac{8}{3}\right\}$ b. $-\dfrac{7}{3} \le x \le 1$ c. $-2 < x < 6$ d. $x < -\dfrac{9}{4}$ or $x > \dfrac{7}{4}$ **6.** a. $\dfrac{3}{2}\sqrt{2}$

b. $\sqrt{85}$ c. $\sqrt{(\gamma - \alpha)^2 + (\delta - \beta)^2}$ d. $\sqrt{(b - a)^2 + [f(b) - f(a)]^2}$ **7.** a.

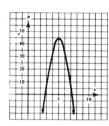

b. 5 shirts   c. $45   d. Less than 3 shirts or more than 7 shirts   **8.** 40 telephones

9. a. $\left\{-5, \dfrac{3}{2}\right\}$ b. $\left\{\dfrac{5 \pm \sqrt{43}}{3}\right\}$ **10.** a. $\left\{\dfrac{5 \pm \sqrt{71}\,i}{8}\right\}$ b. $\left\{\dfrac{2}{3}, -1 - 2t\right\}$

PROBLEM SET 3.1, PAGE 112

1. B   **3.** A   **5.** H   **7.** C   **9.** E   **11.** U   **13.** T   **15.** M   **17.** R
19. Y   **23.** a. $x^2 - x - 2$ b. $y^2 - y - 6$ **25.** a. $c^2 - 6c - 7$ b. $z^2 + 2z - 15$
27. a. $3x^2 + 4x + 1$ b. $3x^2 + 5x + 2$ **29.** a. $x^2 + 2xy + y^2$ b. $x^2 - 2xy + y^2$
31. a. $25x^2 - 16$ b. $9y^2 - 4$ **33.** a. $x^2 + 8x + 16$ b. $y^2 - 6y + 9$ **35.** a. 11   b. 4
c. -10 **37.** a. -4 b. 10   c. -10 **39.** a. $3x^2 + 3$ b. $x^3 - x^2 - 4x - 2$
41. a. $x^3 - 7x^2 + 6x - 6$ b. $15x^3 - 22x^2 + 5x + 2$ **43.** $3x^3 + 8x^2 - 9x + 2$
45. $2x^3 - 3x^2 - 8x - 3$ **47.** $x^3 - 3x^2 + 4$ **49.** $x^3 - 9x^2 + 27x - 27$
51. $12x^6 - 20x^5 + 23x^4 - 34x^3 + 71x^2 - 53x + 12$

PROBLEM SET 3.2, PAGE 119

1. $m(e + i + y)$ **3.** $a^2 + b^2$ **5.** $(a + b)(a^2 - ab + b^2)$ **7.** $(m - n)^2$ **9.** $(a + b)^3$
11. $(d - c)^3$ **13.** $(a + b)(x + y)$ **15.** $(x - 7)(x + 5)$ **17.** $(3x + 1)(x - 2)$
19. $b(4a - 1)(2a + 3)$ **21.** $y(4y^2 + y - 21)$ **23.** $p^2(4p - 3)(3p + 5)$

25. $(x - y - 1)(x - y + 1)$ **27.** $5(a - 1)(5a + 1)$ **29.** $-\dfrac{1}{9}(x + 3y)(5x + 3y)$

31. $\dfrac{1}{y^8}(x^3 - 13y^4)(x^3 + 13y^4)$ **33.** $-3(2m - 1)$ **35.** $(x^n - y^n)(x^{2n} + x^n y^n + y^{2n})$

37. $(x^n - y^n)^2$ **39.** $x(x - 5)$ **41.** $(2x - 1)(2x + 1)(x - 2)(x + 2)$
43. $(z - 2)(z + 2)(z^2 + 2z + 4)(z^2 - 2z + 4)$ **45.** $(z - 2)^2(z + 2)(z^2 + 2z + 4)$

47. $\dfrac{1}{216}(3x + 1)(2x - 1)(9x^2 - 3x + 1)(4x^2 + 2x + 1)$ **49.** $(x^3 + 2)(x - 2)(x^2 + 2x + 4)$

51. $(x - y - a - b)(x - y + a + b)$ **53.** $(x - y - c + b)(x - y + a - b)$
55. $4y(x + 2z)$ **57.** $(2x + 2y + a + b)(x + y - 3a - 3b)$

59. 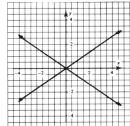   **61.** 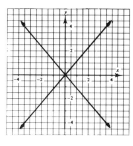   **63.**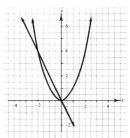

65. 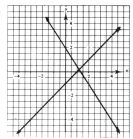   **67.**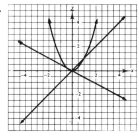

PROBLEM SET 3.3, PAGE 127

1.

$$2 \,\big|\; \begin{array}{cccc} 3 & -9 & 11 & -10 \\ & 6 & -6 & 10 \end{array}$$

$$\begin{array}{cccc} +3 & -3 & 5 & 0 \end{array}$$

$$3x^2 - 3x + 5$$

3.

$$-2 \,\big|\; \begin{array}{cccc} 2 & -4 & -10 & 12 \\ & -4 & 16 & -12 \end{array}$$

$$\begin{array}{cccc} 2 & -8 & 6 & 0 \end{array}$$

$$2x^2 - 8x + 6$$

5.

$$-4 \,\big|\; \begin{array}{cccc} 1 & 7 & 8 & -20 \\ & -4 & -12 & 16 \end{array}$$

$$\begin{array}{cccc} 1 & 3 & -4 & -4 \end{array}$$

$$x^2 + 3x - 4 + \frac{-4}{x + 4}$$

7.

$$1 \,\big|\; \begin{array}{ccccc} 1 & 2 & 0 & -5 & 2 \\ & 1 & 3 & 3 & -2 \end{array}$$

$$\begin{array}{ccccc} 1 & 3 & 3 & -2 & 0 \end{array}$$

$$x^3 + 3x^2 + 3x - 2$$

9.

$$-2 \,\big|\; \begin{array}{cccccc} 1 & 0 & 0 & 0 & 0 & -32 \\ & -2 & 4 & -8 & 16 & -32 \end{array}$$

$$\begin{array}{cccccc} 1 & -2 & 4 & -8 & 16 & -64 \end{array}$$

$$x^4 - 2x^3 + 4x^2 - 8x + 16 + \frac{-64}{x + 2}$$

11. $x^2 + x - 6$ **13.** $x^2 + x - 12$

15. $4x^3 + 8x^2 - 7x - 7$ **17.** $x^3 - 2x^2 + 3x - 4$

19. $3x^3 - 2x^2 - 5$ **21.** $Q(x) = 2x^3 - 6x^2 + 2x - 2; R(x) = 0$ **23.** $Q(x) = x^3 - x + 1;$

$R(x) = 1 - x$ **25.** $Q(x) = 3x^2 - 7x + 5; R(x) = 0$ **27.** $6x^3 + 18x^2 + 53x + 159 + \dfrac{478}{x - 3}$

29. $5x^3 + 12x^2 - 15x - 1 + \dfrac{-13}{x - 1}$ **31.** $x^4 - 1 + \dfrac{4}{x + 5}$ **33.** $x^2 - 3x + 2$ **35.** $2x^2 - 5x + 3$

PROBLEM SET 3.4, PAGE 138

1.   a. $f(1) = -3$ b. $f(-1) = -19$ c. $f(0) = -4$ d. $f(6) = 842$ e. $f(-4) = -448$
3.   a. $P(1) = -824$ b. $P(-1) = -758$ c. $P(0) = -812$ d. $P(3) = -890$
e. $P(7) = -1022$ **5.**   a. $g(0) = -10$ b. $g(-1) = -8$ c. $g(-2) = 68$ d. $g(5) = 2140$
e. $g(-3) = 380$ **7.**   a. $f(0) = -3$ b. $f(1) = 8$ c. $f(\frac{1}{2}) = 0$ d. $f(-\frac{1}{2}) = -\frac{5}{2}$
e. $f(-3) = 840$ **9.**   a. $P(-1) = 0$ b. $P(1) = 42$ c. $P(4) = 0$ d. $P(\frac{3}{2}) = 0$
e. $P(-\frac{5}{2}) = 0$

11.   0 positive roots; 1 negative root

13.   2 or 0 positive roots; 1 negative root

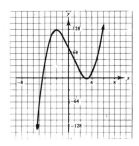

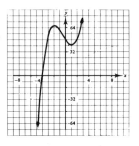

15.   2 or 0 positive roots; 1 negative root

17.   2 or 0 positive roots; 2 or 0 negative roots

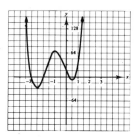

19. 3 or 1 positive roots; 1 negative root

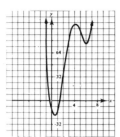

21. 2 or 0 positive roots; 3 or 1 negative roots

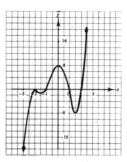

23. 2 or 0 positive roots; 2 or 0 negative roots

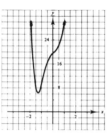

25. 2 or 0 positive roots; 1 negative root

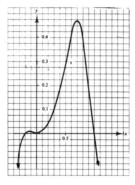

27.

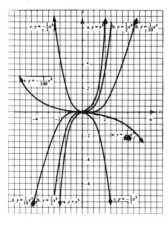

29.

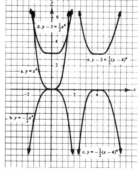

PROBLEM SET 3.5, PAGE 151

1. $\pm 1, \pm 3, \pm 5, \pm 15$ **3.** $\pm 1, \pm 2, \pm 3, \pm 4, \pm 6, \pm 12, \pm \dfrac{1}{2}, \pm \dfrac{3}{2}$ **5.** $\pm 1, \pm 2, \pm 3, \pm 4, \pm 6, \pm 12$

7. $\pm 1, \pm 3, \pm 5, \pm 15, \pm \dfrac{1}{2}, \pm \dfrac{3}{2}, \pm \dfrac{5}{2}, \pm \dfrac{15}{2}$ **9.** $\pm 1, \pm 2, \pm 3, \pm 6, \pm 9, \pm 18$ **11.** $\{1, 2, -2\}$

13. $\{2, 3, -3\}$ **15.** $\{2, -2, -3\}$ **17.** $\left\{-3, 5, -\dfrac{1}{2}\right\}$ **19.** $\{1, -1, 3, -6\}$ **21.** $\{-3, -5, -7\}$

23. $\left\{\dfrac{1}{2}, -\dfrac{5}{2}, \dfrac{7}{2}\right\}$ **25.** $\{-1, 3, -3, -4\}$ **27.** $\{2, -1, 3 \pm i\}$ **29.** $\left\{\dfrac{1}{2}, 1, \dfrac{1 \pm i}{2}\right\}$

31. $\{2, \pm 2i\}$ **33.** $\{2 \pm i, -2 \pm i\}$ **35.** $\left\{\dfrac{3}{2}, 1, \pm i\sqrt{3}\right\}$ **37.** $\left\{-2 \pm \sqrt{5}, \dfrac{2}{3}, \pm i\right\}$

39. $\{2 \pm \sqrt{3}, -1, 4 \pm \sqrt{13}\}$ **41.** $f(0) = -2$ and $f(-1) = 55$ and by the Location Theorem there is at least one root between -1 and 0 **43.** Yes; $f(-1) = 1$, $f(-2) = -5$ so by the Location Theorem there is at least one root between -2 and -1.   **45.** 8 cm   **47.** $\{1.8\}$ **49.** $\{-0.2, 2.6, 0.4, -4.8\}$

REVIEW PROBLEMS, PAGE 154

1. a. $9x^4 + 15x^3 - 101x^2 - 71x - 12$ b. $3t^3 + 4t^2 - 32t - 11$ c. $x^2 + x - 12$
d. $27w^3 + 27w^2 + 9w + 1$ e. 3   **2.** a. -12 b. -40 c. -42 d. 48   e. 0

3.

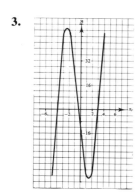

4. $\left\{-\dfrac{1}{3}, -4, 3\right\}$ **5.** a. $3x(3x + 4)$

b. $(x - 1)(x + 1)(x - 5)(x + 5)$

c. $\dfrac{1}{y^2}(2x - 2xy - y^2)(2x + 2xy + y^2)$

d. $(x + 2)(2x - 1)(2x + 1)$

e. $(2x - 1)(4x^2 + 2x + 1)(x + 1)(x^2 - x + 1)$

6. $4x^3 + 6x^2 - 2x + 3 + \dfrac{1}{3x + 1}$

7.

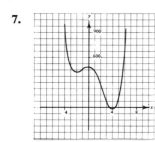

8. $\{1 \pm 2i, 1 \pm \sqrt{2}\}$ **9.** $\left\{1, \dfrac{1 \pm \sqrt{3}\, i}{2}\right\}$

10. Yes; if $f(x) = x^5 - x + 1$, then $f(-1) = 1$ and $f(-2) = -29$. Since these have opposite signs, the Location Theorem says there must be a real root between -2 and -1.

PROBLEM SET 4.1, PAGE 162

1. All reals except $x = \pm2$ **3.** All reals except $x = \pm7, \pm3$ **5.** Yes   **7.** No   **9.** No

11. $\dfrac{y + x}{xy}$ **13.** $\dfrac{2y + 3x}{xy}$ or $x^{-1}y^{-1}(2y + 3x)$ **15.** $\dfrac{2 + 3x + 3y}{x + y}$ **17.** $\dfrac{3x + 2}{x + 2}$ **19.** $\dfrac{-3}{x - y}$

21. $\dfrac{-1}{x - 2}$ **23.** $(y - 3)^2$ or $y^2 - 6y + 9$ **25.** $x^{-1}y^{-1}(x + y)^2$ **27.** $\frac{1}{3}x^{-3}y^{-2}(x^2 + 3x^4y^3 + 3y)$

29. $(x^2 + 1)^{-1}(x - 1)^2(x + 1)^2$ **31.** $(x^2 + 1)^{-2}$ **33.** $2(x^2 - 2)^2(x + 5)^3(5x^2 + 15x - 4)$
35. $-\frac{1}{3}(x^2 - 3)^{-2}(x - 3)(x + 3)$ **37.** $x^{-2}(x^2 + 2)$ **39.** $x^{-2}(-x^2 + 5x + 1)$ or $-x^{-2}(x^2 - 5x - 1)$

41. $x^{-3}(3x^3 - x^2 - 1)$ **43.** $(s + t)^{-3}(3s^2 + 3st - t^2)$ **45.** $\dfrac{-3}{y + 5}$ **47.** 1

49. $\dfrac{2x(3x - 10)}{(2x + 1)(x - 3)}$ **51.** $\dfrac{m(n + m)}{m - mn^2 + n}$ **53.** $\dfrac{x^3 + x^2 + x}{2x^2 + 2x + 1}$

PROBLEM SET 4.2, PAGE 170

1. $A = -5$ **3.** $C = 9$ **5.** $E = 12$ **7.** $G = -2$ **9.** I doesn't exist   **11.** $K = 0$
13. $M = \frac{1}{4}$ **15.** $P = 2$ **17.** $R = \frac{1}{3}$ **19.** $T = 3$ **21.** V doesn't exist   **23.** X doesn't
exist   **25.** Z doesn't exist
27. $f(x) = x - 3,\ x \neq -2$ **29.** $f(x) = 2x - 3,\ x \neq 5$ **31.** $f(x) = 2x + 1,\ x \neq 5, -3$

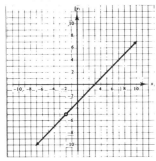
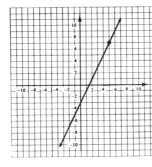
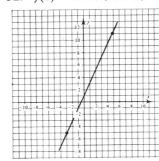

33. $f(x) = x^2 + 4x + 2, x \neq -2$

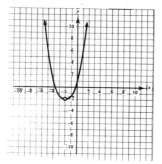

35. $f(x) = x^2 + 10x + 20, x \neq -2$

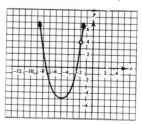

PROBLEM SET 4.3, PAGE 176

1. Continuous   **3.** Continuous   **5.** Discontinuous at $x = 3$ **7.** Discontinuous at $x = -2$
9. Continuous   **11.** Discontinuous at $x = -2$ **13.** Discontinuous everywhere
15. Continuous   **17.** Continuous; $0 < t \leq 24$ (on a 24-hour clock)   **19.** Discontinuous; domain is $1 \leq d \leq 31$, d a positive integer   **21.** Answers vary

PROBLEM SET 4.4, PAGE 192

1. $x = 0; y = 0$ **3.** $x = 0; y = 1$ **5.** $x = 0; y = 2$ **7.** $x = -3; y = 0$ **9.** $x = 4;$
$y = x + 4$ **11.** $x = 1; y = -x - 1$ **13.** None   **15.** $x = -3; y = 1$

17.

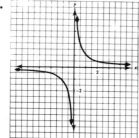

19.

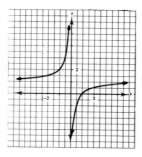

21.

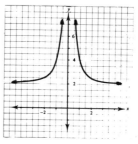

23.

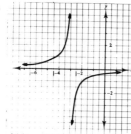

25.

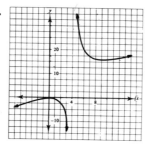

27.

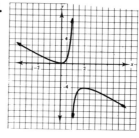

29. 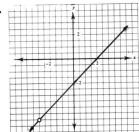   **31.**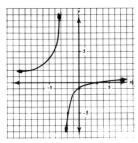

PROBLEM SET 4.5, PAGE 199

1. $\dfrac{1}{x} + \dfrac{2}{x^2} + \dfrac{5}{x^3}$ **3.** $\dfrac{2}{x-2} + \dfrac{3}{(x-2)^2}$ **5.** $\dfrac{2}{x+1} + \dfrac{3}{(x+1)^2} + \dfrac{-3}{(x+1)^3}$ **7.** $\dfrac{4}{x-2} + \dfrac{3}{x-1}$

9. $\dfrac{2}{x} + \dfrac{3}{x-1} + \dfrac{5}{(x-1)^2}$ **11.** $x + 2 + \dfrac{3}{x-1} + \dfrac{1}{(x-1)^2}$ **13.** $\dfrac{1}{x+2} + \dfrac{-1}{x+3}$ **15.** $\dfrac{5}{6(x+5)}$

15. $\dfrac{5}{6(x+5)} + \dfrac{1}{6(x-1)}$ **17.** $\dfrac{1}{x-2} + \dfrac{3}{x+2}$ **19.** $\dfrac{4}{x} + \dfrac{3}{x+1} + \dfrac{-2}{x-1}$ **21.** $\dfrac{3}{x-1} + \dfrac{2x-4}{x^2+1}$

23. $\dfrac{2}{5-x} + \dfrac{-3}{4-x}$ or $\dfrac{-2}{x-5} + \dfrac{3}{x-4}$ **25.** $\dfrac{1}{x+1} + \dfrac{-2}{(x+1)^2} + \dfrac{2x-3}{x^2+5x+1}$

27. $\dfrac{5}{x-1} + \dfrac{-1}{(x-1)^2} + \dfrac{3x}{x^2+1}$ **29.** $\dfrac{3x+1}{x^2+1} + \dfrac{4}{x+1} + \dfrac{-1}{(x+1)^2}$

REVIEW PROBLEMS, PAGE 201

1. a. $\dfrac{x^3 + x^2 + 4x - 1}{x^3}$ b. $\dfrac{x+2}{x} + \dfrac{3-x}{x^2} + \dfrac{x-1}{x^3} = 1 + \dfrac{2}{x} + \dfrac{3}{x^2} - \dfrac{1}{x} + \dfrac{1}{x^2} - \dfrac{1}{x^3} = 1 + \dfrac{1}{x} + \dfrac{4}{x^2} + \dfrac{-1}{x^3}$

c. All real numbers except $x = 0$ d. Yes (don't forget to check to see that the domains are the same)   **2.** a. $x^{-3}(x^3 + x^2 + 4x - 1)$ b. $x^{-1}y^{-1}(y + x + xy)$ c. $-5(x^2+5)^{-4}(x-1)(x+1)$

3. a. $\dfrac{x+2y}{(x+y)^2}$ b. 1   c. $\dfrac{xy - x^2}{xy^2 - y + x}$ **4.** a. 2   b. 2   c. $-\frac{1}{2}$ d. $\frac{3}{2}$ e. 2

5. a. $\dfrac{1}{x-1} + \dfrac{3}{(x-1)^2}$ b. $\dfrac{3}{x-1} + \dfrac{2}{x-2} + \dfrac{-1}{(x-2)^2}$ c. $\dfrac{2}{x-1} + \dfrac{3x+1}{x^2+3} + \dfrac{2x+1}{(x^2+3)^2}$

6. $x = 2$; $y = 2$; discontinuous at $x = 2$.   **7.**   No asymptotes; discontinuous at $x = 1$.

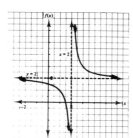

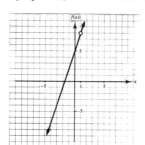

8. $x = 2$; $x = -1$; $y = 2$; discontinuous at $x = 2$; $x = -1$ **9.**   No asymptotes; discontinuous at $x = 2$.

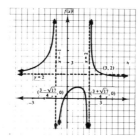

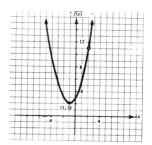

10. $x = 2$; $y = 2x + 1$; discontinuous at $x = 2$.

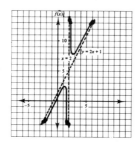

PROBLEM SET 5.1, PAGE 210

1. $A = 5$ **3.** $C = -3$ **5.** $E = -6$ **7.** $G = 16$ **9.** $I = -16$ **11.** $L = 512$
13. $N = 7$ **15.** $S = 10$ **17.** $U = 18$ **19.** $Z = 100$

21.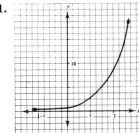

23.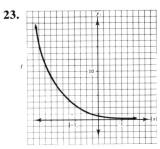

25.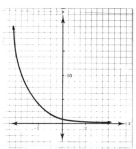

27.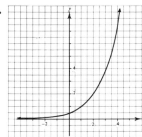

29.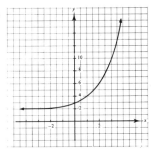

31.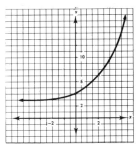

33.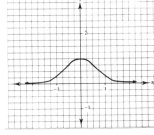

35. $x^2 y^{-1}$ **37.** $x - 2x^{1/2} y^{1/2} + y$ **39.** $x + y$

41.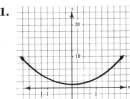

43.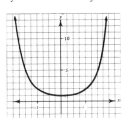

45.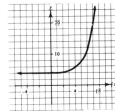

47.   a. If $b = 1$, then $f(x) = 1$; algebraic   b. If $b = 0$, then $f(x) = 0$; algebraic   **49.** $2^{\sqrt{2}} \approx 2.7$

51. $10^{\sqrt{2}} \approx 26.0$ **53.** $\sqrt{2}^{\sqrt{3}} \approx 1.8$

55. 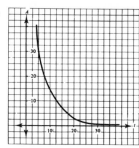 **57.**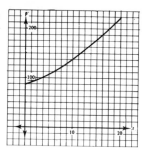

PROBLEM SET 5.2, PAGE 219

1. $x = 7$ **3.** $x = \dfrac{5}{3}$ **5.** $x = \dfrac{2}{3}$ **7.** $x = -\dfrac{4}{3}$ **9.** $x = 3$ **11.** $x = -2$ **13.** $x = \dfrac{1}{4}$

15. $x = -\dfrac{2}{15}$ **17.** $\$1000\left(1 + \dfrac{0.12}{2}\right)^{10(2)} \approx \$3207.13$. From table use 6% compounded

annually for 20 years $\approx \$3207.14$ **19.** About 24 years   **21.** $2226   **23.** About 1%
25. $k = 0.0075$ **27.** About 28 years   **29.** 11.9 psi   **31.** a. About 15 years
b. About 18 years

33. 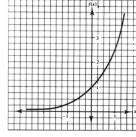   **35.** 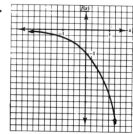   **37.**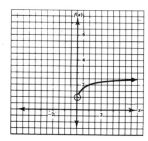

39.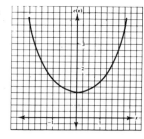

PROBLEM SET 5.3, PAGE 228

1. $\log_2 64 = 6$ **3.** $\log_{10} 1000 = 3$ **5.** $\log_5 125 = 3$ **7.** $\log_n m = p$ **9.** $\log_{1/3} 9 = -2$

11. $\log_9 \dfrac{1}{3} = -\dfrac{1}{2}$ **13.** $10^4 = 10{,}000$ **15.** $10^0 = 1$ **17.** $2^{-3} = \dfrac{1}{8}$ **19.** $e^2 = e^2$

21. $m^p = n$ **23.** $\left(\dfrac{1}{2}\right)^{-4} = 16$ **25.** 2 **27.** 4 **29.** -1 **31.** 2 **33.** 6 **35.** $\dfrac{5}{3}$

37. $\left\{\dfrac{7}{2}\right\}$ **39.** $\{10^5\}$ **41.** x is any positive real number **43.** $\{15\}$ **45.** $\{25\}$ **47.** $\{6\}$

49. $\{5\}$

51. **53.** **55.**

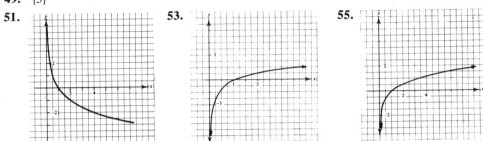

57. Let $A = b^x$; then $\log_b A = x$ (definition of logarithm).

$A^p = (b^x)^p$ raise both sides to the pth power
$A^p = b^{px}$ property of exponents
$\log_b A^p = px$ definition of logarithm
$\log_b A^p = p \log_b A$ substitution

PROBLEM SET 5.4, PAGE 237

1. 0.630 **3.** 0.926 **5.** 2.82 **7.** 1.35 **9.** 2.92 **11.** 4.85 **13.** $0.623 - 2$ or -1.38 **15.** $0.726 - 2$ or -1.274 **17.** 6.91 **19.** $4914 \approx 4900$ (HIGH) **21.** $5517840 \approx 55{,}200{,}000$ (OHBLISS) **23.** $5937 \approx 5940$ (LEGS) **25.** $378193771 \approx 380{,}000{,}000$ (ILLEGIBLE) **27.** $40 \approx 40.0$ (OH) **29.** $57334.4914 \approx 57{,}300$ (HIGH·HEELS) **31.** $0.9405645491 \approx 0.941$

33. about 1.3% **35.** No; should be about 12 years **37.** $t = \dfrac{-5700 \, \ell n \, P}{\ell n \, 2}$ **39.** About 10,500 years old **41.** a. About 12 watts b. 170 days c. About 400 days **43.** a. About 30 days

b. No c. $N = 80(1 - e^{-t/62.5})$ **45.** a. About 108°F b. $k = 0.08$ c. $t = -\dfrac{1}{k} \ell n \left(\dfrac{T - A}{B - A}\right)$

47. a. About 3600 units sold b. About 4700 units sold c. About \$117,000 **49.** It is approximately 316,000 times noisier on the outside than on the inside.

REVIEW PROBLEMS, PAGE 242

1. a. 9   b. 25   c. 72   d. $\sqrt{3}$ e. $x - y$

2. a. 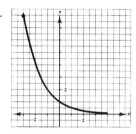   b. 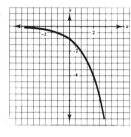   c.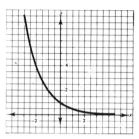

d. 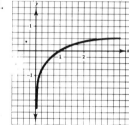   e.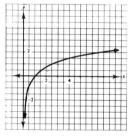

3. a. {3}   b. $\left\{\dfrac{1}{4}\right\}$ c. {−1}   d. {2}   e. {3}

4. a. 3   b. −3   c. −2   d. 5   e. 4/3

5. a. {243}   b. {7}   **6.** a. 5.86   b. −2.67

c. 0.936   d. 9.68   e. 6.44   **7.** $k = 0.026$

8. About 1.6%   **9.** a. $44,600   b. $38,700

c. $185,000   **10.** a. $31,604   b. $n = \dfrac{-\log\left(1 - \dfrac{Ar}{P}\right)}{\log(1 + r)}$

PROBLEM SET 6.1, PAGE 252

1. a. $\pi/6$ b. $\pi/2$ c. $3\pi/2$ d. $\pi/4$ e. 2π f. $\pi/3$ g. π

3. a.   b.   c.   d.   e.

5. a. 40° b. 180° **7.** a. π b. $\pi/6$ **9.** a. 330° b. 160° **11.** a. $7\pi/4$ b. π
13. a. 2.7168 b. 1.2832 **15.** a. 0 b. 3.1416 **17.** a. 5.4978 b. $3\sqrt{5} - 2\pi$ or
0.4250 **19.** a. 0 b. 1 c. -1 d. -1 e. 0 **21.** a. 0 b. -1 c. -1 d. 0
e. -1 **23.** a. 1 b. 1 c. 1 d. 1 e. $(c(\theta), s(\theta))$ is a point on the unit circle $x^2 + y^2 = 1$;
therefore, $c^2(\theta) + s^2(\theta) = 1$ **25.** 65.67° **27.** $-85.33°$ **29.** 315.42° **31.** 29.28°
33. 48.469° **35.** $-94.359°$ **37.** 40° **39.** 6° **41.** $-165°$ **43.** 114.59°
45. $-14.32°$ **47.** 22.92° **49.** $2\pi/9$ or .6981 **51.** $-16\pi/45$ or -1.1170 **53.** $127\pi/90$ or
4.4331 **55.** 1.95 **57.** -1.10 **59.** 1 m **61.** 31.40 m **63.** 70.69 cm **65.** 12.5 ft
67. 3.14 cm **69.** About 4000 miles **71.** 890 km

PROBLEM SET 6.2, PAGE 261

1. 60° **3.** $\pi/4$ **5.** $\pi/2$ **7.** 0 **9.** $\sqrt{2}/2$ **11.** 1 **13.** -1 **15.** $\sqrt{3}/2$

17. 0 **19.** 0 **21.** $\dfrac{1}{2}$ **23.** $\sqrt{2}/2$ **25.** 0 **27** $|x|\sqrt{2}$ **29.** $-x\sqrt{2}$ **31.** $\dfrac{1}{2}$

33. 0 **35.** 1 **37.** 1 **39.** $\dfrac{3}{4}$ **41.** $\dfrac{1}{4}$

PROBLEM SET 6.3, PAGE 268

1. a. 1 b. 1 c. $\dfrac{1}{2}\sqrt{3}$ d. $\dfrac{2}{3}\sqrt{3}$ e. Undefined f. $\dfrac{1}{3}\sqrt{3}$ **3.** a. 1 b. 1 c. 0

d. $\dfrac{2}{3}\sqrt{3}$ e. $\dfrac{1}{2}\sqrt{2}$ f. 0 **5.** a. 1 b. -1 c. -1 d. 0 e. Undefined f. Undefined

7. 1 **9.** 1 **11.** a. $\sqrt{2}/2$ b. $\sqrt{2}/2$ **13.** a. 2 b. $\dfrac{1}{3}\sqrt{3}$ **15.** a. $-\sqrt{3}$ b. $-\sqrt{3}$

17. $\cos\theta = 3/5$; $\sin\theta = 4/5$; $\tan\theta = 4/3$; $\sec\theta = 5/3$; $\csc\theta = 5/4$; $\cot\theta = 3/4$ **19.** $\cos\theta = -3/5$;
$\sin\theta = -4/5$; $\tan\theta = 4/3$; $\sec\theta = -5/3$; $\csc\theta = -5/4$; $\cot\theta = 3/4$ **21.** $\cos\theta = 5/13$; $\sin\theta = 12/13$;
$\tan\theta = 12/5$; $\sec\theta = 13/5$; $\csc\theta = 13/12$; $\cot\theta = 5/12$ **23.** $\cos\theta = 5/13$; $\sin\theta = -12/13$;
$\tan\theta = -12/5$; $\sec\theta = 13/5$; $\csc\theta = -13/12$; $\cot\theta = -5/12$ **25.** $\cos\theta = 2/\sqrt{29}$; $\sin\theta = -5\sqrt{29}$;
$\tan\theta = -5/2$; $\sec\theta = \sqrt{29}/2$; $\csc\theta = -\sqrt{29}/5$; $\cot\theta = -2/5$ **27.** $\cos\theta = -4/\sqrt{41}$; $\sin\theta = -5/\sqrt{41}$;
$\tan\theta = 5/4$; $\sec\theta = -\sqrt{41}/4$; $\csc\theta = -\sqrt{41}/4$; $\cot\theta = 4/5$ **29.** $\cos 450° = 0$; $\sin 450° = 1$;
$\tan 450°$ undefined; $\sec 450°$ undefined; $\csc 450° = 1$; $\cot 450° = 0$

31. $\cos(-\pi/6) = \sqrt{3}/2$; $\sin(-\pi/6) = -\dfrac{1}{2}$; $\tan(-\pi/6) = -\sqrt{3}/3$; $\sec(-\pi/6) = 2/\sqrt{3}$ or $(2/3)\sqrt{3}$;

$\csc(-\pi/6) = -2/\sqrt{3}$; $\cot(-\pi/6) = -3/\sqrt{3}$ or $-\sqrt{3}$ **33.** $\cos(-2\pi) = 1$; $\sin(-2\pi) = 0$;
$\tan(-2\pi) = 0$; $\sec(-2\pi) = 1$; $\csc(-2\pi)$ and $\cot(-2\pi)$ undefined **35.** $\cos 390° = \sqrt{3}/2$;
$\sin 390° = 1/2$; $\tan 390° = \sqrt{3}/3$; $\sec 390° = 2/\sqrt{3}$; $\csc 390° = 2$; $\cot 390° = 3/\sqrt{3}$ or $\sqrt{3}$

37. $\cos(-135°) = -\sqrt{2}/2$; $\sin(-135°) = -\sqrt{2}/2$; $\tan(-135°) = 1$; $\sec(-135°) = -2/\sqrt{2}$ or $-\sqrt{2}$;
$\csc(-135°) = -2/\sqrt{2}$ or $-\sqrt{2}$; $\cot(-135°) = 1$ **39.** $\cos 3\pi = -1$; $\sin 3\pi = 0$; $\tan 3\pi = 0$; $\sec 3\pi = -1$;
$\csc 3\pi$ and $\cot 3\pi$ undefined **41.** 0.6 **43.** 2.0 **45.** -0.4 **47.** -1.0 **49.** -0.6
51. 0.2 **53.** $(47) - 1.015427$; $(48)\,0.342020$; $(49) - 0.642788$; $(50)\,5.671282$; $(51)\,0.176327$
55. $(47) -1.0154$; $(48)\,0.3420$; $(49) -0.6428$; $(50)\,5.6713$; $(51)\,0.1763$ **57.** 0.205271
59. 0.841471 **61.** 0.886995 **63.** 0.366176

PROBLEM SET 6.4, PAGE 277

1. a. Undefined   b. 0   c. 0   d. $\sqrt{3}/2$ e. Undefined   f. 0   **3.** a. Undefined
b. -1 c. 0   d. 1   e. 1   f. 1/2   **5.** a. 1   b. $\sqrt{2}$ c. $\sqrt{3}$ d. 1   e. 0   f. -1

19.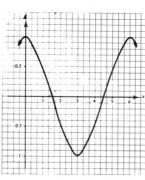

21.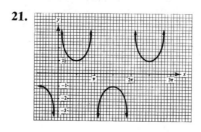

23.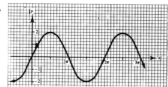

25.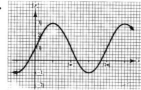

27.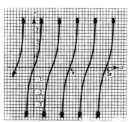

29.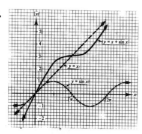

31.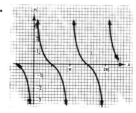

PROBLEM SET 6.5, PAGE 289

1. a. 1   b. Undefined   c. 0   d. Undefined   e. -1 f. 0   **3.** a. $\sqrt{3}/3$ b. 2
c. $\sqrt{2}/2$ d. $\sqrt{2}/2$ e. $\sqrt{3}$ f. Undefined   **5.** a. $\sqrt{3}$ b. 0   c. $2\sqrt{3}/3$ d. $\sqrt{2}$
e. -1 f. $\sqrt{3}/3$

7.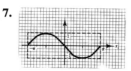

9.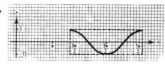

11.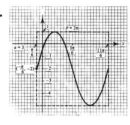

13.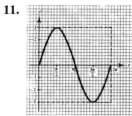

15.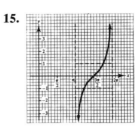

17.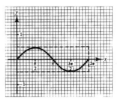

19.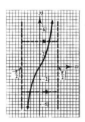

21.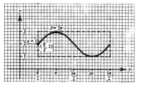

23.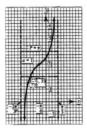

25.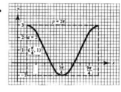

27.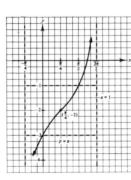

29.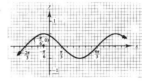

31.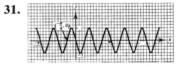

33.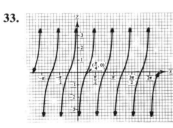

35.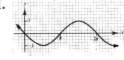

37.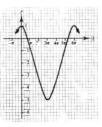

39.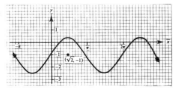

41.

43.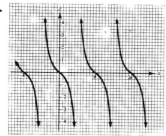

45.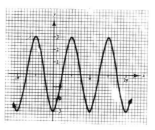

47.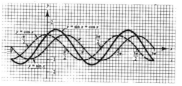

49.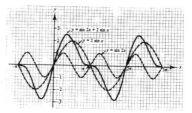

51.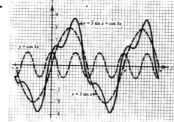

PROBLEM SET 6.6, PAGE 297

1. a. 0   b. 30°   c. 30°   d. 0   **3.** a. 60°   b. 45°   c. 45°   d. 45°   **5.** a. 150°
b. −45°   c. −45°   d. 120°   **7.** a. 0   b. 60°   c. 60°   d. 60°   **9.** 33°   **11.** 85°
13. 19°   **15.** 54°   **17.** 108°   **19.** 69°   **21.** −66°   **23.** 1   **25.** 1/3
27. $2\pi/15$ **29.** 0.4163   **31.** $30° \pm 360°n$; $150° \pm 360°n$ **33.** $60° \pm 180°n$
35. $150° \pm 360°n$; $210° \pm 360°n$ **37.** $23° \pm 360°n$; $157° \pm 360°n$ **39.** $54° \pm 180°n$

41.

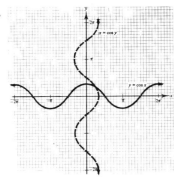

43.

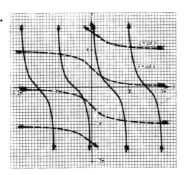

45.

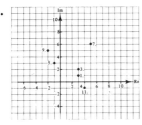

PROBLEM SET 6.7, PAGE 303

1–11.

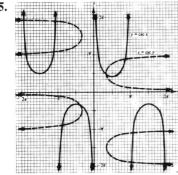

Note: The numbers in Problems 13–43 should also be graphed.

13. $\sqrt{2}$ cis 45°   **15.** 2 cis 330°   **17.** 2 cis 300°

19. 4 cis 330°   **21.** cis 0°   **23.** cis 270°

25. 3 cis 55°   **27.** 4 cis 100°   **29.** $\sqrt{2} + \sqrt{2}i$

31. $2\sqrt{2} - 2\sqrt{2}i$

33. $-\sqrt{3}/2 + \frac{1}{2}i$ **35.** 1   **37.** $-\frac{3}{2} + (3\sqrt{3}/2)i$ **39.** $4.2262 + 9.0631i$ **41.** $-7\sqrt{2}/2 + (7\sqrt{2}/2)i$

43. $-8.8633 - 1.5628i$

PROBLEM SET 6.8, PAGE 307

1. $6 \text{ cis } 210°$ **3.** $48 \text{ cis } 152°$ **5.** $2 \text{ cis } 15°$ **7.** $\frac{5}{2} \text{ cis } 267°$ **9.** $2 \text{ cis } 190°$

11. $18 \text{ cis } 167°$ **13.** $8 \text{ cis } 150°$ **15.** $\text{cis } 330°$ **17.** $64 \text{ cis } 180°$ **19.** $8 \text{ cis } 270°$
21. $(\sqrt{2}/2) \text{ cis } 75°$ **23.** $4 \text{ cis } 330°$ **25.** $12\sqrt{2} \text{ cis } 45°$ **27.** $4\sqrt{2} \text{ cis } 225°$ **29.** $\text{cis } 30°$
31. $12 \text{ cis } 330°$ **33.** $(3\sqrt{3} + 3) + (3\sqrt{3} - 3)i$ **35.** $-3 + 3\sqrt{3}\,i$ **37.** $3 \text{ cis } 270°$

39. $\left(\frac{3}{2}\sqrt{3} - \frac{3}{2}\right) + \left(-\frac{3}{2}\sqrt{3} - \frac{3}{2}\right)i$ **41.** $\frac{3}{4} + \frac{3}{4}\sqrt{3}\,i$ **43.** $-4.3079 + 34.3832i$ **45.** $.3131 + 2.2281\,i$

PROBLEM SET 6.9, PAGE 311

1. $3 \text{ cis } 120°, 3 \text{ cis } 300°$ **3.** $5 \text{ cis } 150°, 5 \text{ cis } 330°$ **5.** $3 \text{ cis } 100°, 3 \text{ cis } 220°, 3 \text{ cis } 340°$
7. $2 \text{ cis } 88°, 2 \text{ cis } 178°, 2 \text{ cis } 268°, 2 \text{ cis } 358°$ **9.** $4 \text{ cis } 56°, 4 \text{ cis } 146°, 4 \text{ cis } 236°, 4 \text{ cis } 326°$
11. $2 \text{ cis } 32°, 2 \text{ cis } 104°, 2 \text{ cis } 176°, 2 \text{ cis } 248°, 2 \text{ cis } 320°$ **13.** $\text{cis } 60°, \text{cis } 180°, \text{cis } 300°$ **15.** $3 \text{ cis } 0°,$
$3 \text{ cis } 120°, 3 \text{ cis } 240°$ **17.** $\sqrt[8]{2} \text{ cis } 56.25°, \sqrt[8]{2} \text{ cis } 146.25°, \sqrt[8]{2} \text{ cis } 236.25°, \sqrt[8]{2} \text{ cis } 326.25°$ **19.** $\text{cis } 54°,$
$\text{cis } 126°, \text{cis } 198°, \text{cis } 270°, \text{cis } 342°$ Note: Problems 21–25 should also be illustrated graphically.

21. $1, -\frac{1}{2} + (\sqrt{3}/2)i, -\frac{1}{2} - (\sqrt{3}/2)i$ **23.** $1 + \sqrt{3}\,i, -2, 1 - \sqrt{3}\,i$ **25.** $-0.6840 + 1.8794i,$

$-1.2856 - 1.5321i, 1.9696 - 0.3473i$ **27.** $2 \text{ cis } 30°, 2 \text{ cis } 90°, 2 \text{ cis } 150°, 2 \text{ cis } 210°, 2 \text{ cis } 270°,$
$2 \text{ cis } 330°$ **29.** $\text{cis } 0°, \text{cis } 40°, \text{cis } 80°, \text{cis } 120°, \text{cis } 160°, \text{cis } 200°, \text{cis } 240°, \text{cis } 280°, \text{cis } 320°$
31. $\text{cis } 9°, \text{cis } 45°, \text{cis } 81°, \text{cis } 117°, \text{cis } 153°, \text{cis } 189°, \text{cis } 225°, \text{cis } 261°, \text{cis } 297°, \text{cis } 333°$ **33.** $4 \text{ cis } 30°,$
$4 \text{ cis } 270°, 4 \text{ cis } 150°$

REVIEW PROBLEMS, PAGE 315

1. a. $\frac{1}{2}$ b. $\frac{1}{2}$ c. 1 d. 1 e. Undefined   f. $\sqrt{3}/3$ g. $\sqrt{2}/2$ h. $\sqrt{3}$ **2.** a. $30°$ or

$\pi/6$ b. $30°$ or $\pi/6$ c. $-60°$ or $-\pi/3$ d. $45°$ or $\pi/4$ e. $90°$ or $\pi/2$ f. $-45°$ or $-\pi/4$
g. $60°$ or $\pi/3$ h. $0°$ or 0 **3.** a. 1.089615865 b. -1.338648128 c. 1.458652446
d. $18.3°$ e. $130.5°$ f. $73°$

4. a. 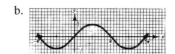   b.   c.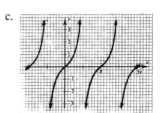

d. 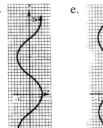   e. 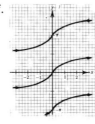   f. 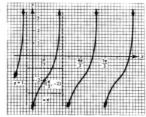   5. a.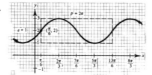

b. 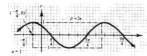   c. 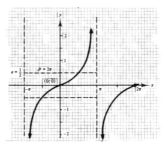   d.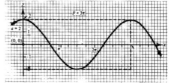

e. 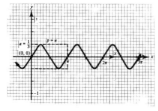   f.

6. a. $-128 + 128\sqrt{3}\, i$ b. $3\sqrt{3} - 3i$ **7.** a. 8 cis 360°   b. cis 154°   **8.** $\sqrt{7}$ cis 165°, $\sqrt{7}$ cis 345°
or $-2.556 + 0.6848i$, $2.556 - 0.6848i$ **9.** cis 0°, cis 90°, cis 180°, cis 270° or ± 1, $\pm i$ **10.** About
157 m

PROBLEM SET 7.1, PAGE 322

3. III, IV   **5.** I, IV   **7.** II   **9.** I   **11.** $\cot(A + B)$ **13.** $\tan(\pi/15)$
15. $\cos(\pi/8)$ **17.** $\cos 127°$ **19.** $\sec^2(\pi/5)$ **21.** 1   **23.** -1 **25.** 1   **31.** $\tan\theta$
$= \tan\theta$; $\cot\theta = 1/\tan\theta$; $\sec\theta = \pm\sqrt{1 + \tan^2\theta}$; $\cos\theta = 1/(\pm\sqrt{1 + \tan^2\theta})$; $\sin\theta = \tan\theta/(\pm\sqrt{1 + \tan^2\theta})$;
$\csc\theta = (\pm\sqrt{1 + \tan^2\theta})/\tan\theta$ **33.** $\sec\theta = \sec\theta$; $\cos\theta = 1/\sec\theta$; $\tan\theta = \pm\sqrt{\sec^2\theta - 1}$; $\cot\theta$
$= 1/(\pm\sqrt{\sec^2\theta - 1})$; $\sin\theta = (\pm\sqrt{\sec^2\theta - 1})/\sec\theta$; $\csc\theta = \sec\theta/(\pm\sqrt{\sec^2\theta - 1})$ **35.** $\sin\theta = -12/13$;
$\tan\theta = -12/5$; $\sec\theta = 13/5$; $\csc\theta = -13/12$; $\cot\theta = -5/12$ **37.** $\cot\theta = 12/5$; $\sec\theta = 13/12$;
$\cos\theta = 12/13$; $\sin\theta = 5/13$; $\csc\theta = 13/5$ **39.** $\cos\theta = \sqrt{5}/3$; $\tan\theta = 2\sqrt{5}/5$; $\cot\theta = \sqrt{5}/2$; $\csc\theta$

$= 3/2; \sec \theta = 3\sqrt{5}/5$ **41.** $\cos \theta = 5\sqrt{34}/34; \tan \theta = -3/5; \cot \theta = -5/3; \sin \theta = -3\sqrt{34}/34; \csc \theta$
$= -\sqrt{34}/3$ **43.** $\sin \theta = -3\sqrt{10}/10; \cos \theta = \sqrt{10}/10; \cot \theta = -1/3; \tan \theta = -3; \sec \theta = \sqrt{10}$

45. $\dfrac{1 - \cos^2 \theta}{\sin \theta} = \dfrac{\sin^2 \theta}{\sin \theta}$ **47.** $\dfrac{\dfrac{\sin \theta}{\cos \theta} + \dfrac{\cos \theta}{\sin \theta}}{\dfrac{1}{\sin \theta \cos \theta}} = \dfrac{\sin^2 \theta + \cos^2 \theta}{\cos \theta \sin \theta} \cdot \dfrac{\sin \theta \cos \theta}{1}$

$\qquad\qquad = \sin \theta$ $\qquad\qquad\qquad\qquad\qquad\qquad = 1$

49. $\dfrac{\cos \theta + \dfrac{\sin^2 \theta}{\cos \theta}}{\sin \theta} = \dfrac{\cos^2 \theta + \sin^2 \theta}{\cos \theta \sin \theta}$ **51.** $\sin \theta + \cot \theta = \sin \theta + \dfrac{\cos \theta}{\sin \theta}$

$\qquad\qquad = \dfrac{1}{\cos \theta \sin \theta}$ $\qquad\qquad\qquad\qquad = \dfrac{\sin^2 \theta + \cos \theta}{\sin \theta}$

53. $\dfrac{\tan \theta + \cot \theta}{\sec \theta \csc \theta} = \dfrac{\dfrac{\sin \theta}{\cos \theta} + \dfrac{\cos \theta}{\sin \theta}}{\dfrac{1}{\cos \theta} \cdot \dfrac{1}{\sin \theta}}$ **55.** $\sec^2 \theta + \tan^2 \theta = \dfrac{1}{\cos^2 \theta} + \dfrac{\sin^2 \theta}{\cos^2 \theta}$

$\qquad = \dfrac{\sin^2 \theta + \cos^2 \theta}{\cos \theta \sin \theta} \cdot \dfrac{\cos \theta \sin \theta}{1}$ $\qquad\qquad\qquad = \dfrac{1 + \sin^2 \theta}{\cos^2 \theta}$

$\qquad = 1$

57. $(\cot \theta - \sec \theta)(\sin \theta \cos \theta) = \left(\dfrac{\cos \theta}{\sin \theta} - \dfrac{1}{\cos \theta}\right)(\sin \theta \cos \theta)$

$\qquad\qquad\qquad\qquad = \dfrac{\cos^2 \theta - \sin \theta}{\sin \theta \cos \theta}(\sin \theta \cos \theta)$

$\qquad\qquad\qquad\qquad = \cos^2 \theta - \sin \theta$

PROBLEM SET 7.2, PAGE 330

1. $\sin^3 \theta + \cos^2 \theta \sin \theta = \sin \theta(\sin^2 \theta + \cos^2 \theta)$
$\qquad\qquad\qquad\qquad = \sin \theta$

3. $\cot \theta \tan^2 \theta = (\cot \theta \tan \theta) \tan \theta$ **5.** $\tan^2 \theta \sin^2 \theta = \tan^2 \theta(1 - \cos^2 \theta)$
$\qquad\qquad = \tan \theta$ $\qquad\qquad\qquad\qquad = \tan^2 \theta - \tan^2 \theta \cos^2 \theta$

$\qquad\qquad\qquad\qquad\qquad\qquad = \tan^2 \theta - \dfrac{\sin^2 \theta}{\cos^2 \theta} \cos^2 \theta$

$\qquad\qquad\qquad\qquad\qquad\qquad = \tan^2 \theta - \sin^2 \theta$

7. $\tan A + \cot A = \dfrac{\sin A}{\cos A} + \dfrac{\cos A}{\sin A}$

$\qquad = \dfrac{\sin^2 A + \cos^2 A}{\cos A \sin A}$

$\qquad = \dfrac{1}{\cos A \sin A}$

$\qquad = \sec A \csc A$

9. $\dfrac{\sec x + \csc x}{\csc x \sec x} = \dfrac{1}{\csc x} + \dfrac{1}{\sec x}$

$\qquad\qquad\qquad\quad = \sin x + \cos x$

11. $\dfrac{1 - \sec^2 t}{\sec^2 t} = \dfrac{1}{\sec^2 t} - 1$

$\qquad\qquad = \cos^2 t - 1$

$\qquad\qquad = -\sin^2 t$

13. $(\sec \theta - \cos \theta)^2 = \sec^2 \theta - 2 \sec \theta \cos \theta - \cos^2 \theta$

$\qquad\qquad\qquad = (\tan^2 \theta + 1) - 2 + (1 - \sin^2 \theta)$

$\qquad\qquad\qquad = \tan^2 \theta - \sin^2 \theta$

15. $\dfrac{1 - \sin^2 2\theta}{1 + \sin 2\theta} = \dfrac{(1 - \sin 2\theta)(1 + \sin 2\theta)}{1 + \sin 2\theta}$

$\qquad\qquad\qquad = 1 - \sin 2\theta$

17. $\dfrac{\sin^2 \lambda + \sin \lambda \cos \lambda + \sin \lambda}{\sin \lambda + \cos \lambda + 1} = \dfrac{\sin \lambda (\sin \lambda + \cos \lambda + 1)}{\sin \lambda + \cos \lambda + 1}$

$\qquad\qquad\qquad\qquad = \sin \lambda$

19. $\sin 2\alpha \cos 2\alpha (\tan 2\alpha + \cot 2\alpha)$

$\qquad = \sin 2\alpha \cos 2\alpha \cdot \left(\dfrac{\sin 2\alpha}{\cos 2\alpha} + \dfrac{\cos 2\alpha}{\sin 2\alpha} \right)$

$\qquad = \sin 2\alpha \cos 2\alpha \cdot \left(\dfrac{\sin^2 2\alpha + \cos^2 2\alpha}{\cos 2\alpha \sin 2\alpha} \right)$

$\qquad = 1$

21. $\csc 3\beta - \cos 3\beta \cot 3\beta = \dfrac{1}{\sin 3\beta} - \dfrac{\cos^2 3\beta}{\sin 3\beta}$

$\qquad\qquad\qquad\qquad = \dfrac{1 - \cos^2 3\beta}{\sin 3\beta}$

$\qquad\qquad\qquad\qquad = \dfrac{\sin^2 3\beta}{\sin 3\beta}$

$\qquad\qquad\qquad\qquad = \sin 3\beta$

23. $\dfrac{\sin^2 B - \cos^2 B}{\sin B + \cos B} = \dfrac{(\sin B - \cos B)(\sin B + \cos B)}{\sin B + \cos B}$

$\qquad\qquad\qquad = \sin B - \cos B$

25. $\tan^2 2\gamma + \sin^2 2\gamma + \cos^2 2\gamma = \tan^2 2\gamma + 1$

$\qquad\qquad\qquad\qquad\qquad = \sec^2 2\gamma$

27. $\dfrac{\tan \theta + \cot \theta}{\sec \theta \csc \theta} = \dfrac{(\sin \theta / \cos \theta) + (\cos \theta / \sin \theta)}{\sec \theta \csc \theta}$

$\qquad = \dfrac{\sin^2 \theta + \cos^2 \theta}{\cos \theta \sin \theta} \cdot \dfrac{1}{\sec \theta \csc \theta}$

$\qquad = \dfrac{1}{\cos \theta \sin \theta} \cdot \dfrac{\sin \theta \cos \theta}{1}$

$\qquad = 1$

29. $\dfrac{1}{\sin\theta + \cos\theta} + \dfrac{1}{\sin\theta - \cos\theta} = \dfrac{\sin\theta - \cos\theta + \sin\theta + \cos\theta}{(\sin\theta + \cos\theta)(\sin\theta - \cos\theta)}$

$$= \frac{2\sin\theta}{\sin^2\theta - \cos^2\theta}$$

$$= \frac{2\sin\theta}{2\sin^2\theta - 1}$$

$$= \frac{\sin\theta}{\sin^2\theta - (1/2)}$$

31. $\dfrac{1 + \tan C}{1 - \tan C} = \dfrac{1 + \tan C}{1 - \tan C} \cdot \dfrac{1 + \tan C}{1 + \tan C}$

$$= \frac{1 + 2\tan C + \tan^2 C}{1 - \tan^2 C}$$

$$= \frac{\sec^2 C + 2\tan C}{1 - (\sec^2 C - 1)}$$

$$= \frac{\sec^2 C + 2\dfrac{\sin C}{\cos C}}{2 - \sec^2 C}$$

$$= \frac{\sec^2 C + 2\tan C}{2 - \sec^2 C}$$

33. $\dfrac{\sin^3 x - \cos^3 x}{\sin x - \cos x} = \dfrac{(\sin x - \cos x)(\sin^2 x + \sin x \cos x + \cos^2 x)}{\sin x - \cos x}$

$$= \sin^2 x + \sin x \cos x + \cos^2 x$$

$$= 1 + \sin x \cos x$$

35. $\sqrt{(3\cos\theta - 4\sin\theta)^2 + (3\sin\theta + 4\cos\theta)^2}$

$\quad = \sqrt{9\cos^2\theta - 24\sin\theta\cos\theta + 16\sin^2\theta + 9\sin^2\theta + 24\sin\theta\cos\theta + 16\cos^2\theta}$

$\quad = \sqrt{9 + 16}$

$\quad = 5$

37. $\dfrac{(\sec^2\gamma + \tan^2\gamma)^2}{\sec^4\gamma - \tan^4\gamma} = \dfrac{(\sec^2\gamma + \tan^2\gamma)^2}{(\sec^2\gamma - \tan^2\gamma)(\sec^2\gamma + \tan^2\gamma)}$

$$= \dfrac{\sec^2\gamma + \tan^2\gamma}{\sec^2\gamma - \tan^2\gamma}$$

$$= \dfrac{\sec^2\gamma + \tan^2\gamma}{(1 + \tan^2\gamma) - \tan^2\gamma}$$

$$= \sec^2\gamma + \tan^2\gamma$$

$$= (1 + \tan^2\gamma) + \tan^2\gamma$$

$$= 1 + 2\tan^2\gamma$$

39. $(\sec 2\theta + \csc 2\theta)^2 = \sec^2 2\theta + 2\sec 2\theta \csc 2\theta + \csc^2 2\theta$

$$= \dfrac{1}{\cos^2 2\theta} + \dfrac{2}{\cos 2\theta \sin 2\theta} + \dfrac{1}{\sin^2 2\theta}$$

$$= \dfrac{\sin^2 2\theta + 2\sin 2\theta \cos 2\theta + \cos^2 2\theta}{\cos^2 2\theta \sin^2 2\theta}$$

$$= \dfrac{1 + 2\sin 2\theta \cos 2\theta}{\cos^2 2\theta \sin^2 2\theta} \quad \cdot$$

41. $\dfrac{1}{\csc\theta - \cot\theta} = \dfrac{1}{\csc\theta - \cot\theta} \cdot \dfrac{\csc\theta + \cot\theta}{\csc\theta + \cot\theta}$

$$= \dfrac{\csc\theta + \cot\theta}{\csc^2\theta - \cot^2\theta}$$

$$= \dfrac{\csc\theta + \cot\theta}{(1 + \cot^2\theta) - \cot^2\theta}$$

$$= \csc\theta + \cot\theta$$

43. $2\csc A - \cot A \cos A + \cos^2 A \csc A = 2\csc A - \dfrac{\cos^2 A}{\sin A} + \dfrac{\cos^2 A}{\sin A}$

$$= 2\csc A$$

45. $\dfrac{\tan\theta}{\cot\theta} - \dfrac{\cot\theta}{\tan\theta} = \dfrac{\sin^2\theta}{\cos^2\theta} - \dfrac{\cos^2\theta}{\sin^2\theta}$

$$= \frac{\sin^4\theta - \cos^4\theta}{\sin^2\theta\cos^2\theta}$$

$$= \frac{(\sin^2\theta + \cos^2\theta)(\sin^2\theta - \cos^2\theta)}{\sin^2\theta\cos^2\theta}$$

$$= \frac{\sin^2\theta - \cos^2\theta}{\sin^2\theta\cos^2\theta}$$

$$= \frac{1}{\cos^2\theta} - \frac{1}{\sin^2\theta}$$

$$= \sec^2\theta - \csc^2\theta$$

47. $\dfrac{1 + \tan^3\theta}{1 + \tan\theta} = \dfrac{(1 + \tan\theta)(1 - \tan\theta + \tan^2\theta)}{1 + \tan\theta}$

$$= 1 - \tan\theta + \tan^2\theta$$

$$= \sec^2\theta - \tan\theta$$

49. $\dfrac{\cos^2\theta - \cos\theta\csc\theta}{\cos^2\theta\csc\theta - \cos\theta\csc^2\theta} = \dfrac{\cos\theta(\cos\theta - \csc\theta)}{\cos\theta\csc\theta(\cos\theta - \csc\theta)}$

$$= \frac{\cos\theta}{\cos\theta\csc\theta}$$

$$= \frac{1}{\csc\theta}$$

$$= \sin\theta$$

51. $\dfrac{\cos\theta + \cos^2\theta}{\cos\theta + 1} = \dfrac{\cos\theta(1 + \cos\theta}{\cos\theta + 1}$

$$= \cos\theta$$

Also, $\dfrac{\cos\theta\sin\theta + \cos^2\theta}{\sin\theta + \cos\theta} = \dfrac{\cos\theta(\sin\theta + \cos\theta)}{\sin\theta + \cos\theta}$

$$= \cos\theta$$

53. $\sin\theta + \cos\theta + 1 = \dfrac{(\sin\theta + \cos\theta + 1)(\sin\theta + \cos\theta - 1)}{\sin\theta + \cos\theta - 1}$

$$= (\sin^2\theta + \sin\theta\cos\theta - \sin\theta + \cos\theta\sin\theta + \cos^2\theta$$

$$- \cos\theta + \sin\theta + \cos\theta - 1)\left(\frac{1}{\sin\theta + \cos\theta - 1}\right)$$

$$= \frac{2\sin\theta\cos\theta}{\sin\theta + \cos\theta - 1}$$

55. $\dfrac{\csc\theta + 1}{\csc\theta - 1} - \dfrac{\sec\theta - \tan\theta}{\sec\theta + \tan\theta} = \dfrac{\dfrac{1}{\sin\theta} + 1}{\dfrac{1}{\sin\theta} - 1} - \dfrac{\dfrac{1}{\cos\theta} - \dfrac{\sin\theta}{\cos\theta}}{\dfrac{1}{\cos\theta} + \dfrac{\sin\theta}{\cos\theta}}$

$$= \frac{1 + \sin\theta}{1 - \sin\theta} - \frac{1 - \sin\theta}{1 + \sin\theta}$$

$$= \frac{(1 + \sin\theta)^2 - (1 - \sin\theta)^2}{1 - \sin^2\theta}$$

$$= \frac{1 + 2\sin\theta + \sin^2\theta - 1 + 2\sin\theta - \sin^2\theta}{\cos^2\theta}$$

$$= \frac{4\sin\theta}{\cos^2\theta}$$

$$= 4\tan\theta\sec\theta$$

57. $\dfrac{\cos\theta + 1}{\cos\theta - 1} + \dfrac{1 - \sec\theta}{1 + \sec\theta} = \dfrac{\cos\theta + 1}{\cos\theta - 1} + \dfrac{1 - \dfrac{1}{\cos\theta}}{1 + \dfrac{1}{\cos\theta}}$

$$= \frac{\cos\theta + 1}{\cos\theta - 1} + \frac{\cos\theta - 1}{\cos\theta + 1}$$

$$= \frac{\cos^2\theta + 2\cos\theta + 1 + \cos^2\theta - 2\cos\theta + 1}{\cos^2\theta - 1}$$

$$= \frac{2\cos^2\theta + 2}{\cos^2\theta - 1}$$

$$= \frac{2\cos^2\theta + 2}{-\sin^2\theta}$$

$$= -2\cot^2\theta - 2\csc^2\theta$$

PROBLEM SET 7.3, PAGE 341

	angle θ	$\cos \theta$	$\sin \theta$	$\tan \theta$
1.	$-15°$	$\dfrac{\sqrt{6} + \sqrt{2}}{4}$	$\dfrac{\sqrt{2} - \sqrt{6}}{4}$	$-2 + \sqrt{3}$
3.	$75°$	$\dfrac{\sqrt{6} - \sqrt{2}}{4}$	$\dfrac{\sqrt{2} + \sqrt{6}}{4}$	$2 + \sqrt{3}$
5.	$345°$	$\dfrac{\sqrt{6} + \sqrt{2}}{4}$	$\dfrac{\sqrt{2} - \sqrt{6}}{4}$	$-2 + \sqrt{3}$

	$\cos \tfrac{1}{2}\theta$	$\sin \tfrac{1}{2}\theta$	$\tan \tfrac{1}{2}\theta$	$\cos 2\theta$	$\sin 2\theta$	$\tan 2\theta$
7.	$3\sqrt{10}/10$	$\sqrt{10}/10$	$1/3$	$7/25$	$24/25$	$24/7$
9.	$-5\sqrt{26}/26$	$\sqrt{26}/26$	$-1/5$	$119/169$	$-120/169$	$-120/199$
11.	$\sqrt{7}/3$	$\sqrt{2}/3$	$\sqrt{14}/7$	$-31/81$	$20\sqrt{14}/81$	$-20\sqrt{14}/31$

13. $\dfrac{\sqrt{3}\cos\theta - \sin\theta}{2}$ **15.** $\dfrac{1 + \tan\theta}{1 - \tan\theta}$ **17.** $\dfrac{\sqrt{3}\cos\theta - \sin\theta}{2}$ **19.** $\cos 40° + \cos 110°$

21. $\cos 11° - \cos 59°$ **23.** $\dfrac{1}{2}\cos 3\theta - \cot 7\theta$ **25.** $2 \sin 53° \cos 10°$ **27.** $2 \sin 80° \cos 1°$

29. $-2\sin(x/2)\cos(3x/2)$ **31.** $\sqrt{2}/2$ **33.** $1/2$ **35.** -1 **37.** $\dfrac{1}{2}\sqrt{2 - \sqrt{2}}$

39. $\sqrt{2} - 1$ **41.** 0.9135 **43.** 0.2079 **45.** 1.1918

47. $\tan(\alpha + \beta) = \dfrac{\sin(\alpha + \beta)}{\cos(\alpha + \beta)}$

$$= \frac{\sin\alpha\cos\beta + \cos\alpha\sin\beta}{\cos\alpha\cos\beta - \sin\alpha\sin\beta} \cdot \frac{\dfrac{1}{\cos\alpha\cos\beta}}{\dfrac{1}{\cos\alpha\cos\beta}}$$

$$= \frac{\dfrac{\sin\alpha}{\cos\alpha} + \dfrac{\sin\beta}{\cos\beta}}{1 - \dfrac{\sin\alpha}{\cos\alpha} \cdot \dfrac{\sin\beta}{\cos\beta}}$$

$$= \frac{\tan\alpha + \tan\beta}{1 - \tan\alpha\tan\beta}$$

49. $\cot(\alpha + \beta) = \dfrac{\cos(\alpha + \beta)}{\sin(\alpha + \beta)}$

$$= \frac{\cos\alpha\cos\beta - \sin\alpha\sin\beta}{\sin\alpha\cos\beta + \cos\alpha\sin\beta} \cdot \frac{\dfrac{1}{\sin\alpha\cos\beta}}{\dfrac{1}{\sin\alpha\cos\beta}}$$

$$= \frac{\cot\alpha - \cot\beta}{1 + \cot\alpha\cot\beta}$$

Another acceptable answer is $\dfrac{\cot\beta\cot\alpha - 1}{\cot\beta + \cot\alpha}$.

51. 33/65 **53.** 56/65

	$\cot 2\theta$	$\cos 2\theta$	$\cos\theta$	$\sin\theta$	$\tan\theta$
55.	$-3/4$	$-3/5$	$\sqrt{5}/5$	$2\sqrt{5}/5$	2
57.	$1/\sqrt{3}$	$1/2$	$\sqrt{3}/2$	$1/2$	$\sqrt{3}/2$
59.	$-4/3$	$-4/5$	$\sqrt{10}/10$	$3\sqrt{10}/10$	3

61. $\dfrac{\cos 5\theta}{\sin\theta} - \dfrac{\sin 5\theta}{\cos\theta} = \dfrac{\cos 5\theta\cos\theta - \sin 5\theta\sin\theta}{\sin\theta\cos\theta}$

$$= \frac{\cos(5\theta + \theta)}{\sin\theta\cos\theta}$$

$$= \frac{\cos 6\theta}{\sin\theta\cos\theta}$$

63. $\dfrac{\sin(\theta + h) - \sin\theta}{h} = \dfrac{\sin\theta\cos h + \cos\theta\sin h - \sin\theta}{h}$

$$= \cos\theta\left(\frac{\sin h}{h}\right) + \sin\theta\left(\frac{\cos h - 1}{h}\right)$$

$$= \cos\theta\left(\frac{\sin h}{h}\right) - \sin\theta\left(\frac{1 - \cos h}{h}\right)$$

65. $\sin(\alpha + \beta + \gamma) = \sin(\alpha + \beta)\cos\gamma + \cos(\alpha + \beta)\sin\gamma$
$= (\sin\alpha\cos\beta + \cos\alpha\sin\beta)\cos\gamma + (\cos\alpha\cos\beta - \sin\alpha\sin\beta)\sin\gamma$
$= \sin\alpha\cos\beta\cos\gamma + \cos\alpha\sin\beta\cos\gamma + \cos\alpha\cos\beta\sin\gamma - \sin\alpha\sin\beta\sin\gamma$

67.
$$\sin(\alpha + \beta) = \sin\alpha\cos\beta + \cos\alpha\sin\beta$$
$$\underline{\sin(\alpha - \beta) = \sin\alpha\cos\beta - \cos\alpha\sin\beta}$$
$$\sin(\alpha + \beta) - \sin(\alpha - \beta) = 2\cos\alpha\sin\beta \qquad \text{(subtracting)}$$

Let $x = \alpha + \beta$ and $y = \alpha - \beta$; then, $\alpha = (x + y)/2$ and $\beta = (x - y)/2$. Therefore,

$$\sin x - \sin y = 2\sin\left(\frac{x - y}{2}\right)\cos\left(\frac{x + y}{2}\right).$$

69. $\cos 2\theta = \cos^2 \theta - \sin^2 \theta$

Let $\theta = 2\theta$, then $\cos 4\theta = \cos^2 2\theta - \sin^2 2\theta$.

71. $\dfrac{\cos 3\theta - \cos \theta}{\sin \theta - \sin 3\theta} = \dfrac{-2 \sin 2\theta \sin \theta}{2 \cos 2\theta \sin (-\theta)}$

$$= \frac{\sin 2\theta}{\cos 2\theta}$$

$$= \tan 2\theta$$

73. $\tan (B/2) = \dfrac{1 - \cos B}{\sin B}$ (from Problem 66)

$$= \frac{1}{\sin B} - \frac{\cos B}{\sin B}$$

$$= \csc B - \cot B$$

75. $\sin 3\theta = \sin (2\theta + \theta)$
$= \sin 2\theta \cos \theta + \cos 2\theta \sin \theta$
$= 2 \sin \theta \cos \theta \cos \theta + (\cos^2 \theta - \sin^2 \theta) \sin \theta$
$= 2 \sin \theta \cos^2 \theta + \cos^2 \theta \sin \theta - \sin^3 \theta$
$= 3 \cos^2 \theta \sin \theta - \sin^3 \theta$
$= 3(1 - \sin^2 \theta) \sin \theta - \sin^3 \theta$
$= 3 \sin \theta - 4 \sin^3 \theta$

PROBLEM SET 7.4, PAGE 348

1. $\{\pi/3, 5\pi/3\}$ **3.** $\{\pi/8, 3\pi/8, 9\pi/8, 11\pi/8\}$ **5.** $\{\pi/12, 5\pi/12, 3\pi/4, 13\pi/12, 17\pi/12, 7\pi/4\}$
7. $\{0, \pi, \pi/2, 3\pi/2\}$ **9.** $\{\pi/3, \pi/2, 5\pi/3\}$ **11.** $\{0, \pi, \pi/3, 4\pi/3\}$ **13.** $\{0, \pi, 1.2661, 5.0171\}$
15. $\{\pi/6, 5\pi/6, 7\pi/6, 11\pi/6\}$ **17.** $\{\pi/4, 5\pi/4\}$ **19.** $\{\pi/4, 3\pi/4, 5\pi/4, 7\pi/4\}$ **21.** $\{3\pi/2\}$
23. $\{\pi/3, 5\pi/3\}$ **25.** $\{0.9046, 5.3786\}$ **27.** $\{0.2846, 2.8510\}$ **29.** $\{\pi/6, \pi/3, 7\pi/6, 2\pi/3\}$

PROBLEM SET 7.5, PAGE 354

1. $\alpha = 30°$; $\beta = 60°$; $\gamma = 90°$; $a = 80$; $b = 140$; $c = 160$ **3.** $\alpha = 39°$; $\beta = 51°$; $\gamma = 90°$; $a = 68$; $b = 83$; $c = 110$ **5.** $\alpha = 25°$; $\beta = 65°$; $\gamma = 90°$; $a = 6.1$; $b = 13$; $c = 14$ **7.** $\alpha = 77°$; $\beta = 13°$; $\gamma = 90°$; $a = 390$; $b = 90$; $c = 400$ **9.** $\alpha = 50°$; $\beta = 40°$; $\gamma = 90°$; $a = 24$; $b = 29$; $c = 45$
11. $\alpha = 69.2°$; $\beta = 20.8°$; $\gamma = 90.0°$; $a = 26.5$; $b = 10.0$; $c = 28.3$ **13.** $\alpha = 32.6°$; $\beta = 57.4°$; $\gamma = 90.0°$; $a = 70.0$; $b = 109$; $c = 130$ **15.** $\alpha = 67°$; $\beta = 23°$; $\gamma = 90°$; $a = 9000$; $b = 3800$; $c = 9800$ **17.** $\alpha = 73.6°$; $\beta = 16.4°$; $\gamma = 90.0°$; $a = 8780$; $b = 2580$; $c = 9140$ **19.** $\alpha = 65.44°$; $\beta = 24.56°$; $\gamma = 90.00°$; $a = 7003$; $b = 3200$; $c = 7700$ **21.** The building is 23 m tall.   **23.** The ship is 200 m away.   **25.** The distance is 350 ft.   **27.** The top of the ladder is 13 ft above the ground.   **29.** The angle of elevation is about 56°.   **31.** The height is 1251 ft.   **33.** The distance is 780 ft.   **35.** The height of the Empire State Building is 1250 ft, and that of the Sears Tower is 1454 ft.   **37.** The distance is 51.8 ft.   **39.** The center is 14.7 ft high.   **41.** The plate should be placed at 12 ft for a pitch of 5/12.   **43.** The distance from the earth to Venus is 63,400,000 (or 6.34×10^7) miles.   **45.** The radius of the inscribed circle is 633.9 ft.   **49.** The shadow will be 12 ft long.

PROBLEM SET 7.6, PAGE 369

1. $\alpha = 54°$; $\beta = 133°$; $\gamma = 7°$; $a = 7.0$; $b = 8.0$; $c = 2.0$ **3.** $\alpha = 82°$; $\beta = 50°$; $\gamma = 48°$; $a = 18$; $b = 14$; $c = 12$ **5.** $\alpha < 90°$; $a < h < b$, no triangle formed   **7.** $\alpha = 80°$; $\beta = 52°$; $\gamma = 48°$; $a = 5.0$; $b = 4.0$; $c = 3.8$ **9.** $h \approx 8.63 < a < b$; ambiguous case   *Solution I:* $\alpha = 47.0°$; $\beta = 57.8°$; $\gamma = 75.2°$; $a = 10.2$; $b = 11.8$; $c = 13.5$ *Solution II:* $\alpha = 47.0°$; $\beta' = 122.2°$; $\gamma' = 10.8°$; $a = 10.2$; $b = 11.8$; $c' = 2.61$ **11.** $h \approx 4.14 < a < b$; ambiguous case   *Solution I:* $\alpha = 56°$; $\beta = 67°$; $\gamma = 57°$; $a = 4.5$; $b = 5.0$; $c = 4.6$ *Solution II:* $\alpha = 56°$; $\beta' = 113°$; $\gamma' = 11°$; $a = 4.5$; $b = 5.0$; $c' = 1.0$ **13.** $\alpha = 65°$; $\beta = 78°$; $\gamma = 37°$; $a = 38$; $b = 41$; $c = 25$ **15.** $\alpha = 110.6°$; $\beta = 21.2°$; $\gamma = 48.2°$; $a = 38.2$; $b = 14.8$; $c = 30.4$ **17.** $\alpha = 14°$; $\beta = 42°$; $\gamma = 142°$; $a = 26$; $b = 71$; $c = 88$ **19.** $\alpha = 60.2°$; $\beta = 84.8°$; $\gamma = 35.0°$; $a = 14.2$; $b = 16.3$; $c = 9.39$ **21.** $\alpha = 147°$; $\beta = 15.0°$; $\gamma = 18.0°$; $a = 49.5$; $b = 23.5$; $c = 28.1$ **23.** $h \approx 43.8 < c < b$; ambiguous case   *Solution I:* $\alpha = 90.8°$; $\beta = 57.1°$; $\gamma = 32.1°$; $a = 98.2$; $b = 82.5$; $c = 52.2$ *Solution II:* $\alpha' = 25.0°$; $\beta' = 122.9°$; $\gamma = 32.1°$; $a' = 41.5$; $b = 82.5$; $c = 52.2$ **33.** He is 5.39 miles from the target.   **35.** The distance from L to T is 454 ft.   **37.** The length of the tower is 179 ft.   **39.** The height of the tower is 985.2 ft.   **41.** The pier is 6.72 miles long.   **47.** 180 sq units   **49.** 20,100 sq units

REVIEW PROBLEMS, PAGE 374

2. $\cos \delta = -4/5$; $\tan \delta = -3/4$; $\csc \delta = 5/3$; $\sec \delta = -5/4$; $\cot \delta = -4/3$

3.
$$\frac{1 + \tan^2 \theta}{\csc \theta} = \frac{\sec^2 \theta}{\csc \theta}$$

$$= \frac{1}{\cos^2 \theta} \cdot \sin \theta$$

$$= \frac{1}{\cos \theta} \cdot \frac{\sin \theta}{\cos \theta}$$

$$= \sec \theta \tan \theta$$

4.
$$\frac{\cos \theta}{\sec \theta} - \frac{\sin \theta}{\cot \theta} = \cos^2 \theta - \frac{\sin^2 \theta}{\cos \theta}$$

$$= \frac{\cos^3 \theta - \sin^2 \theta}{\cos \theta}$$

Also,
$$\frac{\cos \theta \cot \theta - \tan \theta}{\csc \theta} = \frac{\cos \theta \cdot \dfrac{\cos \theta}{\sin \theta} - \dfrac{\sin \theta}{\cos \theta}}{\dfrac{1}{\sin \theta}}$$

$$= \left(\frac{\cos^2 \theta}{\sin \theta} - \frac{\sin \theta}{\cos \theta} \right) \sin \theta$$

$$= \frac{(\cos^3 \theta - \sin^2 \theta) \sin \theta}{\sin \theta \cos \theta}$$

$$= \frac{\cos^3 \theta - \sin^2 \theta}{\cos \theta}$$

5. $2 \sin\left(\dfrac{h}{2}\right) \cos\left(\dfrac{2x + h}{2}\right)$

6. $\dfrac{\sin 5\theta + \sin 3\theta}{\cos 5\theta - \cos 3\theta} = \dfrac{2 \sin 4\theta \cos \theta}{-2 \sin 4\theta \sin \theta}$

$= -\cot \theta$

7.   a. $\{0, \pi\}$ b. $\{3\pi/2\}$ c. $\{1.8798, 2.8085, 5.0214, 5.9501\}$ **8.**   It is 278 ft across the bridge.   **9.**   The angle of elevation is 48.9°, and the distance is 430 ft.   **10.**   At 12:42 P.M. or at 4:40 P.M.

PROBLEM SET 8.1, PAGE 379

1.   17; arithmetic   **3.**   96; geometric   **5.**   85; neither   **7.** pq^5; geometric   **9.**   36; neither   **11.**   2; neither   **13.** $\dfrac{5}{6}$; neither   **15.**   216; neither   **17.**   0, 3, 6   **19.** $a + d$, $a + 2d, a + 3d$ **21.** $0, \dfrac{1}{3}, \dfrac{1}{2}$ **23.** $-2, +3, -4$ **25.** $2, \dfrac{3}{2}, \dfrac{4}{3}$ **27.** $-5, -5, -5$ **29.**   303   **31.**   21   **33.** -7 **35.** $3, 1, \dfrac{1}{3}, \dfrac{1}{9}, \dfrac{1}{27}$ **37.**   1, 2, 3, 5, 8

PROBLEM SET 8.2, PAGE 384

1. $s_n = 1 + 5n$ **3.** $s_n = -15 + 7n$ **5.** $s_n = nx$ **7.**   5, 9, 13, 17; $1 + 4n$ **9.**   100, 95, 90, 85; $105 - 5n$ **11.** $5, 5 + x, 5 + 2x, 5 + 3x$; $5 - x + xn$ **13.** $s_{20} = 101$ **15.** $S_{10} = 845$ **17.** $d = -2$ **19.** $S_{15} = 480$ **21.** $S_{20} = 11{,}500$ **23.**   There are $\dfrac{99 - 41}{2} = 29$ terms so the sum is 2030.   **25.**   There are $\dfrac{136 - 48}{2} = 44$ terms so the sum is 4048.   **27.** $S_n = n(n + 1)$ **29.** $S_{87} = 3828$ **31.**   a. $\dfrac{9}{2}$ b. 4   c. -1 d. 84   e. 48   **33.**   55 blocks

PROBLEM SET 8.3, PAGE 389

1. $5, 15, 45$; $5 \cdot 3^{n-1}$ **3.** $1, -2, 4$; $(-2)^{n-1}$ **5.** $-15, -3, -\dfrac{3}{5}$; $-15\left(\dfrac{1}{5}\right)^{n-1}$ **7.** $8, 8x, 8x^2$; $8x^{n-1}$ **9.** $a = 3, r = 2$ **11.** $a = 1, r = \dfrac{1}{2}$ **13.** $a = x, r = x$ **15.** $s_n = 3 \cdot 2^{n-1}$ **17.** $s_n = 2^{1-n}$ **19.** $s_n = x^n$ **21.** $s_5 = 486$ **23.** $S_5 = 726$ **25.** $S_{10} = 1{,}111{,}111{,}111$ **27.** $S_1 = \dfrac{1}{3}$; $S_{10} \approx 0.499992$ **29.** $S_{15} \approx 38{,}146{,}972{,}670$ (accuracy depends on type of calculator used)   **31.** $S_{10} = 11{,}111{,}111{,}110$ **33.**   a. $2\sqrt{2}$ b. 4   c. $-\sqrt{15}$ d. $-2\sqrt{5}$ e. $4\sqrt{5}$

PROBLEM SET 8.4, PAGE 395

1. 2   **3.** 200   **5.** -67.5 **7.** $\dfrac{4}{9}$ **9.** $\dfrac{3}{11}$ **11.** $\dfrac{27}{11}$ **13.** $\dfrac{1394}{3333}$ **15.** $\dfrac{1}{2}$

17. 13.5   **19.** $4 + 2\sqrt{2}$ **21.** $2 + \dfrac{3}{2}\sqrt{2}$ **23.** $\dfrac{1132}{225}$ **25.** 190 ft   **27.** \$10,000

29. Total area $= \text{I} + \text{II} + \text{III} + \text{IV} + \text{V} + \text{VI} + \cdots$

$$= (\text{I} + \text{II} + \text{III}) + (\text{IV} + \text{V} + \text{VI}) + \cdots$$

$$= 3\left(\frac{a}{2}\right)^2 + 3\left(\frac{1}{2}\cdot\frac{a}{2}\right)^2 + 3\left(\frac{1}{2}\cdot\frac{1}{2}\cdot\frac{a}{2}\right)^2 + \cdots$$

$$= \frac{3}{2^2}a^2 + \frac{3}{2^4}a^2 + \frac{3}{2^6}a^2 + \cdots$$

$$= \frac{3}{4}a^2 + \frac{3}{4}a^2\left(\frac{1}{4}\right) + \frac{3}{4}a^2\left(\frac{1}{4}\right)^2 + \cdots$$

$$a = \frac{3}{4}a^2,\ r = \frac{1}{4};\ \lim_{n\to\infty} S_n = \frac{(3/4)a^2}{1 - (1/4)} = a^2$$

31.   area of $ABCD = a^2$

$$\text{area of } EFGH = \left(\frac{1}{2}a\right)^2 = \frac{1}{4}a^2$$

$$\text{area of next square} = \left(\frac{1}{2}\cdot\frac{1}{2}a\right)^2 = \frac{a^2}{4^2}$$

$$\vdots$$

$$\text{area of } n\text{th square} = \left(\frac{a}{2^{n-1}}\right)^2 = \frac{a^2}{4^{n-1}}$$

$$S_n = \text{total area} = a^2 + \frac{a^2}{4} + \frac{a^2}{4}\left(\frac{1}{4}\right) + \frac{a^2}{4}\left(\frac{1}{4}\right)^2 + \cdots + \frac{a^2}{4}\left(\frac{1}{4}\right)^{n-2} + \cdots$$

First term is a^2; $r = \dfrac{1}{4}$; $\lim\limits_{n\to\infty} S_n = \dfrac{a^2}{1 - (1/4)} = \dfrac{4}{3}a^2.$

PROBLEM SET 8.5, PAGE 404

1. 22   **3.** 2   **5.** 72   **7.** 132   **9.** 220   **11.** 20   **13.** 49   **15.** 8   **17.** 28

19. 56   **21.** 70   **23.** 1326   **27.** $\sum\limits_{k=1}^{7} 2^{-k}$ **29.** $\sum\limits_{k=0}^{4} 6^k$ **31.** $a_1 b_1 + a_2 b_2 + \cdots + a_r b_r$

33. $\sum\limits_{j=1}^{r} ka_j = ka_1 + ka_2 + ka_3 + \cdots + ka_r$

$\qquad = k(a_1 + a_2 + a_3 + \cdots + a_r)$

$\sum\limits_{j=1}^{r} k = \sum\limits_{j=1}^{r} ka_j, \quad \text{where } a_j = 1 \text{ for all } j$

$\qquad = k(a_1 + a_2 + a_3 + \cdots + a_r)$

$\qquad = k\underbrace{(1 + 1 + 1 + \cdots + 1)}_{r \text{ terms}}$

$\qquad = kr$

35. a. 100   b. 200   **37.** $11^0 = 1; 11^1 = 11; 11^2 = 121; 11^3 = 1331$ The powers of 11 are the rows of Pascal's triangle. (For 11^5 or higher powers, look at row 5: 1 5 10 10 5 1. When two digits are in the same column, the tens digit must be carried into the next column, just as in addition: $11^5 = 1\ 6\ 1\ 0\ 5\ 1$.)

PROBLEM SET 8.6, PAGE 409

1. $(x + y)^5 = x^5 + 5x^4y + 10x^3y^2 + 10x^2y^3 + 5xy^4 + y^5$ **3.** $(x + 2)^5 = x^5 + 10x^4 + 40x^3 + 80x^2,$ $+ 80x + 32$ **5.** $(a + b)^4 = a^4 + 4a^3b + 6a^2b^2 + 4ab^3 + b^4$ **7.** 462   **9.** 1001
11. $x^{16} + 32x^{15}y + 480x^{14}y^2 + 4480x^{13}y^3$ **13.** $x^{12} - 24x^{11}y + 264x^{10}y^2 - 1760x^9y^3$
15. $1 - 0.24 + 0.0264 - 0.00176$ **17.** $1 + 0.56 + 0.1372 + 0.019208$ **19.** $1 - 0.14 + 0.0084$
$- 0.00028$ **21.** $5.60 to the nearest cent

PROBLEM SET 8.7, PAGE 414

1. *Step 1.* Prove it true for $n = 1$:

$$1 = \frac{1(1 + 1)}{2} \ \checkmark$$

Step 2. Assume it true for $n = k$:

$$1 + 2 + 3 + \cdots + k = \frac{k(k + 1)}{2}$$

Step 3. Prove it true for $n = k + 1$:

To prove: $1 + 2 + 3 + \cdots + (k + 1) = \dfrac{(k + 1)(k + 2)}{2}$

Proof: by hypothesis (Step 2),

$$1 + 2 + 3 + \cdots + k = \frac{k(k + 1)}{2}.$$

Add $(k + 1)$ to both sides:

$$1 + 2 + 3 + \cdots + k + (k + 1) = \frac{k(k + 1)}{2} + (k + 1)$$

$$= \frac{k(k + 1)}{2} + \frac{2(k + 1)}{2}$$

$$= \frac{(k + 1)(k + 2)}{2}$$

Step 4.   The proposition is true for all positive integers n by PMI.

3.   *Step 1.*   Prove it true for $n = 1$:

$$\sum_{j=1}^{1} (2j - 1)^2 = (2 - 1)^2$$

$$= 1$$

$$\frac{n(2n - 1)(2n + 1)}{3} = \frac{1(2 - 1)(2 + 1)}{3}$$

$$= 1$$

Step 2.   Assume it true for $n = k$:

$$\sum_{j=1}^{k} (2j - 1)^2 = 1^2 + 3^2 + 5^2 + \cdots + (2k - 1)^2$$

So

$$1^2 + 3^2 + 5^2 + \cdots + (2k - 1)^2 = \frac{k(k + 1)(2k + 1)}{3}.$$

Step 3.   Prove it true for $n = k + 1$:
To prove:

$$1^2 + 3^2 + 5^2 + \cdots + [2(k + 1) - 1]^2 = \frac{(k + 1)(k + 1 + 1)(2(k + 1) + 1)}{3}$$

or

$$1^2 + 3^2 + 5^2 + \cdots + (2k + 1)^2 = \frac{(k + 1)(k + 2)(2k + 3)}{3}$$

Proof: by hypothesis,

$$1^2 + 3^2 + 5^2 + \cdots + (2k - 1)^2 = \frac{k(k + 1)(2k + 1)}{3}$$

Add $(2k + 1)^2$ to both sides:

$$1^2 + 3^2 + 5^2 + \cdots + (2k - 1)^2 + (2k + 1)^2 = \frac{k(2k - 1)(2k + 1)}{3} + (2k + 1)^2$$

$$= \frac{k(2k - 1)(2k + 1) + 3(2k + 1)^2}{3}$$

$$= \frac{(2k + 1)[k(2k - 1) + 3(2k + 1)]}{3}$$

$$= \frac{(2k + 1)(2k^2 + 5k + 3)}{3}$$

$$= \frac{(2k + 1)(k + 1)(2k + 3)}{3}$$

Step 4.   The proposition is true for all positive integers n by PMI.

5.   *Step 1.*   Prove it true for $n = 1$:

$$(2 \cdot 1)^2 = 4$$

$$\frac{2 \cdot 1(1 + 1)(2 \cdot 1 + 1)}{3} = \frac{2 \cdot 1 \cdot 2 \cdot 3}{3} = 4$$

Step 2.   Assume it true for $n = k$:

$$2^2 + 4^2 + 6^2 + \cdots + (2k)^2 = \frac{2k(k + 1)(2k + 1)}{3}$$

Step 3.   Prove it true for $n = k + 1$:

To prove: $2^2 + 4^2 + 6^2 + \cdots + (2k + 2)^2 = \frac{2k(k + 1)(k + 2)(2k + 3)}{3}$

Proof: By hypothesis,

$$2^2 + 4^2 + 6^2 + \cdots + (2k)^2 = \frac{2k(k + 1)(2k + 1)}{3}.$$

Add $(2k + 2)^2$ to both sides:

$$2^2 + 4^2 + \cdots + (2k)^2 + (2k + 2)^2 = \frac{2k(k + 1)(2k + 1)}{3} + (2k + 2)^2$$

$$= \frac{2k(k + 1)(2k + 1) + 3 \cdot (2k + 2)^2}{3}$$

$$= \frac{2k(k + 1)(2k + 1) + 3 \cdot 4(k + 1)^2}{3}$$

$$= \frac{2(k + 1)[k(2k + 1) + 3 \cdot 2(k + 1)]}{3}$$

$$= \frac{2(k + 1)(2k^2 + 7k + 6)}{3}$$

$$= \frac{2(k + 1)(k + 2)(2k + 3)}{3}$$

Step 4.   The proposition is true for all positive integers n by PMI.

7. *Step 1.*   Prove it true for $n = 1$:

$$1(1 + 2) = 3$$

$$\frac{1(1 + 1)(2 \cdot 1 + 7)}{6} = \frac{1 \cdot 2 \cdot 9}{6} = 3$$

Step 2.   Assume it true for $n = k$:

$$1 \cdot 3 + 2 \cdot 4 + \cdots + k(k + 2) = \frac{k(k + 1)(2k + 7)}{6}$$

Step 3.   Prove it true for $n = k + 1$:

To prove: $1 \cdot 3 + 2 \cdot 4 + \cdots (k + 1)(k + 3) = \dfrac{(k + 1)(k + 2)(2k + 9)}{6}$

Proof: By hypothesis,

$$1 \cdot 3 + 2 \cdot 4 + \cdots + k(k + 2) = \frac{k(k + 1)(2k + 7)}{6}.$$

Add $(k + 1)(k + 3)$ to both sides:

$$1 \cdot 3 + 2 \cdot 4 + \cdots + k(k + 2) + (k + 1)(k + 3) = \frac{k(k + 1)(2k + 7)}{6} + (k + 1)(k + 3)$$

$$= \frac{k(k + 1)(2k + 7) + 6(k + 1)(k + 3)}{6}$$

$$= \frac{(k + 1)[k(2k + 7) + 6(k + 3)]}{6}$$

$$= \frac{(k + 1)(2k^2 + 13k + 18)}{6}$$

$$= \frac{(k + 1)(k + 2)(2k + 9)}{6}$$

Step 4.   The proposition is true for all positive integers n by PMI.

9. *Step 1.*   Prove it true for $n = 1$:

$1^5 - 1 = 0$, which is divisible by 5.

Step 2.   Assume it true for $n = k$:
That is, $k^5 - k$ is divisible by 5.

Step 3.   Prove it true for $n = k + 1$:
To prove: $(k + 1)^5 - (k + 1)$ is divisible by 5.

$$\text{Now } (k + 1)^5 - (k + 1) = k^5 + 5k^4 + 10k^3 + 10k^2 + 5k + 1 - k - 1$$
$$= (k^5 - k) + (5k^4 + 10k^3 + 10k^2 + 5k).$$

$k^5 - 5$ is divisible by 5 by hypothesis.
$5k^4 + 10k^3 + 10k^2 + 5k = 5(k^4 + 2k^3 + 2k^2 + k)$ is divisible by 5.

Therefore $(k + 1)^5 - (k + 1)$ is divisible by 5.

Step 4. The proposition is true for all positive integers n by PMI.

11. *Step 1.* Prove it true for $n = 1$:

$$(1 + 1)^2 \geq 1 + 1^2$$
$$4 \geq 3; \text{ true}$$

Step 2. Assume it true for $n = k$:
$$(1 + k)^2 \geq 1 + k^2$$

Step 3. Prove it true for $n = k + 1$:

To prove: $(2 + k)^2 \geq 1 + (1 + k)^2$

or
$$4 + 4k + k^2 \geq 1 + 1 + 2k + k^2$$
or
$$2k \geq -2$$
$$k \geq -1$$

That is, we wish to prove $k \geq -1$. By hypothesis,

$$(1 + k)^2 \geq 1 + k^2$$
$$1 + 2k + k^2 \geq 1 + k^2$$
$$2k \geq 0$$
$$k \geq 0$$

If $k \geq 0$, then $k \geq -1$, so the result is proved.

Step 4. The proposition is true for all positive integers n by PMI.

13. Conjecture: $1 + 4 + 7 + \cdots + (3n - 2) = \dfrac{n(3n - 1)}{2}$

Step 1. Prove it true for $n = 1$:

$$1 \overset{?}{=} \frac{1(3 \cdot 1 - 1)}{2}$$

$$1 = 1 \; \checkmark$$

Step 2. Assume it true for $n = k$:

that is, $1 + 4 + 7 + \cdots + (3k - 2) = \dfrac{k(3k - 1)}{2}$

Step 3. Prove it true for $n = k + 1$:

To prove: $1 + 4 + 7 + \cdots + (3k + 1) = \dfrac{(k + 1)(3k + 2)}{2}$

Proof: By hypothesis, $1 + 4 + 7 + \cdots + (3k - 2) = \dfrac{k(3k - 1)}{2}$.

Add $(3k + 1)$ to both sides:

$$1 + 4 + 7 + \cdots + (3k + 2) + (3k + 1) = \frac{k(3k - 1)}{2} + (3k + 1)$$

$$= \frac{k(3k - 1)}{2} + \frac{2(3k + 1)}{2}$$

$$= \frac{3k^2 - k + 6k + 2}{2}$$

$$= \frac{(k + 1)(3k + 2)}{2}$$

Step 4. The proposition is true for all positive integers n by PMI.

15. *Step 1.* Prove it true for $n = 1$:

$(b^m)^1 = b^{m \cdot 1}$; true

Step 2. Assume it true for $n = k$:

$(b^m)^k = b^{mk}$

Step 3. Prove it true for $n = k + 1$:

$$
\begin{aligned}
(b^m)^{k+1} &= b^{m(k+1)} \\
(b^m)^k &= b^{mk} && \text{(by hypothesis)} \\
(b^m)^k b^m &= b^{mk} b^m && \text{(multiply both sides by } b^m) \\
(b^m)^{k+1} &= b^{mk} b^m && \text{(definition)} \\
&= b^{mk+m} && \text{(Problem 14)} \\
&= b^{m(k+1)} && \text{(distributive)}
\end{aligned}
$$

Step 4. The proposition is true for all positive integers n by PMI.

17. *Step 1.* Prove it true for $n = 1$:

$$\left(\frac{a}{b}\right)^1 = \frac{a^1}{b^1} \ \checkmark$$

Step 2. Assume it true for $n = k$:

$$\left(\frac{a}{b}\right)^k = \frac{a^k}{b^k}$$

Step 3. Prove it true for $n = k + 1$:

$$\left(\frac{a}{b}\right)^{(k+1)} = \frac{a^{k+1}}{b^{k+1}} \quad \text{(hypothesis)}$$

$$\left(\frac{a}{b}\right)^k \left(\frac{a}{b}\right) = \frac{a^k}{b^k}\left(\frac{a}{b}\right) \quad \text{(multiply both sides by } a/b)$$

$$\left(\frac{a}{b}\right)^{k+1} = \frac{a^k}{b^k} \cdot \frac{a}{b} \quad \text{(definition)}$$

$$= \frac{a^k \cdot a}{b^k \cdot b} \qquad \text{(multiplication of fractions)}$$

$$= \frac{a^{k+1}}{b^{k+1}} \qquad \text{(definition)}$$

Step 4. The proposition is true for all positive integers n by PMI.

19. *Step 1.* Prove it true for $n = 1$:

$2^1 > 1 \ \checkmark$

Step 2. Assume it true for $n = k$:

that is, $2^k > k$

Step 3. Prove it true for $n = k + 1$:

To prove: $2^{k+1} > k + 1$

Now $2^{k+1} = 2^k \cdot 2$
$\qquad\qquad = 2^k \cdot (1 + 1)$
$\qquad\qquad = 2^k + 2^k$
$\qquad\qquad > 2^k + 1 \qquad \text{(since } 2^k > 1 \text{ for all } k\text{)}$
$\qquad\qquad > k + 1 \qquad \text{(by hypothesis)}$

Step 4. The proposition is true for all positive integers n by PMI.

21. *Step 1.* Prove it true for $n = 1$:

$|a_1| \le |a_1| \ \checkmark$

Step 2. Assume it true for $n = k$:

$|a_1 + a_2 + \cdots + a_k| \le |a_1| + |a_2| + \cdots + |a_k|$

Step 3. Prove it true for $n = k + 1$:

$|a_1 + a_2 + \cdots + a_k| \le |a_1| + |a_2| + \cdots + |a_k| \qquad \text{(by hypothesis)}$

$|a_1 + a_2 + \cdots + a_k| + |a_{k+1}| \le |a_1| + |a_2| + \cdots + |a_k| + |a_{k+1}|$

Add $|a_{k+1}|$ to both sides.

But

$|(a_1 + a_2 + \cdots + a_k) + a_{k+1}| \le |a_1 + a_2 + \cdots + a_k| + |a_{k+1}| \qquad \text{(by the triangle inequality)}$

Therefore,

$|a_1 + a_2 + \cdots + a_k + a_{k+1}| \le |a_1| + |a_2| + \cdots + |a_{k+1}|.$

Step 4. The proposition is true for all positive integers n by PMI.

23. *Step 1.* Prove it true for $n = 1$:

$$(a + b)^1 = \sum_{j=0}^{1} \binom{1}{j} a^{1-j} b^j$$

$$= \binom{1}{0}a + \binom{1}{1}b$$

$$= a + b$$

Step 2. Assume it true for $n = k$:

i.e. $(a + b)^k = \sum_{j=0}^{k} \binom{k}{j} a^{k-j} b^j$

Step 3. Prove it true for $n = k + 1$:

that is, $(a + b)^{k+1} = \sum_{j=0}^{k+1} \binom{k+1}{j} a^{k+1-j} b^j$

Proof: By hypothesis,

$$(ab)^k = \sum_{j=0}^{k} \binom{k}{j} a^{k-j} b^j$$

$$= a^k + \cdots + \binom{k}{r-1} a^{k-r+1} b^{r-1} + \binom{k}{r} a^{k-r} b^r + \cdots + b^k$$

Multiply both sides by $(a + b)$:

$$(a + b)^k (a + b) = \left[a^k + \cdots + \binom{k}{r-1} a^{k-r+1} b^{r-1} + \binom{k}{r} a^{k-r} b^r + \cdots + b^k \right](a + b)$$

$$(a + b)^{k+1} = \left[a^{k+1} + \cdots + \binom{k}{r-1} a^{k-r+1} b^{r-1} + \binom{k}{r} a^{k-r+1} b^r + \cdots + ab^k \right]$$

$$+ \left[a^k b + \cdots + \binom{k}{r-1} a^{k-r+1} b^r + \binom{k}{r} a^{k-r} b^{r+1} + \cdots + b^{k+1} \right]$$

$$= a^{k+1} + \cdots + \left[\binom{k}{r} + \binom{k}{r-1} \right] a^{k-r+1} b^r + \cdots + b^{k+1}$$

From Problem 22 we proved $\binom{k}{r} + \binom{k}{r-1} = \binom{k+1}{r}$, and the result is proved.

Step 4. The Binomial Theorem is proved for all positive integers n by PMI.

REVIEW PROBLEMS, PAGE 418

1. a. Arithmetic; $s_n = -9 + 10n$ b. Neither; 11111, 111111; number of ones increases by 1
c. Geometric; $s_n = 11^{n-1}$ d. Geometric; $s_n = 162 \cdot 3^{-n}$ e. Neither; 25, 36; sequence of square

numbers: n^2 **2.** a. $S_{10} = 460$ c. $S_{10} = \dfrac{1}{10}(11^{10} - 1)$ d. $-3^{-6} + 81$ **3.** a. $s_{10} = 2560$;

$S_5 = 5115$ b. $s_{10} = 5$; $S_5 = 200$ c. $d = 2$; $S_{10} = 110$ d. $r = \dfrac{1}{2}$; $S_{10} = 1023$ **4.** a. $-\dfrac{24}{11}$

b. 200   **5.** a. 2   b. 70   c. $\dfrac{p!}{q!(p-q)!}$ d. 20,300   e. $3^{10} - 1$ or 59,048   **6.** a. $x^5 +$

6. a. $x^5 - 5x^4y + 10x^3y^2 - 10x^2y^3 + 5xy^4 - y^5$ b. $32x^5 + 80x^4y + 80x^3y^2 + 40x^2y^3 + 10xy^4 + y^5$
c. 7920   **7.** 2046   **8.** In 24 hr there will be 72 divisions; $S_{72} = 2^{82} - 2^{10}$
9. Let $y - d$, y, $y + d$ be the numbers. Then

$$(y - d - 1) + y + (y + d + 4) = 21$$
$$y = 6.$$

Also, x, xr, xr^2 is the geometric sequence, where $xr = 6$. So $r = 6/x$. Thus,

$$x + 6 + x\left(\frac{6}{x}\right)^2 = 21$$

$$x^2 + 6x + 36 = 21x$$

$$x^2 - 15x + 36 = 0$$

$$x = 3 \quad \text{or} \quad 12.$$

If $x = 3$, then 3, 6, 12 is the geometric sequence.
If $x = 12$, then 12, 6, 3 is the geometric sequence.
The three numbers are 3, 6, and 12.

10. *Step 1.* Prove it true for $n = 1$:

$$4 = 2 \cdot 1(1 + 1)$$
$$= 2 \cdot 2$$
$$= 4 \quad \surd$$

Step 2. Assume it true for $n = k$:

$$4 + 8 + 12 + \cdots + 4k = 2k(k + 1)$$

Step 3. Prove it true for $n = k + 1$:

$$4 + 8 + 12 + \cdots + 4(k + 1) = 2(k + 1)(k + 2)$$

Proof:		
$4 + 8 + 12 + \cdots + 4k = 2k(k + 1)$		(by hypothesis)
$4 + 8 + 12 + \cdots + 4k + (4k + 4) = 2k(k + 1) + (4k + 4)$		(add $4k + 4$ to both sides)
$4 + 8 + \cdots + (4k + 4) = 2k(k + 1) + 4(k + 1)$		(distributive)
$= (2k + 4)(k + 1)$		(distributive)
$= 2(k + 2)(k + 1)$		(distributive)

Step 4. By Steps 1, 2, and 3 and the principle of mathematical induction, it is true for all positive
integers.

PROBLEM SET 9.1, PAGE 432

1.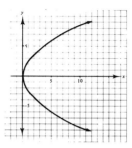

3.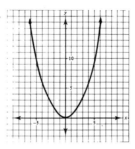

5.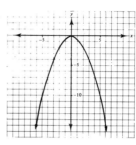

7.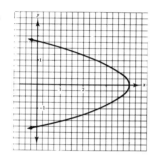

9.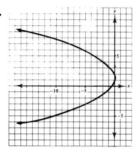

11.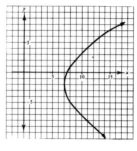

13. $y^2 = 10\left(x - \dfrac{5}{2}\right)$

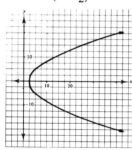

15. $(y - 2)^2 = -16(x + 1)$

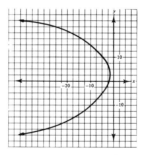

17. $(x + 2)^2 = 24(y + 3)$

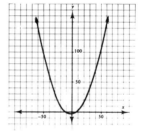

19. $x = 2$

21. $(x + 3)^2 = -\dfrac{1}{3}(y - 2)$

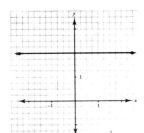

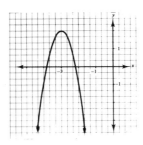

23. $(x - 100)^2 = -200(y - 50)$ **25.** It is 2.25 m from the vertex on the axis of symmetry.   **27.** Let $P(x, y)$ be a point on the parabola. Then the distance of P to the line is $|c - y|$ and $PF = \sqrt{(x + 0)^2 + (y + c)^2}$. So

$$\sqrt{(x + 0)^2 + (y + c)^2} = |c - y|$$
$$x^2 + y^2 + 2cy^2 + c^2 = c^2 - 2cy + y^2$$
$$x^2 = -4cy.$$

29. Let $P(x, y)$ be a point on the parabola. Then the distance of P to the line is $c - x$ and $|PF| = \sqrt{(x + c)^2 + (y - 0)^2}$. So

$$\sqrt{(x + c)^2 + y^2} = c - x$$
$$x^2 + 2cx + c^2 + y^2 = c^2 - 2cx + x^2$$
$$y^2 = -4cx.$$

31. $(-3, 4)$ and $(1, 12)$ **33.** $(2, 3)$ and $(3, 1)$

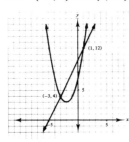

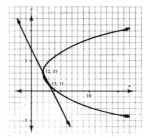

35. $T = \dfrac{1}{40}(A - 20)^2 + 11$; at age 40 her time will be 21.0 sec.

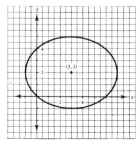

PROBLEM SET 9.2, PAGE 444

1.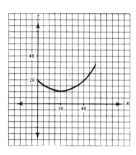

3.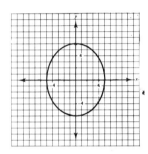

5.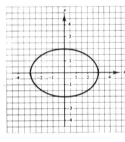

7.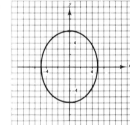

9.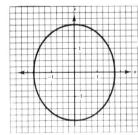

11.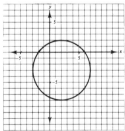

13.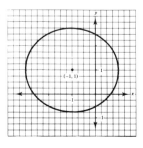

15.

Graphs for Problems 17–23 should also be included.

17. $(x - 4)^2 + (y - 5)^2 = 36$ **19.** $\dfrac{x^2}{24} + \dfrac{y^2}{49} = 1$ **21.** $\dfrac{16(x + 1)^2}{225} + \dfrac{16(y + 3)^2}{81} = 1$

23. $\dfrac{x^2}{89} + \dfrac{16y^2}{89} = 1$ **25.** 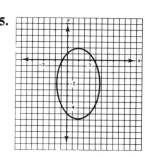   **27.**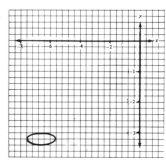

29.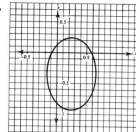

31. Let $P(x, y)$ be any point on the circle. Then, by the distance formula, $\sqrt{(x - h)^2 + (y - k)^2} = r$ or $(x - h)^2 + (y - h)^2 = r^2$.   **33.** Greatest distance is $1.5 \cdot 10^8$; least distance is $1.3 \cdot 10^8$ **35.** The walls are 20 ft from the foci.   **37.** $(-2, 0)$ **39.** Let the coordinates of one focus be $(c, 0)$, where $c^2 = a^2 - b^2$. Then, if $x = c$,

$$\frac{c^2}{a^2} + \frac{y^2}{b^2} = 1$$

$$y^2 = b^2\left(1 - \frac{c^2}{a^2}\right)$$

$$= b^2 - \frac{b^2}{a^2}(a^2 - b^2)$$

$$= \frac{b^4}{a^2}$$

$$y = \pm\frac{b^2}{a^2}$$

Thus, the length of the latus rectum, $2y$, is $2b^2/a^2$.

PROBLEM SET 9.3, PAGE 453

1.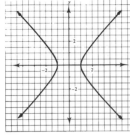

3.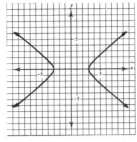

5.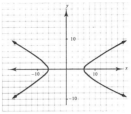

7.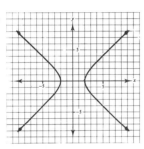

9.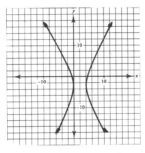

11. $\dfrac{y^2}{25} - \dfrac{x^2}{24} = 1$ **13.** $\dfrac{x^2}{1} - \dfrac{y^2}{24} = 1$ **15.** $\dfrac{x^2}{9} - \dfrac{(y+3)^2}{7} = 1$

17.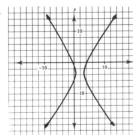

19.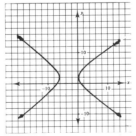

21.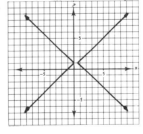

23.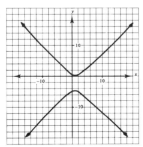

25. Let $P(x, y)$ be a point on the hyperbola.

$$\left|\sqrt{(x + c)^2 + (y - 0)^2} - \sqrt{(x - c)^2 + (y - 0)^2}\right| = 2a$$

$$\sqrt{(x + c)^2 + y^2} = \sqrt{(x - c)^2 + y^2} \pm 2a$$

$$x^2 + 2cx + c^2 + y^2 = x^2 - 2cx + c^2 + y^2 \pm 4a\sqrt{(x - c)^2 + y^2} + 4a^2$$

$$4cx - 4a^2 = \pm 4a\sqrt{(x - c)^2 + y^2}$$

$$cx - a^2 = \pm a\sqrt{(x - c)^2 + y^2}$$

$$c^2x^2 - 2a^2cx + a^4 = a^2[(x^2 - 2cx + c^2) + y^2]$$

$$= a^2x^2 - 2a^2cx + a^2c^2 + a^2y^2$$

$$c^2x^2 - a^2x^2 - a^2y^2 = a^2c^2 - a^4$$

$$(c^2 - a^2)x^2 - a^2y^2 = a^2(c^2 - a^2)$$

$$b^2x^2 - a^2y^2 = a^2b^2$$

$$\frac{x^2}{a^2} - \frac{y^2}{b^2} = 1$$

27. $d = \frac{b}{a}\sqrt{x^2 - a^2} - \frac{b}{a}x$

$$= \frac{b}{a}\sqrt{x^2\left(1 - \frac{a^2}{x^2}\right)} - \frac{b}{a}x$$

$$= \frac{b}{a}x\sqrt{1 - \frac{a^2}{x^2}} - \frac{b}{a}x$$

$$= \frac{b}{a}x\left(\sqrt{1 - \frac{a^2}{x^2}} - 1\right)$$

$$\lim_{|x| \to \infty}\left(\sqrt{1 - \frac{a^2}{x^2}} - 1\right) = 0$$

$$\lim_{|x| \to \infty} d = 0$$

29. The coordinates of two corners are (a, b), $(-a, -b)$. The other diagonal has the same length.

$$d = \sqrt{(a + a)^2 + (b + b)^2}$$

$$= \sqrt{4a^2 + 4b^2}$$

$$= 2\sqrt{a^2 + b^2}$$

$$= 2\sqrt{c^2} \quad (\text{since } c^2 = a^2 + b^2)$$

$$= 2c$$

31. At one focus, $x = c$ so

$$\frac{c^2}{a^2} - \frac{y^2}{b^2} = 1$$

$$\frac{y^2}{b^2} = \frac{c^2}{a^2} - 1$$

$$y^2 = \frac{b^2}{a^2}(c^2 - a^2)$$

$$= \frac{b^2}{a^2} \cdot b^2 \quad (c^2 = a^2 + b^2)$$

$$= \frac{b^4}{a^2}$$

$$y = \pm \frac{b^2}{a}$$

So the length of this segment is $2b^2/a$.

PROBLEM SET 9.4, PAGE 466

1. Parabola; $(y + 1)^2 = 4(x - 5)$ **3.** Parabola; $(x - 3)^2 = \frac{2}{3}(y + 8)$ **5.** Line; $y = -\frac{25}{9}x + 25$

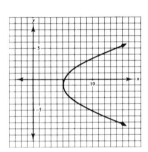 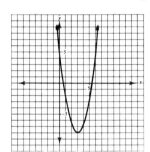 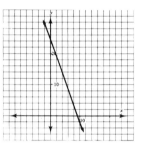

7. Parabola; $(y - 3)^2 = 4(x + 1)$ **9.** Ellipse; $\dfrac{(x - 3)^2}{25} + \dfrac{(y - 4)^2}{9} = 1$

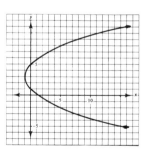 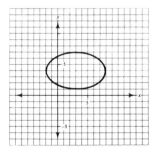

11.   Rotated parabola; $\theta = 45°$; $(x' + 3)^2 = 6(y' + 2)$

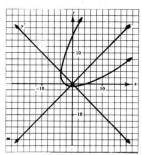

13.   (15) hyperbola; (16) hyperbola; (17) ellipse; (18) hyperbola; (19) parabola; (20) ellipse; (21) hyperbola

15.   a. $\theta = 45°$; $x = \dfrac{1}{\sqrt{2}}(x' - y')$; $y = \dfrac{1}{\sqrt{2}}(x' + y')$ b. $\dfrac{x'^2}{16} - \dfrac{y'2}{16} = 1$

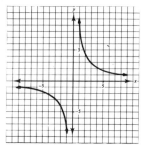

17.   a. $\theta = 45°$; $x = \dfrac{1}{\sqrt{2}}(x' - y')$; $y = \dfrac{1}{\sqrt{2}}(x' + y')$ b. $\dfrac{x'2}{9} + \dfrac{y'^2}{4} = 1$

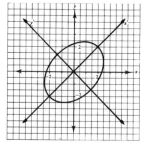

19. a. $\tan \theta = 2$; $x = \dfrac{1}{\sqrt{5}}(x' - 2y')$; $y = \dfrac{1}{\sqrt{5}}(2x' + y')$ b. $(x' + 1)^2 = 4\left(y' + \dfrac{7}{10}\right)$

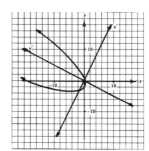

21. a. $\theta = 30°$; $x = \dfrac{1}{2}(\sqrt{3}\,x' - y')$; $y = \dfrac{1}{2}(x' + \sqrt{3}\,y')$ b. $\dfrac{x'^2}{4} - \dfrac{y'^2}{9} = 1$

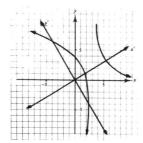

23. a. $\theta = 30°$; $x = \dfrac{1}{2}(\sqrt{3}\,x' - y')$; $y = \dfrac{1}{2}(x' + \sqrt{3}\,y')$ b. $x'^2 = -\dfrac{1}{16}(y' - 4)$

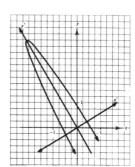

25.   a. $\theta = 60°$; $x = \dfrac{1}{2}(x' - \sqrt{3}\,y')$; $y = \dfrac{1}{2}(\sqrt{3}\,x' + y')$ b. $\dfrac{(x' - 2)^2}{4} + \dfrac{(y' - 1)^2}{1} = 1$

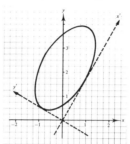

27.   a. $\tan \theta = 3$; $x = \dfrac{1}{\sqrt{10}}(x' - 3y')$; $y = \dfrac{1}{\sqrt{10}}(3x' + y')$ b. $\dfrac{(x' - \sqrt{10})^2}{44} + \dfrac{(y' - 2\sqrt{10})^2}{4} = 1$

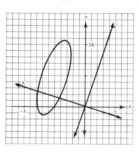

28–38.   The graphs form a picture of a bell.

PROBLEM SET 9.5, PAGE 473

1.

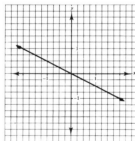

3.

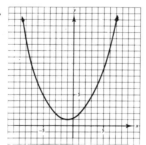

5.

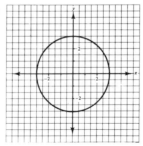

7.

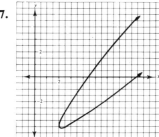

9.

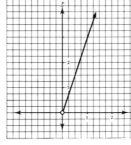

11. $y = -\dfrac{x}{2}$

13. $(x + 1)^2 = 4\left(y - \dfrac{3}{4}\right)$

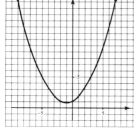

15. $x^2 + y^2 = 9$

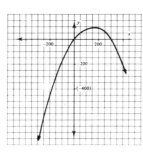

17. $x^2 - 2xy + y^2 - 13x + 12y + 38 = 0$

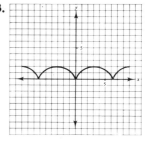

19. $y = 3x, \; x > 0$

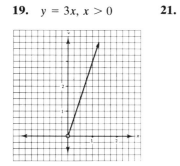

21.

23.

25. 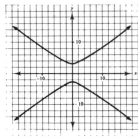 **27.** 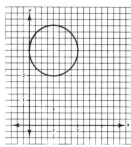 **29.**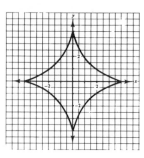

PROBLEM SET 9.6, PAGE 483

In Problems 1–10, polar forms are listed first, rectangular second.
1. $(4, \pi/4) = (-4, 5\pi/4); (2\sqrt{2}, 2\sqrt{2})$ **3.** $(5, 2\pi/3) = (-5, 5\pi/3); (-5/2, 5\sqrt{3}/2)$ **5.** $(3/2, 7\pi/6)$
$= (-3/2, \pi/6); (-3\sqrt{3}/4, -3/4)$ **7.** $(-4, 4) = (4, 0.86); (2.61, 3.03)$ **9.** $(-4, \pi) = (4, 0); (4, 0)$
11. $(5\sqrt{2}, \pi/4) = (-5\sqrt{2}, 5\pi/4)$ **13.** $(4, 5\pi/3) = (-4, 2\pi/3)$ **15.** $(3\sqrt{2}, 7\pi/4) = (-3\sqrt{2}, 3\pi/4)$
17. $(2, 5\pi/6) = (-2, 11\pi/6)$ **19.** $(13, 2.75) = (-13, 5.90)$ **21.** Given (r, θ), find (x, y). By the
definition of the trigonometric functions, $\cos \theta = x/r$ and $\sin \theta = y/r$. Therefore, $x = r \cos \theta$ and
$y = r \sin \theta$. **23.** Yes **25.** No **27.** Yes **29.** Yes **31.** Yes
Answers to Problems 33–43 vary.

45. 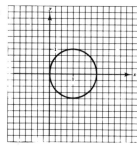 **47.** 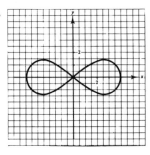 **49.**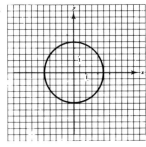

51. 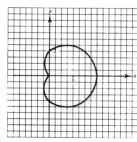 **53.** 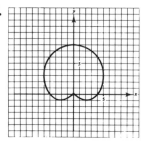 **55.**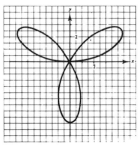

57. 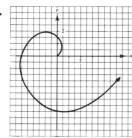   **59.**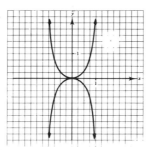

REVIEW PROBLEMS, PAGE 485

1. a. Hyperbola   b. Ellipse   c. Parabola   d. Line   e. Parabola   f. Ellipse
g. Hyperbola   h. Circle   i. Line   j. Hyperbola   **2.** a. $(5, 2.38) = (-5, 5.52)$ b. $(3, 4\pi/3)$
$= (-3, \pi/3)$ c. $(-2, 2) = (2, 5.14)$ d. $(-5, 3.1416) = (5, 0)$ e. $(5, \sqrt{7.5}) = (-3.6085, 3.4610)$;

$(3, -2\pi/3) = (-1.5, -2.5981)$; $(-2, 2) = (0.8323, -1.8186)$; $(-5, 9.4247) = (5, 0)$ **3.** a. $\dfrac{(x-4)^2}{4}$

$+ \dfrac{(y-1)^2}{3} = 1$ b. $\dfrac{(x+5)^2}{1} - \dfrac{(y-4)^2}{3} = 1$ **4.** a. $(y-3)^2 = 20(x-6)$ b. $\dfrac{4(x+5)^2}{25}$

$- \dfrac{4(y-4)^2}{75} = 1$

5. a. 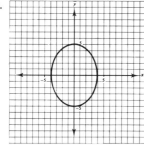   b.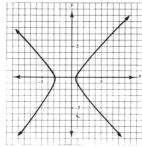

6. a.

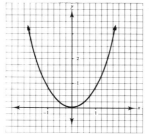

b.

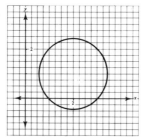

7. a.

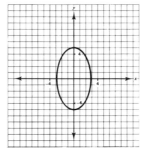

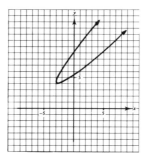

8. a.

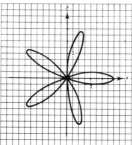

b.

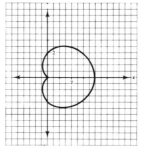

9.

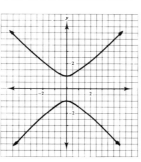

10. a. 45°   b. 1/2

PROBLEM SET A.1, PAGE 494

1. a. 5   b. −4   c. −14   d. 135   **3.** a. 3   b. −3   c. 10   d. 5   **5.** a. 11
b. −12   c. 174   d. 421   **7.** a. −3   b. 42   c. 132   d. −1   **9.** a. −19   b. −1

11. a. 20   b. −52   **13.** a. 4   b. 16   **15.** a. $\dfrac{-92}{105}$ b. $\dfrac{31}{30}$ **17.** a. $\dfrac{1}{9}$ b. $\dfrac{71}{63}$

19. a. $\dfrac{239}{1000}$ b. $\dfrac{4}{5}$ **21.** a. $\dfrac{30}{7}$ b. $\dfrac{141}{26}$ **23.** a. $\dfrac{2}{3}$ b. $\dfrac{-7}{27}$ **25.** a. $\dfrac{-625}{686}$ b. $\dfrac{158}{5}$

27. a. $1 - 6i$ b. $3 + 3i$ **29.** a. −4   b. $8 - 10i$ **31.** a. $-3 + 2i$ b. $2 + 5i$
33. a. 29   b. 89   **35.** a. 1   b. i **37.** a. −1   b. $-i$ **39.** a. $-i$

b. 1   **41.** $32 - 24i$ **43.** $-9 + 40i$ **45.** $-\dfrac{3}{2} + \dfrac{3}{2}i$ **47.** $-2i$ **49.** $-\dfrac{6}{29} + \dfrac{15}{29}i$

51. $1 + i$ **53.** $-\dfrac{35}{37} - \dfrac{12}{37}i$

PROBLEM SET B.1, PAGE 506

1. {2}   **3.** {2}   **5.** ∅   **7.** {5}   **9.** {1}   **11.** $[-10, \infty)$ **13.** $(-2, \infty)$
15. $(-\infty, -1]$ **17.** $(-\infty, 4/3)$ **19.** $(10, \infty)$ **21.** $(-\infty, -2)$ **23.** $[1, 3]$ **25.** $[1, 4]$
27. $[-3, -1]$ **29.** $(-\infty, -3) \cup (1, \infty)$ **31.** $(-4/3, 3/2)$ **33.** $(-\infty, -2] \cup [2, \infty)$

35. $[2, 3]$ **37.** $(-\infty, -8/3) \cup (4, \infty)$ **39.** $(-1 - \sqrt{2}, -1 + \sqrt{2})$ **41.** $\left(-\infty, \dfrac{-3 - \sqrt{37}}{2}\right]$

$\cup \left[\dfrac{-3 + \sqrt{37}}{2}, \infty\right)$ **43.** $[1 - \sqrt{7}, 1 + \sqrt{7}]$ **45.** $68° < F < 86°$ **47.** Must produce more than

eight.   **49.** The height and width must be less than or equal to 12 in., and the length less than or equal
to 48 in.   **51.** It is above 192 ft for 4 sec.   **53.** The minimum width is about 72.47 ft.

Index

APPLICATIONS INDEX

FORMULAS (continued from inside front cover)

$$e = \lim_{n \to \infty} \left(1 + \frac{1}{n}\right)^n$$

Arithmetic sequence: $s_n = a + (n - 1)d$ Arithmetic series: $S_n = \frac{n}{2}[2a + (n - 1)d]$

Geometric sequence: $s_n = ar^{n-1}$ Geometric series: $S_n = \dfrac{a(r^n - 1)}{r - 1}$

$$\lim_{n \to \infty} S_n = \frac{a}{1 - r}$$

Binomial expansion: $(a + b)^n = \sum\limits_{k=0}^{n} \binom{n}{k} a^{n-k}b^k$ where $\binom{n}{k} = \dfrac{n!}{k!(n - k)!}$

Relationship between polar and rectangular coordinate systems: $(x, y) = (r, \theta)$

$$r = \sqrt{x^2 + y^2} \qquad \tan \theta = \frac{y}{x} \qquad x = r \cos \theta \qquad y = r \sin \theta$$

IDENTITIES

Fundamental Identities

1. $\sec \theta = \dfrac{1}{\cos \theta}$

2. $\csc \theta = \dfrac{1}{\sin \theta}$

3. $\cot \theta = \dfrac{1}{\tan \theta}$

4. $\tan \theta = \dfrac{\sin \theta}{\cos \theta}$

5. $\cot \theta = \dfrac{\cos \theta}{\sin \theta}$

6. $\cos^2 \theta + \sin^2 \theta = 1$

7. $\tan^2 \theta + 1 = \sec^2 \theta$

8. $1 + \cot^2 \theta = \csc^2 \theta$

Cofunction Identities

9. $\cos(\pi/2 - \theta)$
   $= \sin \theta$

10. $\sin(\pi/2 - \theta)$
    $= \cos \theta$

11. $\tan(\pi/2 - \theta)$
    $= \cot \theta$

Opposite-Angle Identities

12. $\cos(-\theta) = \cos \theta$

13. $\sin(-\theta) = -\sin \theta$

14. $\tan(-\theta) = -\tan \theta$

Sum and Difference Identities

15. $\cos(\alpha + \beta) = \cos \alpha \cos \beta - \sin \alpha \sin \beta$
16. $\cos(\alpha - \beta) = \cos \alpha \cos \beta + \sin \alpha \sin \beta$
17. $\sin(\alpha + \beta) = \sin \alpha \cos \beta + \cos \alpha \sin \beta$
18. $\sin(\alpha - \beta) = \sin \alpha \cos \beta - \cos \alpha \sin \beta$

19. $\tan(\alpha + \beta) = \dfrac{\tan \alpha + \tan \beta}{1 - \tan \alpha \tan \beta}$

20. $\tan(\alpha - \beta) = \dfrac{\tan \alpha - \tan \beta}{1 + \tan \alpha \tan \beta}$

Double-Angle Identities

21. $\cos 2\theta = \cos^2 \theta - \sin^2 \theta$
    $= 2 \cos^2 \theta - 1$
    $= 1 - 2 \sin^2 \theta$

22. $\sin 2\theta = 2 \sin \theta \cos \theta$

23. $\tan 2\theta = \dfrac{2 \tan \theta}{1 - \tan^2 \theta}$

Half-Angle Identities

24. $\cos \frac{1}{2}\theta = \pm\sqrt{\dfrac{1 + \cos \theta}{2}}$

25. $\sin \frac{1}{2}\theta = \pm\sqrt{\dfrac{1 - \cos \theta}{2}}$

26. $\tan \frac{1}{2}\theta = \dfrac{1 - \cos \theta}{\sin \theta}$

     $= \dfrac{\sin \theta}{1 + \cos \theta}$

Product Identities

27. $2 \cos \alpha \cos \beta = \cos(\alpha - \beta) + \cos(\alpha + \beta)$
28. $2 \sin \alpha \cos \beta = \sin(\alpha + \beta) + \sin(\alpha - \beta)$
29. $2 \sin \alpha \sin \beta = \cos(\alpha - \beta) - \cos(\alpha + \beta)$
30. $2 \cos \alpha \sin \beta = \sin(\alpha + \beta) - \sin(\alpha - \beta)$

Sum Identities

31. $\cos \alpha + \cos \beta = 2 \cos\left(\dfrac{\alpha + \beta}{2}\right) \cos\left(\dfrac{\alpha - \beta}{2}\right)$

32. $\cos \alpha - \cos \beta = -2 \sin\left(\dfrac{\alpha + \beta}{2}\right) \sin\left(\dfrac{\alpha - \beta}{2}\right)$

33. $\sin \alpha + \sin \beta = 2 \sin\left(\dfrac{\alpha + \beta}{2}\right) \cos\left(\dfrac{\alpha - \beta}{2}\right)$

34. $\sin \alpha - \sin \beta = 2 \sin\left(\dfrac{\alpha - \beta}{2}\right) \cos\left(\dfrac{\alpha + \beta}{2}\right)$